D0710516

Dynamics and Vibration of Structures

Dynamics and Vibration of Structures

DEMETER G. FERTIS

Professor of Civil Engineering
The University of Akron

A Wiley-Interscience Publication
JOHN WILEY & SONS, New York · London · Sydney · Toronto

Library of Congress Cataloging in Publication Data:

Fertis, Demeter G.
Dynamics and Vibration of Structures.

"A Wiley-Interscience publication."
Bibliography: p. 475.
1. Structural dynamics. 2. Vibration. I. Title.

TA654.F47 624'.171 72-11788
ISBN 0-471-25777-X

Printed in the United States of America

10–9 8 7 6 5 4 3 2 1

To KIKI, SUE, and LITSA

PREFACE

This book is intended as a text for courses in structural dynamics and vibrations usually offered to seniors and first-year graduate students in civil engineering and engineering mechanics. It can be also used as a professional reference book for practicing civil, mechanical, and aeronautical engineers. Selected chapters can serve as a text for specialized courses in mechanical engineering.

The book is so written as to satisfy both theoretical and practical needs on the subject, with emphasis on applications, clarity, and versatility in methods of analysis. Each method is treated in a way that recognizes its application to problems where it does the most good.

Thus the book, being neither strictly theoretical nor strictly practical, is written to provide a variety of tools needed for the dynamic analysis of structures. These needs were recognized and taken into consideration during the time this material was taught by me at Wayne State University, the University of Iowa, and the University of Akron. At the University of Akron the text is presently used for senior and graduate courses on dynamics and vibration of structures.

Chapter 1 is devoted to fundamentals on structural vibrations that are essential for better understanding of the dynamic response of structures. Chapter 2 provides a rather extensive treatment of the dynamic response of spring-mass systems. Rigorous as well as numerical methods of analysis are included, by taking into consideration also various types of forcing functions and damping. The dynamic response of idealized frames and simple buildings is discussed in Chapter 3, and rigorous methods of analysis for systems with infinite degrees of freedom are treated in Chapter 4.

The method of modal analysis and its application to various structural systems is the topic of Chapter 5. The application of this method requires the computation of natural frequencies and mode shapes. Therefore, a small portion of Chapter 5, as well as Chapter 6, gives well known methods of vibration by including the more recent concepts of transfer matrices and dynamic hinge. Chapter 7 examines the dynamic response of structures with members of variable stiffness. Unique methods of analysis are presented that greatly simplify the dynamic analysis of complex structural problems.

Chapters 8 and 9 review Fourier transforms, Laplace transforms, and variational methods of analysis. These concepts are somewhat more theoretical than others, but their treatment and application to problems on structural dynamics and vibration are clearly illustrated.

Approximate methods of analysis for the dynamic response of structures are discussed in Chapter 10. These methods are somewhat practical, but they provide a simple and reasonable approach for computing the dynamic response of single and multielement elastic and elastoplastic structural systems. The dynamic analysis of structures subjected to blast and earthquake is contained in Chapter 11. Practical as well as rigorous methods of analysis are included.

The last chapter introduces the reader to the subject of random excitations that has received particular attention during the past decade. The stochastic methods of approach introduced in this chapter are particularly applicable to structures subjected to earthquake excitations. Both theory and applications are given.

I express my indebtedness to my colleagues for their kindness and support, and give credit to the works of others that have inspired the work of this text. In particular, I thank Dr. A. L. Simon, Professor and Head of the Department of Civil Engineering of the University of Akron, for his encouragement and support. To Apostolos C. Raptis, Ph.D. student of the Electrical Engineering Department at the University of Akron, I express my gratitude for his suggestions and help. To Mrs. Dorothy Guilliams goes my appreciation for the excellent typing of the manuscript. Thanks are also due to Mrs. Rosemary Phillips for her assistance in the typing. To my wife and children I express my deepest appreciation for their patience and understanding.

DEMETER G. FERTIS

Akron, Ohio
June 1972

CONTENTS

Dynamics and Vibration of Structures

1

FUNDAMENTALS OF STRUCTURAL VIBRATIONS

1-1 INTRODUCTION

The analysis and design of structures to resist the effects of time-dependent forces constitute the field of structural dynamics. For example, the dynamic forces that are transmitted to a structure that supports oscillating machinery must be considered in its analysis and design. The same analogy is also true for a bridge that supports moving loads, a structure that is subjected to suddenly applied dynamic forces such as blast and wind gust, or for a building whose foundation is suddenly disturbed by the presence of some random earthquake activity.

Any structure that is excited by a time-varying force is subject to vibrations. Consequently, knowledge regarding vibration response of structures constitutes an essential part in their analysis for dynamic response. Basic concepts and methods of structural vibrations that are needed in the dynamic analysis of structures are presented in this chapter.

1-2 DEFINITIONS AND FUNDAMENTAL ASPECTS OF PERIODIC MOTION

The field of engineering vibrations can be thought of as a special application of the principles of dynamics to bodies performing some kind of repetitive motion. Therefore, Newton's laws of motion can be used to study such vibratory motions.

Vibrations are time-dependent displacements of a particle, or a system of particles, with respect to an equilibrium position. If these displacements are repetitive and their repetition is executed at equal intervals of time with respect to an equilibrium position, the resulting motion is said to be periodic. The time t required for the motion to be repeated is called the period. The frequency of vibration, f_n, is the number of cycles per unit of time. That is,

$$f_n = \frac{1}{\tau} \tag{1-1}$$

A cycle is defined as a complete motion during the period τ of vibration. In addition to the definition given above, the frequency of vibration is often designated as the number of radians per unit of time and it is usually represented by the Greek letter ω.

The vibration of a body is usually classified as free or forced. In the case of free vibrations, the only forces acting on the body are those owing to the inertia forces developed as a result of its distribution of mass. A body vibrating freely, that is, without impressed vibration from an outside source, does so at one or more of its natural frequencies. The lowest natural frequency of vibration is termed the fundamental frequency.

If an external vibrating force system is applied to a body, the body vibrates at the frequency of the applied force. An alternating external force system may arise as a consequence of many natural phenomena such as waves, wind, sound, blast, or earthquake activity as well as from any mechanically produced causes. In each case, the wave motion of the disturbance will vibrate a structure at the frequency of the oscillating force. A condition of resonance will occur when the frequency of the applied force system coincides with one of the natural frequencies of the body. At the reasonant condition, the amplitude of vibration will approach infinity with time. Practically, however, the amplitude of vibration may exceed allowable values in a rather short period of time, with the subsequent loss of structural integrity.

Fortunately, the free vibration of engineering structures is usually harmonic. The assumption of harmonic vibration permits a problem to be mathematically expressed in terms of sine and cosine functions. Harmonic motion is characteristically periodic, but it is not true that periodic motion is harmonic.

Harmonic motion can be described by the motion of the projection P' of a particle P, Fig. 1-1a, moving at constant angular velocity ω around the circumference of the circle, of radius r, on the vertical diameter of the circle. Similar observations can be obtained by considering the projection of P on the horizontal diameter of the circle. If at time $t=0$ the particle P is at A, the angular displacement θ at any time t is $\theta=\omega t$. Thus the displacement y of the projection P', measured from the origin 0, is

$$y=r\sin\omega t \tag{1-2}$$

The linear velocity of P' is

$$\dot{y}=\frac{dy}{dt}=r\omega\cos\omega t \tag{1-3}$$

and the linear acceleration is

$$\ddot{y}=\frac{d^2y}{dt^2}=-r\omega^2\sin\omega t=-\omega^2 y \tag{1-4}$$

or

$$\ddot{y}+\omega^2 y=0 \tag{1-5}$$

Equation 1-5 is the differential equation of motion of P' along the vertical diameter of the circle with the specified initial conditions. The general solution of this equation is

$$y=A\sin\omega t+B\cos\omega t \tag{1-6}$$

where the constants of integration A and B are determined by utilizing the initial conditions of the motion.

The displacement, velocity, and acceleration of P', as given by Eqs. 1-2,

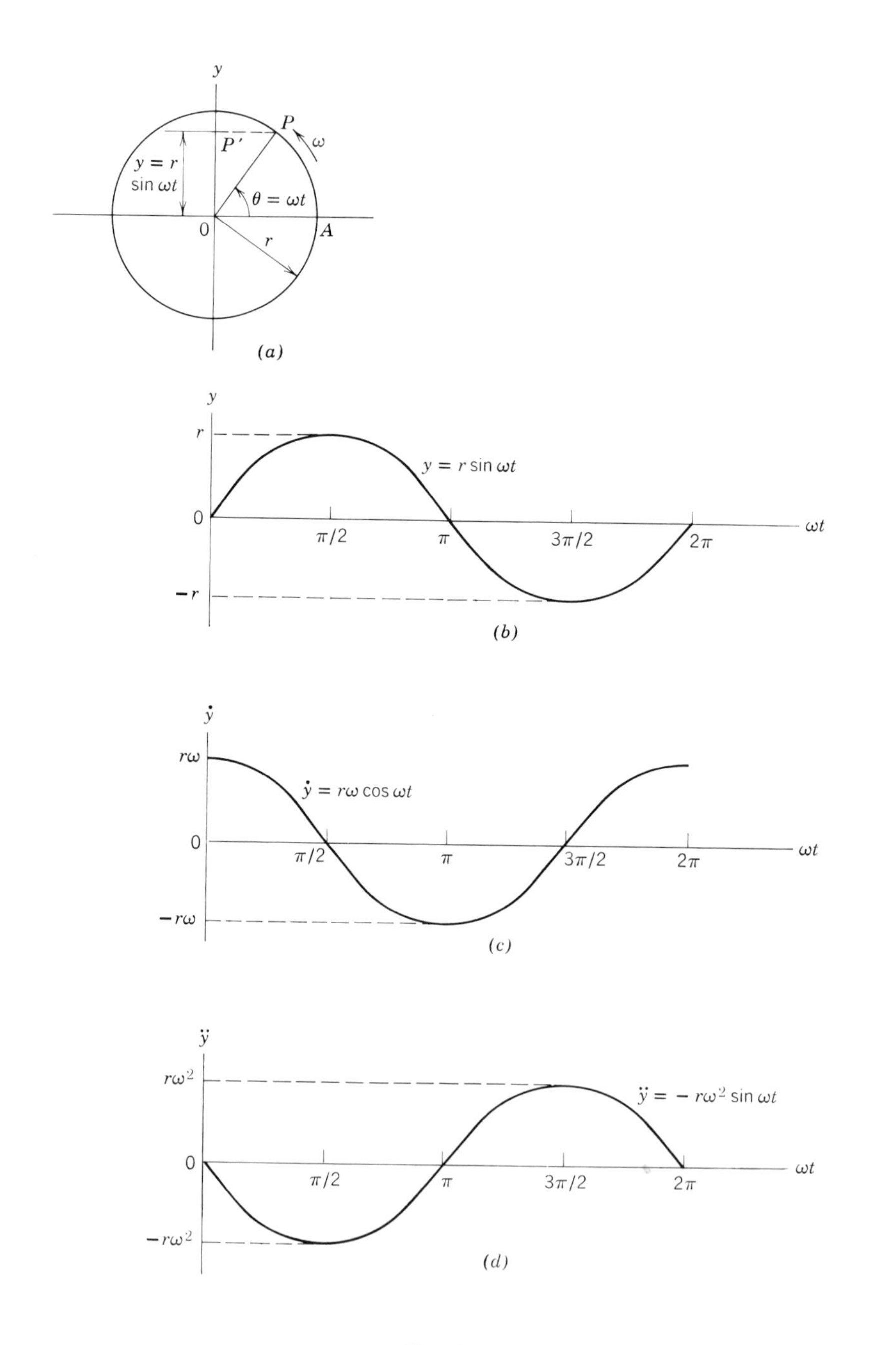
y
P
ω
P'
y = r
sin ωt
θ = ωt
0
A
r
(a)
y
r
y = r sin ωt
0
π/2
π
3π/2
2π
ωt
−r
(b)
ẏ
rω
ẏ = rω cos ωt
0
π/2
π
3π/2
2π
ωt
−rω
(c)
ÿ
rω²
ÿ = − rω² sin ωt
0
π/2
π
3π/2
2π
ωt
−rω²
(d)

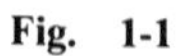

Fig. 1-1

1-3, and 1-4 are shown, respectively, in Figs. 1-1*b*, 1-1*c*, and 1-1*d*. The graphs in this figure show that the velocity is 90° ahead of the velocity vector. Consequently, the acceleration vector is 180° ahead of the displacement vector.

The amplitude of vibration of a damped structure, vibrating freely, may be attenuated by resisting forces developing during motion, since it is not subject to external exciting forces to sustain the vibration. The resisting forces dissipate energy and in time they die out the vibration. This phenomenon is known as damping. The causes for such actions are many, but the associated energy losses are often small and they could be neglected in the analysis of many engineering structures.

The three most common types of damping are usually defined as viscous, coulomb, and hysteresis. Viscous damping is usually associated with bodies moving through fluids at low velocities. In this case, the damping force $F_d(t)$ is assumed to be proportional to the first power of the velocity of the motion. That is,

$$F_d(t) = -c\dot{y} \tag{1-7}$$

where c is the damping constant and $\dot{y}$ is the velocity. The constant c is defined as the constant resistance per unit velocity.

Coulomb or dry friction damping, is associated with the sliding of bodies on dry surfaces, and the resisting force F_f is given by the expression

$$F_f = \pm \mu N \tag{1-8}$$

where μ is the coefficient of kinetic friction of the material and N is the normal force. This type of resistance depends on the normal force, the nature of sliding surfaces, and the characteristics of the material.

Hysteresis damping, often called solid or structural damping, is associated with the internal friction of the material, and it is approximately proportional to the amplitude of the displacement of the deformed body. It is independent of the frequency of vibration, and it is the result of the friction between internal planes that slip and slide during deformation of the body. The energy absorbed in this manner is dissipated in the form of heat.

If a force-displacement curve is plotted as shown in Fig. 1-2, the energy loss ΔU per cycle of motion is the area within the loop. In this figure, F is the spring force, y is the displacement, and Y is the amplitude of vibration. Experiments have shown that ΔU is approximately proportional to the square of the amplitude. That is,

$$\Delta U = k\pi c_0 Y^2 \tag{1-9}$$

where k is the spring stiffness and c_0 is the dimensionless constant of the material for solid damping.

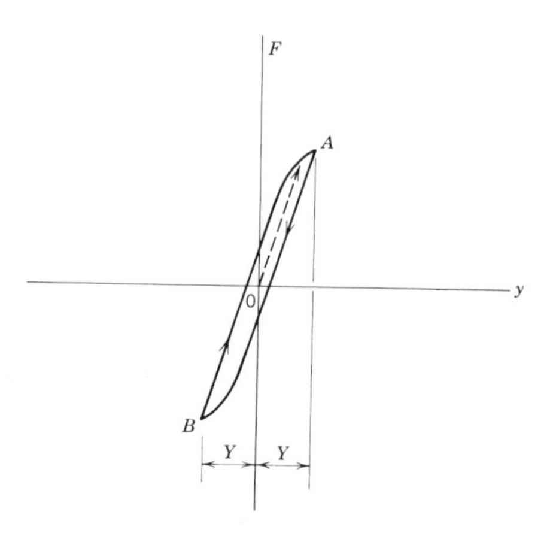

Fig. 1-2

1-3 DIFFERENTIAL EQUATIONS OF MOTION FOR VARIOUS SYSTEMS

The differential equations of motion for systems with various degrees of freedom are derived in this section. The degrees of freedom of an engineering system are defined by the number of independent coordinates necessary to establish the position of its particles at any time t. Consider for example the spring-mass system shown in Fig. 1-3*a*, whose mass m is restricted to move in the vertical direction only and under the influence of viscous damping. The time-varying force acting on m is $F(t)$, k is the spring constant defined as the force required to stretch the spring by an amount equal to unity, y is the vertical displacement of m, and c is the damping constant. The position of m at any time t is completely determined if $y(t)$ is known. Consequently this system has only one degree of freedom.

The differential equation of motion for this system can be derived by using the free-body diagram in Fig. 1-3*b* and applying Newton's second law of motion. That is,

$$m\ddot{y} = F(t) - ky - c\dot{y}$$

or

$$m\ddot{y} + c\dot{y} + ky = F(t) \tag{1-10}$$

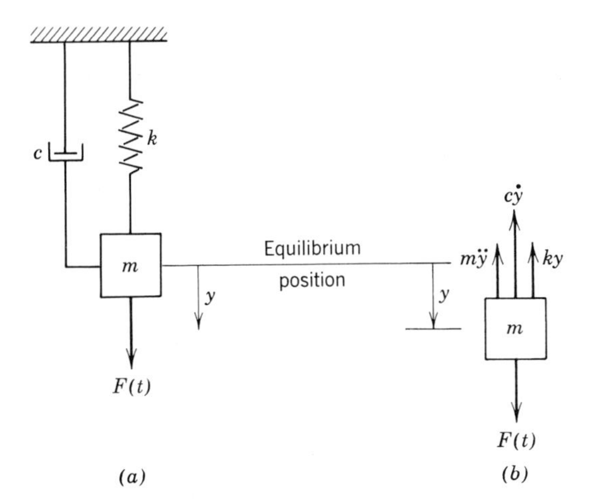
c
k
m
Equilibrium
position
y
F(t)
(a)
cẏ
mÿ
ky
y
m
F(t)
(b)

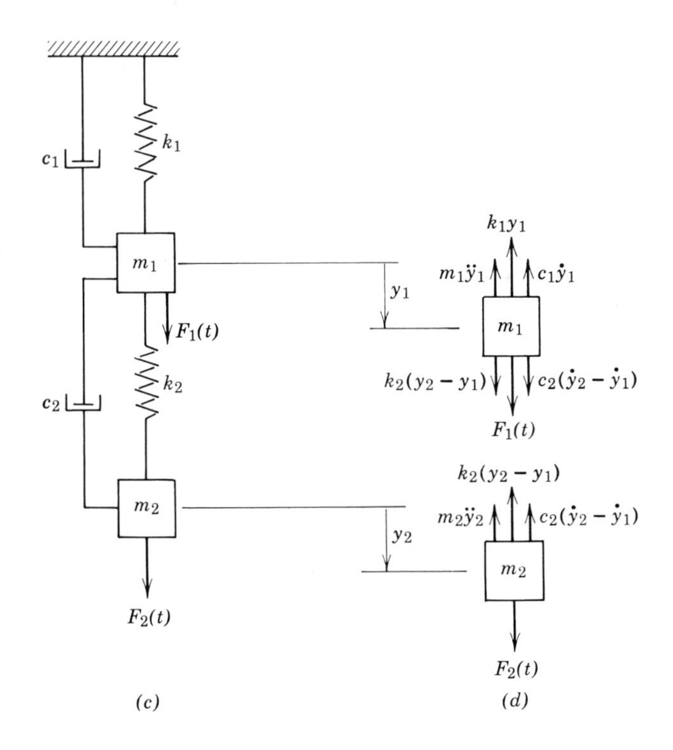
c1
k1
m1
F1(t)
c2
k2
m2
F2(t)
(c)
k1y1
m1ÿ1
c1ẏ1
y1
m1
k2(y2 − y1)
c2(ẏ2 − ẏ1)
F1(t)
k2(y2 − y1)
m2ÿ2
c2(ẏ2 − ẏ1)
y2
m2
F2(t)
(d)

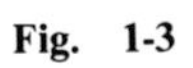
Fig. 1-3

where $\ddot{y}$ denotes second derivative with respect to time. Equation 1-10 is a linear nonhomogeneous second-order differential equation with constant coefficients. Its solution will yield the displacement function $y(t)$.

If the time-varying force $F(t)$ is zero, Eq. 1-10 yields

$$m\ddot{y}+c\dot{y}+ky=0 \tag{1-11}$$

Equation 1-11 is a second-order linear homogeneous equation with constant coefficients and represents the free vibration of the one-degree system under the influence of vicous damping.

In the case where the viscous damping force $c\dot{y}$ is zero,

$$m\ddot{y}+ky=0 \tag{1-12}$$

This homogeneous equation represents the undamped free vibration of the spring-mass system.

The spring-mass system in Fig. 1-3*c*, if restricted to move in the vertical direction only, has two degrees of freedom, because it requires the two independent coordinates y_1 and y_2 to locate the positions of masses m_1 and m_2, respectively, during motion. The system is again assumed to be under the influence of viscous damping, where c_1 and c_2 are the damping constants. The free-body diagrams in Fig. 1-3*d* depict the forces acting on masses m_1 and m_2. By using these free-body diagrams and applying dynamic equilibrium in the vertical direction, the equations of motion are as follows:

$$-m_1\ddot{y}_1-k_1y_1+k_2(y_2-y_1)-c_1\dot{y}_1+c_2(\dot{y}_2-\dot{y}_1)+F_1(t)=0$$

$$-m_2\ddot{y}_2-k_2(y_2-y_1)-c_2(\dot{y}_2-\dot{y}_1)+F_2(t)=0$$

or

$$m_1\ddot{y}_1+c_1\dot{y}_1-c_2(\dot{y}_2-\dot{y}_1)+k_1y_1-k_2(y_2-y_1)=F_1(t) \tag{1-13}$$

$$m_2\ddot{y}_2+c_2(\dot{y}_2-\dot{y}_1)+k_2(y_2-y_1)=F_2(t) \tag{1-14}$$

Equations 1-13 and 1-14 are the differential equations of motion of the two-degree system. Their simultaneous solution will yield the displacement functions $y_1(t)$ and $y_2(t)$ for masses m_1 and m_2, respectively.

The free vibration with viscous damping is obtained when the time-varying forces $F_1(t)$ and $F_2(t)$ are zero. In this case, the system of the homogeneous equations of motion is

$$m_1\ddot{y}_1+c_1\dot{y}_1-c_2(\dot{y}_2-\dot{y}_1)+k_1y_1-k_2(y_2-y_1)=0 \tag{1-15}$$

$$m_2\ddot{y}_2+c_2(\dot{y}_2-\dot{y}_1)+k_2(y_2-y_1)=0 \tag{1-16}$$

If the damping forces are zero, Eqs. 1-15 and 1-16 yield

$$m_1\ddot{y}_1 + k_1 y_1 - k_2(y_2 - y_1) = 0 \tag{1-17}$$

$$m_2\ddot{y}_2 + k_2(y_2 - y_1) = 0 \tag{1-18}$$

which is the case of free undamped vibration.

In the derivations above, the elastic springs connecting the masses are assumed to be linear and of negligible mass compared to the heavy rigid masses m, m_1, and m_2. In addition, the time-varying functions acting on the masses are assumed to be low-frequency forcing functions. If, for example, the frequency of $F(t)$ is of high magnitude that can excite the natural modes of the spring itself, the system in Fig. 1-3*a* can no longer be analyzed as a one-degree system. The distributed mass and elasticity of the spring should be taken into consideration.

Systems with continuous mass and elasticity, such as beams, columns, frames, and shafts, have infinite degrees of freedom, because during motion an infinite number of coordinates are required to locate the position of the particles making up the deformable elastic body. For example, the uniform stiffness beam in Fig. 1-4*a* has infinite degrees of freedom. Under the action of the distributed dynamic force $q(t, x)$, its differential equation can be derived by considering the free-body diagram of an element of length dx as shown in Fig. 1-4*b*. In this diagram, M and V denote the dynamic moment and the dynamic shear force, respectively, of the beam section. The product $m\ddot{y}dx$ is the inertia force acting on the element dx, where m denotes the mass per unit of length of the beam and y the vertical dynamic displacement at a distance x.

Dynamic equilibrium of the element in the vertical direction yields

$$q(t,x) - m\ddot{y} + \frac{dV}{dx} = 0 \tag{1-19}$$

In the equation above,

$$V = \frac{\partial M}{\partial x} = -\frac{\partial^3 y}{\partial x^3}(EI) \tag{1-20}$$

where E and I are the modulus of elasticity and cross-sectional moment of inertia, respectively, of the beam. By substituting Eq. 1-20 into Eq. 1-19, we have

$$EI\frac{\partial^4 y}{\partial x^4} + m\ddot{y} = q(t,x) \tag{1-21}$$

which is the differential equation of motion of the beam in the transverse

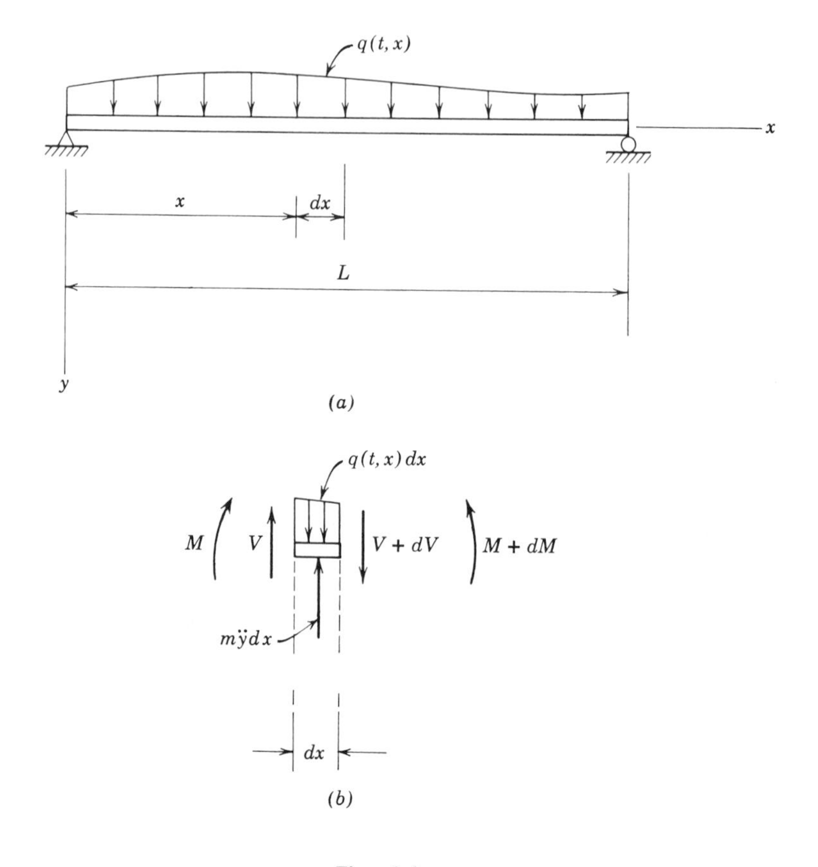

Fig. 1-4

direction. Its solution will yield the time displacement function $y(t, x)$.

The differential equation for the free undamped vibration of the beam is obtained if the time-dependent force $q(t, x)$ in Eq. 1-21 is zero. That is,

$$EI\frac{\partial^4 y}{\partial x^4} + m\ddot{y} = 0 \tag{1-22}$$

These derivations do not take into consideration the effects of deformations due to shear forces and the rotatory inertia forces acting on the element. These are not tremendous limitations because for many structural problems, such as slender beams, these effects are small and can be neglected. For a similar reason, the effects of damping are in many cases neglected.

1-4 LAGRANGE'S EQUATION

In the preceding section, the differential equations of motion were derived either by using Newton's second law of motion or by applying what is known as D'Alambert's dynamic equilibrium. For many structural problems the derivation of such equations is rather straightforward, and application of dynamic equilibrium or Newton's law is usually sufficient. In other cases, however, the analysis of structures for dynamic response becomes more convenient by making use of a powerful tool known as Lagrange's[1] equation. This equation was developed by Lagrange and was published in his book *Mechanics Analytique* in 1788. Since its inception, it enjoyed a wide popularity in the solution of many classical and modern dynamics problems. Lagrange's equation makes use of energy concepts and generalized[2] coordinates.

The generalized coordinates for a multiple degree of freedom system are designated as $q_1, q_2, \ldots, q_i, \ldots, q_n$. The number of such coordinates required to define the deformation configuration of a system during motion should be equal to the number of degrees of freedom of the system. They are independent of one another, and there is one generalized coordinate associated with each degree of freedom.

The derivation of Lagrange's equation is initiated by considering the beam in Fig. 1-5*a* which supports the weights $W_1, \ldots, W_2, \ldots, W_j, \ldots, W_n$ at points $1, 2, \ldots, j, \ldots, n$, respectively. In addition, the member is acted upon by the dynamic forces $F_1, F_2, \ldots, F_k, \ldots, F_m$. It is further assumed that the beam is of negligible weight, and thus the number of weights supported by this member defines its degrees of freedom. Since there are n weights, the system has n degrees of freedom, and n generalized coordinates are needed to define its configuration during motion.

For a deflection configuration such as the one in Fig. 1-5*b*, the n generalized coordinates are $q_1, q_2, \ldots, q_i, \ldots, q_n$, and they are taken at the positions shown in the figure. Other locations could be also selected, because the only requirement in their choice is that they should be equal in number with the number of degrees of freedom of the system. In addition, the deflections $y_1, y_2, \ldots, y_j, \ldots, y_n$ under the weight concentration points of the beam, should be defined by the generalized coordinates. In other words, each deflection y should be a function of $q_1, q_2, \ldots, q_i, \ldots, q_n$. Mathematically these relationships are written as

$$y_1 = f_1(q_1, q_2, \ldots, q_n) \qquad y_2 = f_2(q_1, q_2, \ldots, q_n) \qquad y_3 = f_3(q_1, q_2, \ldots, q_n) \tag{1-23}$$

[1]Additional information regarding the derivation and application of Lagrange's equation can be found in Reference 13.

[2]More information on generalized coordinates and generalized forces appears in Reference 13.

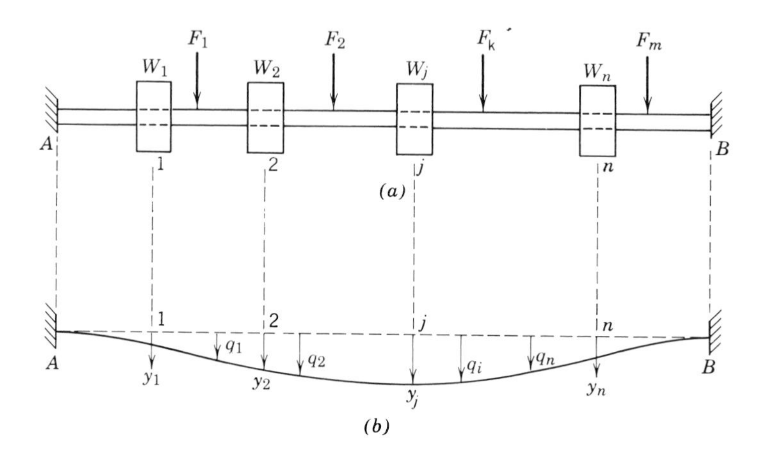
F_1
F_2
F_k
F_m
W_1
W_2
W_j
W_n
A
B
1
2
j
n
(a)
A
B
1
2
j
n
q_1
q_2
q_i
q_n
y_1
y_2
y_j
y_n
(b)

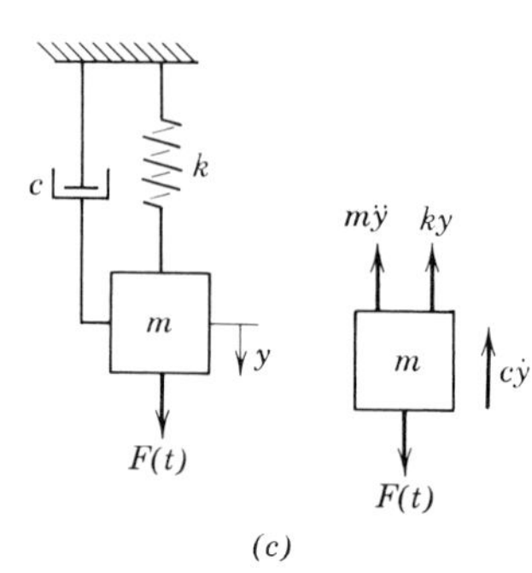
c
k
m
y
F(t)
$m\ddot{y}$
ky
m
$c\dot{y}$
F(t)
(c)

c
k_1
m_1
y_1
$F_1(t)$
c
k_2
m_2
y_2
$F_2(t)$
c
k_3
m_3
y_3
$F_3(t)$
$m_1\ddot{y}_1$
$c\dot{y}_1$
k_1y_1
m_1
$c(\dot{y}_2 - \dot{y}_1)$
$k_2(y_2 - y_1)$
$F_1(t)$
$m_2\ddot{y}_2$
$c(\dot{y}_2 - \dot{y}_1)$
$k_2(y_2 - y_1)$
m_2
$c(\dot{y}_3 - \dot{y}_2)$
$k_3(y_3 - y_2)$
$F_2(t)$
$m_3\ddot{y}_3$
$c(\dot{y}_3 - \dot{y}_2)$
$k_3(y_3 - y_2)$
m_3
$F_3(t)$
(d)

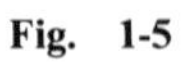
Fig. 1-5

$$\cdots\cdots\cdots\cdots$$

$$y_n = f_n(q_1, q_2, \ldots, q_n)$$

By considering the generalized coordinate q_i and introducing a virtual displacement δq_i, the work δW done by the external forces during the virtual displacement is equal to the change δU in the internal strain energy U of the system. That is,

$$\delta U = \delta W \tag{1-24}$$

The two terms can be written as

$$\delta U = \frac{\partial U}{\partial q_i}\delta q_i \tag{1-25}$$

$$\delta W = \frac{\partial W}{\partial q_i}\delta q_i \tag{1-26}$$

Thus Eq. 1-24 yields

$$\frac{\partial U}{\partial q_i}\delta q_i = \frac{\partial W}{\partial q_i}\delta q_i \tag{1-27}$$

In a dynamic system, forces producing work δW are usually the external forces $F_1, F_2, \ldots, F_m$, the inertia forces $m_j\ddot{y}_j = (W_j / g)\ddot{y}_j$ acting on the m_j masses $(j = 1, 2, \ldots, n)$, and the forces due to damping. If δW_e, δW_d and δW_{in} represent the virtual work done by the external, damping, and inertia forces, respectively, Eq. 1-27 can be written as

$$\frac{\partial U}{\partial q_i}\delta q_i = \frac{\partial W_e}{\partial q_i}\delta q_i + \frac{\partial W_d}{\partial q_i}\delta q_i - \sum_{j=1}^{n}(m_j\ddot{y}_j)\frac{\partial y_i}{\partial q_i}\delta q_i \tag{1-28}$$

In the right side of this equation, the amplitude y_j in the last term is the displacement of the mass m_j.

The kinetic energy T of the system, by definition, is

$$T = \sum_{j=1}^{n}\tfrac{1}{2}m_j\dot{y}_j^2 \tag{1-29}$$

Thus

$$\frac{\partial T}{\partial \dot{q}_i} = \sum_{j=1}^{n} m_j\dot{y}_j\frac{\partial \dot{y}_j}{\partial \dot{q}_i} \tag{1-30}$$

The equation can be written as

$$\frac{\partial T}{\partial \dot{q}_i} = \sum_{j=1}^{n} m_j \dot{y}_j \frac{\partial y_j}{\partial q_i} \tag{1-31}$$

because by Eq. 1-23

$$\dot{y}_j = \frac{\partial y_j}{\partial q_i} \dot{q}_i \tag{1-32}$$

and

$$\frac{\partial \dot{y}_j}{\partial \dot{q}_i} = \frac{\partial y_j}{\partial q_i} \tag{1-33}$$

The last term in the right-hand side of Eq. 1-28 can be expressed as

$$-\sum_{j=1}^{n} (m_j \ddot{y}_j) \frac{\partial y_j}{\partial q_i} \delta q_i = -\frac{d}{dt} \sum_{j=1}^{n} (m_j \dot{y}_j) \frac{\partial y_j}{\partial q_i} \delta q_i + \sum_{j=1}^{n} (m_j \dot{y}_j) \frac{\partial \dot{y}_j}{\partial q_i} \delta q_i \tag{1-34}$$

From Eq. 1-29

$$\frac{\partial T}{\partial q_i} = \sum_{j=1}^{n} (m_j \dot{y}_j) \frac{\partial \dot{y}_j}{\partial q_i} \tag{1-35}$$

Therefore, by substituting Eqs. 1-31 and 1-35 into Eq. 1-34,

$$-\sum_{j=1}^{n} (m_j \ddot{y}_j) \frac{\partial y_j}{\partial q_i} \delta q_i = -\frac{d}{dt} \frac{\partial T}{\partial \dot{q}_i} \delta q_i + \frac{\partial T}{\partial q_i} \delta q_i \tag{1-36}$$

Equations 1-36 and 1-28, by substitution, yield

$$\frac{\partial U}{\partial q_i} \delta q_i = \frac{\partial W_e}{\partial q_i} \delta q_i + \frac{\partial W_d}{\partial q_i} \delta q_i - \frac{d}{dt} \frac{\partial T}{\partial \dot{q}_i} \delta q_i + \frac{\partial T}{\partial q_i} \delta q_i$$

or, by rearranging terms and simplifying,

$$\frac{d}{dt} \frac{\partial T}{\partial \dot{q}_i} - \frac{\partial T}{\partial q_i} + \frac{\partial U}{\partial q_i} - \frac{\partial W_d}{\partial q_i} = \frac{\partial W_e}{\partial q_i} \tag{1-37}$$

Equation 1-37 is known as Lagrange's equation. Its application regarding the dynamic response of structural systems will yield as many differential equations of motion as there are degrees of freedom. The quantities that are required in applying Lagrange's equations are the expressions for T, U, W_d and W_e in terms of the generalized coordinates $q_1, q_2, \ldots, q_n$.

As an application of Lagrange's equations, the one-degree spring-mass system in Fig. 1-5c is considered. The displacement y of the mass m will be taken as the generalized coordinate q_i. One generalized coordinate is needed here because the spring-mass system has only one degree of freedom. The expressions for the energies required to apply Eq. 1-37 are as follows:

$$T = \tfrac{1}{2} m\dot{y}^2 \qquad U = \tfrac{1}{2} ky^2$$

$$W_d = (-c\dot{y})y \qquad W_e = F(t)y$$

where c is the viscous damping factor. By substituting these results into Eq. 1-37 and simplifying,

$$m\ddot{y} + ky + c\dot{y} = F(t) \tag{1-38}$$

Equation 1-38 is the familiar general differential equation given by Eq. 1-10.

Consider now the three-degree spring-mass system in Fig. 1-5*d*. In this case the displacements y_1, y_2, and y_3 of masses m_1, m_2, and m_3, respectively, are chosen as the generalized displacements. The required energy expressions are as follows:

$$T = \tfrac{1}{2} m_1\dot{y}_1^2 + \tfrac{1}{2} m_2\dot{y}_2^2 + \tfrac{1}{2} m_3\dot{y}_3^2$$

$$U = \tfrac{1}{2} k_1 y_1^2 + \tfrac{1}{2} k_2 (y_2 - y_1)^2 + \tfrac{1}{2} k_3 (y_3 - y_2)^2$$

$$W_d = (-c\dot{y}_1)y_1 + [-c(\dot{y}_2 - \dot{y}_1)](y_2 - y_1) + [-c(\dot{y}_3 - \dot{y}_2)](y_3 - y_2)$$

$$W_e = F_1(t)y_1 + F_2(t)y_2 + F_3(t)y_3$$

The required derivatives are

$$\frac{\partial T}{\partial \dot{y}_1} = m_1\dot{y}_1 \qquad \frac{\partial T}{\partial \dot{y}_2} = m_2\dot{y}_2 \qquad \frac{\partial T}{\partial \dot{y}_3} = m_3\dot{y}_3$$

$$\frac{\partial T}{\partial y_1} = \frac{\partial T}{\partial y_2} = \frac{\partial T}{\partial y_3} = 0$$

$$\frac{\partial U}{\partial y_1} = k_1 y_1 - k_2(y_2 - y_1)$$

$$\frac{\partial U}{\partial y_2} = k_2(y_2 - y_1) - k_3(y_3 - y_2)$$

$$\frac{\partial U}{\partial y_3} = k_3(y_3 - y_2)$$

$$\frac{\partial W_d}{\partial y_1} = -c\dot{y}_1 + c(\dot{y}_2 - \dot{y}_1)$$

$$\frac{\partial W_d}{\partial y_2} = -c(\dot{y}_2 - \dot{y}_1) + c(\dot{y}_3 - \dot{y}_2)$$

$$\frac{\partial W_d}{\partial y_3} = -c(\dot{y}_3 - \dot{y}_2)$$

$$\frac{\partial W_e}{\partial y_1} = F_1(t) \qquad \frac{\partial W_e}{\partial y_2} = F_2(t) \qquad \frac{\partial W_e}{\partial y_3} = F_3(t)$$

Thus, by Eq. 1-37, the three differential equations of motion for the three-degree system are

$$m_1\ddot{y}_1 + k_1y_1 - k_2(y_2 - y_1) + c\dot{y}_1 - c(\dot{y}_2 - \dot{y}_1) = F_1(t) \tag{1-39}$$

$$m_2\ddot{y}_2 + k_2(y_2 - y_1) - k_3(y_3 - y_2) + c(\dot{y}_2 - \dot{y}_1) - c(\dot{y}_3 - \dot{y}_2) = F_2(t) \tag{1-40}$$

$$m_3\ddot{y}_3 + k_3(y_3 - y_2) + c(\dot{y}_3 - \dot{y}_2) = F_3(t) \tag{1-41}$$

The solution for other types of problems can be obtained in a similar manner.

1-5 FREE VIBRATION OF ONE-DEGREE SPRING-MASS SYSTEMS

Consider the spring-mass system shown in Fig. 1-3*a*. Its undamped free vibration is represented by Eq. 1-12. That is,

$$m\ddot{y} + ky = 0 \tag{1-42}$$

By combining Eqs. 1-4 and 1-42,

$$-m\omega^2 y + ky = 0$$

or

$$\omega = \sqrt{k/m} \tag{1-43}$$

The expression yields the natural undamped frequency of vibration in units of radians per second (rps). In hertz (Hz) the frequency f_n is

$$f_n = \frac{\omega}{2\pi} = \frac{1}{2\pi}\sqrt{k/m} \tag{1-44}$$

The period of vibration τ is

$$\tau = \frac{1}{f_n} = 2\pi\sqrt{m/k} \tag{1-45}$$

The displacement function $y(t)$ can be determined by solving Eq. 1-42. This is a linear homogeneous equation with constant coefficients, whose solution is assumed to have the form

$$y = Ae^{\psi t} \tag{1-46}$$

where A and ψ are constants. By substituting Eq. 1-46 into Eq. 1-42,

$$m\psi^2 + k = 0$$

and

$$\psi_{1,2} = \pm\sqrt{-k/m} = \pm i\sqrt{k/m} = \pm i\omega \tag{1-47}$$

where $i = \sqrt{-1}$. Thus, by Eqs. 1-46 and 1-47,

$$y = A_1 e^{i\omega t} + A_2 e^{-i\omega t} \tag{1-48}$$

where A_1 and A_2 are constants.

Equation 1-48 can be expressed in trigonometric form by applying Euler's expression

$$e^{\pm i(\omega t)} = \cos\omega t \pm i\sin\omega t \tag{1-49}$$

This yields

$$y = C_1\cos\omega t + C_2\sin\omega t \tag{1-50}$$

where C_1 and C_2 replace the constants $(A_1 + A_2)$ and $i(A_1 - A_2)$, respectively.

If the initial conditions of motion are $y_{t=t_0} = y_0$ and $\dot{y}_{t=t_0} = \dot{y}_0$, then

$$C_1 = y_0\cos\omega t_0 - \frac{\dot{y}_0}{\omega}\sin\omega t_0$$

$$C_2 = y_0\sin\omega t_0 + \frac{\dot{y}_0}{\omega}\cos\omega t_0$$

and Eq. 1-50 yields

$$y(t) = y_0 \cos\omega(t - t_0) + \frac{\dot{y}_0}{\omega} \sin\omega(t - t_0) \tag{1-51}$$

$$\dot{y}(t) = -y_0\omega \sin\omega(t - t_0) + \dot{y}_0 \cos\omega(t - t_0) \tag{1-52}$$

The two terms in the right-hand side of Eq. 1-51 are separately plotted in Figs. 1-6*a* and 1-6*b*.

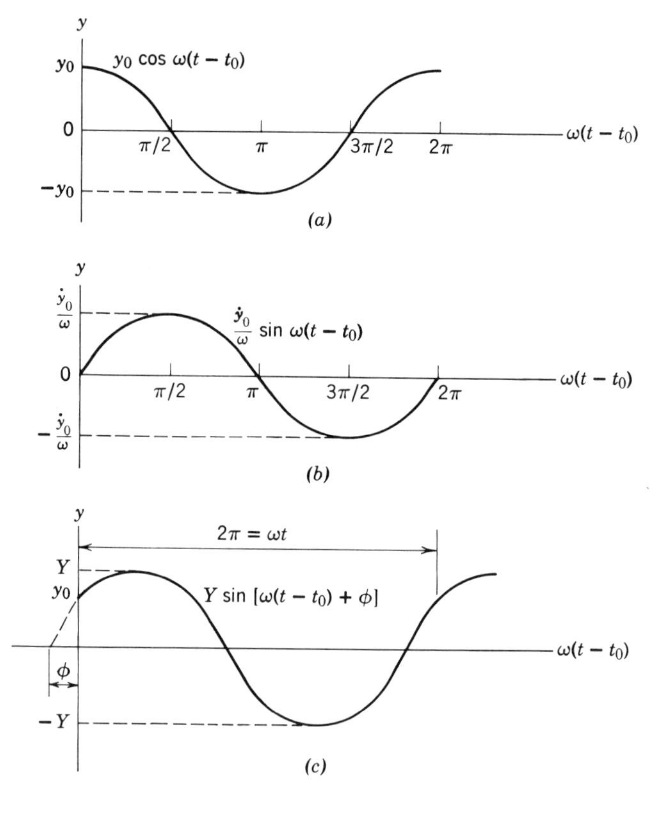

Fig. 1-6

Equations 1-51 and 1-52 can be rearranged differently by substituting for y_0 and $\dot{y}_0/\omega$ the expressions

$$y_0 = Y \sin\phi \qquad \frac{\dot{y}_0}{\omega} = Y \cos\phi$$

where ϕ is the phase angle. On this basis

$$y(t) = Y \sin[\omega(t - t_0) + \phi] \tag{1-53}$$

$$\dot{y}(t) = Y\omega \cos[\omega(t - t_0) + \phi] \tag{1-54}$$

where

$$Y = \sqrt{y_0^2 + (\dot{y}_0/\omega)^2} \tag{1-55}$$

and

$$\tan\phi = \frac{Y \sin\phi}{Y \cos\phi} = \frac{\omega y_0}{\dot{y}_0} \tag{1-56}$$

Equation 1-53 is plotted in Fig. 1-6*c*.

1-6 FREE VIBRATION WITH VISCOUS DAMPING

Consider now the case where the one-degree spring-mass system in Fig. 1-3*a* vibrates freely under the influence of viscous damping. Its differential equation of motion is given by Eq. 1-11, that is,

$$m\ddot{y} + c\dot{y} + ky = 0 \tag{1-57}$$

Its solution is assumed of the form

$$y = Ce^{pt} \tag{1-58}$$

where C and p are constants.

By substituting Eq. 1-58 into Eq. 1-57, we have

$$mp^2 + cp + k = 0 \tag{1-59}$$

and the two roots of this equation are

$$p_{1,2} = -\frac{c}{2m} \pm \sqrt{(c/2m)^2 - k/m} \tag{1-60}$$

Thus the solution given by Eq. 1-58 becomes

$$y(t) = C_1 e^{p_1 t} + C_2 e^{p_2 t} \tag{1-61}$$

where C_1 and C_2 are constants to be determined from the initial conditions of the motion. For example, at $t = 0$ the initial conditions for y and $\dot{y}$ may be taken as y_0 and zero, respectively. The quantities p_1 and p_2 are given by Eq. 1-60.

Three important conclusions can be drawn from Eq. 1-60:

1. If the sum of the terms under the radical is equal to zero, then

$$\left(\frac{c}{2m}\right)^2=\frac{k}{m} \tag{1-62}$$

The term k/m is equal to ω^2, where ω is the undamped natural frequency of the system. Under this condition, Eq. 1-60 yields $p_1=-c/2m$. This form of damping represents a transition from oscillatory to nonoscillatory configuration and is known as the condition of critical damping. For example, if the motion starts from the position y_0 with velocity $\dot{y}_0$, the mass will return to rest without oscillation. This does not usually occur in practice.

2. The condition of overdamping occurs when the terms under the radical satisfy the inequality

$$\left(\frac{c}{2m}\right)^2>\frac{k}{m} \tag{1-63}$$

In this case the motion is aperiodic and p_1 and p_2 are always real values and negative.

3. The third and most usual case is when

$$\left(\frac{c}{2m}\right)^2<\frac{k}{m} \tag{1-64}$$

which is known as the underdamped condition. In this case, the system is said to be lightly damped and it will vibrate with a decreasing amplitude. Damping for most mechanical and structural systems is usually light.

The three conditions can be expressed in somewhat different manner by introducing the critical damping factor c_c. Critical damping is defined as the value of c in Eq. 1-60 that makes the algebraic sum of the terms under the radical equal to zero. Thus, from Eq. 1-62,

$$c_c=2\sqrt{km}=2m\omega \tag{1-65}$$

From this point of view, the damping of the system may be specified in terms of c_c and the damping ratio

$$\zeta=\frac{c}{c_c} \tag{1-66}$$

In this manner, Eq. 1-60 can be written as

$$p_{1,2} = \left(-\zeta \pm \sqrt{\zeta^2 - 1}\right)\omega \tag{1-67}$$

Consequently, the condition of critical damping occurs when $\zeta = 1$, that of overdamping when $\zeta > 1$, and that of light damping when $\zeta < 1$.

By applying Euler's expression, Eq. 1-49, the expression given by Eq. 1-61 becomes

$$y(t) = e^{-\zeta\omega t}\left[A_1 \cos\left(\omega\sqrt{\zeta^2 - 1}\right)t + A_2 \sin\left(\omega\sqrt{\zeta^2 - 1}\right)t\right] \tag{1-68}$$

The natural frequency ω_d of the damped one-degree system is

$$\omega_d = \omega\sqrt{\zeta^2 - 1} \tag{1-69}$$

1-7 FREE VIBRATION WITH COULOMB DAMPING

The resisting force in this case is assumed constant, provided that the surface is uniform and that the difference in its value between starting and moving conditions is small so that it can be neglected. For a one-degree spring-mass system vibrating freely, Fig. 1-7*a*, the differential equation of motion is

$$m\ddot{y} + ky \pm F_f = 0 \tag{1-70}$$

y
k
m
$\pm F_f$

(*a*)

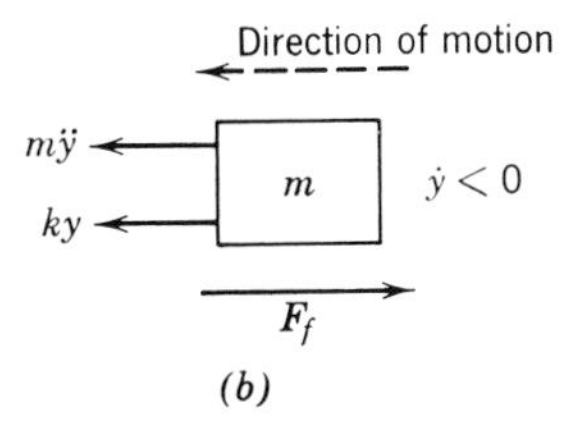

(*b*)

Fig. 1-7

where F_f is the resisting force and is given by Eq. 1-8. The positive sign of F_f applies when the velocity $\dot{y}$ is positive, and the negative sign is used for negative velocity. A solution of Eq. 1-70 is valid only for the time interval in which the velocity remains unchanged with respect to sign.

Consider first the case where the initial displacement of the mass m in Fig. 1-7*a* is y_0 and that the velocity $\dot{y}$ is zero at this point. If the mass is released from this position, it moves to the left, and the velocity is negative. Thus during the first half cycle of motion the differential equation of motion is

$$m\ddot{y}+ky=F_f \tag{1-71}$$

This is a nonhomogeneous differential equation and its solution $y(t)$ is obtained by superimposing the homogeneous or complementary solution $y_c(t)$ and the particular solution $y_p(t)$. That is,

$$y(t)=y_c(t)+y_p(t) \tag{1-72}$$

The solution $y_c(t)$ is given by Eq. 1-50 as

$$y_c(t)=C_1\cos\omega t+C_2\sin\omega t \tag{1-73}$$

The solution $y_p(t)$, since F_f is constant, is taken as

$$y_p=A \tag{1-74}$$

where A is constant. By substituting Eq. 1-74 in Eq. 1-71, we have

$$A=\frac{F_f}{k}$$

Thus

$$y_p=\frac{F_f}{k} \tag{1-75}$$

By utilizing Eqs. 1-72, 1-73, and 1-75, we have the complete solution $y(t)$:

$$y(t)=C_1\cos\omega t+C_2\sin\omega t+\frac{F_f}{k} \tag{1-76}$$

The initial conditions are $y=y_0$ and $\dot{y}=0$ at $t=0$. Thus

$$C_1=y_0-\frac{F_f}{k} \qquad C_2=0$$

and

$$y(t)=\left(y_0-\frac{F_f}{k}\right)\cos\omega t+\frac{F_f}{k} \tag{1-77}$$

During the first half cycle of motion, the mass m is moving to the left and the velocity $\dot{y}$ will become zero when $\omega t=\pi$. This is easily verified by taking the first derivative $\dot{y}(t)$ of Eq. 1-77. Thus, at $t=\pi/\omega$, the first negative peak of Eq. 1-77 is

$$y=-\left(y_0-\frac{2F_f}{k}\right) \tag{1-78}$$

At the end of the first half cycle the mass m will start moving to the right and the velocity will become positive. Consequently, the equation of motion for the second half cycle is

$$m\ddot{y}+ky=-F_f \tag{1-79}$$

The solution of this equation is

$$y(t)=A_1\cos\omega t+A_2\sin\omega t-\frac{F_f}{k} \tag{1-80}$$

The initial conditions are $y=-(y_0-2F_f/k)$ and $\dot{y}=0$ at $t=\pi/\omega$. This yields

$$A_1=y_0-\frac{3F_f}{k} \qquad A_2=0$$

and

$$y(t)=\left(y_0-\frac{3F_f}{k}\right)\cos\omega t-\frac{F_f}{k} \tag{1-81}$$

At the end of the second half cycle of motion the velocity is again zero and the displacement y is obtained by placing $t=2\pi/\omega$ in Eq. 1-81. That is,

$$y_{t=2\pi/\omega}=y_0-\frac{4F_f}{k} \tag{1-82}$$

This shows that in each cycle the amplitude is reducing by an amount equal to $4F_f/k$. It will become zero when

$$y_0=\left(\frac{4F_f}{k}\right)n \tag{1-83}$$

where n is the number of complete cycles.

1-8 FREE VIBRATION WITH HYSTERESIS DAMPING

Hysteresis damping is defined in Section 1-2, where Eq. 1-9 yields the energy loss per cycle of motion. The magnitude of this loss depends on the area within the loop in Fig. 1-2. For some materials, such as rubber, the loop may be large. For others, such as steel, this loop is usually small. For many engineering structures the amount of hysteresis damping is often small and can be neglected.

For a one-degree spring-mass system the effects of hysteresis damping could be taken into consideration by determining an equivalent viscous damping factor ζ_e and an equivalent viscous damping constant c_e. On this basis, Eq. 1-64 yields

$$y(t) = e^{-\zeta_e \omega t}\left[A_1 \cos\left(\omega\sqrt{\zeta_e^2 - 1}\right)t + A_2 \sin\left(\omega\sqrt{\zeta_e^2 - 1}\right)t\right] \quad (1\text{-}84)$$

where A_1 and A_2 are constants depending on the initial conditions of the motion.

An appropriate expression for determining ζ_e and c_e may be derived by considering the consecutive peaks A, B, and C in Fig. 1-8 and applying

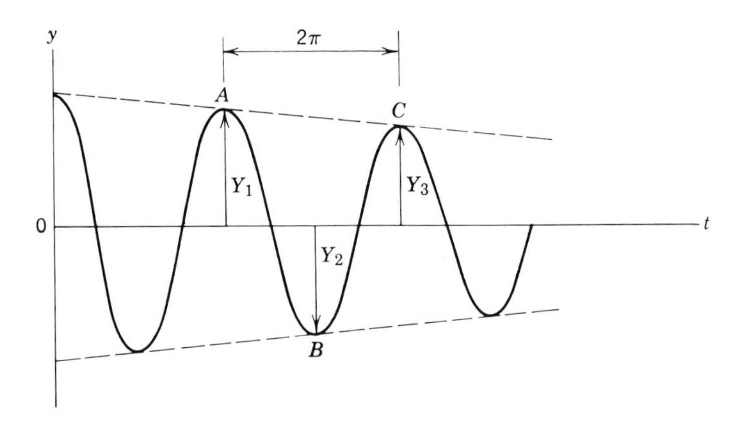

Fig. 1-8

energy principles. That is, if the energy loss per quarter cycle is assumed to be equal to $k\pi c_0 Y^2/4$, the energy equation for the half cycle, A to B in Fig. 1-8, is

$$\frac{kY_1^2}{2} - \frac{k\pi c_0 Y_1^2}{4} - \frac{k\pi c_0 Y_2^2}{4} = \frac{kY_2^2}{2}$$

or

$$\left(\frac{Y_1}{Y_2}\right)^2 = \frac{2+\pi c_0}{2-\pi c_0} \tag{1-85}$$

Similarly, for the next half cycle B to C,

$$\left(\frac{Y_2}{Y_3}\right)^2 = \frac{2+\pi c_0}{2-\pi c_0} \tag{1-86}$$

The product of Eqs. 1-85 and 1-86, by considering c_0 as small, yields the approximate expression

$$\frac{Y_1}{Y_3} \approx 1+\pi c_0 \tag{1-87}$$

and the log decrement δ, defined as

$$\delta = \ln\left(\frac{Y_1}{Y_2}\right) \approx \pi c_0 \tag{1-88}$$

Thus, by utilizing the equation above,

$$2\pi\zeta_e \approx \pi c_0$$

or

$$\zeta_e = \frac{c_0}{2} \tag{1-89}$$

With the help of Eqs. 1-65, 1-66, and 1-89,

$$c_e = c_c\zeta_e = c_0\sqrt{km} = \frac{c_0 k}{\omega} \tag{1-90}$$

For example, if in a vibration test the ratio Y_1 / Y_3 is measured and is found to be equal to 1.04, then, by Eq. 1-87, $c_0 = 0.0128$. Consequently, Eqs. 1-89 and 1-90 yield $\zeta_e = 0.0064$ and $c_e = 0.0128\sqrt{km}$, respectively. With ζ_e known, Eq. 1-84 will yield the amplitude response of the system.

1-9 FREE VIBRATION OF TWO-DEGREE SPRING-MASS SYSTEMS

Consider the two-degree spring-mass system in Fig. 1-3c and let it be assumed that the dynamic forces $F_1(t)$, $F_2(t)$ and the damping constants c_1,

c_2 are all zero. If the system is excited, the masses m_1 and m_2 will vibrate freely about their equilibrium position and the motion is represented by the differential equations (1-17) and (1-18) in Section 1-3. That is,

$$m_1\ddot{y}_1 + k_1 y_1 - k_2(y_2 - y_1) = 0 \tag{1-91}$$

$$m_2\ddot{y}_2 + k_2(y_2 - y_1) = 0 \tag{1-92}$$

The two expressions above are ordinary linear homogeneous differential equations with constant coefficients. If the motion is assumed to be harmonic, the dynamic amplitudes y_1 and y_2 are expressed as

$$y_1 = Y_1 \sin\omega t \qquad y_2 = Y_2 \sin\omega t \tag{1-93}$$

where Y_1 and Y_2 are maximum amplitudes.

By substituting the expressions in Eq. 1-93 into Eqs. 1-91 and 1-92 and rearranging terms,

$$(m_1\omega^2 - k_1 - k_2)Y_1 + k_2 Y_2 = 0 \tag{1-94}$$

$$k_2 Y_1 + (m_2\omega^2 - k_2)Y_2 = 0 \tag{1-95}$$

For a solution other than $Y_1 = Y_2 = 0$, the determinant of the coefficients of Y_1 and Y_2 in Eqs. 1-94 and 1-95 must be zero. That is,

$$\begin{vmatrix} (m_1\omega^2 - k_1 - k_2) & k_2 \\ k_2 & (m_2\omega^2 - k_2) \end{vmatrix} = 0 \tag{1-96}$$

Expansion of the determinant yields the frequency equation

$$\omega^4 - \left(\frac{k_1 + k_2}{m_1} + \frac{k_2}{m_2}\right)\omega^2 + \frac{k_1 k_2}{m_1 m_2} = 0 \tag{1-97}$$

Equation 1-97 yields the two values of ω^2 as

$$\omega_{1,2}^2 = \frac{1}{2}\left(\frac{k_1 + k_2}{m_1} + \frac{k_2}{m_2}\right) \pm \frac{1}{2}\left[\left(\frac{k_1 + k_2}{m_1} + \frac{k_2}{m_2}\right)^2 - 4\frac{k_1 k_2}{m_1 m_2}\right]^{\frac{1}{2}} \tag{1-98}$$

Normally, there are four values of ω that satisfy Eq. 1-97. These roots are $\pm\omega_1$ and $\pm\omega_2$. In practical problems, such as the spring-mass system above, the negative values of ω are disregarded as being of no physical significance. Thus the natural frequencies of the system are the two positive roots ω_1 and ω_2, which are equal in number with the number of degrees of freedom of the system.

Each natural frequency of vibration is associated with a definite deflection configuration of the system, generally known as mode of vibration. Consequently, the number of modes of vibration equals the number of natural frequencies at which the system vibrates, and also equals the number of its degrees of freedom.

For the two-degree spring-mass system, the two modes corresponding to ω_1 and ω_2 can be determined by using either Eq. 1-94 or Eq. 1-95. By utilizing Eq. 1-95,

$$\frac{Y_2}{Y_1}=\frac{k_2}{k_2-m_2\omega^2}$$

or

$$Y_2=CY_1 \tag{1-99}$$

where

$$C=\frac{k_2}{k_2-m_2\omega^2} \tag{1-100}$$

Equation 1-99 yields the relationship between the amplitudes Y_1 and Y_2, thus describing the mode shapes at which the system vibrates. For $\omega=\omega_1$ and $\omega=\omega_2$, the corresponding values of C are

$$C_1=\frac{k_2}{k_2-m_2{\omega_1}^2} \qquad C_2=\frac{k_2}{k_2-m_2{\omega_2}^2} \tag{1-101}$$

and Eq. 1-99 yields,

$$Y_2^{(1)}=C_1Y_1^{(1)} \tag{1-102}$$

$$Y_2^{(2)}=C_2Y_1^{(2)} \tag{1-103}$$

The superscripts (1) and (2) indicate first and second mode, respectively. That is, Eq. 1-102 provides the amplitude relationship in the first mode, and that of the second mode is given by Eq. 1-103. If the position of the one mass is defined, the position of the other mass can be determined. In practice, only the amplitude relationship is usually of importance, and a value of unity is assigned to one of the amplitudes. If this is the case, the modes so defined are said to be normalized to unity. This is valid, because a natural frequency is dependent only on the corresponding mode shape and is independent of the actual values of the amplitudes in that mode.

The same expressions and conclusions regarding frequencies and mode shapes can be drawn by assuming that the solutions of the differential equations (1-91) and (1-92) are given by the expressions

$$y_1 = Y_1 e^{\psi t} \qquad y_2 = Y_2 e^{\psi t} \tag{1-104}$$

and proceeding as above.

It is easily noted in this procedure that the four values of ψ are $\psi_1 = i\omega_1$, $\psi_2 = -i\omega_1$, $\psi_3 = i\omega_2$, and $\psi_4 = -i\omega_2$, where $i = \sqrt{-1}$. Thus, by utilizing these values and the amplitude relationships given by Eqs. 1-102 and 1-103, the displacement functions y_1 and y_2 in Eq. 1-104 yield

$$y_1 = (Y_1)_1 e^{i\omega_1 t} + (Y_1)_2 e^{-i\omega_1 t} + (Y_1)_3 e^{i\omega_2 t} + (Y_1)_4 e^{-i\omega_2 t}$$

$$y_2 = C_1[(Y_1)_1 e^{i\omega_1 t} + (Y_1)_2 e^{-i\omega_1 t}] + C_2[(Y_1)_3 e^{i\omega_2 t} + (Y_1)_4 e^{-i\omega_2 t}]$$

By applying Euler's relation, the equations yield

$$y_1 = A_1 \cos\omega_1 t + A_2 \sin\omega_1 t + A_3 \cos\omega_2 t + A_4 \sin\omega_2 t \tag{1-105}$$

$$y_2 = C_1(A_1 \cos\omega_1 t + A_2 \sin\omega_1 t) + C_2(A_3 \cos\omega_2 t + A_4 \sin\omega_2 t) \tag{1-106}$$

The two equations show that the motion is composed of two harmonic motions corresponding to the natural frequencies ω_1 and ω_2 of the system. When the system vibrates with frequency ω_1 only, Eq. 1-106 yields the amplitude relationship

$$y_2 = C_1 y_1 \tag{1-107}$$

From the same equation, the amplitude relationship with frequency ω_2 only is

$$y_2 = C_2 y_1 \tag{1-108}$$

The same conclusions regarding mode shapes can be again drawn here. The constants A_1, A_2, A_3, and A_4 in Eqs. 1-105 and 1-106 can be determined from given initial conditions of the motion. For example,

$$\text{at } t = t_0: \qquad \begin{aligned} y_1 &= y_1^0 & \dot{y}_1 &= \dot{y}_1^0 \\ y_2 &= y_2^0 & \dot{y}_2 &= \dot{y}_2^0 \end{aligned} \tag{1-109}$$

These conditions can be modified and written in terms of initial conditions that take into account the normal modes. From Eqs. 1-105 and 1-106 it is noted that

$$y_1 = y_1^{(1)} + y_1^{(2)} \tag{1-110}$$

$$y_2 = y_2{}^{(1)} + y_2{}^{(2)} = C_1 y_1{}^{(1)} + C_2 y_1{}^{(2)} \tag{1-111}$$

where the superscripts (1) and (2) denote the first and second mode, respectively. Therefore, the expressions in Eq. 1-109 can be written as

$$\begin{aligned} \text{at} \quad t = t_0: \quad & y_1 = y_1^0 = y_1^{0(1)} + y_1^{0(2)} \\ & y_2 = y_2^0 = y_2^{0(1)} + y_2^{0(2)} = C_1 y_1^{0(1)} + C_2 y_1^{0(2)} \\ & \dot{y}_1 = \dot{y}_1^0 = \dot{y}_1^{0(1)} + \dot{y}_1^{0(2)} \\ & \dot{y}_2 = \dot{y}_2^0 = \dot{y}_2^{0(1)} + \dot{y}_2^{0(2)} = C_1 \dot{y}_1^{0(1)} + C_2 \dot{y}_1^{0(2)} \end{aligned} \tag{1-112}$$

Simultaneous solution of the four expressions will yield the four initial conditions $y_1^{0(1)}$, $y_1^{0(2)}$, $\dot{y}_1^{0(1)}$, and $\dot{y}_1^{0(2)}$ for the two modes.

1-10 TWO-DEGREE SYSTEMS WITH VISCOUS DAMPING

If the time-varying forces $F_1(t)$ and $F_2(t)$ in Fig. 1-3*c* are zero, the two-degree spring-mass system will vibrate freely under the influence of viscous damping. The differential equations of motion are given by Eqs. 1-15 and 1-16. That is,

$$m_1 \ddot{y}_1 + c_1 y_1 - c_2 (\dot{y}_2 - \dot{y}_1) + k_1 y_1 - k_2 (y_2 - y_1) = 0 \tag{1-113}$$

$$m_2 \ddot{y}_2 + c_2 (\dot{y}_2 - \dot{y}_1) + k_2 (y_2 - y_1) = 0 \tag{1-114}$$

The type of motion to be examined here is the one where y_1 and $\dot{y}_1$ are assumed to be greater than y_2 and $\dot{y}_2$, respectively. With this in mind, the terms of Eqs. 1-113 and 1-114 are rearranged as follows:

$$m_1 \ddot{y}_1 + (c_1 + c_2) \dot{y}_1 - c_2 \dot{y}_2 + (k_1 + k_2) y_1 - k_2 y_2 = 0 \tag{1-115}$$

$$m_2 \ddot{y}_2 - c_2 \dot{y}_1 + c_2 \dot{y}_2 - k_2 y_1 + k_2 y_2 = 0 \tag{1-116}$$

The solutions for y_1 and y_2 are taken as

$$y_1 = A_1 e^{\Psi t} \quad \text{and} \quad y_2 = A_2 e^{\Psi t} \tag{1-117}$$

where Ψ, A_1, and A_2 are constants.

By substituting these solutions into Eqs. 1-115 and 1-116 and rearranging items,

$$[m_1 \Psi^2 + (c_1 + c_2) \Psi + k_1 + k_2] A_1 - (c_2 \Psi + k_2) A_2 = 0 \tag{1-118}$$

$$-(c_2 \Psi + k_2) A_1 + (m_2 \Psi^2 + c_2 \Psi + k_2) A_2 = 0 \tag{1-119}$$

For a nontrivial solution the determinant of the coefficients of A_1 and A_2 must be zero. That is

$$\begin{vmatrix} m_1\Psi^2+(c_1+c_2)\Psi+k_1+k_2 & -(c_2\Psi+k_2) \\ -(c_2\Psi+k_2) & (m_2\Psi^2+c_2\Psi+k_2) \end{vmatrix}=0$$

Expansion of the determinant yields the characteristic equation

$$m_1m_2\Psi^4+[m_1c_2+(c_1+c_2)m_2]\Psi^3+[m_1k_2+(k_1+k_2)m_2+c_1c_2]\Psi^2 + (c_1k_2+c_2k_1)\Psi+k_1k_2=0 \tag{1-120}$$

This is a fourth degree equation, and four values of Ψ satisfy it. When these values are properly substituted into Eq. 1-117, the solutions of Eqs. 1-115 and 1-116 are obtained.

An examination of Eq. 1-120 reveals that the coefficients of Ψ^4, Ψ^3, Ψ^2, and Ψ, as well as the product k_1k_2, are all positive. Thus a solution of this equation could be the one where all four roots are complex. Since there is damping, the real part of each root will be negative. Consequently,

$$\begin{aligned} \Psi_1&=-a_1+ib_1 \qquad & \Psi_3&=-a_2+ib_2 \\ \Psi_2&=-a_1-ib_1 \qquad & \Psi_4&=-a_2-ib_2 \end{aligned} \tag{1-121}$$

In the equation above, a_1, a_2, b_1, and b_2 are positive

By substituting the values of Ψ from Eq. 1-121 into the solutions given by Eq. 1-117, we have

$$y_1=(A_1)_1e^{(-a_1+ib_1)t}+(A_1)_2e^{(-a_1-ib_1)t} + (A_1)_3e^{(-a_2+ib_2)t}+(A_1)_4e^{(-a_2-ib_2)t} \tag{1-122}$$

$$y_2=(A_2)_1e^{(-a_1+ib_1)t}+(A_2)_2e^{(-a_1-ib_1)t} + (A_2)_3e^{(-a_2+ib_2)t}+(A_2)_4e^{(-a_2-ib_2)t} \tag{1-123}$$

or, by applying the Eulerian relation and simplifying,

$$y_1=e^{-a_1t}(C_1\cos b_1t+C_2\sin b_1t)+e^{-a_2t}(C_3\cos b_2t+C_4\sin b_2t) \tag{1-124}$$

$$y_2 = e^{-a_1 t}(C_5 \cos b_1 t + C_6 \sin b_1 t) + e^{-a_2 t}(C_7 \cos b_2 t + C_8 \sin b_2 t) \tag{1-125}$$

The equations above show that the decay of the motion is exponential and is defined by the real parts a_1 and a_2 of the roots in Eq. 1-121. On the other hand, the vibration of the system is characterized by the imaginary parts b_1 and b_2. If damping is very light, then a_1 and a_2 will be small and b_1 and b_2 will be closely the same as the natural frequencies ω_1 and ω_2 of the system.

There are eight constants in the pair of Eqs. 1-122 and 1-123, or the pair (1-124) and (1-125), that would have to be determined. In determining these constants, the first step includes the use 1-121 and the auxiliary equation (1-118) or (1-119). For example, by substituting the value of Ψ_1 from Eq. 1-121 into Eq. 1-119, the ratio $(A_2)_1/(A_1)_1$ is determined. In a similar manner, the ratios $(A_2)_2/(A_1)_2$, $(A_2)_3/(A_1)_3$, and $(A_2)_4/(A_1)_4$ are obtained by substituting into Eq. 1-119 the values of Ψ_2, Ψ_3, and Ψ_4. From these ratios, the ratios C_5/C_1, C_6/C_2, C_7/C_3, and C_8/C_4 can be also obtained, because Eqs. 1-124 and 1-125 have been derived from Eqs. 1-122 and 1-123, respectively, by applying the Eulerian relation. The additional four conditions needed to complete the evaluation of all eight constants are the initial conditions of the motion. For example,

$$\text{at} \quad t = t_0: \quad \begin{matrix} y_1 = y_1^0 & \dot{y}_1 = \dot{y}_1^0 \\ y_2 = y_2^0 & \dot{y}_2 = \dot{y}_2^0 \end{matrix} \tag{1-126}$$

By returning to Eq. 1-120, another set of four roots of Ψ can be obtained by substituting $-\Psi$ for Ψ in Eq. 1-120. On this basis, four negative real roots are obtained. Proceeding as before and examining the motion, it will be found that it is aperiodic. This solution represents the case of large damping where the system, if disturbed, will return to its equilibrium position without oscillation.

The final set of four roots of Ψ that can be obtained from Eq. 1-120 includes two real and negative roots and a complex conjugate pair. Examination of this solution reveals that for each mass of the system, an aperiodic part of the motion is superimposed on a damped vibration.

In practice, however, structures are usually subjected to light damping and only the first case, if any, would be of concern.

1-11 FREE VIBRATION OF UNIFORM BEAMS

The partial differential equation of motion for the undamped free vibration of beams is given by Eq. 1-22 in Section 1-3. That is,

$$EI\frac{\partial^4 y}{\partial x^4}+m\ddot{y}=0 \tag{1-127}$$

by using Eq. 1-4 and

$$\lambda^4=\frac{m\omega^2}{EI} \tag{1-128}$$

Eq. 1-127 yields

$$\frac{d^4y}{dx^4}-\lambda^4 y=0 \tag{1-129}$$

Total derivatives have been used in the equation above because y varies only with x.

The solution of Eq. 1-129, with EI assumed constant, is taken as

$$y=Ae^{\psi x} \tag{1-130}$$

where A and ψ are constants. This yields

$$\psi^4=\lambda^4 \tag{1-131}$$

Therefore, with $i=\sqrt{-1}$, the four roots of Eq. 1-131 are

$$\psi_1=\lambda \qquad \psi_3=i\lambda$$

$$\psi_2=-\lambda \qquad \psi_4=-i\lambda$$

and Eq. 1-130 yields

$$y=A_1e^{\lambda x}+A_2e^{i\lambda x}+A_3e^{-\lambda x}+A_4e^{-i\lambda x} \tag{1-132}$$

By applying the Eulerian relation as in previous sections, Eq. 1-132 takes the form

$$y=C_1\cosh\lambda x+C_2\sinh\lambda x+C_3\cos\lambda x+C_4\sin\lambda x \tag{1-133}$$

Equation 1-133 is the general solution of Eq. 1-129 and applies to uniform m and EI spans of any boundary conditions. The constants C_1, C_2, C_3, and C_4 can be determined from the boundary conditions of the given problem. At the two ends of a beam span there always exist four end conditions that can be used to determine these constants or ratios between these constants. Utilization of these end restraints permits one to draw conclusions that lead to the evaluation of the values of λ and the corresponding mode shapes of the beam. With known λ's, the natural frequencies ω and the corresponding periods of vibration τ can be obtained from the expressions

$$\omega=\lambda^2\sqrt{EI/m} \tag{1-134}$$

$$\tau=\frac{2\pi}{\omega}=\frac{2\pi}{\lambda^2}\sqrt{m/EI} \tag{1-135}$$

Since a beam has infinite degrees of freedom, the natural frequencies of vibration and the corresponding mode shapes will be infinite in number.

As an illustration, assume that it is required to determine the natural frequencies and mode shapes of a simply supported beam of constant mass and stiffness. The end conditions of the member are:

$$\begin{aligned} \text{at} \quad x=0: \quad & y=0, \quad \frac{d^2y}{dx^2}=0 \\ x=L: \quad & y=0, \quad \frac{d^2y}{dx^2}=0 \end{aligned} \tag{1-136}$$

Application of these four conditions yields

$$C_1+C_3=0 \tag{1-137}$$

$$C_1-C_3=0 \tag{1-138}$$

$$C_1\cosh\lambda L+C_2\sinh\lambda L+C_3\cos\lambda L+C_4\sin\lambda L=0 \tag{1-139}$$

$$C_1\lambda^2\cosh\lambda L+C_2\lambda^2\sinh\lambda L-C_3\lambda^2\cos\lambda L-C_4\lambda^2\sin\lambda L=0 \tag{1-140}$$

Equations 1-137 and 1-138 suggest that C_1 and C_3 should be both zero, because C_1 cannot be equal to both $-C_3$ and $+C_3$. Consequently, Eqs. 1-139 and 1-140 become

$$C_2\sinh\lambda L+C_4\sin\lambda L=0 \tag{1-141}$$

$$C_2\sinh\lambda L-C_4\sin\lambda L=0 \tag{1-142}$$

Algebraic addition and subtraction of Eqs. 1-141 and 1-142 yields the expressions

$$C_2\sinh\lambda L=0 \tag{1-143}$$

$$C_4\sin\lambda L=0 \tag{1-144}$$

In Eq. 1-143 the constant C_2 must be zero because $\sinh\lambda L$ cannot be zero. In Eq. 1-144 the constant C_4 cannot be zero, because if it is zero a trivial solution of no vibration is obtained. Thus the frequency equation is

$$\sin \lambda L = 0 \qquad (1\text{-}145)$$

The roots of λL satisfying Eq. 1-145 are

$$\lambda L = n\pi \qquad n = 1, 2, 3, \ldots \qquad (1\text{-}146)$$

and

$$\lambda = \frac{n\pi}{L} \qquad (1\text{-}147)$$

By applying Eq. 1-134 the natural frequencies of the beam are

$$\omega_n = \frac{n^2\pi^2}{L^2}\sqrt{EI/m} \qquad n = 1, 2, 3, \ldots \qquad (1\text{-}148)$$

The corresponding mode shapes, by Eq. 1-133, are given by the expression

$$y = C_4 \sin \lambda x \qquad (1\text{-}149)$$

where λ is given by Eq. 1-147. Since there are n values of λ, Eq. 1-149 is written as

$$y_n = C_n \sin \frac{n\pi x}{L} \qquad n = 1, 2, 3, \ldots \qquad (1\text{-}150)$$

Equation 1-150, for $n = 1$, yields the mode shape corresponding to the fundamental frequency of the beam. For $n = 2, 3, \ldots$, Eq. 1-150 yields the mode shapes corresponding to the higher frequencies 2, 3, ...

The constant C_n in Eq. 1-150 is arbitrary and can be taken equal to unity. This means that the natural frequencies of a beam depend only on the shape of the corresponding modes and not on the actual amplitudes of the mode. Consequently, the mode shape equation can be written as

$$\beta_n(x) = \sin \frac{n\pi x}{L} \qquad (1\text{-}151)$$

Equations 1-148 and 1-151 could be also obtained by setting equal to zero the determinant of the coefficients of the C's in Eqs. 1-137 to 1-140 inclusive. Expansion of this determinant will yield the expression

$$\sinh \lambda L \sin \lambda L = 0, \qquad (1\text{-}152)$$

and the same conclusions regarding natural frequencies and mode shapes can be drawn from this equation.

1-12 ORTHOGONALITY PROPERTIES OF NORMAL MODES

In many cases the analysis of structural systems for dynamic response requires the computation of its natural frequencies and the corresponding modes of vibration. An important property that helps the determination of these quantities is the orthogonality relation that exists between any two modes of a system with any number of degrees of freedom. For the two-degree spring-mass system discussed in Section 1-10, the amplitudes $Y_1^{(1)}$, $Y_2^{(1)}$, $Y_1^{(2)}$, and $Y_2^{(2)}$ satisfy the important equation

$$m_1 Y_1^{(1)} Y_1^{(2)} + m_2 Y_2^{(1)} Y_2^{(2)} = 0, \tag{1-153}$$

known as the orthogonality property of the amplitudes of the two normal modes of the system.

For a system of any number of degrees of freedom, an orthogonality relationship such as the one given by Eq. 1-153 will exist between modes 1 and 2, 2 and 3, 3 and 4, 1 and 3, and so on. Consequently, Eq. 1-153 can take a more general form that can be used conveniently in a given situation. The derivation of this general expression is as follows:

Consider the nth mode of a system and let the inertia forces of this mode move through the displacements of the mth mode. Then consider the mth mode and let the inertia forces of this mode move through the displacements of the nth mode. In accordance with Betti's[3] law, the following expression can be written:

$$\sum_{i=1}^{i=j} (\omega_n^2 m_i Y_i^{(n)}) Y_1^{(m)} = \sum_{i=1}^{i=j} (\omega_m^2 m_i Y_i^{(m)}) Y_i^{(n)} \tag{1-154}$$

In the equation above, $\omega_n^2 m_i Y_i^{(n)}$ and $\omega_m^2 m_i Y_i^{(m)}$ are the inertia forces of mass m_i for the nth and mth modes, respectively. By rearranging terms, Eq. 1-154 yields

$$(\omega_m^2 - \omega_n^2) \sum_{i=1}^{i=j} m_i Y_i^{(n)} Y_i^{(m)} = 0 \tag{1-155}$$

The superscripts (n) and (m) designate nth and mth modes, respectively. For $m \neq n$, Eq. 1-155 yields

$$\sum_{i=1}^{i=j} m_i Y_i^{(n)} Y_i^{(m)} = 0 \tag{1-156}$$

This relationship between the amplitudes of the nth and mth modes is

[3]See Reference 78, page 24.

known as the orthogonality property of these modes. A condition like the one given by Eq. 1-156 exists between any two modes of the system. One may easily verify that for a two-degree system, that is, $i=1$ and 2, $n=1$, and $m=2$, Eq. 1-156 yields Eq. 1-153.

1-13 THE FLEXIBILITY MATRIX

In many cases the dynamic analysis of structures becomes convenient if the displacements are expressed in terms of flexibility coefficients. A flexibility coefficient a_{ij} is defined as the deflection at point i due to a unit load acting at point j. Figure 1-9a depicts a two-span continuous beam acted upon by loads $P_1, P_2, \ldots, P_n$, producing the total deflections $y_1, y_2, \ldots, y_n$ at points 1, 2, ..., n, respectively. A unit load at point 1, Fig. 1-9b, yields the flexibility coefficients $a_{11}, a_{21}, \ldots, a_{n1}$ at points 1, 2, ..., n, respectively.

The total deflection $y_1, y_2, \ldots, y_n$, in terms of flexibility coefficients, are as follows:

$$
\begin{aligned}
y_1 &= a_{11}P_1 + a_{12}P_2 + \ldots + a_{1n}P_n \\
y_2 &= a_{21}P_1 + a_{22}P_2 + \ldots + a_{2n}P_n \\
&\ldots\ldots\ldots\ldots\ldots\ldots \\
y_n &= a_{n1}P_1 + a_{n2}P_2 + \ldots + a_{nn}P_n
\end{aligned}
\tag{1-157}
$$

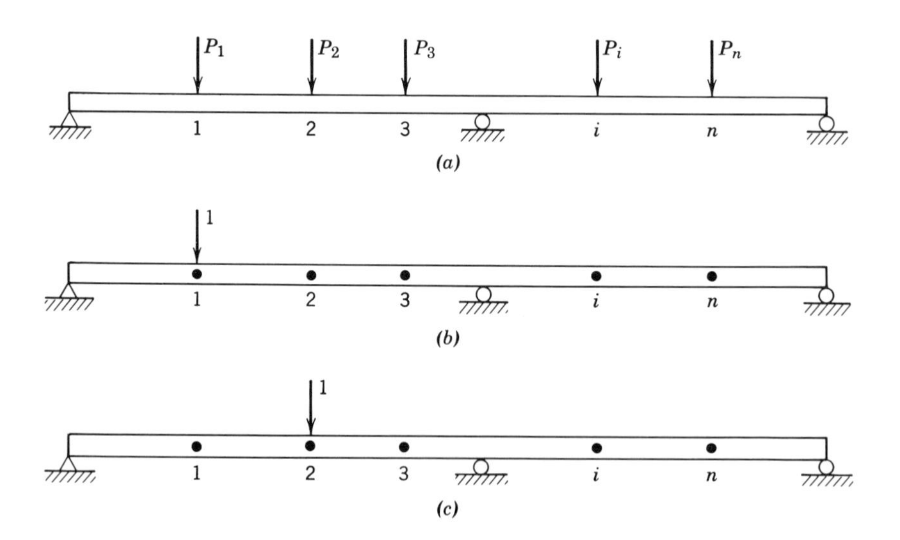

Fig. 1-9

For a given problem, the evaluation of the a_{ij} coefficients can be made by using known methods of strength of materials. It should be noted, however, that by Maxwell's[4] law the coefficient a_{ij} is equal to a_{ji}. That is,

$$a_{ij} = a_{ji} \tag{1-158}$$

In matrix[5] form, the expressions in Eq. 1-157 are written as follows:

$$\begin{bmatrix} y_1 \\ y_2 \\ \vdots \\ y_n \end{bmatrix} = \begin{bmatrix} a_{11} & a_{12} & \cdots & a_{1n} \\ a_{21} & a_{22} & \cdots & a_{2n} \\ \cdots & \cdots & \cdots & \cdots \\ a_{n1} & a_{n2} & \cdots & a_{nn} \end{bmatrix} \begin{bmatrix} P_1 \\ P_2 \\ \vdots \\ P_n \end{bmatrix} \tag{1-159}$$

or, in short form,

$$\{y\} = [M]\{P\} \tag{1-160}$$

The square matrix

$$[M] = \begin{bmatrix} a_{11} & a_{12} & \cdots & a_{1n} \\ a_{21} & a_{22} & \cdots & a_{2n} \\ \cdots & \cdots & \cdots & \cdots \\ a_{n1} & a_{n2} & \cdots & a_{nn} \end{bmatrix} \tag{1-161}$$

is known as the flexibility matrix.

1-14 THE STIFFNESS MATRIX

For certain engineering structures, such as frames, the computation of flexibility coefficients becomes rather laborious and it would be convenient to use stiffness coefficients in their analysis. The symbol k_{ij} is used to denote a stiffness coefficient and is defined as the force required to be applied at point j to produce a deflection equal to unity in the direction of the force, while point i is restraint against translation. That is, the coefficients k_{ij} represent a force system that is capable of translating point j

[4]See Reference 78, page 28.

[5]Additional information on matrices can be found in the appendix.

by an amount equal to unity while preventing the translation of point i.

Consider for example the simply supported beam in Fig. 1-10*a*, loaded as shown. The total displacements $y_1, y_2, \ldots, y_n$ under the load concentration points are also shown in the same figure. For a unit displacement at point 1, the stiffness coefficients constitute the force system in Fig. 1-10*b* that maintains a displacement equal to unity at point 1, and zero displacements at points 2, 3, ..., *n*.

The expressions relating the forces $P_1, P_2, \ldots, P_n$ with the displacements $y_1, y_2, \ldots, y_n$, are as follows:

$$\begin{aligned} P_1 &= k_{11}y_1 + k_{12}y_2 + \ldots + k_{1n}y_n \\ P_2 &= k_{21}y_1 + k_{22}y_2 + \ldots + k_{2n}y_n \\ &\cdots\cdots\cdots\cdots\cdots \\ P_n &= k_{n1}y_1 + k_{n2}y_2 + \ldots + k_{nn}y_n \end{aligned} \qquad (1\text{-}162)$$

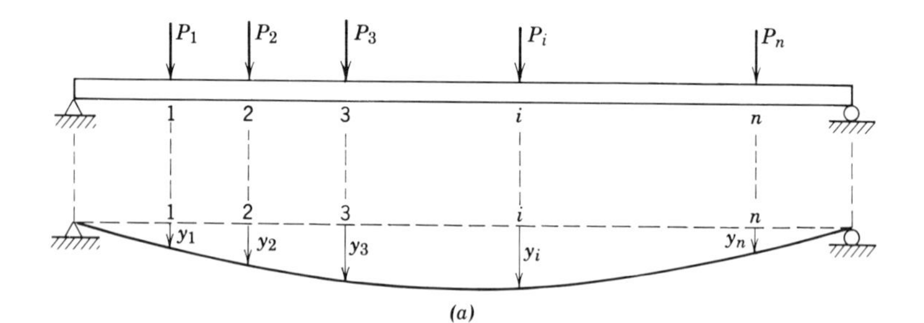

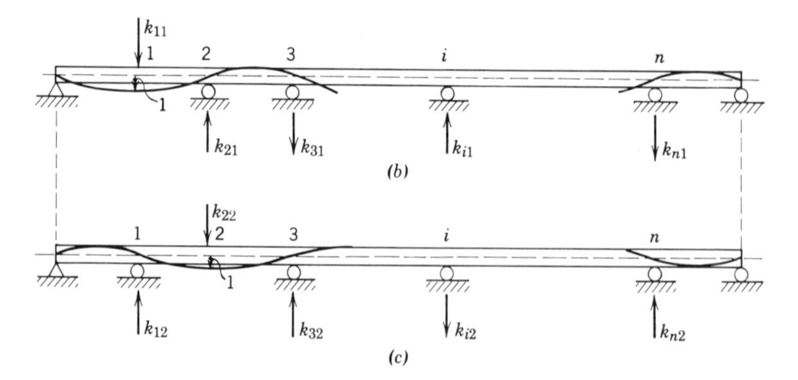

Fig. 1-10

or, in matrix form,

$$\begin{bmatrix} P_1 \\ P_2 \\ \vdots \\ P_n \end{bmatrix} = \begin{bmatrix} k_{11} & k_{12} & \cdots & k_{1n} \\ k_{21} & k_{22} & \cdots & k_{2n} \\ \cdots & \cdots & \cdots & \cdots \\ k_{n1} & k_{n2} & \cdots & k_{nn} \end{bmatrix} \begin{bmatrix} y_1 \\ y_2 \\ \vdots \\ y_n \end{bmatrix} \qquad (1\text{-}163)$$

or

$$\{P\} = [N]\{y\} \qquad (1\text{-}164)$$

The square matrix

$$[N] = \begin{bmatrix} k_{11} & k_{12} & \cdots & k_{1n} \\ k_{21} & k_{22} & \cdots & k_{2n} \\ \cdots & \cdots & \cdots & \cdots \\ k_{n1} & k_{n2} & \cdots & k_{nn} \end{bmatrix} \qquad (1\text{-}165)$$

is known as the stiffness matrix. The computation of the k_{ij} coefficients can be carried out by using known methods in structural analysis. It should be noted, however, that because of Maxwell's law,

$$k_{ij} = k_{ji} \qquad (1\text{-}166)$$

In many cases the analysis of structures for dynamic response requires the use of flexibility coefficients although they may be more difficult to determine. In such cases, the stiffness coefficients can be first computed to obtain the stiffness matrix $[N]$. The inverse[6] of this matrix, designated by $[N^{-1}]$, yields the flexibility matrix $[M]$. Thus Eq. 1-160 can be written as

$$\{y\} = [N^{-1}]\{P\} \qquad (1\text{-}167)$$

The inverse $[N^{-1}]$ of $[N]$ always exists and is unique, provided that $[N]$ is nonsingular. Computer programs are usually available for such computations. It should be noted, however, that the inverse $[M^{-1}]$ of the flexibility matrix $[M]$ yields the stiffness matrix $[N]$.

[6]See the appendix.

1-15 COMPUTATION OF STIFFNESS COEEFICIENTS

Consider the simply supported beam in Fig. 1-11*a* and let it be assumed that it is required to determine the stiffness coefficients at points 1, 2, and 3. The force system in Fig. 1-11*b* consists of the coefficients, k_{11}, k_{21}, and k_{31} which provide unit displacement at point 1 and zero displacements at points 2 and 3.

The computation of these coefficients can be carried out by assuming that the member in Fig. 1-11*b* is a three-span continuous beam with rigid supports at points A, 2, 3, and B, and loaded at point 1 by a force $P = k_{11}$ that is capable of producing unit vertical displacement at this point.

By applying the method of moment distribution as shown in Fig. 1-11*c*, the moments at supports 2 and 3 are $2.613P$ and $0.650P$, respectively. The coefficient k_{11} is determined by using the free-body diagram of span A 2 and the condition of unit displacement at point 1. By applying handbook formulas,

$$y_1 = 1 = \frac{101.40(1728)P}{EI}$$

or

$$P = k_{11} = \frac{EI}{101.40(1728)}$$

When $EI = 30 \times 10^6$ kip-in.^{2},

$$k_{11} = 171.0 \text{ kips/in.}$$

The supporting forces k_{21} and k_{31} can be easily computed by using the free-body diagrams of the beam spans and applying statics in the usual way. This procedure yields

$$k_{21} = -162.2 \text{ kips/in.} \quad k_{31} = 66.6 \text{ kips/in.}$$

The procedure to determine the remaining stiffness coefficients is the same. The continuous beams to use are shown in Figs. 1-11*d* and 1-11*e*. The results are

$$k_{12} = k_{32} = k_{23} = -162.2 \text{ kips/in.}$$

$$k_{13} = 66.6 \text{ kips/in.}$$

$$k_{22} = 235.0 \text{ kips/in.}$$

$$k_{33} = 171.0 \text{ kips/in.}$$

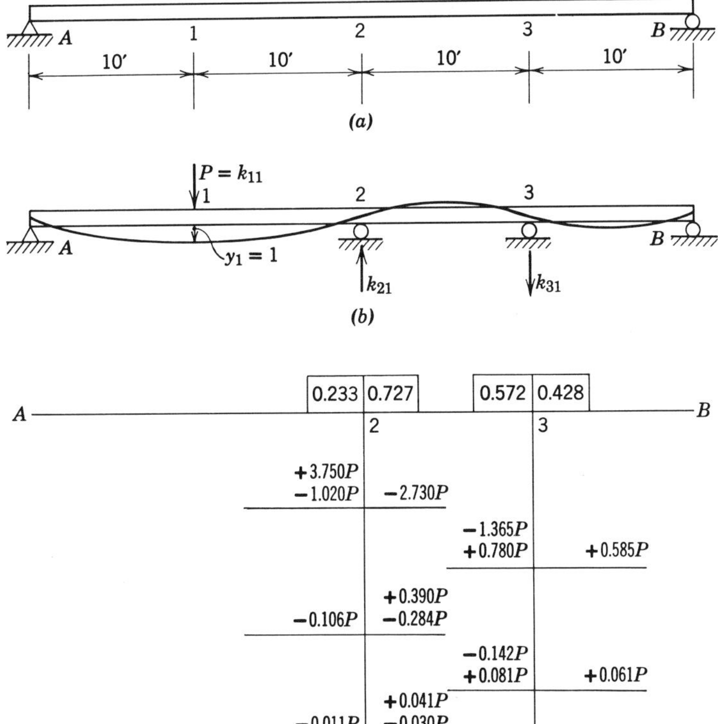

A
1
2
3
B
10′
10′
10′
10′
(a)
P = k11
1
2
3
A
y1 = 1
k21
k31
B
(b)
0.233
0.727
0.572
0.428
A
B
2
3
+3.750P
−1.020P
−2.730P
−1.365P
+0.780P
+0.585P
+0.390P
−0.106P
−0.284P
−0.142P
+0.081P
+0.061P
+0.041P
−0.011P
−0.030P

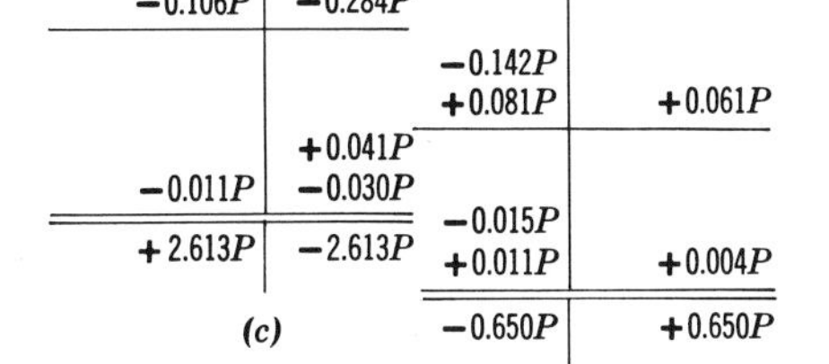

−0.015P
+2.613P
−2.613P
+0.011P
+0.004P
−0.650P
+0.650P

(c)

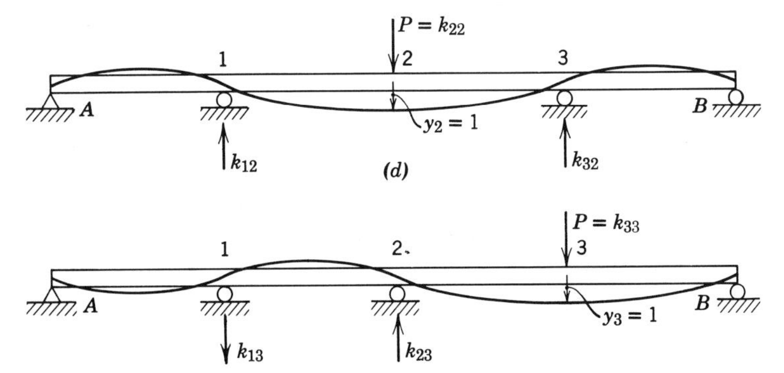

P = k22
1
2
3
A
y2 = 1
B
k12
k32
(d)
P = k33
1
2
3
A
B
y3 = 1
k13
k23
(e)

Fig. 1-11

The results above show that $k_{13} = k_{31}$ and $k_{23} = k_{32}$, well in accordance with the Maxwell's law.

The stiffness matrix $[N]$ is

$$[N] = \begin{bmatrix} k_{11} & k_{12} & k_{13} \\ k_{21} & k_{22} & k_{23} \\ k_{31} & k_{32} & k_{33} \end{bmatrix} = \begin{bmatrix} 171.0 & -162.2 & 66.6 \\ -162.2 & 235.0 & -162.2 \\ 66.6 & -162.2 & 171.0 \end{bmatrix}$$

With the help of the appendix, the flexibility matrix $[M]$ is

$$[M] = [N^{-1}] = \frac{1}{0.31(10)^6} \begin{bmatrix} 13{,}800 & 16{,}900 & 10{,}650 \\ 16{,}900 & 24{,}650 & 16{,}900 \\ 10{,}650 & 16{,}900 & 13{,}800 \end{bmatrix}$$

or

$$[M] = \begin{bmatrix} 0.0445 & 0.0545 & 0.0344 \\ 0.0545 & 0.0795 & 0.0545 \\ 0.0344 & 0.0545 & 0.0445 \end{bmatrix}$$

The units of the elements of $[M]$ are inches. It is interesting to note that direct computation of flexibility coefficients for this problem is easier than computation of stiffness coefficients. In other cases, such as frames, direct computations of stiffness coefficients would be easier.

Consider for example the two-story frame in Fig. 1-12*a* whose columns are made up of 10 WF 25 steel sections and girders of 21 WF 62 steel sections. It is required to determine the horizontal stiffness coefficients at the floor levels, points 1 and 2 in Fig. 1-12*a*. These coefficients are shown in Figs. 1-12*b* and 1-12*f*. The ones in Fig. 1-12*b* are determined here by applying *(a)* a unit horizontal displacement at point 2, Fig. 1-12*c*, *(b)* calculating the fixed-end moments for members *BC* and *DE*, *(c)* balancing the fixed-end moments by using the moment distribution method, and *(d)* determining the stiffness coefficients k_{12} and k_{22} by using the free-body diagram in Fig. 1-12*e* and applying static equilibrium. In this diagram, only the horizontal forces are shown, because they are the only ones needed for the evaluation of the stiffness coefficients.

The fixed-end moments are determined by using the expression

$6EIy_2/L_2{}^2$ and they are shown in Fig. 1-12*c*. The units are kip-in. The final balanced moments are shown in Fig. 1-12*d*. From the free-body diagram in Fig. 1-12*e*,

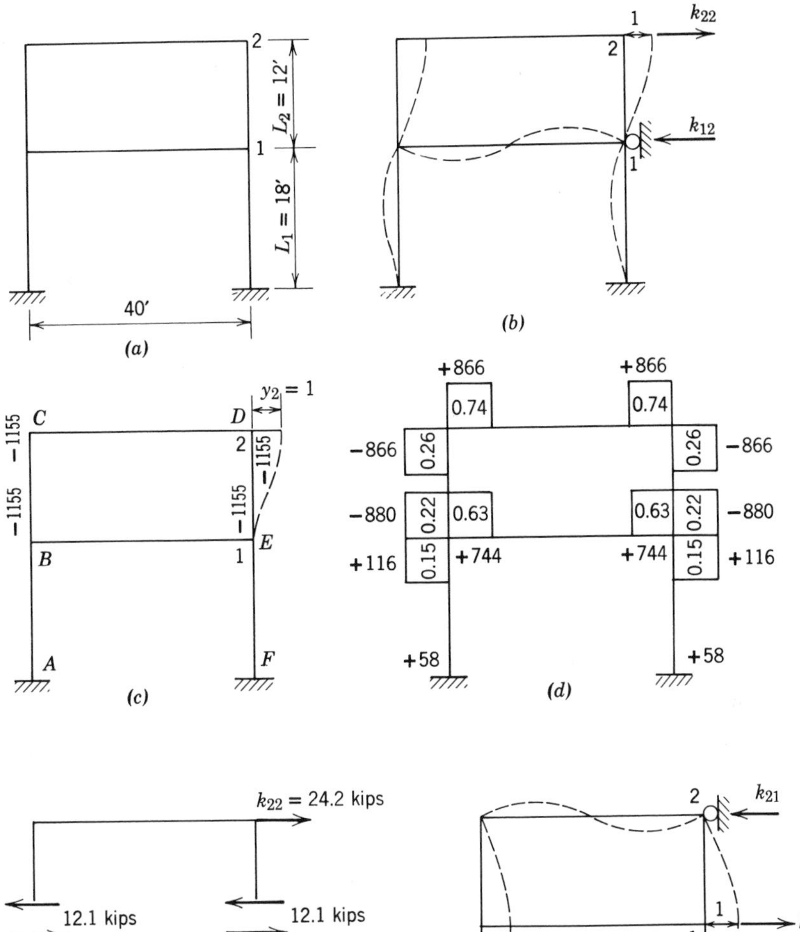

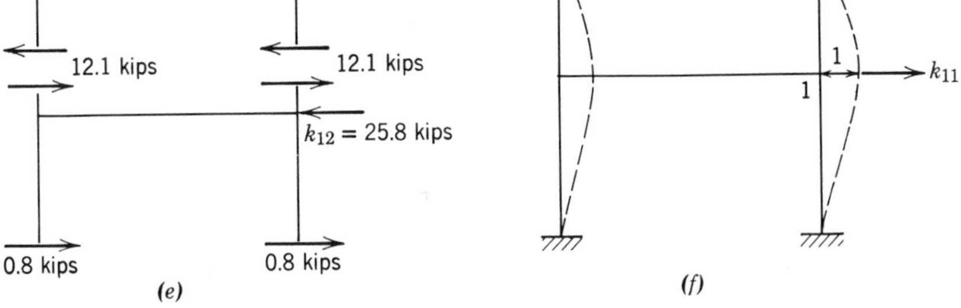

Fig. 1-12

$$k_{12} = 12.1 + 12.1 + 0.8 + 0.8 = 25.8 \text{ kips/in.}$$

$$k_{22} = 12.1 + 12.1 = 24.2 \text{ kips/in.}$$

In a similar manner the coefficients k_{11} and k_{21}, Fig. 1-12*f*, are determined and they are as follows:

$$k_{11} = 36.4 \text{ kips/in.} \quad k_{21} = k_{12} = 25.8 \text{ kips/in.}$$

The stiffness matrix $[N]$ is

$$[N] = \begin{bmatrix} k_{11} & k_{12} \\ k_{21} & k_{22} \end{bmatrix} = \begin{bmatrix} 36.4 & -25.8 \\ -25.8 & 24.2 \end{bmatrix}$$

The inverse $[N^{-1}]$ of $[N]$, as discussed above, should yield the flexibility matrix $[M]$ for this frame.

PROBLEMS

1-1 For each of the spring-mass systems shown below in Fig. P1-1, derive the differential equations of motion.

1-2 By applying Lagrange's equation, determine the differential equations of motion for the spring-mass systems in Problem 1-1.

1-3 A weight of 1.93 lb is suspended by a spring of constant $k = 2.5$ lb/in. Determine the natural frequency and period of vibration of the system.

1-4 If in Problem 1-3 the maximum velocity of the oscillatory motion of the mass is 15 in./sec, determine the amplitude and acceleration.

1-5 A 10-lb weight attached to a spring causes a static elongation of 0.63 in. If the system is displaced from its equilibrium position and then released, determine the natural frequency and period of oscillation.

1-6 A weight of 40 lb is suspended from a spring of constant $k = 15$ lb/in. A dashpot is attached between the weight and the ground and has a resistance of 0.2 lb at a velocity of 3 in./sec. Determine the natural frequency of the system and the critical damping factor of the dashpot.

1-7 Determine the constants A_1 and A_2 in Eq. 1-68 by using the initial conditions $y = y_0$ and $\dot{y} = 0$ at $t = 0$.

1-8 By using Eq. 1-61 and applying the appropriate Eulerian relation, derive Eq. 1-68.

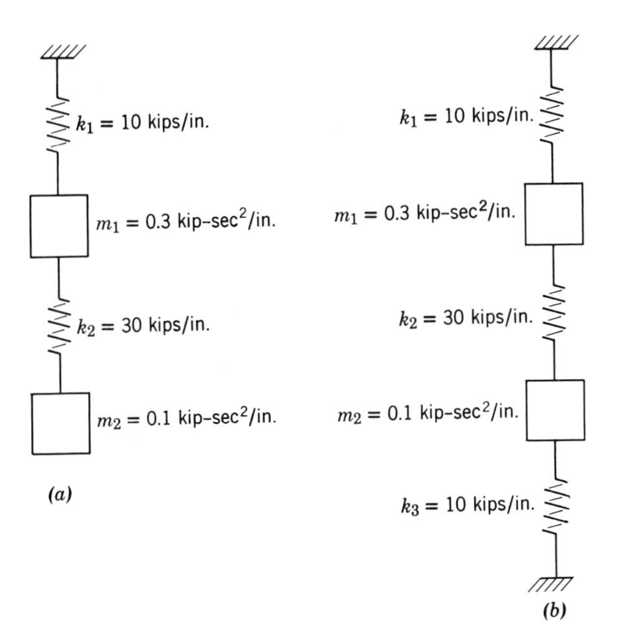
k_1 = 10 kips/in.
m_1 = 0.3 kip-sec^2/in.
k_2 = 30 kips/in.
m_2 = 0.1 kip-sec^2/in.
(a)
k_1 = 10 kips/in.
m_1 = 0.3 kip-sec^2/in.
k_2 = 30 kips/in.
m_2 = 0.1 kip-sec^2/in.
k_3 = 10 kips/in.
(b)

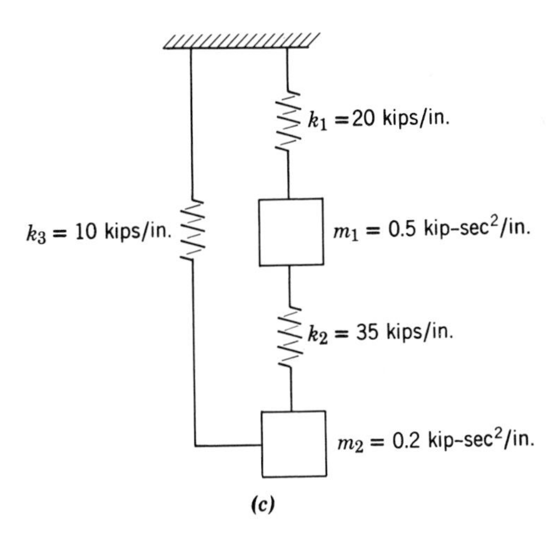
k_1 =20 kips/in.
k_3 = 10 kips/in.
m_1 = 0.5 kip-sec^2/in.
k_2 = 35 kips/in.
m_2 = 0.2 kip-sec^2/in.
(c)

Fig. P1-1

1-9 A weight of 35 lb is attached to a spring of constant $k = 100$ lb/in. and slides back and forth on a dry surface. The coefficient of friction between the weight and surface is 0.4, the initial velocity is zero, and the weight is initially released when the spring is stretched 8 in. What is the displacement of the weight at the end of the fifth cycle of vibration?

1-10 The sliding weight in Fig. 1-7*a* is 100 lb and the coefficient of friction μ between the weight and surface is 0.22. The spring constant k is 40 lb/in., and the weight is initially released with zero velocity when the spring is stretched 5 in. Determine the displacement of the weight at the end of the first and second cycles. Also determine the position at which the weight will stop, measured from the undeformed position of the spring.

1-11 If the ratio Y_1 / Y_3 in Fig. 1-8 is measured experimentally and is found to be equal to 2.5, determine the equivalent viscous damping factor ζ_e and the equivalent viscous damping constant c_e.

1-12 By using the results from Problem 1-11 and Eq. 1-84, write the solution $y(t)$ for the displacement of a mass $m = 0.0777$ kip-sec^2/in. when it is attached to a spring of constant $k = 258$ kips/in. The initial conditions of the mass are $y = y_0 = 1.2$ in. and $\dot{y} = \dot{y}_0 = 0$ at $t = 0$. Also plot $y(t)$ versus time.

1-13 For each of the spring-mass systems in Problems 1-1*a* and 1-1*b*, compute the natural frequencies of vibration and the corresponding mode shapes.

1-14 By utilizing the initial conditions given in Eq. 1-109, determine the constants A_1, A_2, A_3, and A_4 in Eqs. 1-105 and 1-106 and write the general solutions for the displacements y_1 and y_2.

1-15 Determine the constants $C_1, C_2, C_3, \ldots, C_8$ in Eqs. 1-124 and 1-125 by utilizing the initial conditions in Eq. 1-126 and the values of Ψ given by Eq. 1-121. Write the general expression for the amplitudes y_1 and y_2.

1-16 For a two-degree spring-mass system with viscous damping, determine the auxiliary equations and the characteristic equation when $k_1 = 80$ lb/in., $k_2 = 10$ lb/in., $m_1 = 0.15$ lb-sec^2/in., $m_2 = 0.1$ lb-sec^2/in., $c_1 = 0.004$ lb-sec/in., and $c_2 = 0.03$ lb-sec/in.

1-17 By using the results of Problem 1-13, verify the orthogonality property of normal modes given by Eq. 1-156.

1-18 For the beam problems shown below in Fig. P1-18, compute the vertical deflections under the load concentration points by making use of flexibility coefficients and Maxwell's law of reciprocal deflections.

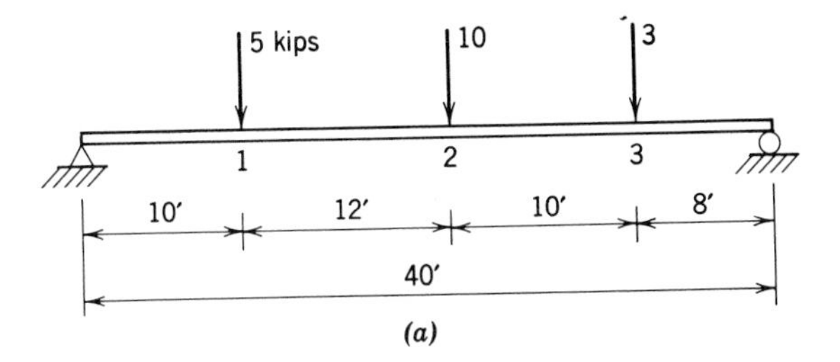

(a)

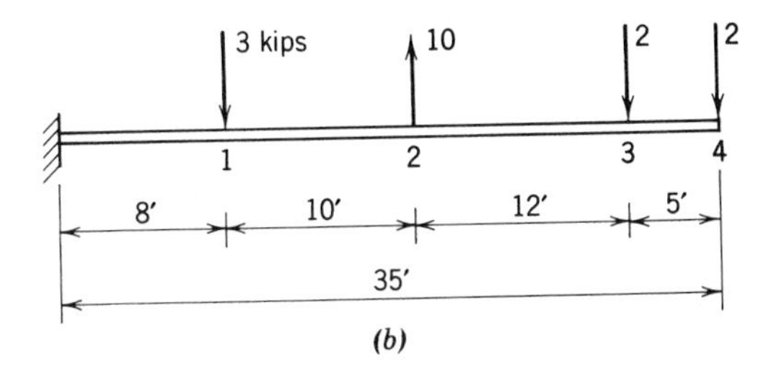

(b)

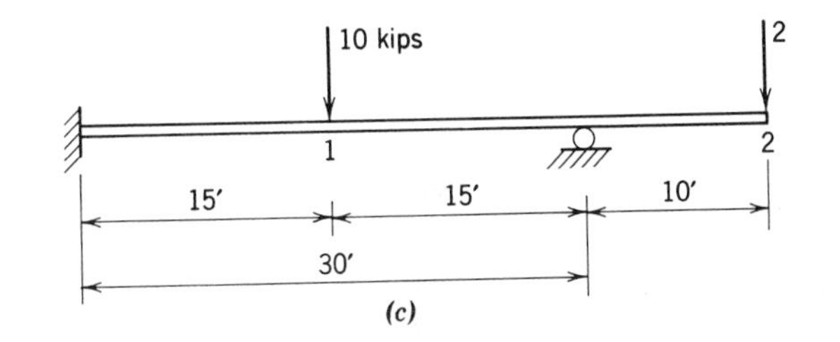

(c)

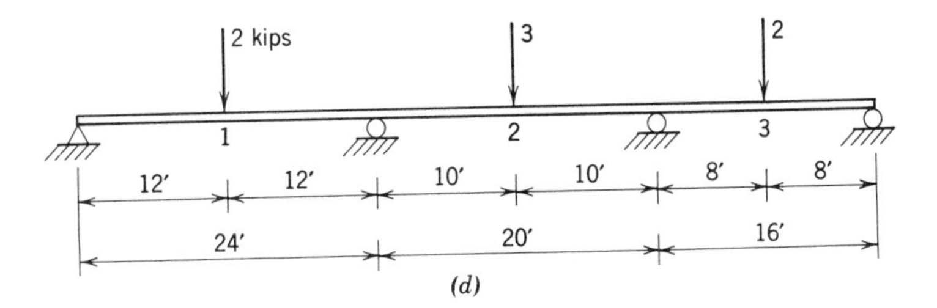

(d)

Fig. P1-18

2

DYNAMIC RESPONSE OF SPRING-MASS SYSTEMS

2-1 INTRODUCTION

The dynamic response of spring-mass systems with one or more degrees of freedom are discussed in this chapter. These problems have found many practical applications in structural dynamics, because many multielement structural systems and machines can be idealized conveniently into spring-mass systems with one or more degrees of freedom, thus making possible an approximate solution to the actual multiple or infinite degree of freedom system.

In the discussion to follow, the spring-mass systems are assumed to be acted upon by various kinds of time-varying forcing functions, and the solutions are found by using rigorous as well as numerical methods. The subject treatment is primarily concentrated on linear elastic systems, except in Section 2-8 where elastoplastic bilinear systems are analyzed. The Fourier method of analysis is also presented in this chapter. It is hoped that this discussion will create a physical feel on dynamic response that is essential in the dynamic analysis of the more complicated structures.

2-2 UNDAMPED HARMONIC EXCITATIONS

The impressed force $F(t)$ acting on the mass m of the one-degree system in Fig. 1-3*a* is assumed to be harmonic and equal to $F \sin \omega_f t$, where F is the peak amplitude and ω_f is the frequency of the force in units of radians per second. By neglecting damping, the differential equation of motion is

$$m\ddot{y} + ky = F \sin \omega_f t \tag{2-1}$$

The two parts of the solution $y(t)$ are

$$Y(t) = y_c(t) + y_p(t) \tag{2-2}$$

where $y_c(t)$ is the complementary solution satisfying the homogeneous part of Eq. 2-1, and $y_p(t)$ is the particular solution based on the forcing function and satisfying Eq. 2-1.

The solution $y_c(t)$ is given by Eq. 1-50, that is,

$$y_c(t) = C_1 \cos \omega t + C_2 \sin \omega t \tag{2-3}$$

Consideration of the nature of the forcing term in Eq. 2-1 suggests that $y_p(t)$ can be taken as

$$y_p(t) = Y \sin \omega_f t \tag{2-4}$$

where Y is the peak amplitude. By substituting Eq. 2-4 into Eq. 2-1 and simplifying,

$$-m\omega_f^2 Y + kY = F \tag{2-5}$$

and

$$Y=\frac{F}{k-m\omega_f^2}=\frac{F}{k}\frac{1}{1-(m/k)\omega_f^2}$$

or

$$Y=\frac{F}{k}\frac{1}{1-(\omega_f/\omega)^2} \tag{2-6}$$

By using Eqs. 2-3, 2-4, and 2-6, Eq. 2-2 yields

$$y(t)=C_1\cos\omega t+C_2\sin\omega t+\frac{F}{k}\frac{1}{1-(\omega_f/\omega)^2}\sin\omega_f t \tag{2-7}$$

$$\dot{y}(t)=-C_1\omega\sin\omega t+C_2\omega\cos\omega t+\frac{F\omega_f}{k}\frac{1}{1-(\omega_f/\omega)^2}\cos\omega_f t \tag{2-8}$$

If the initial conditions of the motion are $y=y_0$ and $\dot{y}=\dot{y}_0$ at $t=0$, then

$$C_1=y_0 \qquad C_2=\frac{\dot{y}_0}{\omega}-\frac{F\omega_f}{k\omega}\frac{1}{1-(\omega_f/\omega)^2}$$

and

$$y(t)=y_0\cos\omega t+\left[\frac{\dot{y}_0}{\omega}-\frac{F\omega_f}{k\omega}\frac{1}{1-(\omega_f/\omega)^2}\right]\sin\omega t$$

$$+\frac{F}{k}\frac{1}{1-(\omega_f/\omega)^2}\sin\omega_f t \tag{2-9}$$

$$\dot{y}(t)=-y_0\omega\sin\omega t+\left[\frac{\dot{y}_0}{\omega}-\frac{F\omega_f}{\omega}\frac{1}{1-(\omega_f/\omega)^2}\right]\omega\cos\omega t$$

$$+\frac{F\omega_f}{k}\frac{1}{1-(\omega_f/\omega)^2}\cos\omega_f t \tag{2-10}$$

The first two terms in each of the two equations prescribe the free vibration and the third term gives the steady-state forced vibration.

Certain important conclusions can be drawn from the third term of Eq. 2-9.When $\omega_f/\omega=1$, the forced frequency ω_f coincides with the free natural frequency ω of the system and the phenomenon of resonance occurs. Under this condition, the amplitude Y in Eq. 2-6 becomes infinitely large with time. If on the other hand $\omega_f < \omega$, then $y_p(t)$ and $F(t)$ are in phase. They are in opposite phase when $\omega_f > \omega$. At resonance, the amplitude increases gradually and will require infinite time to become infinite. Materials that are commonly used in practice, however, are subject to strength limitations, and failures occur long before infinite amplitude is attained

2-3 DAMPED HARMONIC EXCITATIONS

Consider now the case where $F(t)$ in Fig. 1-3a is equal to $F \sin \omega_f t$ and the one-degree system vibrates under the influence of light viscous damping. The differential equation of motion is

$$m\ddot{y} + c\dot{y} + ky = F \sin \omega_f t \tag{2-11}$$

and its complete solution again consists of the homogeneous solution $y_c(t)$ and the particular one $y_p(t)$.

The solution $y_c(t)$ is given by Eq. 1-68. That is,

$$y_c(t) = e^{-\zeta \omega t}\left[A_1 \cos\left(\omega\sqrt{\zeta^2 - 1}\right)t + A_2 \sin\left(\omega\sqrt{\zeta^2 - 1}\right)t\right] \tag{2-12}$$

The solution is transient and is based on the free vibration of the system. Because of damping, it will eventually die out and the only solution left will be the particular one $y_p(t)$.

The solution $y_p(t)$ is a steady-state harmonic oscillation at the frequency of the impressed force, and its amplitude vector lags the force vector by a phase angle ϕ. With this in mind, it is reasonable to assume that $y_p(t)$ is of the form

$$y_p(t) = Y \sin(\omega_f t - \phi) \tag{2-13}$$

By substituting Eq. 2-13 into Eq. 2-11, the amplitude Y and the phase ϕ can be easily found. They are as follows:

$$Y = \frac{F}{\left[(k - m\omega_f^2)^2 + (c\omega_f)^2\right]^{\frac{1}{2}}} \tag{2-14}$$

$$\tan \phi = \frac{c\omega_f}{k - m\omega_f^2} \tag{2-15}$$

These expressions, by making some minor manipulations, are written as

$$Y=\frac{F/k}{\left[(1-m\omega_f^2/k)^2+(c\omega_f/k)^2\right]^{\frac{1}{2}}} \tag{2-16}$$

$$\tan\phi=\frac{c\omega_f/k}{1-m\omega_f^2/k} \tag{2-17}$$

or

$$\Gamma=\frac{Y}{Y_{st}}=\frac{1}{\left\{\left[1-(\omega_f/\omega)^2\right]^2+(2\zeta\omega_f/\omega)^2\right\}^{\frac{1}{2}}} \tag{2-18}$$

$$\tan\phi=\frac{2\zeta\omega_f/\omega}{1-(\omega_f/\omega)^2} \tag{2-19}$$

In Eq. 2-18, $Y_{st}=F/k$ is the deflection of the mass m that is produced by the steady force F, by assuming that F is applied as static load at zero frequency.

The ratio Y/Y_{st}, designated by the Greek letter Γ, is known as the magnification factor, and provides the factor by which Y_{st} should be multiplied to obtain Y. The factor Γ depends only on the ratio ω_f/ω and ζ, and it may be plotted against the ratio ω_f/ω for various values of ζ ($\zeta=0$, 0.05, 0.1, 0.2, ..., 1). The curves so obtained illustrate the influence of ζ at resonance. A plot of Γ versus ω_f/ω for $\zeta=0$, 0.25 and 0.5 is shown in Fig. 2-1.

The complete solution of Eq. 2-11 is obtained by superimposing Eqs. 2-12 and 2-13, where Y in Eq. 2-13 is given by Eq. 2-18. That is,

$$y(t)=e^{-\zeta\omega t}\left[A_1\cos\left(\omega\sqrt{\zeta^2-1}\right)t+A_2\sin\left(\omega\sqrt{\zeta^2-1}\right)t\right]$$

$$+\frac{F/k}{\left\{\left[1-(\omega_f/\omega)^2\right]^2+(2\zeta\omega_f/\omega)^2\right\}^{\frac{1}{2}}}\sin(\omega_f t-\phi) \tag{2-20}$$

In the equation above, ϕ can be found by using Eq. 2-19, and the constants A_1 and A_2 can be obtained by using the initial conditions of the motion.

2-4 IMPULSE

In this case, the forcing function *F(t)* acting on an undamped one-degree spring-mass system is assumed to be an impulse that is suddenly applied at time t_0 when the system is in its equilibrium position $y(t_0)=0$. Under these

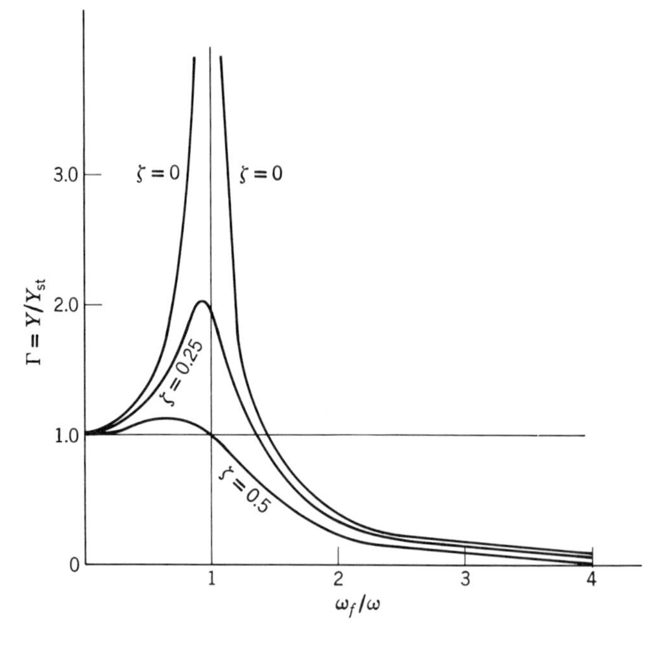

Fig. 2-1

conditions, the resulting motion can be analyzed by using the solution given by Eq. 1-51.

If the impulse is designated by the symbol F_{imp}, the initial velocity $\dot{y}_0$ at time t_0 is F_{imp}/m and the initial displacement is zero. Therefore, Eq. 1-51 yields

$$y(t) = \frac{\dot{y}_0}{\omega} \sin \omega (t - t_0)$$

or

$$y(t) = \frac{F_{imp}}{m\omega} \sin \omega (t - t_0) \qquad \text{for} \qquad t \geqslant t_0 \tag{2-21}$$

2-5 DYNAMIC FORCE OF GENERAL TYPE

The undamped one-degree spring-mass system in this case is subjected to a general-type forcing function $F(t)$ such as that shown in Fig. 2-2. The idea of impulse as expressed by Eq. 2-21 can be used to study the resulting motion. For example at time $t_0 = T$, the quantity $F(T)\,dT$ represented by the shaded

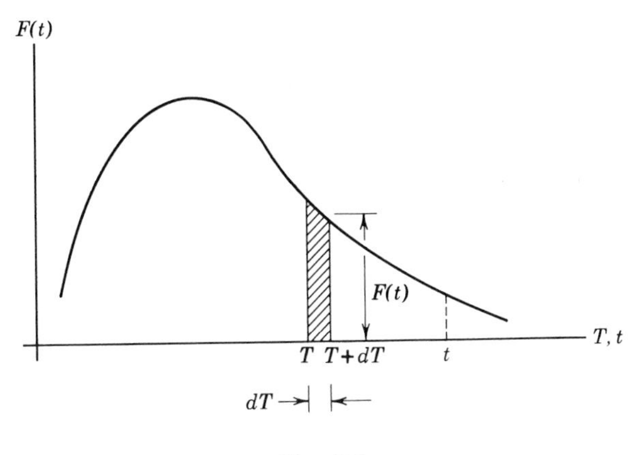

Fig. 2-2

area between times T and $T+dT$ in the force-time plot of Fig. 2-2, can be considered as an infinitesimal impulse that yields the incremental velocity

$$d\dot{y} = \frac{F(T)\,dT}{m} \tag{2-22}$$

By Eq. 2-21, the motion of the mass due to this impulse is given by

$$\frac{F(T)\,dT}{m\omega}\sin\omega(t-T) \quad \text{for } t \geqslant T \tag{2-23}$$

If the effects of all infinitesimal impulses between times $T=0$ and $T=t$ are added, the motion of the mass m due to $F(t)$ is given by the expression

$$y = \frac{1}{m\omega}\int_0^t F(T)\sin\omega(t-T)\,dT \tag{2-24}$$

The complete solution can be obtained by superimposing Eqs. 1-151 and 2-24. Thus, with $t_0=0$, the complete solution is

$$y = y_0\cos\omega t + \frac{\dot{y}_0}{\omega}\sin\omega t + \frac{1}{m\omega}\int_0^t F(T)\sin\omega(t-T)\,dT \tag{2-25}$$

Equation 2-25 is a general expression that can be used to determine the dynamic response of undamped single-degree of freedom systems. Any variation of $F(T)$ will yield a closed-form solution, provided that the integral in this equation can be evaluated. Applications of Eq. 2-25 are discussed in the following section.

Consider now the case where the spring-mass system, in addition to the general-type dynamic force $F(t)$, is also subjected to viscous damping. In this case the homogeneous solution is given by Eq. 1-68. That is,

$$y_c(t) = e^{-\mu t}[A_1 \cos\omega_d t + A_2 \sin\omega_d t] \tag{2-26}$$

where $\omega_d = \omega\sqrt{\zeta^2 - 1}$ and $\mu = \zeta\omega$. By applying the initial conditions $y_c(0) = y_0$ and $\dot{y}_c(0) = \dot{y}_0$, the constants A_1 and A_2 are

$$A_1 = y_0 \qquad A_2 = \frac{\dot{y}_0 + \mu y_0}{\omega_d}$$

and

$$y_c(t) = e^{-\mu t}\left(y_0 \cos\omega_d t + \frac{\dot{y}_0 + \mu y_0}{\omega_d} \sin\omega_d t \right) \tag{2-27}$$

The solution due to $F(t)$ can be obtained by dividing $F(t)$ into an infinite number of infinitesimal impulses as before. Thus, for the damped case, the response due to an element of impulse is

$$\frac{F(T)\,dT}{m\omega_d} e^{-\mu(t-T)} \sin\omega_d(t-T) \tag{2-28}$$

When the effects of all impulses are added, the motion of the mass m due to $F(t)$ is

$$\frac{1}{m\omega_d}\int_0^t F(T)e^{-\mu(t-T)} \sin\omega_d(t-T)\,dT \tag{2-29}$$

Thus superposition of Eqs. 2-27 and 2-29 yields the complete solution

$$y = e^{-\mu t}\left(y_0 \cos\omega_d t + \frac{\dot{y}_0 + \mu y_0}{\omega_d} \sin\omega_d t \right)$$
$$+ \frac{1}{m\omega_d}\int_0^t F(T)e^{-\mu(t-T)} \sin\omega_d(t-T)\,dT \tag{2-30}$$

2-6 SPECIAL TYPES OF FORCING FUNCTION

In the preceding sections, one-degree spring-mass systems were subjected to general and particular types of time-varying force functions, and their dynamic response was obtained by also considering the effects of viscous damping. In this section, some special types of forcing functions that have gained wide interest in the field of structural dynamics are considered. In each case, the effects of damping are neglected. If desired, they can be taken into consideration as discussed earlier in the chapter.

Suddenly Applied Constant Force

Consider the case where the spring-mass system in Fig. 2-3*a* is subjected to a suddenly applied constant force $F(t) = F$, whose plot against time is shown

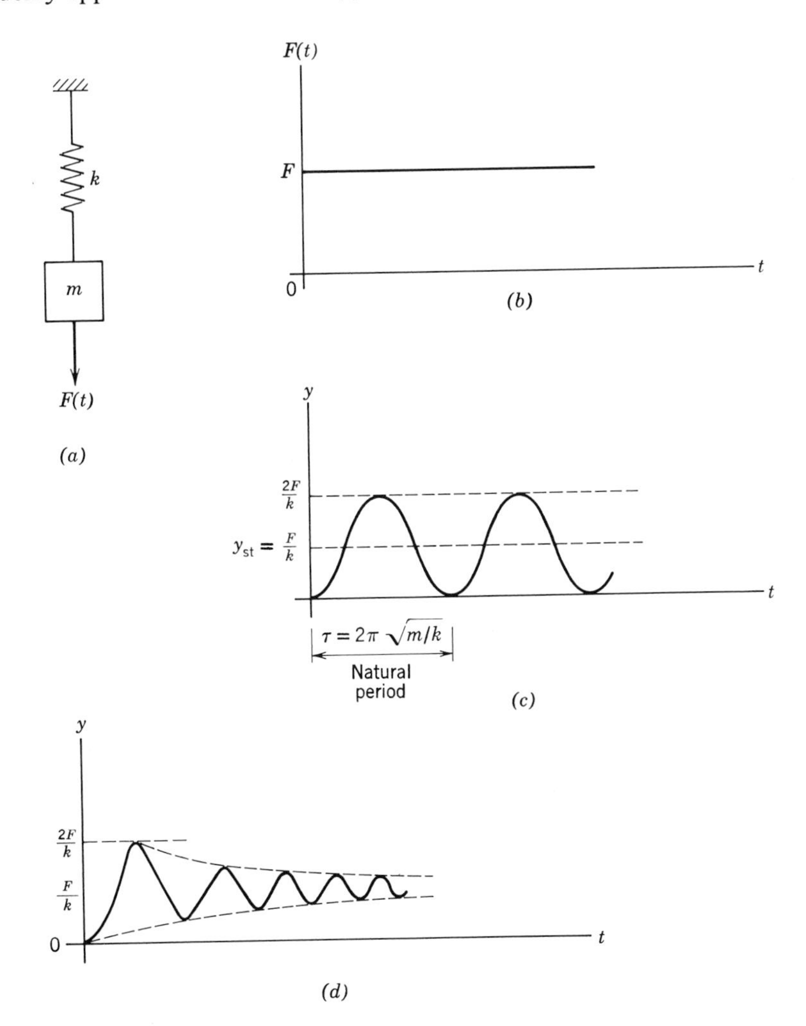

Fig. 2-3

in Fig. 2-3*b*. The magnitude of this force is F and remains constant indefinitely. A free-body diagram of the mass m reveals that the differential equation of motion is

$$m\ddot{y}+ky=F \tag{2-31}$$

The general solution of this equation may be found in a way similar to that used in preceding sections:

$$y(t)=C_1\cos\omega t+C_2\sin\omega t+\frac{F}{k} \tag{2-32}$$

where ω is the free undamped vibration of the spring-mass system.

Let it now be assumed that the initial conditions of the motion are $y_0=\dot{y}_0=0$ at $t=0$. This yields

$$C_1=-\frac{F}{k} \qquad C_2=0$$

Thus

$$y(t)=-\frac{F}{k}\cos\omega t+\frac{F}{k}$$

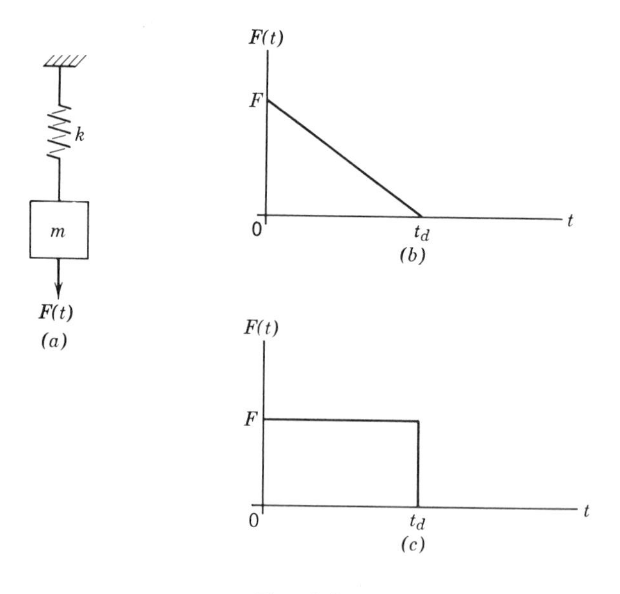

Fig. 2-4

or

$$y(t)=\frac{F}{k}(1-\cos\omega t) \tag{2-33}$$

Here, the quantity F/k is the zero frequency displacement y_{st}, that is, the displacement of the system when F is assumed to be gradually applied like a static load. The ratio of the dynamic deflection $y(t)$ to the static deflection y_{st} is the magnification factor Γ. Thus

$$\Gamma=\frac{y(t)}{y_{st}}=1-\cos\omega t \tag{2-34}$$

The maximum value of Γ is 2, which shows that the effect of a suddenly applied constant force is to double the static deflection. The force F will make the system vibrate about the equilibrium position $y(t)=y_{st}$ as shown in Fig. 2-3*c*. If damping is present, the vibration will eventually die out and the system will come to rest at $y(t)=y_{st}$ as shown in Fig. 2-3*d*.

Triangular Impulsive Load

In this case, the spring-mass system in Fig. 2-4*a* is subjected to a suddenly applied force $F(t)$ whose variation with time is shown in Fig. 2-4*b*. Its value at time $t=0$ is F and decreases linearly to zero at $t=t_d$. Thus the expression for $F(t)$ can be written as

$$F(t)=F\left(1-\frac{T}{t_d}\right) \tag{2-35}$$

where the symbol T is used for time for proper clarification in carrying out the required integrations.

The general differential equation for this problem is

$$m\ddot{y}+ky=F\left(1-\frac{T}{t_d}\right) \tag{2-36}$$

Its solution can be obtained directly from the solution given by Eq. 2-25. The response of the system, however, will have to be determined in two stages.

Stage 1 $\quad T\leqslant t_d$

By utilizing the initial conditions $y_0=\dot{y}_0=0$ and the forcing function given by Eq. 2-35, Eq. 2-25, by carrying out the required integration, yields

$$y(t) = \frac{F}{k}(1-\cos\omega t) + \frac{F}{kt_d}\left(\frac{\sin\omega t}{\omega} - t\right) \tag{2-37}$$

The magnification factor Γ is

$$\Gamma = 1 - \cos\omega t + \frac{\sin\omega t}{\omega t_d} - \frac{t}{t_d} \tag{2-38}$$

Stage 2 $T \geqslant t_d$

For this time interval, the initial conditions are determined by substituting $t = t_d$ in Eq. 2-37. Thus, for $t = t_d$,

$$y_0 = \frac{F}{k}\left(\frac{\sin\omega t_d}{\omega t_d} - \cos\omega t_d\right) \tag{2-39}$$

and

$$\dot{y}_0 = \frac{F}{k}\left(\omega\sin\omega t_d + \frac{\cos\omega t_d}{t_d} - \frac{1}{t_d}\right) \tag{2-40}$$

By substituting Eqs. 2-39 and 2-40 into Eq. 2-25 and making $F(T) = 0$ and $t = t - t_d$, we obtain

$$y(t) = \frac{F}{k\omega t_d}[\sin\omega t_d - \sin\omega(t - t_d)] - \frac{F}{k}\cos\omega t \tag{2-41}$$

The magnification factor Γ is

$$\Gamma = \frac{1}{\omega t_d}[\sin\omega t_d - \sin\omega(t - t_d)] - \cos\omega t \tag{2-42}$$

Rectangular Impulsive Load

In this case the spring-mass system in Fig. 2-4*a* is acted upon by a suddenly applied constant force $F(t) = F$, of duration $t = t_d$. The variation of this load is shown in Fig. 2-4*c*. The procedure to determine the dynamic response of the system is similar to the one followed in the preceding problem. The results are as follows:

$$y(t) = \frac{F}{k}(1 - \cos\omega t) \qquad t \leqslant t_d \tag{2-43}$$

$$\Gamma = 1 - \cos\omega t \qquad t \leqslant t_d \tag{2-44}$$

and

$$y(t) = \frac{F}{k}\left[\cos\omega(t - t_d) - \cos\omega t\right] \qquad t \geqslant t_d \tag{2-45}$$

$$\Gamma = \cos\omega(t - t_d) - \cos\omega t \quad t \geqslant t_d \tag{2-46}$$

The dynamic response due to other types of forcing functions can be derived in a similar manner. The magnification factors for six time-dependent forces are tabulated in Table 2-1. In these derivations damping was not considered and the initial conditions used are $y_0 = \dot{y}_0 = 0$ at $t = 0$. In all cases, the static deflection y_{st} is taken equal to F/k.

It may be noted, however, that the magnification factors in Table 2-1 are functions of time. Maximum values for these factors can be computed by maximizing the respective time functions. These equations can be also expressed in terms of the natural period τ of the system by substituting $2\pi/\tau$ for ω. This will emphasize the fact that the ratio of the time duration t_d to the natural period τ of the system, that is, the ratio t_d/τ, is a parameter of particular importance. This is self explained by the results shown plotted in Figs. 2-5 and 2-6. In these figures, the maximum dynamic response of undamped one-degree elastic systems is shown. Each figure contains two graphs. The first one is a plot of Γ_{max} versus the ratio t_d/τ, and the second one gives the variation of the ratio t_m/τ for various values of the ratio t_d/τ, where t_m is the time of maximum response.

These graphs are of great practical importance for design purposes, because for these types of forcing functions only the natural period of the system is needed in order to read from the graphs the maximum magnification factor and the time of maximum response. In addition, it should be noted that the ratio of the maximum dynamic stress to that of the static one is also represented by the maximum magnification factor. The time t_m at which the maximum stress occurs is the t_m given by these plots.

The maximum dynamic responses for an additional forcing function is given in Fig. 2-7. The fact that damping was neglected in the derivations may not be a serious matter. Usually the maximum magnification factor for a structural system corresponds to the first peak of response where the amount of damping is rather small.

As an illustration regarding the use of the graphs, let it be assumed that the member in Fig. 2-8*a*, supporting the attached load $W = 30$ kips, is subjected to a suddenly applied impulsive force *F(t)* as shown. The variation of this force with time is shown in Fig. 2-8*b*, and its maximum value F is 50 kips. The moment of inertia I of the steel beam is constant at every cross section and it is equal to 2096.4 in.^{4} The section modulus S is 175.4 in.^{3} It is

TABLE 2-1

Forcing Functions	Magnification Factor Γ
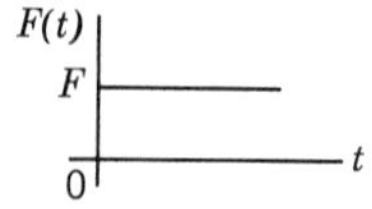 Step: $F(t) = F$ for $t \geq 0$	$\Gamma = 1 - \cos\omega t$
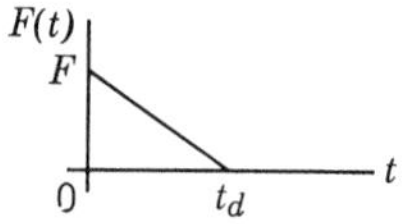 Triangular decreasing from F at 0 to 0 at t_d	$\Gamma = 1 - \cos\omega t + \sin\omega t/\omega t_d - t/t_d \quad t \leqslant t_d$ $\Gamma = (1/\omega t_d)[\sin\omega t_d - \sin\omega(t - t_d)] - \cos\omega t \quad t \geqslant t_d$
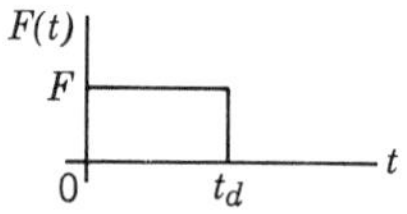 Rectangular pulse F from 0 to t_d	$\Gamma = 1 - \cos\omega t \quad t \leqslant t_d$ $\Gamma = \cos\omega(t - t_d) - \cos\omega t \quad t \geqslant t_d$
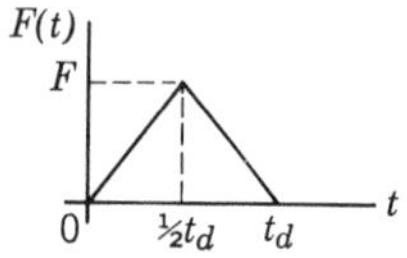 Triangular pulse, peak F at $\frac{1}{2}t_d$, zero at t_d	$\Gamma = (2/t_d)(t - \sin\omega t/\omega) \quad 0 \leqslant t \leqslant t_d/2$ $\Gamma = (2/t_d)\{t_d - t + (1/\omega)[2\sin\omega(t - t_d/2) - \sin\omega t]\} \quad t_d/2 \leqslant t \leqslant t_d$ $\Gamma = (2/\omega t_d)[2\sin\omega(t - t_d/2) - \sin\omega t - \sin\omega(t - t_d)] \quad t \geqslant t_d$
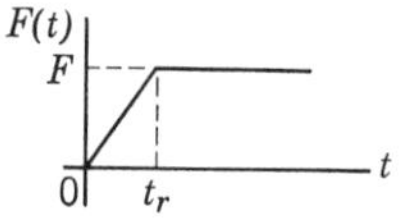 Ramp rising to F at t_r, then constant	$\Gamma = (1/t_r)(t - \sin\omega t/\omega) \quad t \leqslant t_r$ $\Gamma = 1 + (1/\omega t_r)[\sin\omega(t - t_r) - \sin\omega t] \quad t \geqslant t_r$
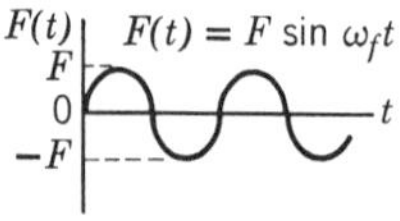 $F(t) = F\sin\omega_f t$	$\Gamma = (1 - \omega_f^2/\omega^2)^{-1}[\sin\omega_f t - (\omega_f/\omega)\sin\omega t]$ where ω_f = frequency of forcing function and ω = natural frequency of the system

Note: $\Gamma = y(t)/y_{st}$ where $y_{st} = F/k$. In the expressions above, ω is the natural frequency of the one degree of freedom system, except as otherwise stated.

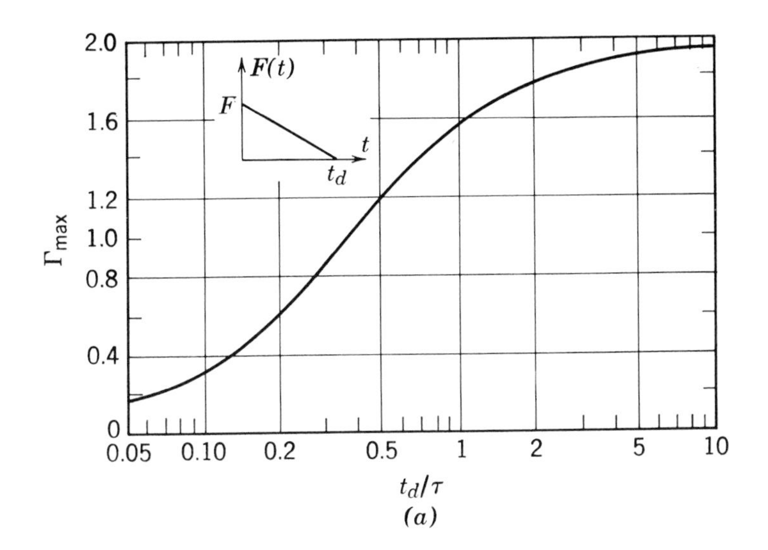

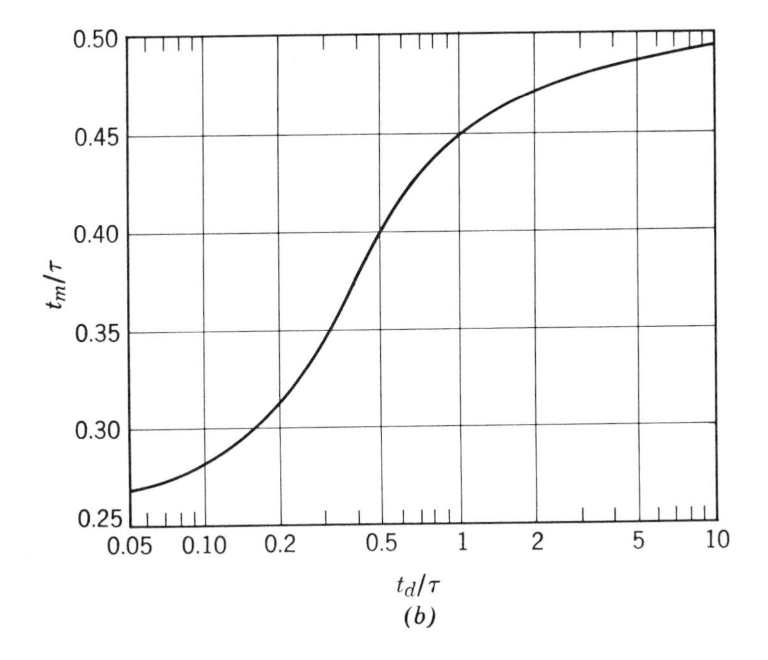

Fig. 2-5. (Reference 41.)

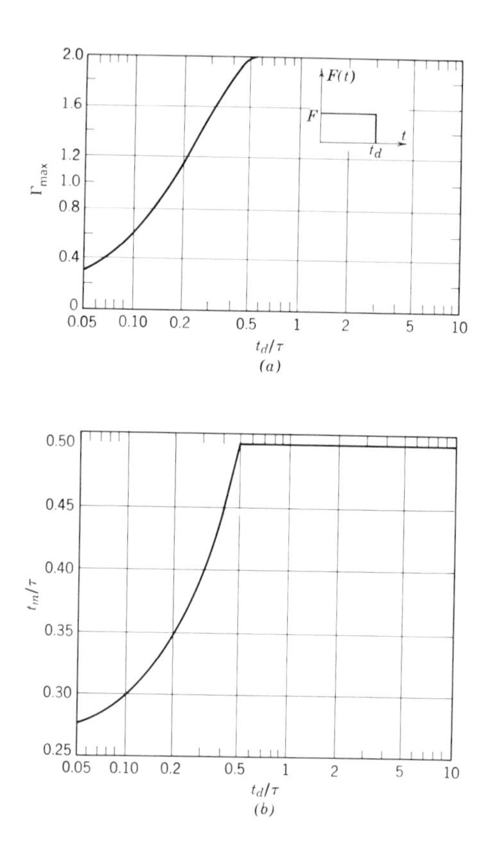

Fig. 2-6. (Reference 41.)

required to determine the maximum dynamic bending stress by assuming that the weight of the member is negligible.

The assumption of negligible beam weight permits one to treat the member as a one degree of freedom elastic system, provided that the maximum stress does not exceed the elastic limit of the material.

The idealized one-degree system is shown in Fig. 2-8*c*. The spring constant k is determined by applying to the center of the beam a force P that is capable of producing the vertical displacement $y=1$ at this point.

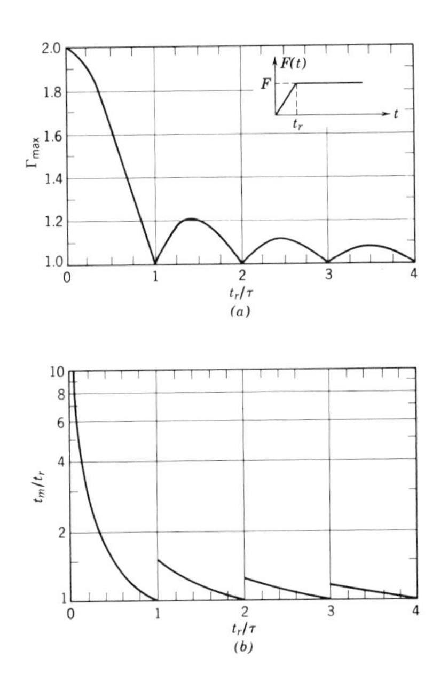

Fig. 2-7. (Reference 41.)

This is given by the expression

$$\frac{PL^3}{192EI} = y = 1$$

or, by solving for P,

$$k = P = \frac{192EI}{L^3} = \frac{(192)(30)(10)^6(2096.4)}{(30)^3(12)^3} = 258.0 \text{ kips/in.}$$

The natural period τ of the system is

$$\tau = 2\pi\sqrt{m/k} = 2\pi\sqrt{(30)/(386)(258)} = 0.109 \text{ sec}$$

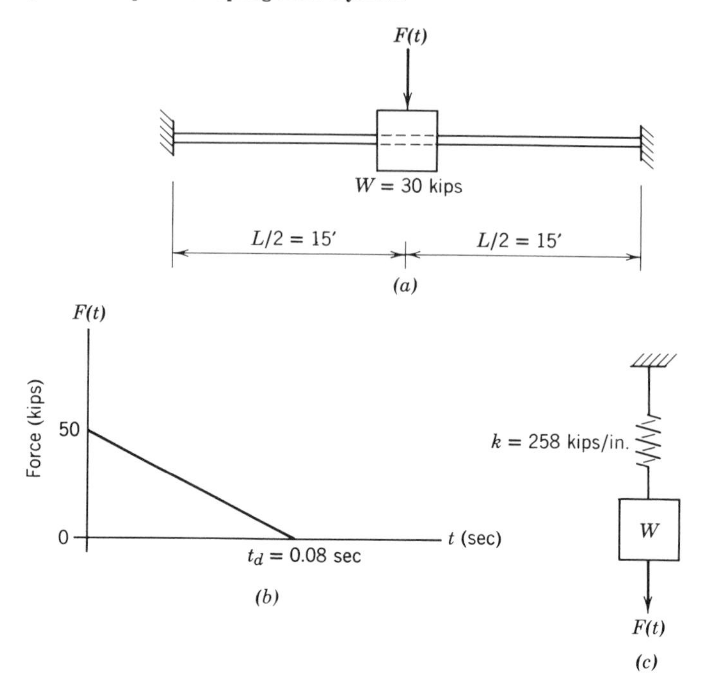

Fig. 2-8

Thus

$$\frac{t_d}{\tau} = \frac{0.08}{0.109} = 0.735$$

From the graphs in Fig. 2-5,

$$(\Gamma)_{\max} = 1.40$$

$$\frac{t_m}{\tau} = 0.44$$

The maximum dynamic stress $\sigma_{\max}$ is equal to the static stress σ_{st} caused by the 50-kip force, multiplied by the maximum magnification factor. That is,

$$\sigma_{\max} = \sigma_{st} (\Gamma)_{\max}$$

or

$$\sigma_{\max} = \frac{M_{st}}{S}(1.40) = \frac{FL/8}{S}(1.40)$$

$$= \frac{(50)(30)(12)}{(8)(175.4)}(1.40) = 17.90 \text{ ksi}$$

This will occur at time

$$t_m = (0.44)\tau = (0.44)(0.109) = 0.048 \text{ sec}$$

It should be noted, however, that in the calculations only the maximum dynamic stress due to $F(t)$ is found. It does not include the stress due to the weight $W = 30$ kips. This is a static stress and it can be added to the results above directly.

2-7 NUMERICAL ANALYSIS

In the preceding sections, rigorous approaches were used to determine the dynamic response of one-degree systems. Such solutions, however, can become very complicated when the loading and resistance functions are relatively complex mathematical expressions. For many practical problems in structural dynamics it could become a serious limitation, because the practicing engineer is often interested in an approximate solution to his problem that satisfies design criteria. Under these conditions, a numerical solution of the differential equations of motion is usually sufficient.

Consider the differential equation of motion

$$m\ddot{y} + ky = F(t) \tag{2-47}$$

This equation yields

$$\ddot{y} = \frac{F(t) - ky}{m} \tag{2-48}$$

where $\ddot{y}$ is the acceleration of the mass of the system. The evaluation of y by a numerical method is what is called here numerical analysis.

In the rigorous solution, the displacement y is obtained as a continuous function of time. In a numerical method, however, the value of y is determined at discrete values of time. Several approaches have been suggested in the past for this purpose. The one selected here is known as the acceleration impulse extrapolation method. In this method, the numerical computations are straightforward and do not involve trial-and-error procedures. On the other hand, the simplicity of this method is not restricted to one degree of freedom systems only. Multidegree[7] systems as well can be treated with equal ease.

Acceleration Impulse Extrapolation Method

In this method, the acceleration curve in Fig. 2-9*a* is replaced by a series of equally spaced impulses occurring at times $t_0, t_1, t_2, \ldots, t_i, \ldots, t_n$. The magnitude of the acceleration impulse at t_i is

$$\ddot{y}_i(\Delta t) \tag{2-49}$$

[7]Applications of this method to multidegree systems can be found in Chapter (3).

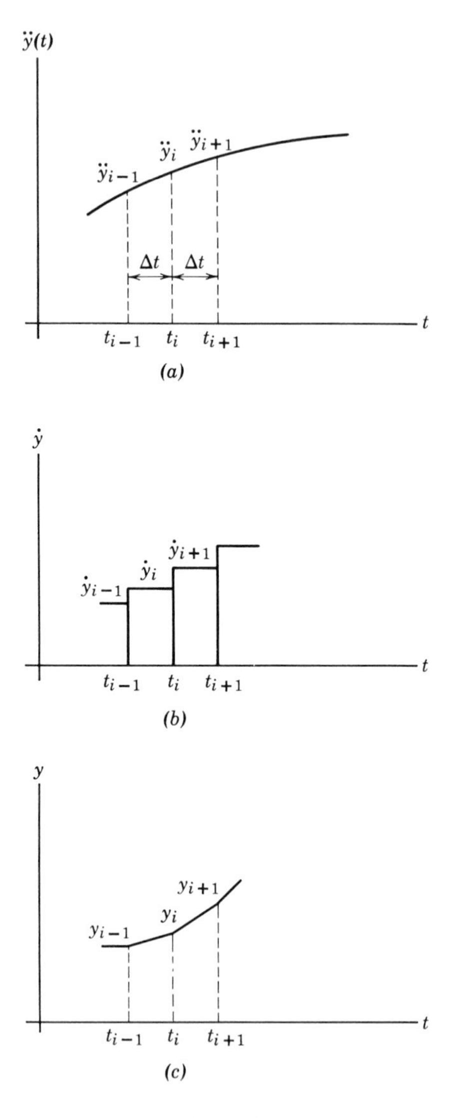

Fig. 2-9

where

$$(\Delta T) = t_1 - t_0 = t_2 - t_1 = \cdots = t_i - t_{i-1} \cdots = t_n - t_{n-1} \qquad (2\text{-}50)$$

Between time intervals, the velocity $\dot{y}$ is constant and the displacement y varies linearly with time as shown in Figs. 2-9*b* and 2-9*c*, respectively.

In the acceleration impulse extrapolation method, the differential equation of motion (2-47) is solved step by step by starting at time $t=0$ where the displacement and velocity are known. Then, by dividing the time scale into discrete intervals, the displacement can be extrapolated from one time station to the next.

Let it be assumed, for example, that the displacements $y^{(i)}$ and $y^{(i-1)}$ at time stations i and $i-1$, respectively, are known from previous computations. The acceleration $\ddot{y}^{(i)}$ at station i can be determined from Eq. 2-48. The average velocity $\dot{y}_{av}$ between time stations i and $i+1$ is given by the approximate relation

$$\dot{y}_{av} = \frac{y^{(i)} - y^{(i-1)}}{\Delta t} + \ddot{y}^{(i)} \Delta t \qquad (2\text{-}51)$$

Thus the displacement $y^{(i+1)}$ at time station $i+1$ may be calculated from the expression

$$y^{(i+1)} = y^{(i)} + \dot{y}_{av} \Delta t \qquad (2\text{-}52)$$

By substituting Eq. 2-51 into Eq. 2-52,

$$y^{(i+1)} = 2y^{(i)} - y^{(i-1)} + \ddot{y}^{(i)} (\Delta t)^2 \qquad (2\text{-}53)$$

This equation can be used to determine the displacement at time station $i+1$, provided that the displacements at time stations $i-1$ and i, as well as the acceleration at time station i, are known.

Equation 2-53, however, cannot be used for the first time interval because at $t=0$ the value of $y^{(i-1)}$ does not exist. For this interval one of the following equations should be used:

$$y^{(1)} = \tfrac{1}{6} (2\ddot{y}^{(0)} + \ddot{y}^{(1)}) (\Delta t)^2 \qquad (2\text{-}54)$$

or

$$y^{(1)} = \frac{\ddot{y}^{(0)}}{2} (\Delta t)^2 \qquad (2\text{-}55)$$

In Eq. 2-54 the acceleration is assumed to vary linearly up to the first time station, and in Eq. 2-55 the acceleration is assumed to be constant during the first time interval and equal to the initial value. For the remaining stations, Eq. 2-53 should be used.

The accuracy of this method is dependent on the smallness of the interval Δt that is used during extrapolation. In many cases, sufficient accuracy can be obtained by using time intervals equal to one-tenth of the smallest natural period of vibration of the system. If better accuracy is required, smaller Δt intervals should be used. The exact solution will be approached as Δt approaches zero.

Consider now the situation where the dynamic response of a system is influenced by the presence of viscous damping. Furthermore, let it be assumed that this system is a one-degree system. Its differential equation of motion is

$$m\ddot{y}+c\dot{y}+ky=F(t) \tag{2-56}$$

and $\ddot{y}$ is

$$\ddot{y}=\frac{F(t)-c\dot{y}-ky}{m} \tag{2-57}$$

In this case, it is required to determine the velocity y at each time station in order to be able to apply the numerical method discussed above. This velocity, at a time station i, can be found by the approximate expression

$$\dot{y}^{(i)}=\frac{y^{(i)}-y^{(i-1)}}{\Delta t}+\ddot{y}^{(i)}\left(\frac{\Delta t}{2}\right) \tag{2-58}$$

The second term on the right-hand side of Eq. 2-58 gives an estimate of the amount by which the average velocity of the preceding time interval should be increased, in order to obtain the velocity at the next time station.

The computation procedure becomes more convenient if Eqs. 2-57 and 2-58 are combined into one equation. This is accomplished by substituting Eq. 2-58 into Eq. 2-57, yielding

$$\ddot{y}^{(i)}=F(t)\frac{2}{2m-c\Delta t}-\frac{2c[y^{(i)}-y^{(i-1)}]}{\Delta t(2m-c\Delta t)}-\frac{2ky^{(i)}}{2m-c\Delta t} \tag{2-59}$$

In this manner, Eq. 2-59 can be used to determine the acceleration $\ddot{y}^{(i)}$ at station i, and Eq. 2-53 should be used to determine the displacement $y^{(i+1)}$ at station $i+1$.

Specific applications of the acceleration impulse extrapolation method are discussed in the sections that follow. At this time, as an illustration, let it be assumed that it is required to find the maximum displacement y of the idealized spring-mass system in Fig. 2-8*c*. By applying Eq. 2-48,

$$\ddot{y}=\frac{F(t)-ky}{m}=\frac{F(t)-258y}{0.0777}$$

or

$$\ddot{y}=12.85F(t)-3320y \tag{2-60}$$

TABLE 2-2

t (sec)	$12.85F(t)$ (in./sec²)	$3320y$ (in./sec²)	$\ddot{y}$ Eq. 2-60 (in./sec²)	$\ddot{y}(\Delta t)^2$ (in.)	y Eq. 2-53
0.00	643.0	0	643.0	0.06430	0
0.01	550.0	106.7	443.3	0.04433	0.03215*
0.02	482.0	360.0	122.0	0.01220	0.10863
0.03	401.0	655.0	−254.0	−0.02540	0.19731
0.04	296.0	865.0	−569.0	−0.05690	0.26059
0.05	241.0	885.0	−644.0	−0.06440	0.26697
0.06	161.0	680.0	−519.0	−0.05190	0.20445
0.07	80.0	298.0	−218.0	−0.02180	0.09003
0.08	0	−153.0	153.0	0.01530	−0.04619
0.09	0	−554.0	554.0	0.05540	−0.16711
0.10	0	−770.0	770.0	0.07740	−0.23191
0.11					−0.21931

*Equation 2-55 is used for this station.

The natural period of the system was calculated in the preceding section and was found to be 0.109 sec. The dynamic displacements of the system will be calculated by using the time interval $\Delta t = 0.01$ sec, approximately equal to one-tenth of the natural period of the system. The results are tabulated in Table 2-2.

For the first time interval, Eq. 2-55 is used, yielding

$$y(t=0.01) = \frac{\ddot{y}^{(0)}}{2}(\Delta t)^2 = \frac{643}{2}(0.01)^2 = 0.03215 \text{ in.}$$

For the remaining intervals, Eq. 2-53 is used. For example,

$$y(t=0.02) = 2(0.03215) - 0 + (443.3)(0.01)^2 = 0.10863 \text{ in.}$$

Table 2-2 shows that the maximum deflection is 0.26697 in. and that it occurs at the time t_m equal to approximately 0.05 sec. From the graph in Fig. 2-5*b*, the time of maximum response is $t_m = 0.048$ sec, which is approximately the same as the one found here. This gives an indication of the accuracy of this numerical method when Δt is one-tenth of the natural period of vibration of the system. In Table 2-2 another peak for y of smaller magnitude occurs at $t = 0.10$ sec. Subsequent peaks may be found by continuing the calculations in Table 2-2. The forcing function $F(t)$ used in the calculations is the one given in Fig. 2-8*b*.

It was found above that the maximum deflection occurred at $t=0.05$ sec and that it is equal to 0.267 in. The spring force F corresponding to this deflection is

$$F=ky=(258)(0.267)=68.9 \text{ kips}$$

This value of the dynamic force is equal to the one acting at the center of the beam in Fig. 2-8*a*, because the spring-mass system in Fig. 2-8*c* is an idealization of this beam with spring stiffness k equal to that at the center of the beam. With this in mind, the maximum dynamic stress σ_{max} at the center of the member is

$$\sigma_{max}=\frac{M}{S}=\frac{FL/8}{S}=\frac{(68.9)(30)(12)}{(8)(175.4)}=17.70 \text{ ksi}$$

The maximum dynamic stress that is found in the preceding section by using the graphs in Fig. 2-5 is 17.90 ksi.

2-8 ELASTOPLASTIC SYSTEMS WITH ONE DEGREE OF FREEDOM[8]

In the preceding sections, the dynamic response of one-degree structural systems was investigated by using *(a)* a rigorous method in which the solution was obtained directly by solving the differential equation of motion and *(b)* by applying the numerical method known as the acceleration impulse extrapolation method. In both approaches, the structural system was assumed to be linearly elastic, thus limiting their application to dynamic responses that do not exceed the elastic limit of the material. In practice, however, the elastic limit for structures that are designed for continuous operating conditions is seldom exceeded. On the other hand, there are cases where it becomes appropriate to consider responses that exceed this elastic limit. For example, if a structure is subjected to a severe dynamic load such as the one due to a blast or an earthquake, but only a few times in its life span, it becomes uneconomical to design this structure to behave entirely in the elastic range. Many ductile materials are characterized by a large yielding range that permits them to absorb large amounts of energy before complete failure. In such special cases, the important decision to be made by a designer is the amount of energy that a structure should be permitted to absorb for a still-safe design. The analysis of this section is limited to inelastic behavior of structures that can be treated as one-degree of freedom systems.

The resistance R of a structure, or a structural component, is defined as the internal force tending to restore the structure to its unloaded static

[8] Additional information on this subject may be found in Reference 8.

position and it can have a variety of forms. For example, if a structural element is made up of brittle material, the resistance is represented by curve 1 in Fig. 2-10*d*. It will have the form of curve 2 if the element is made up of a ductile material, and it will follow the shape of curve 3 if the structural element is of plane concrete. In all these cases, the resistance functions are often idealized in order to simplify the analysis involved. For the majority of structural problems, it is permissible to use the bilinear form shown in Fig. 2-10*c*.

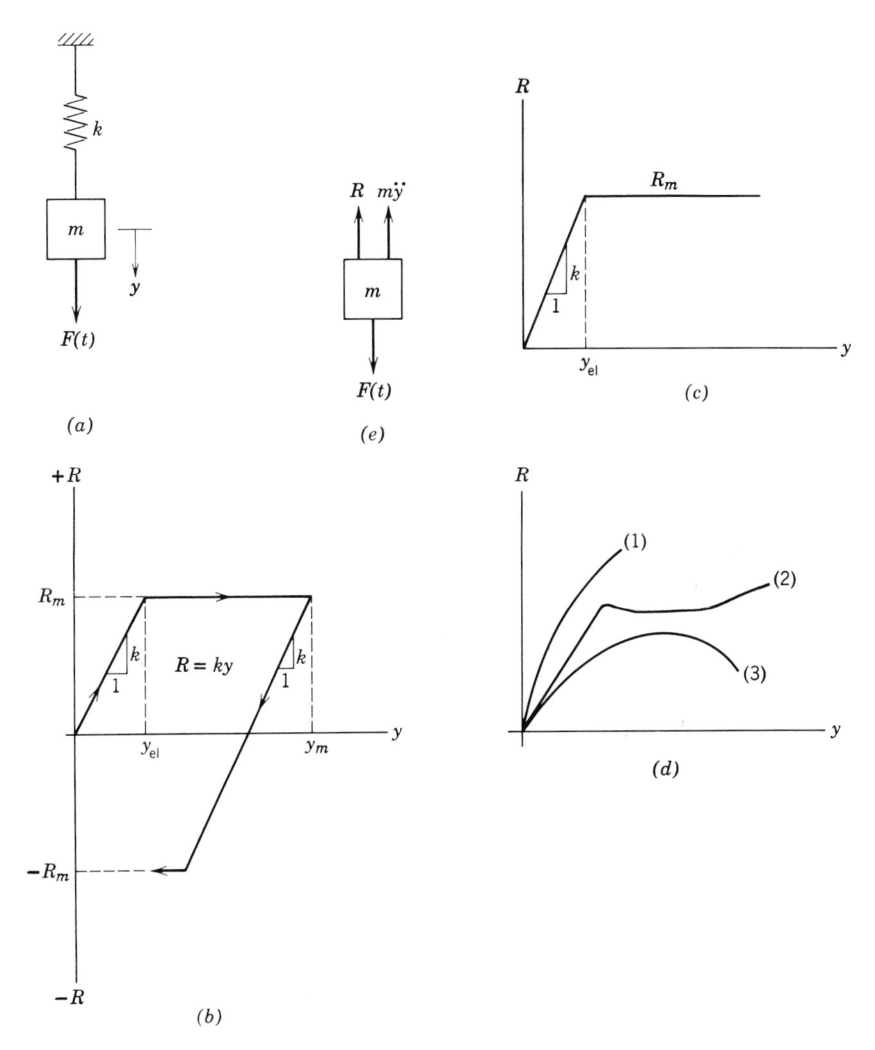

Fig. 2-10

The discussion on inelastic behavior is initiated by considering the displacement y and the resistance function R of the spring-mass system in Fig. 2-10*a*. The resistance R is assumed to have the bilinear form shown in Fig. 2-10*b*, whose slope is equal to the spring constant k as shown. In words, R increases linearly until the elastic limit displacement y_{el} is attained. At this point, the resistance function attains its maximum value R_m. For any further increase of y, R_m will maintain its maximum value. With increasing y and R_m constant, the following two situations are assumed to be developing. It was pointed out earlier that material behavior under loading is characterized by a large yielding range that permits the material to absorb large amounts of energy after it is stressed beyond its elastic limit. The extent of this yielding range is dependent on the ductility limit of the material.

If the displacement y reaches its maximum value y_m before the ductility limit is attained, while R_m is constant, the structural system is said to rebound, and during rebound the resistance R is assumed to decrease linearly along a line parallel to its initial elastic line as shown in Fig. 2-10*b*. This decrease will continue until R becomes equal to $-R_m$. The other situation, however, is for y to continue to increase, with R_m remaining constant, until the ductility limit is attained. In this case, the resistance function is assumed to be composed of the two lines shown in Fig. 2-10*c*.

The resistance function R of the spring-mass system in Fig. 2-10*a* is the force ky in the spring. By applying dynamic equilibrium to the free body in Fig. 2-10*e*, the differential equation of motion is

$$m\ddot{y}+R-F(t)=0 \tag{2-61}$$

If R is assumed to have the form shown in Fig. 2-10*b*, then Eq. 2-61 will take the following forms for the various intervals of the displacement y:

In the elastic range the resistance R is equal to ky and Eq. 2-61 yields

$$m\ddot{y}+ky-F(t)=0 \qquad 0\leqslant y\leqslant y_{el} \tag{2-62}$$

When R becomes R_m and R_m remains constant for values of y between y_{el} and y_m, Eq. 2-61 becomes

$$m\ddot{y}+R_m-F(t)=0 \qquad y_{el}\leqslant y\leqslant y_m \tag{2-63}$$

When R_m starts to decrease linearly until it becomes $-R_m$, the displacement y will attain values between y_m and y_m-2y_{el}. Thus, Eq. 2-61 takes the form

$$m\ddot{y}+R_m-k(y_m-y)-F(t)=0 \qquad (y_m-2y_{el})\leqslant y\leqslant y_m \tag{2-64}$$

Additional equations for the negative plastic range may be written in a similar manner. This is not usually necessary because it seldom becomes of interest to the engineer in practice.

For a given load function $F(t)$, the solutions of the differential equations above can be obtained by using either a rigorous approach or a numerical method such as the acceleration impulse extrapolation method discussed in Section 2-7. When a rigorous solution is used, the initial conditions of Eq. 2-62 are the initial conditions of the given problem. The initial conditions of Eq. 2-63, however, should be the final displacement and the final velocity from the solution of Eq. 2-62. For Eq. 2-64, the final displacement and final velocity from the solution of Eq. 2-63 should be taken as its initial conditions. This procedure is sufficient if the forcing function $F(t)$ is a continuous function of time that does not reduce to zero before the analysis for y is completed. For example, a suddenly applied constant force of infinite duration can be solved by using these equations. If, however, the forcing function $F(t)$ is of finite duration or has discontinuities, additional stages of the equations, one for each discountinuity, should be included in the solution. For the triangular load in Fig. 2-4*b*, for example, the following equations should be used:

$$m\ddot{y} + ky - f(t) = 0 \qquad t < t_d, \quad y < y_{\text{el}} \tag{2-65}$$

$$m\ddot{y} + R_m - F(t) = 0 \qquad t < t_d, \quad y > y_{\text{el}} \tag{2-66}$$

$$m\ddot{y} + ky = 0 \qquad t > t_d, \quad y < y_{\text{el}} \tag{2-67}$$

$$m\ddot{y} + R_m = 0 \qquad t > t_d, \quad y > y_{\text{el}} \tag{2-68}$$

$$m\ddot{y} + R_m - k(y_m - y) - F(t) = 0 \qquad t_d > t > t_m \tag{2-69}$$

$$m\ddot{y} + R_m - k(y_m - y) = 0 \qquad t > t_m, \quad t > t_d \tag{2-70}$$

The first four of the equations express the motion before the maximum displacement y_m is attained, while the last two express the motion when the system is in elastic rebound. For a given problem, the sequence in using these equations to evaluate y depends on the time t_{el}, when the maximum elastic displacement is reached, and on t_m, the time at which the maximum deflection y_m is attained. It is needless to say that rigorous solutions are often complicated and laborious, except for simple types of problems. In such cases, it is advisable to use a numerical method, such as the acceleration impulse extrapolation method.

As an illustration, consider the beam in Fig. 2-8*a* that supports the weight $W = 30.0$ kips, and also carries the dynamic force $F(t)$ whose variation is shown in Fig. 2-11*a*. The weight of the beam is assumed to be negligible, and its idealized one-degree system is shown in Fig. 2-8*c*. Its resistance function is assumed to have the form shown in Fig. 2-11*b*. It is required to determine the maximum deflection y_m by applying the acceleration impulse extrapolation method.

The steel beam in Fig. 2-8*a* has a 24 W 76 cross section, whose plastic section modulus[9] Z_p is 200.1 in.^{3} The plastic[10] moment M_p, based on a yield-point stress of 30.0 ksi, is

$$M_P = (30)(200.1) = 6000.0 \text{ kip-in.}$$

This is the ultimate moment that can be carried by the cross section of the member. For the given loading conditions of the beam, the maximum resistance R_m is equal to the load at the center of the member that produces the ultimate moment. Knowing that plastic hinges will develop simultaneously at the ends and at the center of the member, the resistance

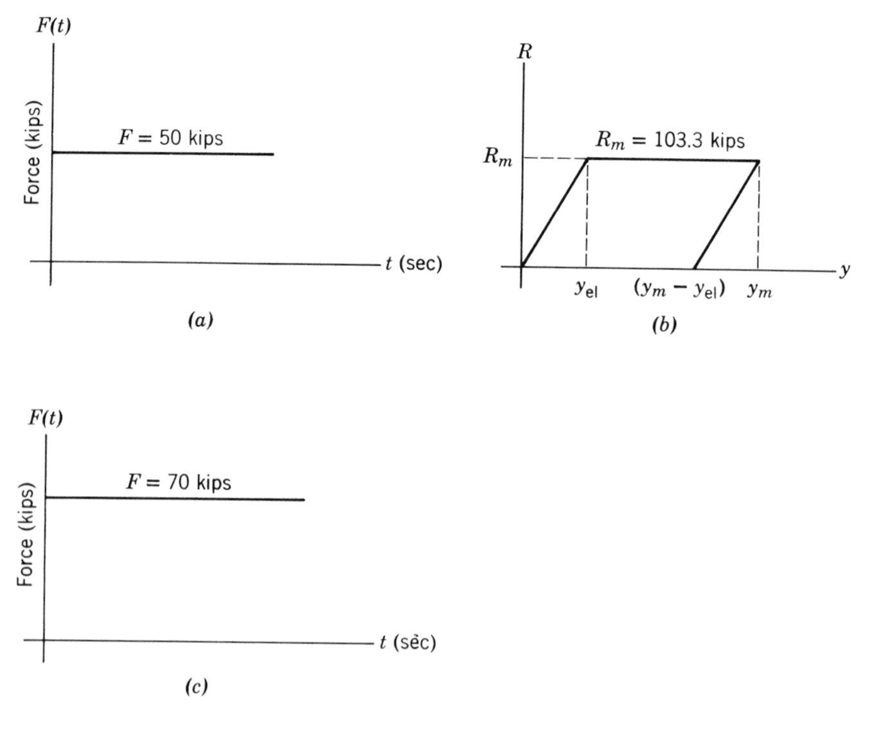

Fig. 2-11

[9]The plastic section modulus for various sizes of beam sections is given in the *Manual of Steel Construction*, 6th. ed. American Institute of Steel Construction, 1964.
[10]Ample information on this subject can be found in Reference 9,

R_m is

$$R_m = \frac{8M_P}{L} = \frac{(8)(6000)}{(30)(12)} = 133.3 \text{ kips}$$

Since the beam carries also the weight $W = 30.0$ kips, the maximum force available to resist the dynamic loading is

$$R_m = 133.3 - 30.0 = 103.3 \text{ kips}$$

Returning now to the idealized one-degree system in Fig. 2-8*c* of spring constant $k = 258$ kips/in., its elastic deflection y_{el} is

$$y_{el} = \frac{R_m}{k} = \frac{103.3}{258.0} = 0.40 \text{ in.}$$

Applying Eq. 2-61, we have the differential equation of motion

$$\frac{W}{g}\ddot{y} + R - F(t) = 0 \qquad (2\text{-}71)$$

Substituting the given values for *F(t)* and W/g and solving for $\ddot{y}$, Eq. 2-71 yields

$$\ddot{y} = 643.5 - 12.87R \qquad (2\text{-}72)$$

For the various intervals of y, Eq. 2-72 yields

$$\ddot{y} = 643.5 - 3325y \qquad 0 \leqslant y \leqslant 0.40 \qquad (2\text{-}73)$$

$$\ddot{y} = -686.5 \qquad 0.40 \leqslant y \leqslant y_m \qquad (2\text{-}74)$$

$$\ddot{y} = 686.5 + 3325(y_m - y) \qquad (y_m - 0.80) \leqslant y \leqslant y_m \qquad (2\text{-}75)$$

TABLE 2-3

t (sec)	$\ddot{y}$ (in./sec^2)	$\ddot{y}(\Delta t)^2$ (in.)	y Eq. 2-53 (in.)
0	643.5	0.0644	0
0.01	536.5	0.0537	0.0322
0.02	251.0	0.0251	0.1181
0.03	−117.5	−0.0118	0.2291
0.04	−446.5	−0.0447	0.3283
0.05	−629.5	−0.0630	0.3828
0.06	−601.5	−0.0602	0.3743
0.07			0.3056

TABLE 2-4

t (sec)	$\ddot{y}$ (in./sec^2)	$\ddot{y}(\Delta t)^2$ (in.)	y Eq. 2-53 (in.)
0	901.0	0.0901	0
0.01	751.0	0.0751	0.0451
0.02	352.0	0.0352	0.1652
0.03	−164.0	−0.0164	0.3205
0.04	−429.0	−0.0429	0.4594
0.05	−429.0	−0.0429	0.5554
0.06	−429.0	−0.0429	0.6085
0.07	−429.0	−0.0429	0.6187
0.08	−320.5	−0.0321	0.5860
0.09	−105.5	−0.0106	0.5212
0.10	144.0	0.0144	0.4458
0.11	348.0	0.0348	0.3848
0.12	436.0	0.0436	0.3586
0.13	378.0	0.0378	0.3760
0.14	194.0	0.0194	0.4312
0.15	−54.0	−0.0054	0.5058
0.16			0.5750

By applying the acceleration impulse extrapolation method and using a time interval $\Delta t = 0.01$ sec, the dynamic deflections y are evaluated and they are shown in Table 2-3. The maximum deflection occurs at time $t = 0.05$ sec and is equal to 0.3828 in. Since this displacement never reached the value $y_{el} = 0.40$ in., the deformation of the system is all elastic. Thus only Eq. 2-73 was used.

Let it now be assumed that the suddenly applied force $F(t)$ has the magnitude shown in Fig. 2-11*c*. In this case, Eq. 2-71 yields

$$\ddot{y} = 901 - 12.87R \tag{2-76}$$

and

$$\ddot{y} = 901 - 3325y \qquad 0 \leqslant y \leqslant y_{el} \tag{2-77}$$

$$\ddot{y} = -429 \qquad 0.04 \leqslant y \leqslant y_m \tag{2-78}$$

$$\ddot{y} = -429 + 3325(y_m - y) \qquad (y_m - 0.80) \leqslant y \leqslant y_m \tag{2-79}$$

By using again the same numerical method and $\Delta t = 0.01$ sec, we obtain the values of y given in Table 2-4. In this case, Eq. 2-77 was used up to $t = 0.04$ sec, Eq. 2-78 for $0.4 \leqslant t \leqslant 0.08$, and Eq. 2-79 for the remaining computations.

The computations, however, could go on indefinitely. The first peak of deflection occurs at t=0.07 sec and is equal to 0.6187 in. Other peaks of deflection will be obtained as the computations continue. It should be noted that the maximum deflection of the elastoplastic solution is greater than $y_{el}=0.40$ in.

If in the solution above the system was assumed to behave only elastically and y_m was determined by using Eq. 2-77 for all times, the elastic vibration would have taken place about the static equilibrium position F/k, where

$$\frac{F}{k}=\frac{70}{258}=0.271 \text{ in.}$$

In the elastoplastic case, however, the system will undergo a certain amount of permanent deformation y_p, which is equal to $(y_m - y_{el})$ as shown in Fig. 2-11*b*. In this expression, y_{el} is the amount by which y_m will be reduced if all loads are removed from the system. The amount of y_p in the solution is

$$y_p = y_m - y_{el} = 0.6187 - 0.4000 = 0.2187 \text{ in.}$$

When the elastoplastic system attains its maximum deflection y_m in Fig. 2-11*b* and starts to rebound, that is, when Eq. 2-79 is applied, it vibrates harmonically for an indefinite period of time, since there is no damping. This residual vibration is elastic, and takes place about an equilibrium position y, where y is given by the expression

$$y = y_m - \frac{R_m - F}{k} \qquad (2\text{-}80)$$

In the equation, the quantity $(R_m - F)/k$ is the peak amplitude Y of the residual vibration, which is the amount by which the maximum displacement y_m should be reduced to obtain the position about which the system vibrates.

For the elastoplastic system discussed above, Eq. 2-80 yields

$$y = 0.6187 - \frac{103.3-70}{258} = 0.4897 \text{ in.}$$

and the peak amplitude Y of the residual vibration is

$$Y = \frac{R_m - F}{k} = \frac{103.3-70}{258} = 0.129 \text{ in.}$$

In other words, the residual vibration takes place about a position y equal to $y_p + y_{st}$, where $y_{st} = F/k$ is the static equilibrium position about which the system would vibrate if its response, at all times, was considered to be elastic only.

2-9 SYSTEMS WITH TWO OR MORE DEGREES OF FREEDOM

The steady-state forced vibration of a two-degree spring-mass system is discussed here. First, the system is assumed to vibrate under the action of harmonic forces without damping. When this case is completed, the analysis continues by including the effects of viscous damping.

Consider the two-degree spring-mass system in Fig. 2-12*a*, and let it be assumed that the harmonic forces $F_1 \cos \omega_f t$ and $F_2 \cos \omega_f t$ are acting on masses m_1 and m_2, respectively. Under these conditions, the system will vibrate harmonically with the frequency ω_f of the exciting forces. At the present, it is further assumed that there is no damping.

At the extreme configuration of the system, the maximum amplitudes of the masses m_1 and m_2 are designated by Y_1 and Y_2, respectively. Since the motion is harmonic, it is reasonable to assume that at this position the harmonic forces attain their maximum magnitudes F_1 and F_2. The free-body diagrams of the masses m_1 and m_2 in Figs. 2-12*b* and 2-12*c*, respectively, depict the forces acting on them at their maximum amplitude position. By applying dynamic equilibrium, the following equations are obtained:

$$m_1\omega_f^2 Y_1 + k_2(Y_2 - Y_1) + F_1 - k_1 Y_1 = 0 \qquad (2\text{-}81)$$

$$m_2\omega_f^2 Y_2 - k_2(Y_2 - Y_1) + F_2 = 0 \qquad (2\text{-}82)$$

or, by rearranging terms,

$$(k_1 + k_2 - m_1\omega_f^2)\,Y_1 - k_2 Y_2 = F_1 \qquad (2\text{-}83)$$

$$-k_2 Y_1 + (k_2 - m_2\omega_f^2)\,Y_2 = F_2 \qquad (2\text{-}84)$$

By solving the two equations simultateously for Y_1 and Y_2,

$$Y_1 = \frac{F_2 k_2 - F_1(m_2\omega_f^2 - k_2)}{(m_1\omega_f^2 - k_1 - k_2)(m_2\omega_f^2 - k_2) - k_2{}^2} \qquad (2\text{-}85)$$

$$Y_2 = \frac{F_1 k_2 - F_2(m_1\omega_f^2 - k_1 - k_2)}{(m_1\omega_f^2 - k_1 - k_2)(m_2\omega_f^2 - k_2) - k_2{}^2} \qquad (2\text{-}86)$$

In the two equations, one can easily notice that the denominators are identical. If these denominators are equal to zero, both amplitudes Y_1 and Y_2 become infinite, thus indicating that a phenomenon of resonance occurs. Therefore, the values of ω_f at resonance can be found from the expression

$$(m_1\omega_f^2 - k_1 - k_2)(m_2\omega_f^2 - k_2) - k_2{}^2 = 0$$

or

$$\omega_f^4-\left(\frac{k_1+k_2}{m_1}+\frac{k_2}{m_2}\right)\omega_f^2+\frac{k_1k_2}{m_1m_2}=0 \tag{2-87}$$

Equation 2-87 is identical to Eq. 1-97, Section 1-9, describing the free undamped vibration of a two-degree spring-mass system. Consequently, the phenomenon of resonance will occur when the frequency ω_f of the harmonic forces coincides with either one of the natural frequencies of the spring-mass system.

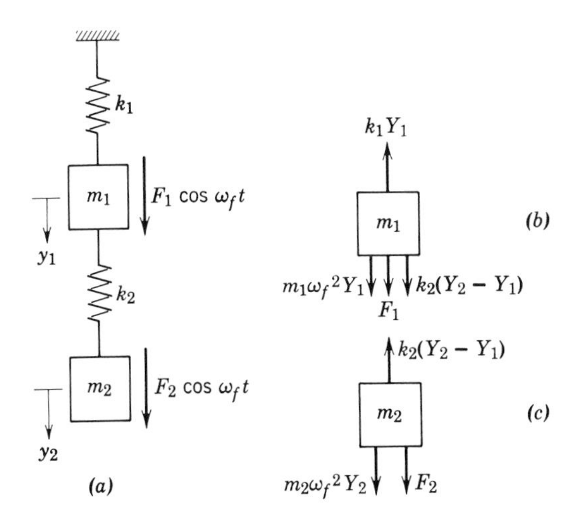

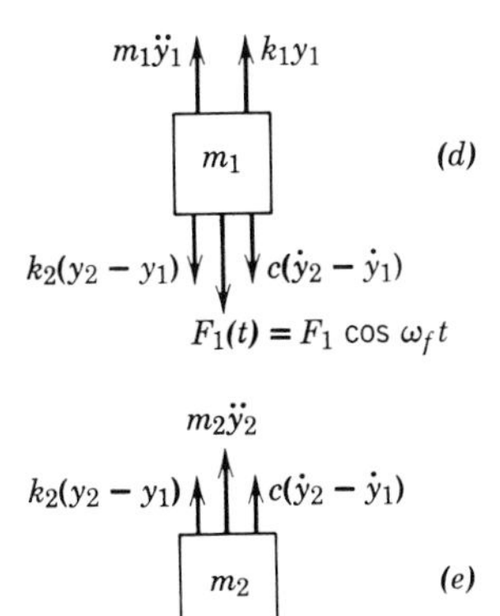

Fig. 2-12

Let it now be assumed that the system in Fig. 2-12a is under the influence of viscous damping. It is further assumed that viscous damping exists only between masses m_1 and m_2 and that the harmonic force $F_2 \cos \omega_f t$ is zero at all times. The free-body diagrams of m_1 and m_2 are shown in Figs. 2-12d and 2-12e, respectively. By applying dynamic equilibrium, the two equations of motion are:

$$-m_1\ddot{y}_1 - k_1 y_1 + k_2(y_2 - y_1) + c(\dot{y}_2 - \dot{y}_1) + F_1(t) = 0 \qquad (2\text{-}88)$$

$$-m_2\ddot{y}_2 - k_2(y_2 - y_1) - c(\dot{y}_2 - \dot{y}_1) = 0 \qquad (2\text{-}89)$$

By rearranging terms, the expressions become,

$$m_1\ddot{y}_1 + c\dot{y}_1 - c\dot{y}_2 + (k_1 + k_2)y_1 - k_2 y_2 = F_1 \cos \omega_f t \qquad (2\text{-}90)$$

$$m_2\ddot{y}_2 - c\dot{y}_1 + c\dot{y}_2 - k_2 y_1 + k_2 y_2 = 0 \qquad (2\text{-}91)$$

It is now assumed that the excitation and the steady-state response are both harmonic and of the same frequency ω_f. It is also assumed that the excitation force $F_1 \cos \omega_f t$ is represented by the expression

$$F_1(t) = F_1 e^{i\omega_f t} \qquad (2\text{-}92)$$

If the response lags the excitation by a phase angle ϕ, the solution[11] of Eqs. 2-90 and 2-91 can be taken as

$$y_1 = Y_1 e^{i(\omega_f t - \phi_1)} \qquad (2\text{-}93)$$

$$y_2 = Y_2 e^{i(\omega_f t - \phi_2)} \qquad (2\text{-}94)$$

In Eqs. 2-92, 2-93, and 2-94 the harmonic force and the harmonic motion are defined by the real parts of these equations.

Equations 2-93 and 2-94 are rewritten as follows:

$$y_1 = \overline{Y}_1 e^{i\omega_f t} \qquad (2\text{-}95)$$

$$y_2 = \overline{Y}_2 e^{i\omega_f t} \qquad (2\text{-}96)$$

where

$$\overline{Y}_1 = Y_1 e^{-i\phi_1} \qquad (2\text{-}97)$$

$$\overline{Y}_2 = Y_2 e^{-i\phi_2} \qquad (2\text{-}98)$$

are the complex amplitudes of y_1 and y_2. By substituting Eqs. 2-92, 2-95,

[11]In vibration analysis, this solution is usually called the method of mechanical impedance. For more information on this method see References 10 and 11.

and 2-96 into the differential equations (2-90) and (2-91), the following system of equations is obtained:

$$(-m_1\omega_f^2+ic\omega_f+k_1+k_2)\,\overline{Y}_1-(ic\omega_f+k_2)\,\overline{Y}_2=F_1 \qquad (2\text{-}99)$$

$$-(ic\omega_f+k_2)\,\overline{Y}_1+(-m_2\omega_f^2+ic\omega_f+k_2)\,\overline{Y}_2=0 \qquad (2\text{-}100)$$

By applying Cramer's rule, the solution of the two equations yields

$$\overline{Y}_1=\frac{\begin{vmatrix} F_1 & -(ic\omega_f+k_2) \\ 0 & (-m_2\omega_f^2+ic\omega_f+k_2) \end{vmatrix}}{\begin{vmatrix} (-m_1\omega_f^2+ic\omega_f+k_1+k_2) & -(ic\omega_f+k_2) \\ -(ic\omega_f+k_2) & (-m_2\omega_f^2+ic\omega_f+k_2) \end{vmatrix}}$$

$$=Y_1e^{-i\phi_1} \qquad (2\text{-}101)$$

and

$$\overline{Y}_2=\frac{\begin{vmatrix} (-m_1\omega_f^2+ic\omega_f+k_1+k_2) & F_1 \\ -(ic\omega_f+k_2) & 0 \end{vmatrix}}{\begin{vmatrix} (-m_1\omega_f^2+ic\omega_f+k_1+k_2) & -(ic\omega_f+k_2) \\ -(ic\omega_f+k_2) & (-m_2\omega_f^2+ic\omega_f+k_2) \end{vmatrix}}$$

$$=Y_2e^{-i\phi_2} \qquad (2\text{-}102)$$

where ϕ_1 and ϕ_2 are the phase angles of the complex amplitudes $\overline{Y}_1$ and $\overline{Y}_2$, respectively.

By expanding the determinants in Eqs. 2-101 and 2-102 and applying known rules of complex numbers, the amplitudes Y_1 and Y_2 are as follows:

$$Y_1=\frac{F_1[(k_2-m_2\omega_f^2)^2+(c\omega_f)^2]^{\frac{1}{2}}}{\{[(k_2-m_2\omega_f^2)(k_1-m_1\omega_f^2)-k_2m_2\omega_f^2]^2+(m_1\omega_f^2+m_2\omega_f^2-k_1)(c\omega_f)^2\}^{\frac{1}{2}}} \qquad (2\text{-}103)$$

$$Y_2=\frac{F_1[k_2{}^2+(c\omega_f)^2]^{\frac{1}{2}}}{\{[(k_2-m_2\omega_f^2)(k_1-m_1\omega_f^2)-k_2m_2\omega_f^2]^2+(m_1\omega_f^2+m_2\omega_f^2-k_1)(c\omega_f)^2\}^{\frac{1}{2}}} \qquad (2\text{-}104)$$

With Y_1 and Y_2 given by Eqs. 2-103 and 2-104, the steady-state solutions are

$$y_1 = Y_1 \cos(\omega_f t - \phi_1) \quad y_2 = Y_2 \cos(\omega_f t - \phi_2) \tag{2-105}$$

The forcing function producing this motion is the harmonic force $F_1 \cos \omega_f t$. The amplitudes y_1 and y_2 lag this force by different amounts indicating that y_1 and y_2 are not in phase during motion.

The solution of spring-mass systems having more than two degrees of freedom could be obtained in some similar manner. It should be noted, however, that a rigorous approach is often complicated and time consuming for practical engineering problems. A satisfactory solution can be usually obtained by using the numerical method acceleration impulse extrapolation method discussed in Section 2-7. Applications of this method to multidegree systems are given in Chapter 3.

2-10 FOURIER SERIES

Fourier has shown[12] that any periodic function can be expressed in terms of combinations of sine and cosine functions. Examples of periodic functions are given in Fig. 2-13. In structural dynamics, time-varying forces *F(t)* are often considered as periodic, and the concept of Fourier series can be used conveniently as a method of analysis.

Any forcing function *F(t)* that is repeated every T seconds can be written as an infinite sum of sines and cosines, with the frequency of the component terms given as $\omega_f = 2\pi/T$, $2\omega_f$, $3\omega_f, \ldots, n\omega_f$. That is,

$$F(t) = \frac{a_0}{2} + a_1 \cos \omega_f t + a_2 \cos 2\omega_f t + \cdots + a_n \cos n\omega_f t + \cdots$$

$$+ b_1 \sin \omega_f t + b_2 \sin 2\omega_f t + \cdots + b_n \sin n\omega_f t + \cdots$$

or

$$F(t) = \frac{a_0}{2} + \sum_{n=1}^{\infty} (a_n \cos n\omega_f t + b_n \sin n\omega_f t) \tag{2-106}$$

The evaluation of the constants a_0, a_n, and b_n to fit a given periodic function *F(t)* can be made by using the expressions

$$a_0 = \frac{2}{T} \int_d^{d+T} F(t)\, dt \tag{2-107}$$

[12] R. V. Churchill, *Fourier Series and Boundary Value Problems*, McGraw-Hill Book Co., Inc, New York, 1941, pp. 70–94.

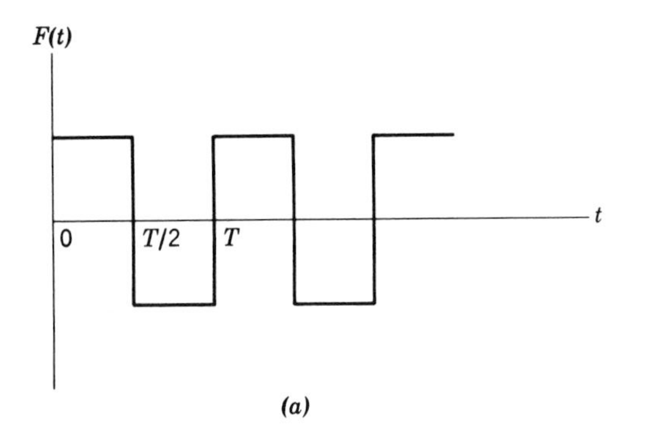

(a)

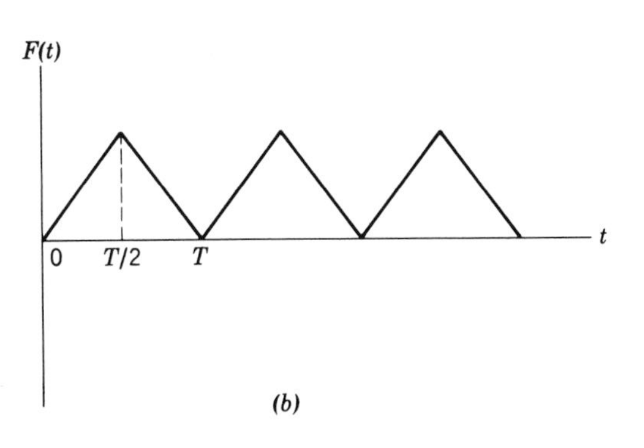

(b)

Fig. 2-13

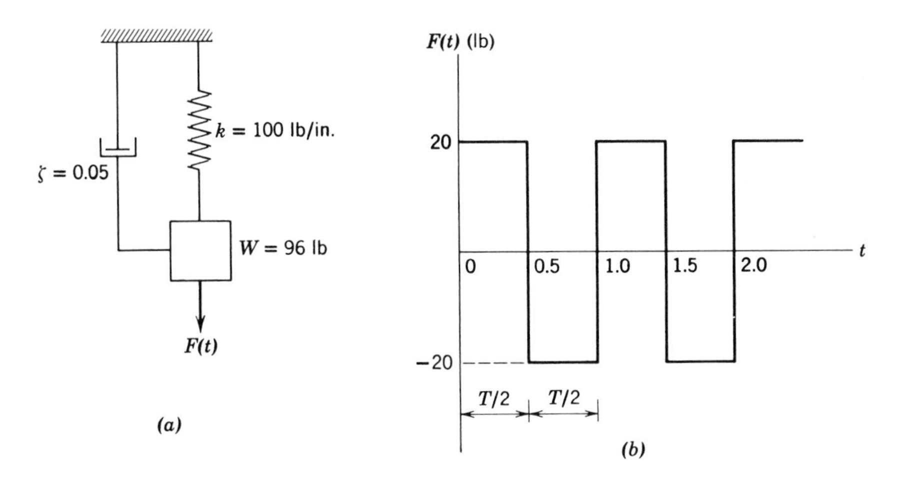

(a)
(b)

Fig. 2-14

$$a_n = \frac{2}{T} \int_d^{d+T} F(t) \cos n\omega_f t \, dt \tag{2-108}$$

$$b_n = \frac{2}{T} \int_d^{d+T} F(t) \sin n\omega_f t \, dt \tag{2-109}$$

where d in the integrals above will usually be equal to either $-T/2$ or zero. The constant a_0 in the series represents the offsetting of the remaining terms of the series.

As an application regarding the Fourier series concept, consider the spring-mass system in Fig. 2-14*a* that is acted upon by the time-varying force shown in Fig. 2-14*b*. It is required to determine the steady-state response of the system.

The first step is to determine the Fourier series expansion of $F(t)$. The forcing function $F(t)$ is clearly an odd function of t and no cosine terms are present. Therefore, the only constant to determine is b_n. By using Eq. 2-109,

$$b_n = \frac{4}{1} \int_0^{0.5} 20 \sin n\omega_f t \, dt$$

$$= 80\left[-\frac{\cos n\omega_f t}{n\omega_f} \right]_0^{0.5}$$

$$= 80\left[-\frac{\cos n\omega_f/2}{n\omega_f} + \frac{1}{n\omega_f} \right]$$

or, since $\omega_f = 2\pi/T$,

$$b_n = (40)\frac{1 - \cos n\pi}{n\pi} \tag{2-110}$$

Equation 2-110 shows that for n even the constant b_n is zero, and it is equal to $80/n\pi$ for n odd. Therefore, the Fourier expansion of $F(t)$ is

$$F(t) = \frac{80}{\pi}\left(\sin 2\pi t + \frac{1}{3} \sin 6\pi t + \frac{1}{5} \sin 10\pi t + \cdots \right) \tag{2-111}$$

The steady-state response can be now easily obtained by using the ideas of Section 2-3. It is only necessary to apply the proper magnification factor and phase angle to each component term of the forcing function $F(t)$, Eq. 2-111, and add the results. The static deflection y_{st} that is produced in the system by the steady forces having the magnitudes of the various terms of

F(t) is as follows:

$$(y_{st})_n = \frac{F_n}{k} = \frac{80/n\pi}{100} = \frac{4}{5n\pi} \qquad n=1, 3, 5, \ldots$$

The undamped natural frequency ω is

$$\omega = \sqrt{k/m} = \sqrt{kg/W} = 20 \text{ rps}$$

The expressions that are required to determine the magnification factors Γ and phase angles ϕ are given by Eqs. 2-18 and 2-19, respectively. The steady-state response for the first three terms of *F(t)*, Eq. 2-111, are given in Table 2-5.

TABLE 2-5

Term	y_{st}	ω_f/ω	Γ Eq. 2-18	ϕ Eq. 2-19	$y_{st}\Gamma \sin(\omega_f t - \phi)$
1	$4/5\pi$	$2\pi/20$	1.11	2°	$0.28 \sin(2\pi t - 2°)$
2	$4/15\pi$	$6\pi/20$	6.83	40°	$0.58 \sin(6\pi t - 40°)$
3	$4/25\pi$	$10\pi/20$	0.68	174°	$0.03 \sin(10\pi t - 174°)$

By substituting the exponential equivalents of the cosine and sine terms into the original form of the series, Eq. 2-106, and making certain manipulations, the Fourier series of *F(t)* can be written in complex form as follows:

$$F(t) = \sum_{n=-\infty}^{\infty} c_n e^{in\omega_f t} \tag{2-112}$$

$$c_0 = \frac{1}{T} \int_d^{d+T} F(t)\, dt \tag{2-113}$$

$$c_n = \frac{1}{T} \int_d^{d+T} F(t)\, e^{-in\omega_f t}\, dt \tag{2-114}$$

where d in the integrals above is usually equal to either $-T/2$ or zero. This form of the Fourier series can be also used as before to determine the dynamic response of structural systems.

PROBLEMS

2-1 The single-degree spring-mass system in Fig. 1-3*a* is subjected to a harmonic force $F_1 \cos \omega_f t$, where $F_1 = 10$ lb and $\omega_f = 100$ cpm. By neglecting damping and assuming that $k = 50$ lb/in. and $m = 0.15$ lb-sec^2/in., determine the maximum forced amplitude of the mass m.

2-2 Repeat Problem 2-1 for the impressed frequencies $\omega_f = 120$, 200, 300, and 600 cpm.

2-3 Repeat Problem 2-1 for an impressed force $F(t) = F_1 \sin \omega_f t$, where $F_1 = 20$ lb and $\omega_f = 200$ Hz.

2-4 Repeat Problem 2-1 by including the effects of viscous damping with damping ratio $\zeta = 0.10$.

2-5 Repeat Problem 2-2 by including the effects of viscous damping with damping ratio $\zeta = 0.20$.

2-6 A weight of 120 lb is suspended by a spring of stiffness $k = 20$ lb/in. If it is forced to vibrate harmonically by a harmonic force of 10 lb with viscous damping of $c = 0.320$ lb/(in.)(sec), determine (*a*) the resonant frequency, (*b*) the amplitude at resonance, and (*c*) the phase angle at resonance.

2-7 Determine the constants A_1 and A_2 in Eq. 2-20 by assuming that at $t = 0$ the initial displacement and initial velocity are y_0 and $\dot{y}_0$, respectively. Write the general expression for $y(t)$ and $\dot{y}(t)$.

2-8 Check the expressions given by Eqs. 2-37 and 2-41 by using Eqs. 2-25 and 2-35 and the initial conditions $y_0 = \dot{y}_0 = 0$.

2-9 Derive Eqs. 2-43 and 2-45 by using Eq. 2-25 and appropriate initial time conditions.

2-10 The spring-mass system in Fig. 2-4*a* is subjected to a suddenly applied force of 30 kips that decreased linearly to zero at $t_d = 0.10$ sec. Determine the maximum dynamic displacement of the mass m due to this force.The spring constant k is 150 kips/in. and $m = 0.08$ kip-sec^2/in.

2-11 For the spring-mass system in Problem 2-10, plot the variation of $y(t)$ versus time for the first 0.40 sec of the motion.

2-12 Repeat Problem 2-10 for a suddenly applied force of 30 kips that is released suddenly at $t = 0.10$ sec.

2-13 For the spring-mass system and loading as given in Problem 2-12, plot the variation of $y(t)$ versus time for the first 0.40 sec of the motion.

2-14 Repeat Problem 2-10 for a suddenly applied force of 30 kips and of infinite duration.

2-15 Repeat Problem 2-10 by utilizing appropriate response charts and also indicate the time the maximum displacement is attained.

2-16 Repeat Problem 2-12 by using appropriate response charts and also indicating the time the maximum displacement is attained.

2-17 Repeat Problem 2-14 by using appropriate response charts and including the time the maximum displacement is attained.

2-18 The system in Fig. 2-8*a* is idealized as shown in Fig. 2-8*c*. Determine the maximum dynamic displacement and maximum dynamic bending stress for a force $F(t)$ of 40 kips applied at $t=0$ and removed suddenly at $t=0.08$ sec. Use appropriate response charts and also indicate the time of maximum response.

2-19 Repeat Problem 2-10 by using the acceleration impulse extrapolation method.

2-20 Repeat Problem 2-12 by applying the acceleration impulse extrapolation method.

2-21 Repeat Problem 2-14 by utilizing the acceleration impulse extrapolation method.

2-22 Repeat Problem 2-18 by applying the acceleration impulse extrapolation method.

2-23 The system in Fig. 2-8*a* is idealized as shown in Fig. 2-8*c*. If the dynamic force $F(t)$ is a suddenly applied constant force of infinite duration and has a magnitude of 90 kips, determine the maximum displacement of the weight W by following appropriate elastoplastic analysis. Neglect the weight of the beam and assume that there is no damping.

2-24 Repeat Problem 2-23 by assuming that the force $F(t)$ is a suddenly applied force of 140 kips at $t=0$ and decreased linearly to zero at $t=0.20$ sec.

2-25 Repeat Problem 2-23 by assuming that the force $F(t)$ is a suddenly applied force of 70 kips at $t=0$ and removed suddenly at $t=0.20$ sec.

2-26 The two-degree spring-mass system in Fig. 1-3*c* is subjected to the harmonic forces $F_1(t)=F_1 \cos \omega_f t$ and $F_2(t)=0$, where $F_1=10$ lb and $\omega_f=100$ cpm. The damping constant $c_1=0$, the damping ratio ζ for the damping element between masses m_1 and m_2 is 0.10, $k_1=40$ lb/in., $k_2=8$ lb/in., $m_1=0.05$ lb-sec^2/in., and $m_2=0.025$ lb-sec^2/in. Determine the forced amplitude of the masses m_1 and m_2.

2-27 Repeat Problem 2-26 for the impressed frequencies $\omega_f=131$ cpm, $\omega_f=286$ cpm, and $\omega_f=572$ cpm.

2-28 The spring-mass system in Fig. 2-14*a* is acted upon by the time-varying force shown in Fig. 2-13*a* of maximum magnitude of 30 lb and $T/2=0.4$ sec. Determine the steady-state response of the system by applying Fourier series analysis. Use only four terms of the Fourier series expansion of $F(t)$.

2-29 Repeat Problem 2-28 by assuming that the spring-mass system is acted upon by the time-varying force shown in Fig. 2-13*b* of maximum magnitude of 20 lb and $T/2=0.5$ sec.

2-30 Repeat Problem 2-28 for $T/2=0.1$, 0.2, and 0.3 sec, and compare results.

2-31 Repeat Problem 2-29 for $T/2=0.1$ sec and compare results.

3

IDEALIZED BEAMS, FRAMES, AND SIMPLE BUILDINGS

3-1 INTRODUCTION

The concepts and methods discussed in the preceding two chapters can be easily used to obtain a reasonable solution regarding the dynamic response of certain types of engineering structures that have infinite degrees of freedom. For the design of many structural problems, stresses and displacements are usually required for points where these values are maximum. For example, if the cantilever beam in Fig. 3-1*a* is used to support the heavy weight W with the dynamic force $F(t)$ acting on it, a reasonable solution for the maximum dynamic vertical displacement and maximum dynamic stress can be obtained by utilizing the idealized one-degree system in Fig. 3-1*b*. In this manner, a system with infinite degrees of freedom is reduced to a simple one-degree spring-mass system. By similar reasoning, the beam in Fig. 3-1*c* and the frame in Fig. 3-1*e*, could be analyzed dynamically by using the spring-mass systems in Figs. 3-1*d* and 3-1*f*, respectively.

The purpose in this chapter is to determine the dynamic response of beams, frames, and building by using idealizations that permit one to apply the concepts and methods discussed in Chapters 1 and 2. These solutions are approximate, but they provide a satisfactory answer to many practical engineering problems. It should be pointed out, however, that such idealizations can be also applied to other types of structures, including plates. Additional approximate methods, as well as additional applications, are discussed later in this book.

3-2 IDEALIZED BEAMS

Approximate dynamic solutions to certain types of beam problems can be obtained by treating them as one-degree spring-mass systems. Such idealizations are usually executed by following logical assumptions that permit one to compute critical stresses and displacements of beams with reasonable accuracy. For example, a satisfactory design for the simply supported beam in Fig. 3-1*c* can be obtained by using the one-degree system in Fig. 3-1*d*. Idealizations of this kind are often permissible, because a successful design can be obtained if the maximum dynamic stress and the maximum dynamic displacement of the beam are known. For the simply supported beam loaded as shown in Fig. 3-1*c*, it is easy to verify that the maximum stress and the maximum displacement will occur at the center section of the beam. Consequently, reasonable accuracy regarding these quantities can be obtained by idealizing the beam as shown in Fig. 3-1*d*. In this figure, k is the stiffness of the beam at center length and is defined as the force at this point that is required to displace it vertically by an amount equal to unity. That is, by using formulas of strength of materials, we have

$$y = \frac{PL^3}{48EI} = \frac{kL^3}{48EI} = 1$$

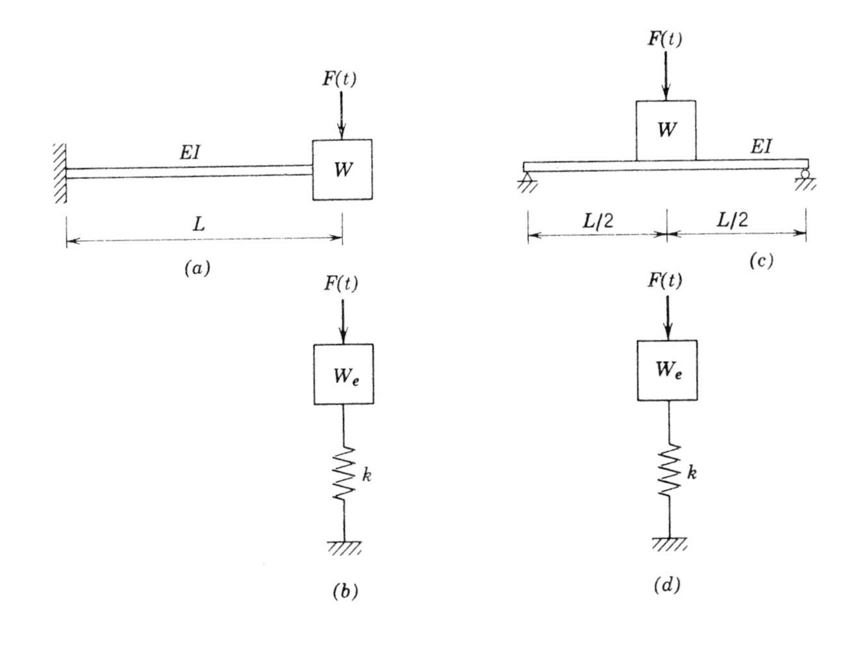
F(t)
EI
W
L
(a)
F(t)
W
EI
L/2
L/2
(c)
F(t)
W_e
k
(b)
F(t)
W_e
k
(d)

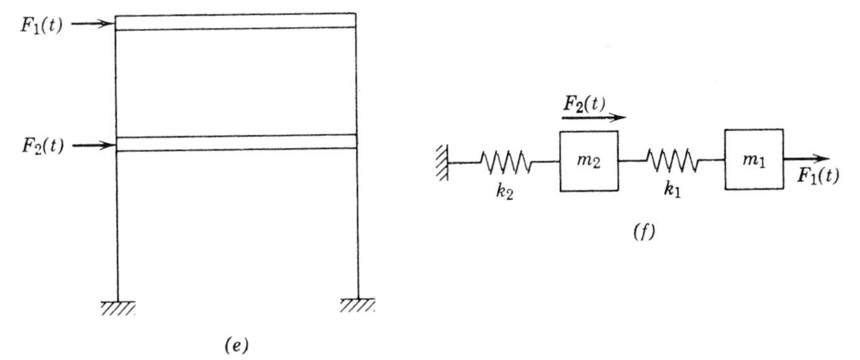
$F_1(t)$
$F_2(t)$
(e)
$F_2(t)$
m_2
m_1
$F_1(t)$
k_2
k_1
(f)

Fig. 3-1

or

$$k = \frac{48EI}{L^3} \tag{3-1}$$

If the weight of the beam is small compared to the heavy concentrated weight W, it can be neglected in the analysis without appreciable loss of accuracy. Therefore, the weight W_e of the idealized system in Fig.3-1*d* would be equal to the weight W supported by the beam. The dynamic force acting on the one-degree system is as shown. On this basis, the vertical dynamic displacement y of the beam's center is represented by the vertical displacement y of the weight W of the idealized one-degree system. The spring force ky is equal to the dynamic force acting at the beam center and $(ky/2)(L/2) = kLy/4$ is the dynamic bending moment acting at the center beam section. With known moment, the maximum dynamic bending stress can be computed by using known formulas of strength of materials.

When a beam is idealized as a spring-mass system as above, the dynamic analysis can be carried out by using the methods discussed in Chapters 1 and 2. To avoid tedious computations, numerical analysis is well in order, since the solution is approximate anyway because of the idealizations involved. Similar procedures and reasoning can be used for the solution of other types of beam problems. See for example Figs. 3-1*a* and 3-1*b*, and the solution obtained for the beam problem in Section 2-7, Fig. 2-8.

3-3 IDEALIZED ONE-STORY RIGID FRAMES AND BUILDINGS

The dynamic response of one-story rigid frames that are subjected to horizontal excitations may be determined with reasonable accuracy by treating the frame as an idealized one-degree spring-mass system.

Consider the rigid frame in Fig.3-2*a* which is acted upon by the time-varying force $F(t)$ as shown. This frame could be represented dynamically by the idealized one-degree system in Fig. 3-2*c*. If the weight of the columns is negligible compared to the weight of the girders, then the weight W in Fig. 3-2*c* would be the total weight of the girders. If the weight of the columns is significant, then it could be taken into account. A reasonable approximation would be to increase the total weight of the girders by half the total weight of the columns.

The spring constant k in Fig. 3-2*c* is taken equal to the static force that should be applied to the frame in the direction of x, in order to produce a deflection x equal to unity. This constant can be determined by using known methods of strength of materials. On this basis, the idealized system in Fig. 3-2*c* represents approximate dynamic behavior of the frame in the

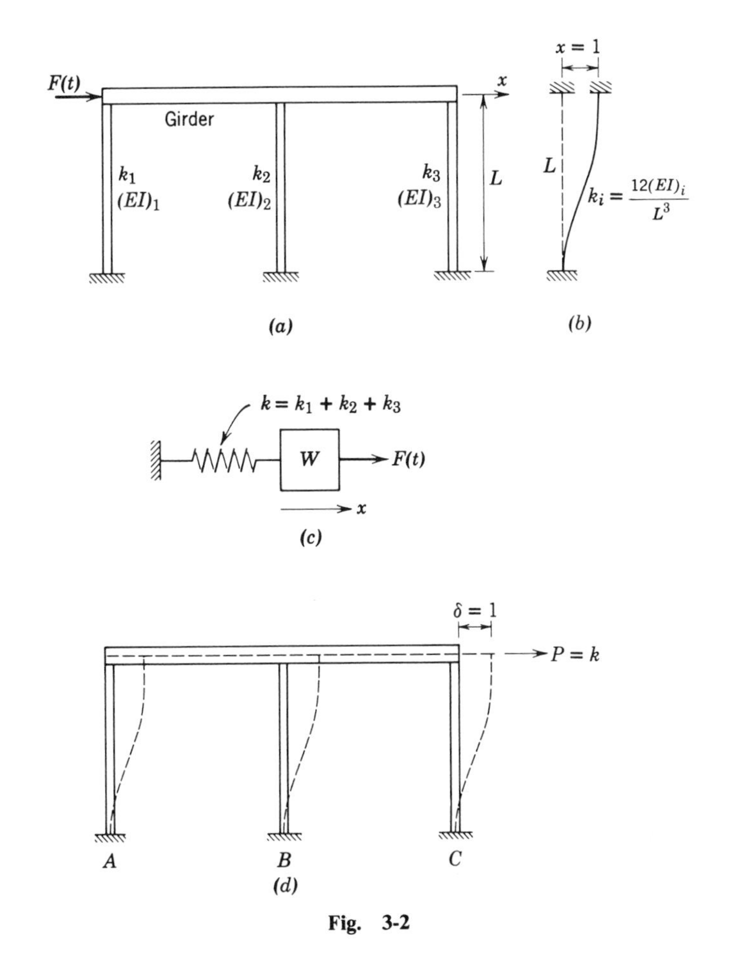

Fig. 3-2

horizontal direction. The deflection of the spring under the action of the force function *F(t)* is approximately equal to the horizontal deflection of the top of the frame.

The idealized system for one-story buildings consisting of rigid frames can be derived in a similar manner. In this case, the weight W in Fig. 3-2*c* will consist of the weight of the roof, with the girders included, and half the weight of the walls. The spring constant k is defined as above. If the building consists of a series of equally spaced frames such as that in Fig. 3-2*a* and also has uniform structural properties and loading along its length, the entire building could be analyzed with reasonable accuracy by considering only an interior frame.

In the following analysis, two cases are investigated. In case 1 the girders are assumed to be infinitely rigid compared to the columns while in case 2 the flexibility of the girders is taken into consideration. In both cases, however, the deformations are assumed not to be dependent on the axial forces present in the columns.

Case 1. Infinitely Rigid Girders

The girders of the structure will be assumed infinitely rigid as compared to the columns. This situation will be approximately true for structures where the stiffness of the girders relative to that of the columns is sufficiently large so that significant rotations at the tops of the columns are prevented. In this manner, the building will behave like a cantilever beam deflected by shear forces only, a condition which is often called the shear building.

The idealized one-degree system for the frame in Fig 3-2*a* is shown in Fig. 3-2*c*. The spring constant k in this case is

$$k = \sum_{i=1}^{3} k_i, \qquad (3\text{-}2)$$

where k_i is given by the expression

$$k_i = \frac{12(EI)i}{L^3} \qquad (3\text{-}3)$$

Here k_i represents the stiffness of column i, defined as the force required to produce unit horizontal displacement at the top of the column as shown in Fig. 3-2*b*. Thus k in Fig. 3-2*c* is the sum of the k 's of the individual columns in Fig. 3-2*a*. The constant k_i would be equal to $3(EI)i/L^3$ if the lower end of the column in Fig. 3-2*b* were hinged.

The natural frequency of vibration f_n of the frame, in hertz, is that of the one-degree system and is given by the expression

$$f_n = \frac{1}{2\pi}\sqrt{kg/W} \qquad (3\text{-}4)$$

and the natural period τ is

$$\tau = \frac{1}{f_n} = 2\pi\sqrt{W/kg} \qquad (3\text{-}5)$$

For a given load function $F(t)$, the dynamic response of the system can be calculated by solving the differential equation

$$m\ddot{x}+kx=F(t), \tag{3-6}$$

where $m=W/g$. In this equation the effects of damping are not included, but they can be taken into consideration as discussed in earlier chapters.

The solution of Eq. 3-6 can be obtained by using any one of the methods already discussed. The displacement of the spring at any time t is equal to the horizontal deflection x of the top of the frame in Fig. 3-2*a*. The force at a time t that develops internally in the spring is equal to the total shear in the three columns of the frame. The maximum bending moment in a column can be determined from the expression[13]

$$\frac{6EIx}{L^2} \tag{3-7}$$

Thus, with the column moments known, the dynamic stresses can be easily found by using known formulas of strength of materials.

The total shear force V in the three columns in Fig. 3-2*a* is

$$V=kx \tag{3-8}$$

and it is distributed among the columns in proportion to their k's. If, for example, the three columns are of equal stiffness, the shear force in each column is equal to one-third of V, and its maximum bending moment M can be calculated from the expression

$$M=\frac{V}{3}\frac{L}{2}=\frac{VL}{6} \tag{3-9}$$

Thus the column moments can be calculated in this case by using either Eq. 3-7 or Eq. 3-9.

As an illustration, let it be assumed that it is required to determine the response of a one-story building whose top view is shown in Fig. 3-3*a*. This building consists of a series of five identical steel frames such as that shown in Fig. 3-3*b*. The roof is a 4-in.-thick reinforced concrete slab with no composite action between slab and girders. The time variation of the applied force *F(t)* for each frame is identical and is as shown in Fig. 3-3*c*. Under these conditions, the building will have uniform properties and loading throughout its length.

In addition, it will be assumed that the weight of the walls and columns is negligible compared to that of the roof and girders. With these conditions in mind, the dynamic response of the building will be calculated by utilizing any one of the interior frames, and deriving the idealized one-degree spring-weight system in Fig. 3-3*d*. The weight W, consisting of the associated

[13]This expression can be found in books of structural analysis. See for example, Reference 10.

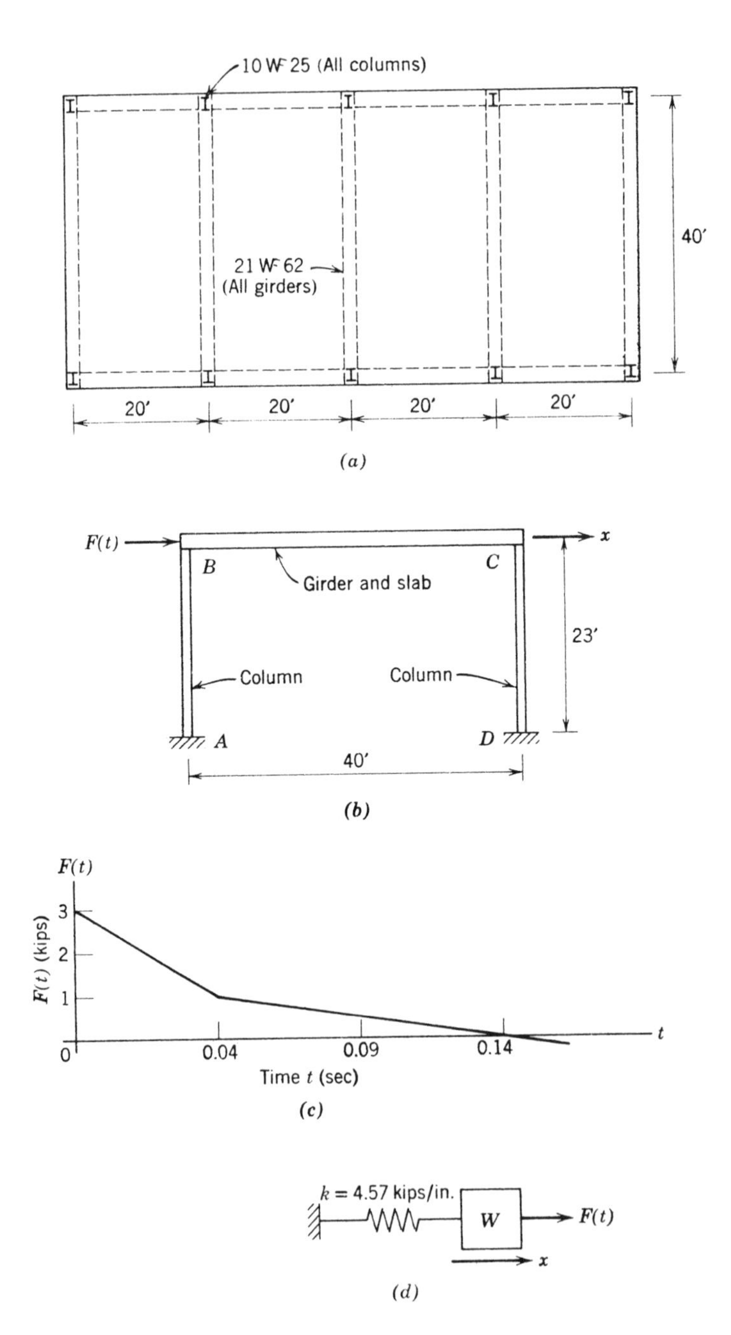

10 W 25 (All columns)
21 W 62
(All girders)
40′
20′
20′
20′
20′
(a)
F(t)
B
Girder and slab
C
x
23′
Column
Column
A
D
40′
(b)
F(t)
3
2
1
F(t) (kips)
0
0.04
0.09
0.14
t
Time t (sec)
(c)
k = 4.57 kips/in.
W
F(t)
x
(d)

Fig. 3-3

weight of the roof slab and girder, is

$$W=(62)(40)+(50)(20)(40)=42{,}480\text{ lb}=42.48\text{ kips}$$

where the weight of concrete is taken equal to 150 lb/ft^3.

The spring constant k, by using Eq. 3-3, is

$$k=\frac{(2)(12)(EI)}{L^3}=\frac{(24)(30)(10)^6(133.2)}{(23)^3(12)^3}$$

$$=4{,}570\text{ lb/in.}=4.57\text{ kips/in.}$$

The dynamic displacements x of the top of the frame are determined by using the acceleration impulse extrapolation method to solve the differential equation of motion of the one-degree system. From Eq. 3-6,

$$\ddot{x}=\frac{g}{W}F(t)-\frac{kg}{W}x$$

or

$$\ddot{x}=9.09F(t)-41.40x \tag{3-10}$$

TABLE 3-1

t (sec)	$9.09F(t)$ (in./sec^2)	$41.40x$ (in./sec^2)	$\ddot{x}$ Eq. 3-10 (in./sec^2)	$\ddot{x}(\Delta t)^2$ (in.)	x Eq. 2-53
0	27.27	0	27.70	0.01108	0
0.02	18.18	0.23	17.95	0.00720	0.00554*
0.04	9.09	0.76	8.33	0.00344	0.01828
0.06	7.25	1.43	5.82	0.00233	0.03446
0.08	5.45	2.19	3.26	0.00131	0.05297
0.10	3.64	3.02	0.62	0.00024	0.07279
0.12	1.82	3.84	−2.02	−0.00081	0.09285
0.14	0	4.65	−4.65	−0.00186	0.11210
0.16	0	5.37	−5.37	−0.00214	0.12949
0.18	0	6.00	−6.00	−0.00224	0.14474
0.20	0	6.53	−6.53	−0.00262	0.15775
0.22	0	6.96	−6.96	−0.00278	0.16814
0.24	0	7.28	−7.28	−0.00291	0.17575
0.26	0	7.48	−7.48	−0.00300	0.18045
0.28	0	7.53	−7.53	−0.00301	0.18215
0.30					0.18084

*Equation 2-55 is used for this station.

From Eq. 3-5,

$$\tau = 2\pi\sqrt{42.48/(4.57)(386)} \quad = 0.975 \text{ sec}$$

In this case, the time interval $\Delta t = 0.02$ sec will be used. For the first time interval Eq. 2-55 yields

$$x(t=0.02) = \frac{27.27}{2}(0.02)^2 = 0.00554 \text{ in.}$$

For the remaining intervals Eq. 2-53 is used. For example,

$$\begin{aligned} x(t=0.04) &= 2(0.00554) - 0 + (18.18)(0.02)^2 \\ &= 0.01828 \text{ in.} \end{aligned}$$

The results are shown in Table 3-1. The maximum dynamic deflection occurs at $t = 0.28$ sec and is equal to 0.18215 in. The shear force V in each column is

$$V = \frac{kx}{2} = \frac{(4.57)(0.18215)}{2} = 0.417 \text{ kips}$$

and the bending moment M at the top of the column is

$$M = \frac{VL}{2} = \frac{(0.417)(23)(12)}{2} = 57.5 \text{ kip-in.}$$

The maximum dynamic stress σ_{max} is

$$\sigma_{max} = \frac{M}{S} = \frac{57.5}{26.4} = 2.18 \text{ ksi}$$

If the structure is assumed to be under the influence of viscous damping with damping constant $c = 0.10$, then by Eq. 1-10

$$\ddot{x} = 9.09F(t) - 0.91\dot{x} - 41.40x, \qquad (3\text{-}11)$$

where the velocity $\dot{x}^{(i)}$ at station i is given by Eq. 2-58. The acceleration $\ddot{x}^{(i)}$ at station i can also be determined directly by applying Eq. 2-59. In the following computations Eqs. 2-58 and 3-11 are used.

At $t = 0$ the velocity $\dot{x}$ is zero and Eq. 3-11 yields

$$\ddot{x} = 9.09F(t) = (9.09)(3) = 27.27 \text{ in./sec}^2$$

Thus, by Eq. 2-55, the displacement at the end of the first time interval is

$$x(t=0.02)=0.00554 \text{ in.}$$

At the end of the same interval, the velocity equation (2-58) yields

$$\dot{x}(t=0.02)=\frac{0.00554-0}{0.02}+\ddot{x}\left(\frac{0.02}{2}\right)$$

or

$$\dot{x}(t=0.02)=0.277+0.01\ddot{x} \tag{3-12}$$

By substituting Eq. 3-12 into Eq. 3-11, we have

$$\begin{aligned}\ddot{x}(t=0.02) &= (9.09)(2.0)-(0.91)(0.277+0.01\ddot{x})-(41.40)(0.00554)\\ &=18.18-0.25-0.0091\ddot{x}-0.23\end{aligned}$$

or, by solving for $\ddot{x}$,

$$\ddot{x}(t=0.02)=\frac{17.70}{1.0091}=17.55 \text{ in./sec}^2$$

Thus, having determined the acceleration at $t = 0.02$ sec, the displacement $x^{(2)}$ at time station 2 can be determined from Eq. 2-53. That is,

$$\begin{aligned}x(t=0.04) &=2(0.00554)-0+(17.55)(0.02)^2\\ &=0.01808 \text{ in.}\end{aligned}$$

In the case of no damping, this displacement was found to be equal to 0.01828 in. The displacements at the remaining time stations can be obtained in the same manner. Tabulation of the results is usually helpful.

Case 2. Flexible Girders

In this case, the flexibility of the frame girders will be taken into account. The idealized one-degree system is again the one shown in Fig. 3-2*c*, but the spring constant k is defined as the force P in Fig. 3-2*d* that is capable of producing a horizontal displacement δ equal to unity. This constant can be determined by a simple moment distribution sidesway procedure or by applying any acceptable method of structural mechanics. If moment distribution is used, the first step is to apply a horizontal displacement $\delta = 1$ as shown in Fig. 3-2*d*. Next, the fixed-end moments at the top and bottom of the columns should be calculated from the relation $6EI\,\delta/L^2$, where $\delta = 1$

and L is the length of the column. Finally, by computing first the moment distribution factors, the moment distribution procedure can be carried out in the usual way. If H_A, H_B, H_C are, respectively, the horizontal shear forces at A, B, and C in Fig. 3-2d, then

$$k = H_A + H_B + H_C$$

By using this value of k in the idealized one-degree system in Fig. 3-2c, the dynamic analysis of the frame takes into consideration the flexibility of the girders.

For the frame in Fig. 3-3b, the moment distribution calculations when a unit horizontal displacement is applied at the top of the columns are shown in Table 3-2. From these computations, the spring stiffness k is

$$k = (2)\frac{298.6 + 282.2}{(23)(12)} = 4.22 \text{ kips/in.}$$

In the case where the girders were assumed to be infinitely rigid, the value of k was 4.57 kips/in. The difference between these two values is rather small, as expected, because the girders are about ten times stiffer than the columns.

With $k = 4.22$ kips/in. in the idealized one-degree system in Fig. 3-3d, the computations regarding the dynamic displacements can be carried out as in case 1. The actual moments at any time t, however, are the moments found in Table 3-2 multiplied by the actual dynamic displacement

TABLE 3-2

A	*B* .149	*B* .851	*C* .851	*C* .149	*D*
−315.0	−315.0			−315.0	−315.0
+23.5	←+47.0	+268.0→	+134.0		
		+77.0	←+154.0	+27.0→	+13.5
−5.8	←−11.5	−65.5→	−32.8		
		+14.0	←+27.9	+4.9→	+2.5
−1.1	←−2.2	−11.8→	−5.9		
		+2.5	←+5.0	+0.9→	+0.5
−0.2	←−0.4	−2.1→	−1.1		
−298.6 kip-in.		+0.5	←+0.9	+0.2→	+0.1
	−0.1	−0.4	+282.0 kip-in.	−282.0 kip-in.	−298.4 kip-in
	−282.2 kip-in.	+282.2 kip-in.			

corresponding to time t, because the moments in this table were due to a unit horizontal displacement. With known moments, the stresses can be calculated in the usual way.

3-4 TWO-STORY RIGID FRAMES AND BUILDINGS

In this section, two-story rigid frames and two-story buildings consisting of rigid frames are considered. The same type of analysis as in the preceding section is applied here, but the girders are assumed to be infinitely rigid as compared to the columns. For many practical problems this assumption is approximately true, because the girders are much more rigid than the columns. There are cases, however, where the flexibility of the girders should be taken into consideration. This subject is discussed in detail in the chapters that follow.

Consider the two-story rigid frame in Fig. 3-4*a*, and let it be assumed that the girders are infinitely rigid compared to the columns. By following the ideas of the preceding section, the dynamic response of this frame will be determined by utilizing the idealized two-degree system in Fig. 3-4*b*. In this figure, W_1 and W_2 represent the total weight of the girders at the floor and roof levels, respectively. The weight of the columns is assumed small compared to that of the girders and is neglected. If this weight is not small, it can be taken into account as discussed in the preceding section. For example, a reasonable approximation would be to increase the weight W_1 of the girders by the tributory weight of the columns of length $(L_1 + L_2)/2$, Fig. 3-4*a*, and W_2 by the tributory amount of length $L_2/2$.

The spring constants k_1 and k_2 are the sum of the column stiffnesses in each floor. For a column of length L_i, Fig. 3-4*c*, the stiffness k_i is

$$k_i = \frac{12(EI)i}{L_i^{\,3}} \tag{3-13}$$

which is the same as Eq. 3-3 in Section 3-3. If the lower end of the column is hinged, then

$$k_i = \frac{3(EI)i}{L_i^{\,3}} \tag{3-14}$$

The displacements x_1 and x_2 of the idealized system in Fig. 3-4*b* are the horizontal displacements x_1 and x_2 of the frame as shown in Fig. 3-4*a*.

The idealized two-degree system for a two-story building consisting of rigid frames such as that in Fig. 3-4*a* is derived in the same way as above. The weights W_1 and W_2 in this case represent the total weight of the floor and roof, respectively, increased by the weight of the tributory wall areas of heights $(L_1 + L_2)/2$ and $L_2/2$, respectively.

By utilizing the free-body diagrams in Fig. 3-4*d* and applying dynamic equilibrium,

$$\frac{W_1}{g}\ddot{x}_1 + k_1 x_1 - k_2(x_2 - x_1) = F_1(t)$$

$$\frac{W_2}{g}\ddot{x}_2 + k_2(x_2 - x_1) = F_2(t)$$

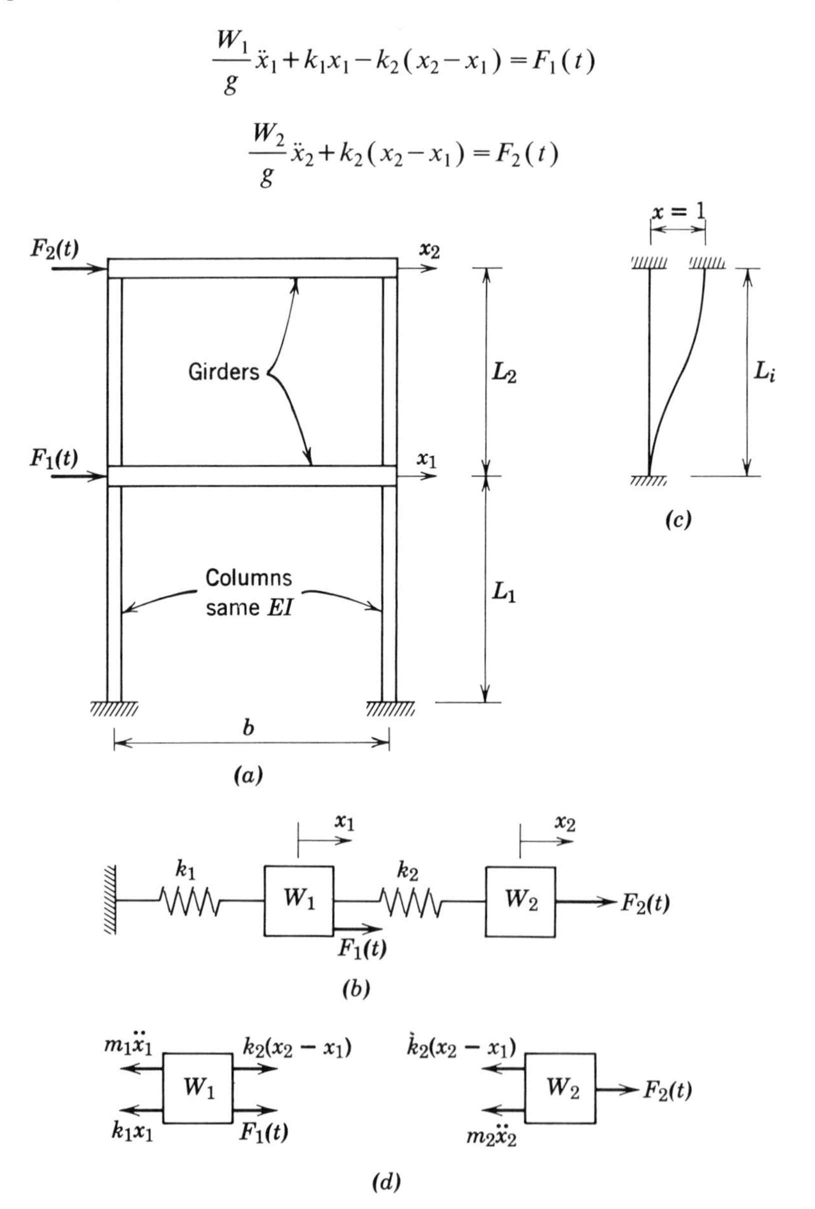

Fig. 3-4

or

$$m_1\ddot{x}_1+k_1x_1-k_2(x_2-x_1)=F_1(t) \tag{3-15}$$

$$m_2\ddot{x}_2+k_2(x_2-x_1)=F_2(t) \tag{3-16}$$

where $m_1=W_1/g$ and $m_2=W_2/g$.

Simultaneous solution of the two differential equations will yield the dynamic displacements $x_1(t)$ and $x_2(t)$ of the frame. The force V_1 in the spring of stiffness k_1 is

$$V_1=k_1x \tag{3-17}$$

This force is equal to the total shear force V_1 at the tops of the columns of the first floor. The force V_2 in the spring of stiffness k_2 is

$$V_2=k_2(x_2-x_1) \tag{3-18}$$

This is equal to the total shear force V_2 at the top of the columns of the second floor. The total shear force of the columns in each floor is distributed to each column in proportion to their stiffness. If V is the shear force in a column, its maximum bending moment M is calculated from the expression

$$M=\frac{VL}{2} \tag{3-19}$$

where L is the length of the column. The maximum bending stresses in each column are determined as before by using elementary formulas of strength of materials.

The natural frequencies of vibration of the system in Fig. 3-4*b* are determined from the two homogeneous differential equations

$$m_1\ddot{x}_1+k_1x_1-k_2(x_2-x_1)=0 \tag{3-20}$$

$$m_2\ddot{x}_2+k_2(x_2-x_1)=0 \tag{3-21}$$

These expressions are identical to Eqs. 1-91 and 1-92 in Section 1-9, and their solution is given in the same section. Thus the two natural undamped frequencies can be determined from the equation

$$\omega_{1,2}^2=\frac{1}{2}\left(\frac{k_1+k_2}{m_1}+\frac{k_2}{m_2}\right)\pm\frac{1}{2}\left[\left(\frac{k_1+k_2}{m_1}+\frac{k_2}{m_2}\right)^2-4\frac{k_1k_2}{m_1m_2}\right]^{\frac{1}{2}} \tag{3-22}$$

The two normal mode shapes can be determined from Eqs. 1-102 and 1-103.

The simultaneous solution of Eqs. 3-15 and 3-16 can be made easier by using the acceleration impulse extrapolation method discussed in Section

2-7. The numerical calculations can be carried out simultaneously in a manner similar to the one used for the single-degree systems discussed in the same section. The displacements $x_1^{(1)}$ and $x_2^{(1)}$ of the first time station are determined by using either Eq. 2-54 or Eq. 2-55. In the first equation the acceleration is assumed to vary linearly up to the first time station, and in the second one it is assumed to be constant during the first time interval and equal to the initial value. For the remaining time stations, the displacements $x_1^{(i+1)}$ and $x_2^{(i+1)}$ are determined by using Eq. 2-53.

The accelerations $\ddot{x}_1^{(i)}$ and $\ddot{x}_2^{(i)}$ at time station i are given by the expressions

$$\ddot{x}_1 = \frac{k_2}{m_1}(x_2 - x_1) - \frac{k_1}{m_1}x_1 + \frac{F_1(t)}{m_1} \tag{3-23}$$

$$\ddot{x}_2 = -\frac{k_2}{m_2}(x_2 - x_1) + \frac{F_2(t)}{m_2} \tag{3-24}$$

which are derived from Eqs. 3-15 and 3-16. The acceleration calculations do not present any problem because the displacements x_1 and x_2 of the ith time station are already known from the extrapolation process.

It was pointed out in Section 2-7 that the accuracy of this method depends on the number of Δt time intervals used in the extrapolation. In this case, sufficient accuracy is usually obtained by using a time interval Δt equal to one-tenth of the smallest natural period of vibration of the system.

As an illustration, consider the two-story rigid frame building whose top view is shown in Fig. 3-5*a*. The five rigid frames forming the building have the dimensions and type shown in Fig. 3-5*b*. The material properties, loading, and so on, are made uniform throughout the length of the building; hence the entire building may be analyzed with reasonable accuracy by using only an interior frame. The idealized two-degree system is shown in Fig. 3-5*c*. The weights W_1 and W_2, by including the tributory wall areas, are as follows:

$$W_1 = (90)(40)(20) + (2)(20)(15)(20) = 84{,}000 \text{ lb}$$
$$= 84.0 \text{ kips}$$

$$W_2 = (60)(40)(20) + (2)(20)(6)(20) = 52{,}800 \text{ lb}$$
$$= 52.8 \text{ kips}$$

By applying Eq. 3-13, the spring constants k_1 and k_2 are

$$k_1 = \frac{(2)(12)(30)(10)^6(133.2)}{(18)^3(12)^3} = 9{,}530 \text{ lb/in.}$$
$$= 9.53 \text{ kips/in.}$$

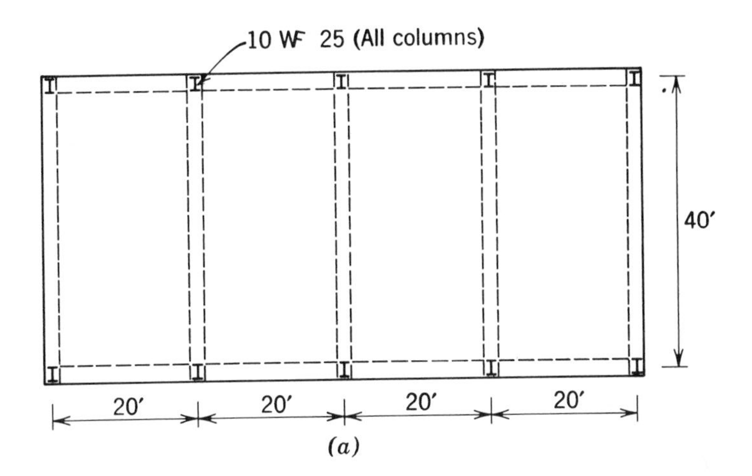
10 WF 25 (All columns)
40′
20′
20′
20′
20′
(a)

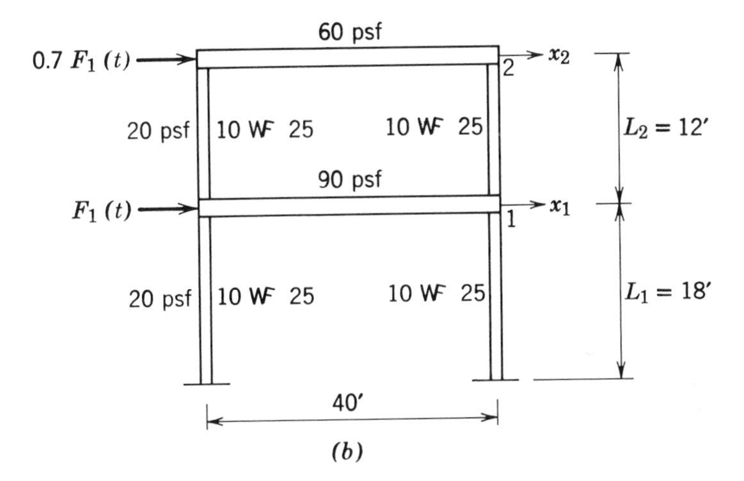
60 psf
0.7 $F_1(t)$
x_2
2
20 psf
10 WF 25
10 WF 25
$L_2 = 12'$
90 psf
$F_1(t)$
x_1
1
20 psf
10 WF 25
10 WF 25
$L_1 = 18'$
40′
(b)

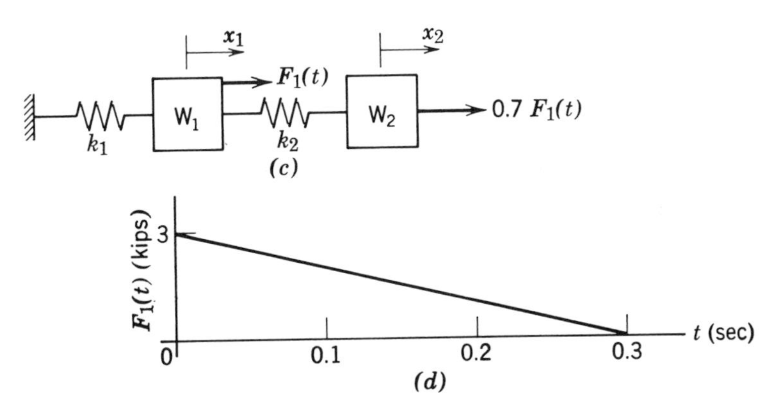
x_1
x_2
$F_1(t)$
W_1
W_2
0.7 $F_1(t)$
k_1
k_2
(c)
$F_1(t)$ (kips)
3
0
0.1
0.2
0.3
t (sec)
(d)

Fig. 3-5

$$k_2 = \frac{(2)(12)(30)(10)^6(133.2)}{(12)^3(12)^3} = 32{,}100 \text{ lb/in.}$$

$$= 32.1 \text{ kips/in.}$$

Thus

$$m_1 = \frac{W_1}{g} = 0.217 \text{ kip-sec}^2\text{/in.}$$

$$m_2 = \frac{W_2}{g} = 0.137 \text{ kip-sec}^2\text{/in.}$$

By applying Eq. 3-22,

$$\omega_1 = 5.10 \text{ rps}$$

$$\omega_2 = 20.00 \text{ rps}$$

or, in hertz, the natural frequencies f_1 and f_2 are

$$f_1 = \frac{\omega_1}{2\pi} = 0.812 \text{ Hz}$$

$$f_2 = \frac{\omega_2}{2\pi} = 3.185 \text{ Hz}$$

The two natural periods τ_1 and τ_2 are

$$\tau_1 = \frac{1}{f_1} = 1.232 \text{ sec}$$

$$\tau_2 = \frac{1}{f_2} = 0.314 \text{ sec}$$

The dynamic displacements x_1 and x_2 are determined by using the acceleration impulse extrapolation method. The time interval Δt is taken equal to 0.03 sec, which is about one-tenth of the natural period τ_2. The time variation of the dynamic force $F_1(t)$ is shown in Fig. 3-5*d*.

By applying Eqs. 3-23 and 3-24, the expressions for the accelerations $\ddot{x}_1$ and $\ddot{x}_2$ are as follows:

$$\ddot{x}_1 = 148(x_2 - x_1) - 43.90x_1 + 4.65F_1(t) \qquad (3\text{-}25)$$

$$\ddot{x}_2 = -234(x_2 - x_1) + 5.10F_1(t) \qquad (3\text{-}26)$$

The results for the time intervals between zero and 0.45 sec are tabulated in Table 3-3. The displacements $x_1^{(1)}$ and $x_2^{(1)}$ for the first time interval, $t = 0.03$ sec, were determined by using Eq. 2-55. For the remaining time stations, Eq. 2-53 was used. The accelerations $\ddot{x}_1^{(0)}$ and $\ddot{x}_2^{(0)}$ at $t = 0$ were determined by using Eqs. 3-25 and 3-26, respectively. The same equations were also used for the computation of the required accelerations at the other time stations.

Table 3-3 shows that the peak values of x_1 and x_2 are 0.3838 and 0.4308 in., respectively, and occur at a time $t = 0.42$ sec. By applying Eqs. 3-17 and 3-18, the total maximum shear force at the top of the columns of each floor are as follows:

$$V_1 = k_1 x_1 = (9.53)(0.3838) = 3.66 \text{ kips}$$

$$V_2 = k_2(x_2 - x_1) = (32.10)(0.4308 - 0.3838) = 1.51 \text{ kips}$$

Since the stiffnesses of the columns are equal, the shear forces are distributed equally among the columns of each floor.

By applying Eq. 3-19, the maximum bending moments M_1 and M_2 in the columns of the first and second floors, respectively, are

$$M_1 = \frac{VL_1}{2} = \frac{(1.83)(18)(12)}{2} = 198 \text{ kip-in.}$$

$$M_2 = \frac{VL_2}{2} = \frac{(0.755)(12)(12)}{2} = 54.2 \text{ kip-in.}$$

Thus the maximum dynamic stresses σ_1 and σ_2 are

$$\sigma_1 = \frac{M_1}{S} = \frac{198}{26.4} = 7.50 \text{ ksi}$$

$$\sigma_2 = \frac{M_2}{S} = \frac{54.2}{26.4} = 2.05 \text{ ksi}$$

The maximum dynamic stress is given by σ_1 and occurs in the columns of the first floor.

3-5 MULTISTORY RIGID FRAMES AND BUILDINGS

The dynamic response of two-story rigid frames and two-story rigid buildings consisting of rigid frames was examined in the preceding section. In this section, these ideas are extended to multistory rigid frame buildings, and their dynamic response is investigated by using idealized multiple degree of freedom systems. As before, the mass of the frame or building is

TABLE 3-3

t (sec)	$148(x_2 - x_1)$	$43.90x_1$	$4.65F_1(t)$	$\ddot{x}_1$ Eq. 3-25 (in./sec^2)	$\ddot{x}_1(\Delta t)^2$ (in.)	x_1 Eq. 2-53 (in.)	$234(x_2 - x_1)$	$5.10F_1(t)$	$\ddot{x}_2$ Eq. 3-26 (in./sec^2)	$\ddot{x}_2(\Delta t)^2$ (in.)	x_2 Eq. 2-53 (in.)
0	0	0	13.95	13.95	0.0126	0	0	15.30	15.30	0.0138	0
0.03	0.086	0.277	12.55	12.36	0.0111	0.0063*	0.141	13.75	13.61	0.0122	0.0069*
0.06	0.340	1.040	11.15	10.45	0.0094	0.0237	0.538	12.25	11.71	0.0105	0.0260
0.09	0.755	2.220	9.75	8.29	0.0075	0.0505	1.193	10.70	9.51	0.0086	0.0556
0.12	1.330	3.720	8.35	5.96	0.0054	0.0848	2.105	9.18	7.07	0.0064	0.0938
0.15	2.060	5.470	6.95	3.54	0.0032	0.1245	3.250	7.65	4.40	0.0040	0.1384
0.18	2.900	7.350	5.58	1.13	0.0010	0.1674	4.580	6.12	1.54	0.0014	0.1870
0.21	3.810	9.270	4.18	−1.28	−0.0012	0.2113	6.020	4.58	−1.44	−0.0013	0.2370
0.24	4.690	11.150	2.79	−3.67	−0.0033	0.2540	7.420	3.06	−4.36	−0.0039	0.2857
0.27	5.490	12.900	1.40	−6.01	−0.0054	0.2934	8.680	1.56	−7.15	−0.0064	0.3305
0.30	6.130	14.380	0	−8.25	−0.0074	0.3274	9.720	0	−9.72	−0.0087	0.3689
0.33	6.600	15.570	0	−8.97	−0.0081	0.3540	10.000	0	−10.00	−0.0090	0.3986
0.36	6.950	16.350	0	−9.40	−0.0085	0.3725	10.950	0	−10.95	−0.0099	0.4193
0.39	7.100	16.800	0	−9.70	−0.0087	0.3825	11.230	0	−11.23	−0.0101	0.4301
0.42	6.950	16.850	0	−9.90	−0.0089	0.3838	11.000	0	−11.00	−0.0099	0.4308
0.45						0.3762					0.4216

*Equation 2-55 has been used to determine this amplitude.

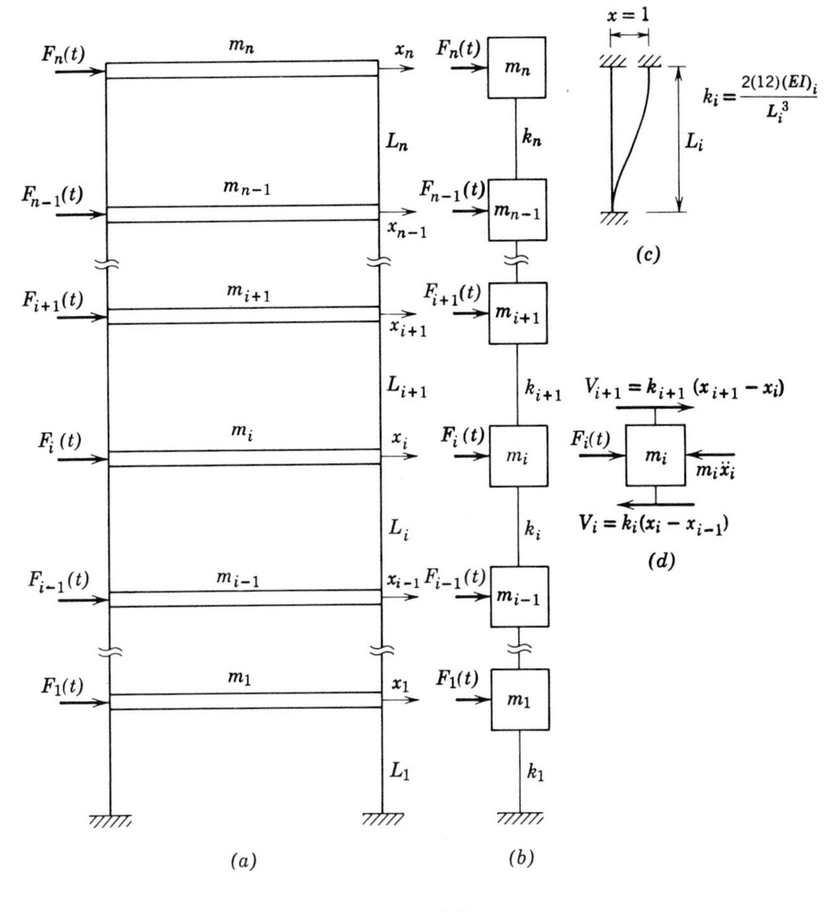

Fig. 3-6

concentrated at the floor and roof levels, and the girders are assumed to be infinitely rigid as compared to the columns. The horizontal motions of the building are assumed independent of its vertical motions, and the dynamic response for these motions is studied by using appropriate idealized systems.

Consider a multistory rigid frame building and let it be assumed that it consists of a series of rigid frames such as that shown in Fig. 3-6*a*. If all the bays of the building are identical, the dynamic response of the entire structure can be determined with reasonable accuracy by considering an interior frame only. By assuming that this frame has the form shown in Fig. 3-6*a*, its idealized multidegree system is as shown in Fig. 3-6*b*. The n masses of this system, that is, $m_1, m_2, \ldots, m_i, \ldots, m_n$, represent the mass of the frame and associated wall and floor areas concentrated at the corresponding floor levels. These mass concentrations are executed in the same way as in Section 3-4. The columns are represented in Fig. 3-6*b* by springs of constants $k_1, k_2, k_3, \ldots, k_i, \ldots, k_n$, that connect and support the masses $m_1, m_2, m_3, \ldots, m_i, \ldots, m_n$. The k 's are determined as in the preceeding section. In Fig. 3-6*c*, for example, the spring stiffness k_i of the columns of the ith story is

$$k_i = \frac{(2)(12)(EI)i}{L_i^{\ 3}} \tag{3-27}$$

The differential equations of motion are derived by considering the free-body diagrams of the masses and applying dynamic equilibrium. For the mass m_i, Fig. 3-6*d*, the differential equation of motion is

$$-m_i\ddot{x}_i + V_{i+1} - V_i + F_i(t) = 0 \tag{3-28}$$

where V_{i+1} and V_i are the total shear forces in the columns of the adjacent stories and $F_i\,(t\,)$ is the external dynamic force.

The shear forces V_i and V_{i+1} are given by the expressions

$$V_i = k_i(x_i - x_{i-1}) \tag{3-29}$$

$$V_{i+1} = k_{i+1}(x_{i+1} - x_i) \tag{3-30}$$

Thus, by substituting Egs. 3-29 and 3-30 into Eq. 3-28, we have

$$-m_i\ddot{x}_i + k_{i+1}(x_{i+1} - x_i) - k_i(x_i - x_{i-1}) + F_i(t) = 0 \tag{3-31}$$

Equation 3-31 is the differential equation of motion for the mass m_i. In a similar manner, the differential equations of motion for the other masses are obtained, which are as follows:

$$\begin{aligned}
&-m_1\ddot{x}_1 + V_2 - V_1 + F_1(t) = 0 \\
&-m_2\ddot{x}_2 + V_3 - V_1 + F_2(t) = 0 \\
&\ldots\ldots\ldots\ldots\ldots \\
&-m_i\ddot{x}_i + V_{i+1} - V_i + F_i(t) = 0 \qquad (3\text{-}32) \\
&\ldots\ldots\ldots\ldots\ldots \\
&-m_n\ddot{x}_n - V_n + F_n(t) = 0
\end{aligned}$$

By expressing the shear forces in terms of the appropriate spring constants and mass displacements, as was done in Eq. 3-28, the following set of equations is obtained:

$$-m_1\ddot{x}_1+k_2(x_2-x_1)-k_1x_1+F_1(t)=0$$

$$-m_2\ddot{x}_2+k_3(x_3-x_2)-k_2(x_2-x_1)+F_2(t)=0$$

$$\cdots\cdots\cdots\cdots\cdots\cdots$$

$$-m_i\ddot{x}_i+k_{i+1}(x_{i+1}-x_i)-k_i(x_i-x_{i-1})+F_i(t)=0 \qquad (3\text{-}33)$$

$$\cdots\cdots\cdots\cdots\cdots\cdots$$

$$-m_n\ddot{x}_n-k_n(x_n-x_{n-1})+F_n(t)=0$$

In the case of a two-story frame building, Eq. 3-33 yields

$$-m_1\ddot{x}_1+k_2(x_2-x_1)-k_1x_1+F_1(t)=0 \qquad (3\text{-}34)$$

$$-m_2\ddot{x}_2-k_2(x_2-x_1)+F_2(t)=0 \qquad (3\text{-}35)$$

These two expressions are identical to Eqs. 3-15 and 3-16 in Section 3-4.

The dynamic displacements $x_1, x_2,\ldots,x_n$ of the frame can be determined by solving simultaneously the differential equations given by Eq. 3-33. In some cases, rigorous solutions are found to be useful and quite rewarding. In other cases, the acceleration impulse extrapolation method will provide a practical solution to a very complicated problem. In applying this numerical method, the accelerations $\ddot{x}_1$, $\ddot{x}_2,\ldots,\ddot{x}_n$ at the various time stations can be determined from the expressions

$$\ddot{x}_1=\frac{k_2}{m_1}(x_2-x_1)-\frac{k_1}{m_1}x_1+\frac{F_1(t)}{m_1}$$

$$\ddot{x}_2=\frac{k_3}{m_2}(x_3-x_2)-\frac{k_2}{m_2}(x_2-x_1)+\frac{F_2(t)}{m_2} \qquad (3\text{-}36)$$

$$\cdots\cdots\cdots\cdots\cdots\cdots$$

$$\ddot{x}_n=-\frac{k_n}{m_n}(x_n-x_{n-1})+\frac{F_n(t)}{m_n}$$

The total shear forces in the columns of each story are given by Eqs. 3-29 and 3-30 and are distributed among the columns in proportion to the k's. If V is the shear force in a column, its maximum bending moment is equal to $VL/2$, where L is the length of the column. With known bending moments, the dynamic stresses are calculated in the usual way.

If the frame vibrates freely without damping, the forcing functions $F_1(t)$, $F_2(t),\ldots,F_n(t)$ in Eq. 3-33 are all zero, and the following system of differential equations is obtained:

$$
\begin{aligned}
-m_1\ddot{x}_1+k_2(x_2-x_1)-k_1x_1&=0\\
-m_2\ddot{x}_2+k_3(x_3-x_2)-k_2(x_2-x_1)&=0\\
\cdots\cdots\cdots\cdots\cdots\cdots\cdots\cdots&\\
-m_i\ddot{x}_i+k_{i+1}(x_{i+1}-x_i)-k_i(x_i-x_{i-1})&=0\\
\cdots\cdots\cdots\cdots\cdots\cdots\cdots\cdots&\\
-m_n\ddot{x}_n-k_n(x_n-x_{n-1})&=0
\end{aligned}
\tag{3-37}
$$

With solutions of the form

$$x_i=X_i\sin\omega t \tag{3-38}$$

Eq. 3-37 yields

$$
\begin{aligned}
\left(\omega^2-\frac{k_2+k_1}{m_1}\right)X_1+\frac{k_2}{m_1}X_2&=0\\
\frac{k_2}{m_2}X_1+\left(\omega^2-\frac{k_3+k_2}{m_2}\right)X_2+\frac{k_3}{m_2}X_3&=0\\
\cdots\cdots\cdots\cdots\cdots\cdots\cdots\cdots&\\
\frac{k_i}{m_i}X_{i-1}+\left(\omega^2-\frac{k_{i+1}+k_i}{m_i}\right)X_i+\frac{k_{i+1}}{m_i}X_{i+1}&=0\\
\cdots\cdots\cdots\cdots\cdots\cdots\cdots\cdots&\\
\frac{k_n}{m_n}X_{n-1}+\left(\omega^2-\frac{k_n}{m_n}\right)X_n&=0
\end{aligned}
\tag{3-39}
$$

This system is a set of n homogeneous algebraic equations containing the n unknown amplitudes X of the system and the unknown natural frequencies ω. The n frequencies ω are determined by setting equal to zero the determinant of the coefficients of X_1, $X_2,\ldots,X_n$ in Eq. 3-39. The characteristic equation that is obtained by expanding this frequency determinant will yield the n natural frequencies of the system. When a frequency is substituted into Eq. 3-39, a relationship between the amplitudes X_1, $X_2,\ldots,X_n$ is obtained. This relationship defines the normal mode corresponding to that frequency. Thus, for n natural frequencies, there correspond n normal modes of vibration.

In the case of two masses, Eq. 3-39 yields

$$\left(\omega^2 - \frac{k_2 + k_1}{m_1}\right)X_1 + \frac{k_2}{m_1}X_2 = 0 \tag{3-40}$$

$$\frac{k_2}{m_2}X_1 + \left(\omega^2 - \frac{k_2}{m_2}\right)X_2 = 0 \tag{3-41}$$

The frequency determinant is

$$\begin{vmatrix} \left(\omega^2 - \dfrac{k_2 + k_1}{m_1}\right) & \dfrac{k_2}{m_1} \\ \dfrac{k_2}{m_2} & \left(\omega^2 - \dfrac{k_2}{m_2}\right) \end{vmatrix} = 0 \tag{3-42}$$

yielding the characteristic equation

$$\omega^4 - \left(\frac{k_1 + k_2}{m_1} + \frac{k_2}{m_2}\right)\omega^2 + \frac{k_1 k_2}{m_1 m_2} = 0 \tag{3-43}$$

Equations 3-40, 3-41, 3-42, and 3-43 are identical, respectively, to Eqs. 1-94, 1-95, 1-96, and 1-97 in Section 1-9.

PROBLEMS

3-1 A weightless steel cantilever beam 20-in. long has a rectangular cross section 2-in. deep and $\frac{1}{2}$-in. wide, and supports a weight of 200 lb at the free end. Determine its natural lateral frequency of vibration.

3-2 Repeat Problem 3-1 by assuming that the system has 20% viscous damping.

3-3 The uniform steel beams in Fig. P3-3 support a weight W located as shown. By assuming that the beams are weightless, determine in each case the idealized one-degree spring-mass system and the natural frequency of vibration. Assume that there is no damping.

3-4 The weight W of the systems in Problem 3-3 is subjected to a suddenly applied dynamic force of 2 kips at $t = 0$ and removed suddenly at $t = 0.3$ sec. Compute the displacement at $t =$ 0.2, 0.3, and 0.5 sec. Use a rigorous solution to solve the idealized system.

3-5 Repeat Problem 3-4 when a force of 2 kips is suddenly applied at $t = 0$ and decreased linearly to zero at $t = 0.3$ sec.

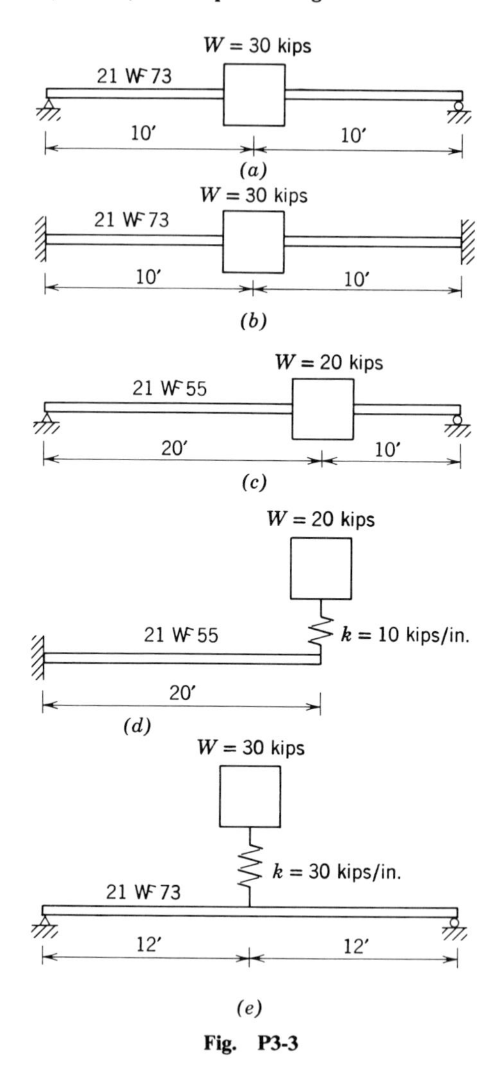

Fig. P3-3

3-6 Determine the time of maximum displacement and the maximum displacement for the systems and loading of Problem 3-4. Make use of appropriate charts.

3-7 Repeat Problem 3-6 for the systems and loading of Problem 3-5. Make use of appropriate charts.

3-8 Repeat Problem 3-4 when the dynamic force $F(t)$ applied at $t=0$ is equal to $F \sin \omega_f t$ where $F=2$ kips and $\omega_f = 100$ rps.

3-9 Repeat Problem 3-4 by assuming that the idealized systems have 20% viscous damping.

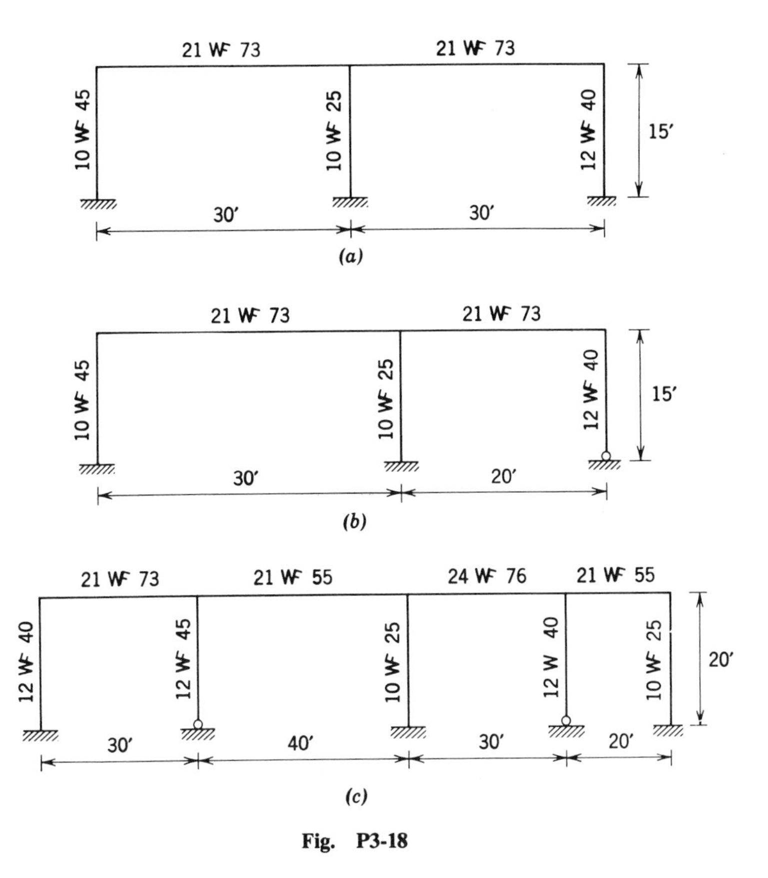

Fig. P3-18

3-10 If the weight of the structural systems in Problem 3-3 is subjected to a suddenly applied force of 2 kips at $t=0$ and removed suddenly at $t=0.3$ sec, determine the magnification factor at $t=0.2$, 0.3, and 0.5 sec.

3-11 By using the acceleration impulse extrapolation method, determine the maximum vertical displacement of the weight in Problem 3-1 when it is subjected to a force of 300 lb applied suddenly at $t=0$ and removed suddenly at $t=0.2$ sec. Neglect damping.

3-12 Repeat Problem 3-11 by assuming that a force of 300 lb is applied suddenly at $t=0$ and decreased linearly to zero at $t=0.3$ sec.

3-13 Repeat Problem 3-6 by utilizing the acceleration impulse extrapolation method.

3-14 Repeat Problem 3-7 by utilizing the acceleration impulse extrapolation method.

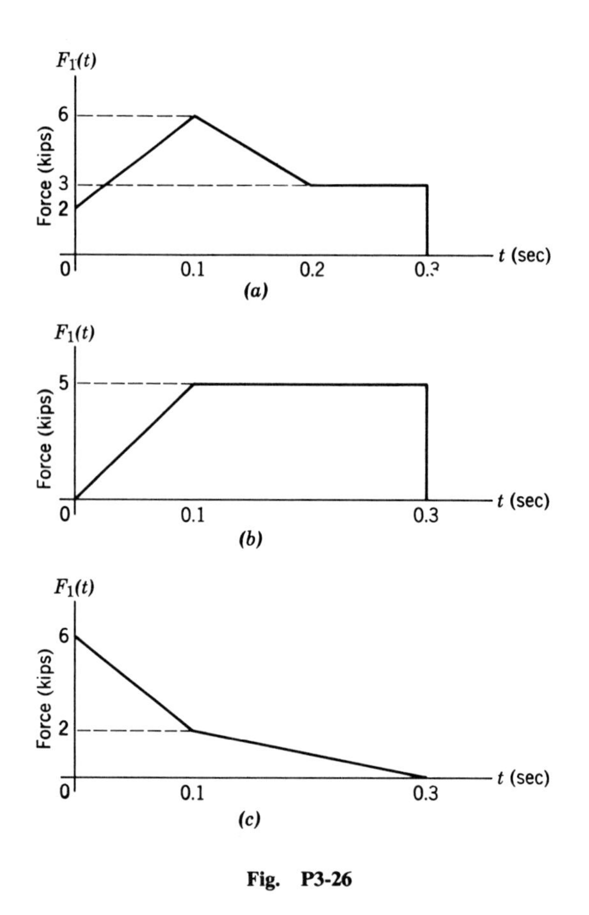

Fig. P3-26

3-15 Repeat Problem 3-11 by assuming that the system has 20% damping.

3-16 Repeat Problem 3-12 by assuming that the system has 10% damping.

3-17 Repeat Problem 3-14 by assuming that the systems have 25% damping.

3-18 For the steel frames shown in Fig. P3-18, determine in each case the idealized single degree of freedom system by assuming that the girders are rigid compared to the columns. Neglect the weight of the columns.

3-19 Repeat Problem 3-18 by taking into consideration the flexibility of the girders.

3-20 By using the appropriate idealized single degree of freedom system, determine in each case the undamped natural frequency of vibration and the period of vibration in the horizontal direction for the steel

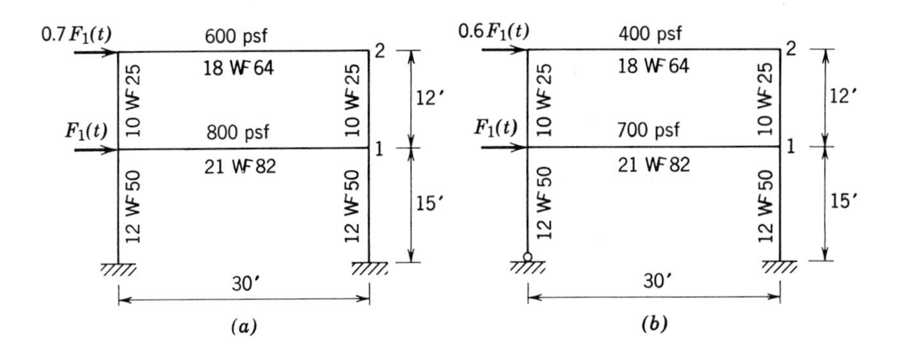

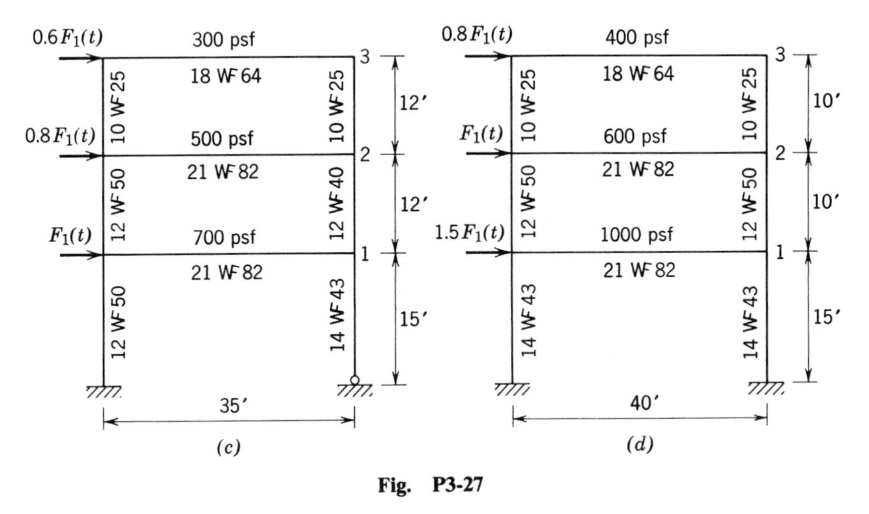

Fig. P3-27

frames in Problem 3-18. Neglect the weight of the columns and assume that the girders are rigid.

3-21 Repeat Problem 3-20 by taking into consideration the flexibility of the girders.

3-22 The steel frames in Problem 3-18 are subjected to a concentrated horizontal dynamic force $F(t)$ at the girder level whose magnitude and time variation is as shown in Fig. 3-3*c*. The girders of each frame are assumed to be infinitely stiff compared to the columns. By using the appropriate idealized one-degree system in each case and applying the acceleration impulse extrapolation method, determine the maximum horizontal displacement at the girder level. Also determine the maximum shear force, moment, and bending stress in each column. Neglect the weight of the columns and assume that there is no damping.

3-23 Repeat Problem 3-22 by taking into consideration the flexibility of the girders.

3-24 Repeat Problem 3-22 by assuming that the structural systems have 15% damping.

3-25 If the dynamic force $F_1(t)$ acting on the two-story building in Fig. 3-5*b* is a force of 5 kips applied suddenly at $t=0$ and then released suddenly at $t=0.3$ sec, determine the maximum horizontal displacements at the girder levels and the maximum bending stresses in the columns. Assume that the girders of the frame are rigid and apply the acceleration impulse extrapolation method.

3-26 Repeat Problem 3-25 when the dynamic force $F_1(t)$ has the variations shown in Fig. P3-26.

3-27 By applying the acceleration impulse extrapolation method and assuming that $F_1(t)$ varies as in Problem 3-26, determine the maximum horizontal displacements at the girder levels for the steel frames shown below. Assume that the girders are rigid and neglect the weight of the columns. (Fig. P3-27.)

3-28 By using the results in Problem 3-27, determine the maximum bending stresses in the columns of each frame.

3-29 Determine the natural frequencies and the corresponding mode shapes in the horizontal direction for the steel frames given in Problem 3-27. Assume that the girders are rigid and neglect damping.

4

SYSTEMS WITH INFINITE DEGREES OF FREEDOM

4-1 INTRODUCTION

In Section 1-3, the differential equation of motion for systems with infinite degrees of freedom was derived and in Section 1-11 the vibration response of some simple beam cases was investigated. In this chapter these ideas are extended to other types of beam problems and plates, and the dynamic response of such structures is obtained by using rigorous solutions. The methodologies presented here could be made more general, but practical considerations limit the analysis to members and plates of constant stiffness. For beams, the stiffness can vary between spans, but is assumed to remain constant within the span. Beams and frames with members of variable stiffness are discussed in Chapter 7.

It should be pointed out, however, that the rigorous solutions suggested in this chapter are quite rewarding for the simpler cases of dynamic loading and boundary conditions, as well as in research. For the more complicated cases, the approximate solutions discussed in later chapters would be more appropriate.

4-2 VIBRATION OF SINGLE-SPAN BEAMS

The natural frequencies and the corresponding mode shapes for a simply supported beam of constant stiffness were obtained in Section 1-11. The general solution used for this problem is given by Eq. 1-133. That is,

$$y = C_1 \cosh\lambda x + C_2 \sinh\lambda x + C_3 \cos\lambda x + C_4 \sin\lambda x \tag{4-1}$$

where

$$\lambda^4 = \frac{m\omega^2}{EI} \tag{4-2}$$

Equation 4-1 is the general solution of Eq. 1-129 and can be used for any beam span of uniform mass and stiffness. The constants C_1, C_2, C_3, and C_4 can be evaluated from the boundary conditions of the span. Thus utilization of the end restraints permits one to draw conclusions that lead to the determination of the values of λ and the corresponding mode shapes of the beam. Computation of natural frequencies and mode shapes for single-span beams are discussed in this section.

Consider now the case where the single-span beam is fixed at the left end and is free at the other. The boundary conditions are

$$\begin{aligned}
&\text{at} \quad x=0: && y = 0 && \text{(a)} \\
& && \frac{dy}{dx} = 0 && \text{(b)} \\
&\text{at} \quad x=L: && \frac{d^2y}{dx^2} = 0 && \text{(c)} \\
& && \frac{d^3y}{dx^3} = 0 && \text{(d)}
\end{aligned} \tag{4-3}$$

By utilizing Eq. 4-1, the end conditions given by Eqs. 4-3*a* and 4-3*b* yield,

$$C_1 + C_3 = 0 \qquad \text{or} \qquad C_1 = -C_3$$

$$C_2 + C_4 = 0 \qquad \text{or} \qquad C_2 = -C_4$$

In a similar manner, Eq. 4-1 and boundary conditions (4-3*c*) and (4-3*d*) yield

$$(\cosh\lambda L + \cos\lambda L)C_3 + (\sinh\lambda L + \sin\lambda L)C_4 = 0 \tag{4-4}$$

$$(\sin\lambda L - \sinh\lambda L)C_3 - (\cosh\lambda L + \cos\lambda L)C_4 = 0 \tag{4-5}$$

where L is the length of the beam. For a solution other than the trivial one, the determininant of the coefficients of C_3 and C_4 in the two equations should be equal to zero. That is

$$\begin{vmatrix} (\cosh\lambda L + \cos\lambda L) & (\sinh\lambda L + \sin\lambda L) \\ (\sin\lambda L - \sinh\lambda L) & -(\cosh\lambda L + \cos\lambda L) \end{vmatrix} = 0 \tag{4-6}$$

By expanding the determinant and simplifying the resulting expression, the following equation is obtained:

$$\cosh\lambda L \cos\lambda L + 1 = 0 \tag{4-7}$$

The values of λL that satisfy Eq. 4-7 yield the natural frequencies ω of the member. The smallest value of λL that satisfies Eq. 4-7 is 1.88, yielding

$$\lambda = \frac{1.88}{L} \tag{4-8}$$

By substituting Eq. 4-8 into Eq. 4-2,

$$\left(\frac{1.88}{L}\right)^4 = \frac{m\omega_0^2}{EI}$$

or

$$\omega_0 = \frac{3.53}{L^2}\sqrt{EI/m} \tag{4-9}$$

This is the fundamental natural frequency of the beam. The next higher value of λL is 4.69, yielding

$$\omega_1 = \frac{22.0}{L^2}\sqrt{EI/m} \tag{4-10}$$

In a similar manner, all the values of λL and the corresponding natural frequencies ω can be determined. The values of λL that satisfy Eq. 4-7 can be also obtained graphically by plotting the curves $y_1 = \cos\lambda L$ and $y_2 = -1/\cosh\lambda L$ for arbitrary values of λL. The intersects of these two curves yield the roots λL of Eq. 4-7. A good approximation to the roots λL of Eq. 4-7, except for the smallest one, is given by the expression

$$(\lambda L)_n = \left(n + \tfrac{1}{2}\right)\pi \quad n = 1, 2, 3, \ldots \tag{4-11}$$

Thus

$$\omega_n = \frac{[(n+0.5)\pi]^2}{L^2}\sqrt{EI/m} \quad n = 1, 2, 3, \ldots \tag{4-12}$$

The fundamental frequency ω_0 should be determined from Eq. 4-9.

The mode shapes corresponding to the frequencies above can be determined from the expressions obtained from the application of the four boundary conditions. From Eq. 4-4,

$$\frac{C_4}{C_3} = -\frac{\cosh\lambda L + \cos\lambda L}{\sinh\lambda L + \sin\lambda L} \tag{4-13}$$

Thus, with $C_1 = -C_3$, $C_2 = -C_4$, and C_4/C_3 as given by the expression above, Eq. 4-1 yields,

$$y = C_3\left[(\cos\lambda x - \cosh\lambda x) - \frac{\cosh\lambda L + \cos\lambda L}{\sinh\lambda L + \sin\lambda L}(\sin\lambda x - \sinh\lambda x)\right] \tag{4-14}$$

The mode shapes for the various values of λ are determined from Eq. 4-14. This expression can be also rewritten as

$$\beta_n(x) = C_n\left[(\cos\lambda_n x - \cosh\lambda_n x) - \frac{\cosh\lambda_n L + \cos\lambda_n L}{\sinh\lambda_n L + \sin\lambda_n L}(\sin\lambda_n x - \sinh\lambda_n x)\right] \tag{4-15}$$

The mode shape corresponding to the fundamental frequency is obtained by substituting $\lambda_n = 1.88/L$ in Eq. 4-15. The mode shapes for the higher frequencies can be determined in a similar manner by using the values of λ given by Eq. 4-11. The natural frequencies and mode shapes for other types of single-span beams can be determined in the same way.

It is interesting to note at this point that the governing differential equation regarding the free vibration of a beam is linear. With this in mind, the most general solution for a given problem can be obtained by superposition of solutions. For example, the deflection shape of the nth mode of a freely vibrating, simply supported beam is given by Eq. 1-150. The general solution y is the superposition of the contributions of all modes. That is,

$$y = \sum_n C_n \sin \frac{n\pi x}{L} \qquad n = 1, 2, 3, \ldots \tag{4-16}$$

In the equation above, there corresponds a different value of C for each mode. This value can only be determined when the initial time conditions are defined. Therefore, mode shapes can only be determined within an arbitrary constant C that is usually taken equal to unity as in Eq. 1-151.

4-3 INITIAL TIME CONDITIONS FOR BEAM MOTIONS

To introduce initial time conditions into the problems discussed in Sections 1-11 and 4-2, Eq. 1-127 should be solved. This equation is written again below for convenience.

$$EI\frac{\partial^4 y}{\partial x^4} + m\ddot{y} = 0 \tag{4-17}$$

Let it now be assumed that the solution of Eq. 4-17 is

$$y(x,t) = Y(x)g(t) \tag{4-18}$$

where $Y(x)$ is a function of x only and $g(t)$ is a function of t only. By substituting Eq. 4-18 into Eq. 4-17 and separating variables,

$$\frac{[\partial^4 Y(x)/\partial x^4](EI)}{mY(x)} = -\frac{\ddot{g}(t)}{g(t)} \tag{4-19}$$

The right-hand side of this equation depends only on t, while the left-hand side is a function of x only. The two sides will be equal only if both are equal to the same constant, which in this case is taken to be equal to $+\omega^2$. Thus, by equating each side of Eq. 4-19 to $+\omega^2$, the following two differential equations are found:

$$EI\frac{\partial^4 Y(x)}{\partial x^4} - m\omega^2 Y(x) = 0 \tag{4-20}$$

$$\ddot{g}(t) + \omega^2 g(t) = 0 \tag{4-21}$$

Equation 4-20 is the same as Eq. 1-129 and its solution is given by Eq. 4-1.

The solution of Eq. 4-21 is that of a structural system having only one degree of freedom and is given by the expression

$$g(t) = C_1 \cos \omega t + C_2 \sin \omega t \tag{4-22}$$

or by the expression

$$g(t) = A \sin(\omega t + \phi) \tag{4-23}$$

Thus Eq. 4-18 yields

$$y(x,t) = Y(x) \sin(\omega t + \phi) \tag{4-24}$$

Here, $Y(x)$ is given by Eq. 4-1, and the constant A of Eq. 4-23 is absorbed by the constants of $Y(x)$. The unknowns in Eq. 4-24 are the four constants of $Y(x)$, ω, and ϕ. These six constants can be determined by using the four end conditions of the beam and the initial time condition such as displacement and velocity at an initial time $t = t_0$.

By considering, for example, a simply supported beam and the deflection expression of its nth mode, Eqs. 4-16 and 4-24 yield the expression

$$y_n = C_n \sin \frac{n\pi x}{L} \sin(\omega_n t + \phi_n) \tag{4-25}$$

This is the deflection equation of the beam for the nth mode. Thus the general equation for the total deflection of the member is

$$y(x,t) = \sum_n C_n \sin \frac{n\pi x}{L} \sin(\omega_n t + \phi_n) \qquad n = 1, 2, 3, \ldots \tag{4-26}$$

The constants ϕ_n and C_n can be determined by applying the initial time conditions.

Let it be assumed, for example, that the initial time conditions are

$$\left.\begin{aligned} \text{at} \quad t=0: \qquad y(x,0) &= y_0 \sin \frac{\pi x}{L} && \text{(a)} \\ \dot{y}(x,0) &= 0 && \text{(b)} \end{aligned}\right\} \tag{4-27}$$

where y_0 is the deflection of the beam's center. By using Eq. 4-26 and applying the conditions above, the following two equations are written:

$$y_0 \sin \frac{\pi x}{L} = \sum_n C_n \sin \frac{n\pi x}{L} \sin \phi_n \tag{4-28}$$

$$-\sum_n C_n \omega_n \sin \frac{n\pi x}{L} \cos \phi_n = 0 \tag{4-29}$$

Equation 4-29 will be satisfied for all values of x if $\cos\phi_n=0$, or $\phi_n=\pi/2$. Thus Eq. 4-28 yields

$$y_0 \sin\frac{\pi x}{L} = \sum_n C_n \sin\frac{n\pi x}{L}$$

or, by expanding the right-hand term,

$$y_0 \sin\frac{\pi x}{L} = C_1 \sin\frac{\pi x}{L} + C_2 \sin\frac{2\pi x}{L} + C_3 \sin\frac{3\pi x}{L} + \cdots \tag{4-30}$$

If the coefficients of the terms in the two sides of this equation are matched, then $C_1 = y_0$ and $C_2 = C_3 = \cdots = C_n = 0$. Thus Eq. 4-26 yields the deflection expression

$$y(x,t) = y_0 \sin\frac{\pi x}{L} \sin\left(\omega_1 t + \frac{\pi}{2}\right) \tag{4-31}$$

This equation satisfies the boundary conditions of the member and shows that for this type of initial time conditions only the first mode influences the motion of the beam. For other types of initial time conditions the solution will have a different form.

4-4 VIBRATION OF CONTINUOUS BEAMS

The solution given by Eq. 4-1 can be used for continuous beams, because this equation is applicable to any beam span of uniform stiffness and arbitrary boundary conditions. Two cases are examined here. In case 1 the ends of the continuous beam are assumed to be simply supported and each beam span is made up of the same uniform mass m and stiffness EI. In case 2 the ends of the beam can be other than simply supported and the mass and stiffness can vary from span to span, but they are kept constant within the span. In both cases the beam spans can be of different lengths.

Case 1.

Consider the n-span continuous beam in Fig. 4-1a that has simply supported ends and the same m and EI for all spans. For an intermediate span L_j, Fig. 4-1b, Eq. 4-1 can be rewritten as

$$y_j = A_j \cosh\lambda_j x + B_j \sinh\lambda_j x + C_j \cos\lambda_j x + D_j \sin\lambda_j x \tag{4-32}$$

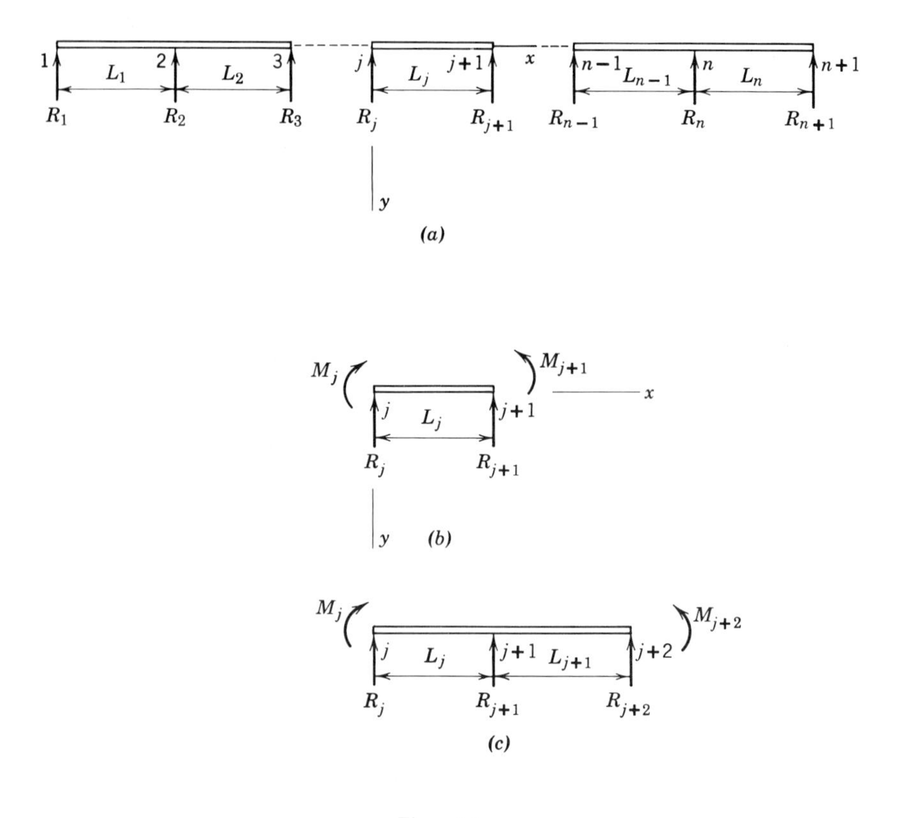

Fig. 4-1

For convenience in writing, the symbols A, B, C, and D are used to represent the constants. It should be also noted that the λ's are given by Eq. 4-2 and that $\lambda_1=\lambda_2=\cdots=\lambda_j=\cdots\lambda_n$.

By applying the boundary condition $y_j=0$ at $x=0$, Eq. 4-32 yields

$$C_j=-A_j \tag{4-33}$$

Thus

$$y_j=B_j\sinh\lambda_j x+D_j\sin\lambda_j x+C_j(\cos\lambda_j x-\cosh\lambda_j x) \tag{4-34}$$

In a similar manner, for span L_{j+1},

$$y_{j+1}=B_{j+1}\sinh\lambda_{j+1}x+D_{j+1}\sin\lambda_{j+1}x+C_{j+1}(\cos\lambda_{j+1}x-\cosh\lambda_{j+1}x) \tag{4-35}$$

The first and second derivatives of y_j with respect to x are

$$\frac{dy_j}{dx}=B_j\lambda_j\cosh\lambda_jx+D_j\lambda_j\cos\lambda_jx-C_j\lambda_j(\sin\lambda_jx+\sinh\lambda_jx)\qquad(4\text{-}36)$$

$$\frac{d^2y_j}{dx^2}=B_j\lambda_j^2\sinh\lambda_jx-D_j\lambda_j^2\sin\lambda_jx-C_j\lambda_j^2(\cos\lambda_jx+\cosh\lambda_jx)\qquad(4\text{-}37)$$

At $x=0$ the two equations yield

$$\frac{dy_j}{dx}=\lambda_j(B_j+D_j)\qquad(4\text{-}38)$$

$$\frac{d^2y_j}{dx^2}=-2C_j\lambda_j^2\qquad(4\text{-}39)$$

Equation 4-38 shows that the rotation of the end j of span L_j is proportional to (B_j+D_j). At the same end, by Eq. 4-39, the constant C_j is proportional to the end curvature d^2y_j/dx^2. Since the ends of the member are simply supported, it may be concluded that C_1 and C_{n+1} are both zero. The first and second derivatives of Eq. 4-35 are the same as those given by Eqs. 4-36 and 4-37, respectively, provided that j is replaced by $j+1$.

In Fig. 4-1*b*, the deflection at support $j+1$ is zero. Thus Eq. 4-34 yields

$$B_j\sinh\lambda_jL_j+D_j\sin\lambda_jL_j+C_j(\cos\lambda_jL_j-\cosh\lambda_jL_j)=0\qquad(4\text{-}40)$$

In addition, Fig. 4-1*c*, two continuity conditions can be written for any two consecutive spans L_j and L_{j+1}. These conditions suggest that the rotations and moments of the two spans at support $j+1$ should be equal. That is,

$$\left(\frac{dy_j}{dx}\right)_{x=L_j}=\left(\frac{dy_{j+1}}{dx}\right)_{x=0}\qquad(4\text{-}41)$$

$$\left(\frac{d^2y_j}{dx^2}\right)_{x=L_j}=\left(\frac{d^2y_{j+1}}{dx^2}\right)_{x=0}\qquad(4\text{-}42)$$

Application of these two conditions yields the following two expressions:

$$B_j\cosh\lambda_jL_j+D_j\cos\lambda_jL_j-C_j(\sin\lambda_jL_j+\sinh\lambda_jL_j)=B_{j+1}+D_{j+1}\qquad(4\text{-}43)$$

$$B_j\sinh\lambda_jL_j-D_j\sin\lambda_jL_j-C_j(\cos\lambda_jL_j+\cosh\lambda_jL_j)=-2C_{j+1}\qquad(4\text{-}44)$$

Addition of Eqs. 4-40 and 4-44 yields

$$C_j \cosh\lambda_j L_j - B_j \sinh\lambda_j L_j = C_{j+1} \tag{4-45}$$

Subtraction of Eqs. 4-40 and 4-44 gives

$$C_j \cos\lambda_j L_j + D_j \sin\lambda_j L_j = C_{j+1} \tag{4-46}$$

From Eqs. 4-45 and 4-46

$$B_j = \frac{C_j \cosh\lambda_j L_j - C_{j+1}}{\sinh\lambda_j L_j} \tag{4-47}$$

$$D_j = \frac{C_{j+1} - C_j \cos\lambda_j L_j}{\sin\lambda_j L_j} \tag{4-48}$$

Thus

$$B_j + D_j = \frac{C_j \cosh\lambda_j L_j - C_{j+1}}{\sinh\lambda_j L_j} + \frac{C_{j+1} - C_j \cos\lambda_j L_j}{\sin\lambda_j L_j}$$

or, by simplifying,

$$B_j + D_j = C_j(\coth\lambda_j L_j - \cot\lambda_j L_j) - C_{j+1}(\operatorname{csch}\lambda_j L_j - \csc\lambda_j L_j) \tag{4-49}$$

In a more compact form, Eq. 4-49 is written as

$$B_j + D_j = C_j \theta_j - C_{j+1} \Psi_j \tag{4-50}$$

where

$$\theta_j = \coth\lambda_j L_j - \cot\lambda_j L_j \tag{4-51}$$

$$\Psi_j = \operatorname{csch}\lambda_j L_j - \csc\lambda_j L_j \tag{4-52}$$

In a similar manner,

$$B_{j+1} + D_{j+1} = C_{j+1} \theta_{j+1} - C_{j+2} \Psi_{j+1} \tag{4-53}$$

By substituting Eqs. 4-47, 4-48, and 4-53 into Eq. 4-43 and simplifying,

$$-C_j(\csc\lambda_j L_j - \operatorname{csch}\lambda_j L_j) + C_{j+1}(\cot\lambda_j L_j - \coth\lambda_j L_j) = C_{j+1}\theta_{j+1} - C_{j+2}\Psi_{j+1}$$

or

$$C_j \Psi_j - C_{j+1}(\theta_j + \theta_{j+1}) + C_{j+2} \Psi_{j+1} = 0 \qquad (4\text{-}54)$$

Equation 4-54 applies to any two consecutive spans of the beam. Consequently, for the n-span beam in Fig. 4-1a, the following equations can be written for each pair of adjacent spans:

$$\begin{aligned} -C_2(\theta_1+\theta_2) + C_3 \Psi_2 &= 0 \\ C_2 \Psi_2 - C_3(\theta_2+\theta_3) + C_4 \Psi_3 &= 0 \\ C_3 \Psi_3 - C_4(\theta_3+\theta_4) + C_5 \Psi_4 &= 0 \\ \cdots\cdots\cdots\cdots\cdots \\ C_{n-1} \Psi_{n-1} - C_n(\theta_{n-1}+\theta_n) &= 0 \end{aligned} \qquad (4\text{-}55)$$

For a beam of two spans and simply supported ends, Eq. 4-55 yields

$$\theta_1 + \theta_2 = 0 \qquad (4\text{-}56)$$

which is known as the frequency equation of the two-span beam. In this expression

$$\begin{aligned} \theta_1 &= \coth \lambda_1 L_1 - \cot \lambda_1 L_1 \\ \theta_2 &= \coth \lambda_2 L_2 - \cot \lambda_2 L_2 \end{aligned} \qquad (4\text{-}57)$$

With $\lambda_1 = \lambda_2 = \lambda$, the product λL_2 in Eq. 4-57 can be expressed in terms of λL_1. In this manner, Eq. 4-56 will become a function of λL_1 only. Usually a rigorous solution of Eq. 4-56 is difficult. An easier way to determine the values of λL_1 that satisfy Eq. 4-56 is to plot[14] λL_1 versus $(\theta_1 + \theta_2)$ for various values of λL_1. With known values of λL_1, the corresponding frequencies can be obtained from the equation

$$\omega = \lambda^2 \sqrt{EI/m} \qquad (4\text{-}58)$$

If the beam in Fig. 4-1a consists of three spans, Eq. 4-55 yields

$$\begin{aligned} -(\theta_1+\theta_2) C_2 + \Psi_2 C_3 &= 0 \\ \Psi_2 C_2 - (\theta_2+\theta_3) C_3 &= 0 \end{aligned} \qquad (4\text{-}59)$$

For a nontrivial solution, the determinant of the coefficients of C_2 and C_3 in

[14]The plotting of these graphs is more convenient if the tables of Reference 18 are used.

Eq. 4-59 must be equal to zero. That is,

$$\begin{vmatrix} -(\theta_1+\theta_2) & \Psi_2 \\ \Psi_2 & -(\theta_2+\theta_3) \end{vmatrix} = 0 \tag{4-60}$$

Equation 4-60 is known as the frequency determinant. Expansion of this determinant yields the frequency equation

$$(\theta_1+\theta_2)(\theta_2+\theta_3)-\Psi_2^2=0 \tag{4-61}$$

Again, the products λL_2 and λL_3 can be expressed in terms of λL_1, thus making Eq. 4-61 a function of λL_1 only. The values of λL_1 that satisfy Eq. 4-61 can be determined graphically by plotting λL_1 versus $[(\theta_1+\theta_2)(\theta_2+\theta_3) - \Psi_2^2]$ for various values of λL_1, and the corresponding natural frequencies ω can be calculated by using Eq. 4-58.

Case 2.

Consider now the more general case where the end supports of the n-span continuous beam can be other than simply supported, and where the λ's of the individual spans are not equal. That is, $\lambda_1 \neq \lambda_2 \neq \lambda_3 \neq \cdots \neq \lambda_n$. By proceeding as in case 1, the equation for any two consecutive spans of the continuous beam is

$$C_j \Psi_j - C_{j+1}\left[\frac{\lambda_{j+1}^2}{\lambda_j^2}\frac{(EI)_{j+1}}{(EI)_j}\theta_j + \frac{\lambda_{j+1}}{\lambda_j}\theta_{j+1}\right] + C_{j+2}\frac{\lambda_{j+2}^2}{\lambda_j\lambda_{j+1}}\frac{(EI)_{j+2}}{(EI)_{j+1}}\Psi_{j+1} = 0 \tag{4-62}$$

This equation is analogous to Eq. 4-54 of the preceding case. The θ's and Ψ's are given, respectively, by Eqs. 4-51 and 4-52.

For an n-span continuous beam, the following equations are written:

$$C_1\Psi_1 - C_2\left[\frac{\lambda_2^2}{\lambda_1^2}\frac{(EI)_2}{(EI)_1}\theta_1 + \frac{\lambda_2}{\lambda_1}\theta_2\right] + C_3\frac{\lambda_3^2}{\lambda_1\lambda_2}\frac{(EI)_3}{(EI)_2}\Psi_2 = 0$$

$$C_2\Psi_2 - C_3\left[\frac{\lambda_3^2}{\lambda_2^2}\frac{(EI)_3}{(EI)_2}\theta_2 + \frac{\lambda_3}{\lambda_2}\theta_2\right] + C_4\frac{\lambda_4^2}{\lambda_2\lambda_3}\frac{(EI)_4}{(EI)_3}\Psi_3 = 0$$

$$C_3\,\Psi_3 - C_4\left[\frac{\lambda_4^2}{\lambda_3^2}\frac{(EI)_4}{(EI)_3}\theta_3 + \frac{\lambda_4}{\lambda_3}\theta_4\right] + C_5\frac{\lambda_5^2}{\lambda_3\lambda_4}\frac{(EI)_5}{(EI)_4}\Psi_4 = 0$$

$$\cdots\cdots\cdots\cdots\cdots\cdots\cdots\cdots \tag{4-63}$$

$$C_{n-1}\,\Psi_{n-1} - C_n\left[\frac{\lambda_n^2}{\lambda_{n-1}^2}\frac{(EI)_n}{(EI)_{n-1}}\theta_{n-1} + \frac{\lambda_n}{\lambda_{n-1}}\theta_n\right]$$

$$+ C_{n+1}\frac{\lambda_{n+1}^2}{\lambda_{n-1}\lambda_n}\frac{(EI)_{n+1}}{(EI)_n}\Psi_n = 0$$

It is interesting to note that if $\lambda_1 = \lambda_2$, $(EI)_1 = (EI)_2$, and the continuous beam has two spans with simply supported ends, Eq. 4-63 yields Eq. 4-56.

Equation 4-62 can be also written in terms of the three moments M_j, M_{j+1}, and M_{j+2} shown in Figs. 4-1*b* and 4-1*c*. This is justified, because from Eq. 4-39,

$$(EI)_j\frac{d^2y_j}{dx^2} = -2(EI)_j\lambda_j^2 C_j = -M_j$$

or

$$C_j = \frac{M_j}{2(EI)_j\lambda_j^2} \tag{4-64}$$

In a similar manner,

$$C_{j+1} = \frac{M_{j+1}}{2(EI)_{j+1}\lambda_{j+1}^2} \tag{4-65}$$

$$C_{j+2} = \frac{M_{j+2}}{2(EI)_{j+2}\lambda_{j+2}^2} \tag{4-66}$$

Thus Eq. 4-62 can be written in terms of the three moments M_j, M_{j+1}, and M_{j+2} by using Eqs. 4-64, 4-65, and 4-66. This form of Eq. 4-62 is known as the three-moment equation. For each support moment an equation of this type can be written, thus yielding a set of expressions similar to those given by Eq. 4-63. If the end supports are fixed, a three-moment equation for each end support can be written by assuming that there is a span outside the end support whose stiffness EI is infinite. When the required equations for a given problem are written, the frequency determinant and the frequency equation are obtained as discussed earlier in this section.

The values of λL that satisfy the frequency equation are determined as discussed above. When these roots are found, the mode shapes are determined by substituting each root of λL into Eq. 4-32 and applying boundary conditions. The existing boundary conditions are always sufficient for the computation of the required constants in a given problem.

As an illustration, let it be assumed that it is required to determine the fundamental frequency of the two-span continuous beam in Fig. 4-2*a*. The stiffness *EI* is constant and is the same for both spans. The uniform mass m of the beam is also the same for both spans and it is equal to 0.216 lb-sec^2/in. per inch of length. The frequency equation of this problem is given by Eq. 4-56. With $\lambda L_2 = 1.33\lambda L_1$, the frequency equation becomes a function of λL_1 only. By plotting λL_1 versus $(\theta_1 + \theta_2)$, the smallest value of λL_1 that satisfies Eq. 4-56 is 2.57. Thus $\lambda = 2.57/L_1$ and Eq. 4-58 yields the fundamental frequency of vibration ω_1 as

$$\omega_1 = \frac{(2.57)^2}{L_1^{\ 2}} \sqrt{EI/m}$$

or

$$\omega_1 = 10.63 \times 10^{-5} \sqrt{EI} \text{ rps}$$

The units used for EI are lb-in.^{2} The higher natural frequencies can be obtained in the same manner. The mode shape corresponding to the fundamental frequency ω_1 is shown in Fig. 4-2*b*.

If the two spans are of equal length, Eq. 4-56 yields two sets of roots. The first set is obtained from the frequency equation $\theta_1 = \theta_2 = \pm\infty$, and the second set from the equation $\theta_1 = \theta_2 = 0$. In the first case, each span will vibrate as a simply supported beam and the roots of λL are $(\lambda L)_n = n\pi$, where $n = 1, 2, 3, \ldots$. In the second case the rotation of the beam at the

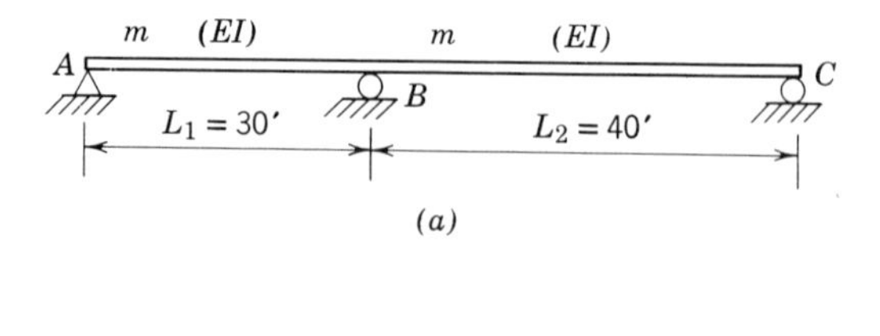

(*a*)

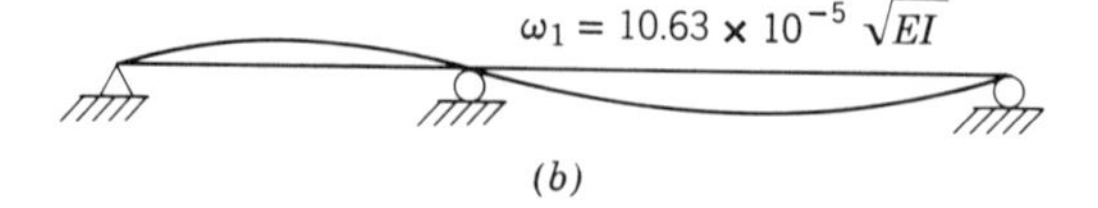

(*b*)

Fig. 4-2

intermediate support during vibration is zero, and each span can be assumed to vibrate like a beam fixed at the one end and simply supported at the other. The natural frequencies and mode shapes in this case can be determined as discussed in Section 4-2.

4-5 DYNAMIC RESPONSE OF BEAMS

The dynamic response of uniform beams acted upon by time-varying forces are investigated in this section.

The discussion is initiated by considering the simply supported beam in Fig. 4-3a, whose stiffness and mass is uniform throughout its length. The time-varying force $q(t,x)$ is uniformly distributed over its length, and the time variation is given by the expression

$$q(t,x) = q\sin\omega_f t \tag{4-67}$$

where ω_f is the frequency of the forcing function.

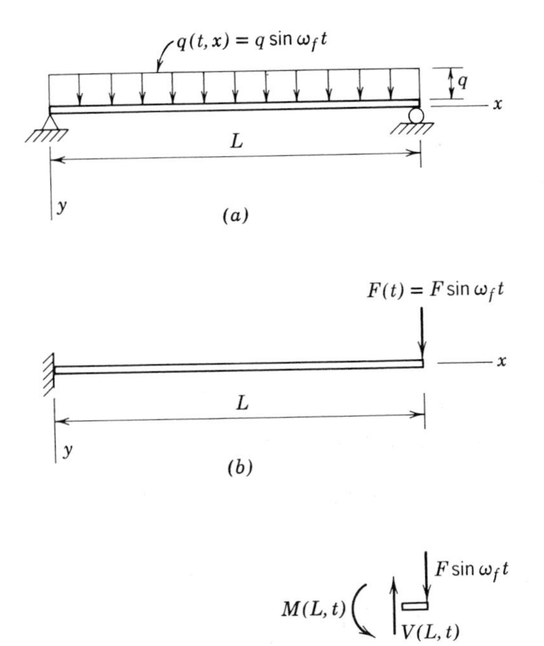

Fig. 4-3

The differential equation of motion is given by Eq. 1-21. By substituting Eq. 4-67 into Eq. 1-21,

$$EI\frac{\partial^4 y}{\partial x^4} + m\ddot{y} = q \sin \omega_f t \tag{4-68}$$

At this point, it is reasonable to conclude that the transient free vibration of the beam will die out because of damping. Thus, by neglecting the transient condition and assuming that the only motion left in the beam is the steady-state one produced by the time-varying force, it is reasonable to assume that a particular solution of Eq. 4-68 is

$$y = Y \sin \omega_f t \tag{4-69}$$

where Y is a function of x that needs to be determined.

By substituting Eq. 4-69 into Eq. 4-68, the following ordinary differential equation is obtained:

$$\frac{d^4 Y}{dx^4} - \lambda^4 Y = \frac{q}{EI} \tag{4-70}$$

where

$$\lambda^4 = \frac{m\omega_f^2}{EI} \tag{4-71}$$

The homogeneous solution of Eq. 4-70 has the form given by Eq. 4-1. The particular solution Y_p, since q/EI is constant, is taken equal to a constant C_0. By substituting $Y_p = C_0$ in Eq. 4-70, C_0 is found to be equal to $-q/m\omega_f^2$, and thus $Y_p = -q/m\omega_f^2$. Therefore, the complete solution of Eq. 4-70 is

$$Y = A \cosh \lambda x + B \sinh \lambda x + C \cos \lambda x + D \sin \lambda x - \frac{q}{\omega_f^2 m} \tag{4-72}$$

where A, B, C, and D are constants.

The end boundary conditions of the beam are

$$\text{at} \quad x = 0: \qquad Y = 0, \quad \frac{d^2 Y}{dx^2} = 0 \tag{4-73}$$

$$\text{at} \quad x = L: \qquad Y = 0, \quad \frac{d^2 Y}{d^2 x} = 0 \tag{4-74}$$

By substituting the conditions given by Eq. 4-73 into Eq. 4-72, the constants A and C are

$$A = C = \frac{q}{2\omega_f^2 m} \tag{4-75}$$

The end conditions given by Eq. 4-74 yield

$$B = -\frac{q}{2\omega_f^2 m} \tanh \frac{\lambda L}{2} \tag{4-76}$$

$$D = \frac{q}{2\omega_f^2 m} \tan \frac{\lambda L}{2} \tag{4-77}$$

By substituting the values of A, B, C, and D into Eq. 4-72

$$Y = \frac{q}{\omega_f^2 m}\left[\frac{\cos\lambda(L/2-x)}{2\cos\lambda L/2} + \frac{\cosh\lambda(L/2-x)}{2\cosh\lambda L/2} - 1\right] \tag{4-78}$$

Thus Eq. 4-69 yields

$$y = \frac{q}{\omega_f^2 m}\left[\frac{\cos\lambda(L/2-x)}{2\cos\lambda L/2} + \frac{\cosh\lambda(L/2-x)}{2\cosh\lambda L/2} - 1\right]\sin\omega_f t \tag{4-79}$$

Equation 4-79 gives the time variation, as well as the variation with respect to x, of the dynamic amplitude of the steady-state forced vibration of the simply supported member. When λL becomes equal to π, 3π, 5π, ... the first term in the bracket of Eq. 4-79 approaches infinity. This shows that the phenomenon of resonance occurs whenever the forcing frequency ω_f coincides with one of the odd-numbered natural frequencies of the member.

The dynamic bending moment M and shear V at any time t and any distance x are determined from the expressions

$$M = -EI\frac{\partial^2 y}{\partial x^2} \tag{4-80}$$

$$V = \frac{\partial M}{\partial x} \tag{4-81}$$

where y is given by Eq. 4-79. Having determined the dynamic moments and shear forces, the dynamic bending and shear stresses can be calculated by using known formulas of strength of materials.

Consider now the case of a cantilever beam loaded at the free end by the forcing function

$$F(t) = F\sin\omega_f t \tag{4-82}$$

where ω_f is the frequency of the force. The beam is shown in Fig. 4-3b and has uniform mass and stiffness throughout its length. The initial displacements and initial velocities at $t=0$ are assumed to be zero.

The governing differential equation of motion for the length L of the member extending between the fixed support and just to the left of the forcing function $F(t)$ is given by Eq. 4-17. For convenience, this equation is written again below.

$$EI\frac{\partial^4 y}{dx^4}+m\ddot{y}=0 \tag{4-83}$$

There are two types of motion that take place. The first one is the transient motion due to the free vibration of the beam and the solution is given by Eq. 4-24. If n modes are considered and the symbol $y_c(x,t)$ is used for this solution, Eq. 4-24 yields

$$y_c(x,t)=\sum_{j=1}^{n} Y_j(x)\sin(\omega_j t+\phi_j) \tag{4-84}$$

The unknowns in Eq. 4-84 are the four constants of the shape function $Y_j(x)$, the natural frequencies ω_j and the phase angles ϕ_j. These unknowns can be determined by using the four boundary conditions of the beam and the two initial time conditions as discussed in Section 4-3. In this problem, the transient motion will be neglected, based on the assumption that it will soon die out because of the influence of some kind of light damping.

The second type of motion is the one produced by the forcing function $F(t)$ whose time variation is given by Eq. 4-82. The solution $y(x,t)$ for this motion must satisfy Eq. 4-83 as well as prescribed boundary conditions of the member. This solution is considered to be of the form.

$$y(x,t)=Y(x)g(t) \tag{4-85}$$

where $Y(x)$ is a function of x and $g(t)$ is a function of time t. By substituting Eq. 4-85 into Eq. 4-83 and separating variables,

$$\frac{EI[\partial^4 Y(x)/\partial x^4]}{mY(x)}=-\frac{\ddot{g}(t)}{g(t)} \tag{4-86}$$

The two sides of the differential equation above will be equal to each other only if both are equal to the same constant, which in this case is designated by the Greek letter Ψ^2. Thus,

$$\frac{d^4Y(x)}{dx^4}-\frac{m\Psi^2}{EI}Y(x)=0 \tag{4-87}$$

$$\ddot{g}(t)+\Psi^2 g(t)=0 \tag{4-88}$$

At this point, it is reasonable to assume that the time variation of the displacements $y(x, t)$ is the same as that of the force $F(t)$. That is,

$$g(t)=\sin \omega_f t \tag{4-89}$$

By substituting Eq. 4-89 into Eq. 4-88,

$$\Psi^2=\omega_f^2 \tag{4-90}$$

Therefore, by using the result above, Eq. 4-87 is written as

$$\frac{d^4Y(x)}{dx^4}-\lambda^4 Y(x)=0 \tag{4-91}$$

where

$$\lambda^4=\frac{m\omega_f^2}{EI} \tag{4-92}$$

The solution of Eq. 4-91 is

$$Y(x)=A\cosh\lambda x+B\sinh\lambda x+C\cos\lambda x+D\sin\lambda x \tag{4-93}$$

Thus, by substituting Eq. 4-93 into Eq. 4-85 and also using Eq. 4-89,

$$y(x,t)=(A\cosh\lambda x+B\sinh\lambda x+C\cos\lambda x+D\sin\lambda x)\sin\omega_f t \tag{4-94}$$

The constants A, B, C, and D can be determined by using the following boundary conditions of the member:

$$\begin{aligned}
&\text{at } x=0: && Y=0 && \text{(a)}\\
& && \frac{dY}{dx}=0 && \text{(b)}\\
&\text{at } x=L: && \frac{d^2Y}{dx^2}=0 && \text{(c)}\\
& && -EI\frac{d^3Y}{dx^3}=F && \text{(d)}
\end{aligned} \tag{4-95}$$

The condition given by Eq. 4-95d is based on the vertical equilibrium of the forces acting on the free-body diagram in Fig. 4-3c, where $V(L, t)$ is the dynamic shear force just to the left of the force $F\sin\omega_f t$.

By applying the boundary conditions at $x=0$, Eq. 4-93 yields

$$\begin{aligned} A &= -C \quad &(a) \\ B &= -D \quad &(b) \end{aligned} \tag{4-96}$$

Therefore,

$$Y(x) = C(-\cosh\lambda x + \cos\lambda x) + D(-\sinh\lambda x + \sin\lambda x) \tag{4-97}$$

By using Eq. 4-97 and applying the boundary conditions at $x=L$,

$$C = \frac{F}{2\lambda^3 EI} \frac{\sinh\lambda L + \sin\lambda L}{(1+\cosh\lambda L \cos\lambda L)} \tag{4-98}$$

$$D = -\frac{F}{2\lambda^3 EI} \frac{\cosh\lambda L + \cos\lambda L}{(1+\cosh\lambda L \cos\lambda L)} \tag{4-99}$$

With A, B, C, and D as given above, Eq. 4-93 yields

$$Y(x) = \frac{F}{2\lambda^3 EI(1+\cosh\lambda L \cos\lambda L)} [(\sinh\lambda L + \sin\lambda L)(\cosh\lambda x - \cos\lambda x)$$

$$- (\cosh\lambda L + \cos\lambda L)(\sinh\lambda x - \sin\lambda x)] \tag{4-100}$$

Thus the displacements $y(x,t)$ due to the forcing function $F(t) = F\sin\omega_f t$ are determined from the expression

$$y(x,t) = Y(x)\sin\omega_f t \tag{4-101}$$

where $Y(x)$ is given by Eq. 4-100.

At $x=L$, that is, at the free end, Eq. 4-101 yields

$$y(L,t) = \frac{F}{\lambda^3 EI} \frac{(\cosh\lambda L \sin\lambda L - \sinh\lambda L \cos\lambda L)}{(1+\cosh\lambda L \cos\lambda L)} \sin\omega_f t \tag{4-102}$$

It should be pointed out, however, that when λL becomes equal to any of the values given by Eqs. 4-8 and 4-11, the displacements given by Eqs. 4-101 and 4-102 become infinite. When the expression for the displacements has been determined, the dynamic moments and shear forces can be computed by applying Eqs. 4-80 and 4-81, respectively. The dynamic stresses can be determined by using known formulas of strength of materials.

4-6 DYNAMIC RESPONSE DUE TO SUPPORT MOTION

There are cases where the supports of a member are subjected to time-varying motions. A beam and girder system acted upon by dynamic forces, earthquake ground motions, or excitations produced by an explosion are examples where support motions may have to be considered in the analysis for dynamic response.

A particular situation that is examined in this section is the case where the end supports of a simply supported beam of uniform mass and stiffness are subjected to a vertical harmonic displacement $y_s(t)$ given by the expression

$$y_s(t) = y_0 \sin \omega_f t \tag{4-103}$$

where, y_0 is the maximum amplitude and ω_f is the frequency of the support motion.

The differential equation of motion for this problem is again Eq. 4-83. The response due to free vibration will be neglected as before, and only the particular solution of Eq. 4-83 will be investigated. This solution is taken as

$$y(x,t) = Y(x)g(t) \tag{4-104}$$

where $Y(x)$ and $g(t)$ are functions of x and t, respectively. It is reasonable to assume that $g(t)$ in Eq. 4-104 is of the form

$$g(t) = \sin \omega_f t \tag{4-105}$$

Thus Eq. 4-104 yields

$$y(t,x) = Y(x) \sin \omega_f t \tag{4-106}$$

By substituting Eq. 4-106 into Eq. 4-83, the following differential equation is obtained:

$$\frac{d^4Y(x)}{dx^4} - \lambda^4 Y(x) = 0 \tag{4-107}$$

where

$$\lambda^4 = \frac{m\omega_f^2}{EI} \tag{4-108}$$

The solution of Eq. 4-107 is given by Eq. 4-93. That is,

$$Y(x) = A \cosh \lambda x + B \sinh \lambda x + C \cos \lambda x + D \sin \lambda x \tag{4-109}$$

The end boundary conditions of the simply supported beam are

$$\text{at} \quad x=0: \qquad Y = y_0 \quad (a) \qquad \frac{d^2Y}{dx^2} = 0 \quad (b) \tag{4-110}$$

$$\text{at} \quad x=L: \qquad Y = y_0 \quad (a) \qquad \frac{d^2Y}{dx^2} = 0 \quad (b) \tag{4-111}$$

The conditions given by Eq. 4-110 yield

$$A = \frac{y_0}{2} \qquad C = \frac{y_0}{2} \tag{4-112}$$

Thus Eq. 4-109 becomes

$$Y(x) = \frac{y_0}{2}(\cosh\lambda x + \cos\lambda x) + B\sinh\lambda x + D\sin\lambda x \tag{4-113}$$

The second derivative of this equation with respect to x is

$$\frac{d^2Y(x)}{dx^2} = \frac{y_0\lambda^2}{2}(\cosh\lambda x - \cos\lambda x) + \lambda^2(B\sinh\lambda x - D\sin\lambda x) \tag{4-114}$$

The conditions given by Eq. 4-111 yield

$$y_0 = \frac{y_0}{2}(\cosh\lambda L + \cos\lambda L) + B\sinh\lambda L + D\sin\lambda L \tag{4-115}$$

$$0 = \frac{y_0}{2}(\cosh\lambda L + \cos\lambda L) + B\sinh\lambda L - D\sin\lambda L \tag{4-116}$$

Addition of Eq. 4-115 and 4-116 yields

$$B = \frac{y_0}{2}\frac{(1-\cosh\lambda L)}{\sinh\lambda L} \tag{4-117}$$

Subtraction of Eqs. 4-115 and 4-116 gives

$$D = \frac{y_0}{2}\frac{(1-\cos\lambda L)}{\sin\lambda L} \tag{4-118}$$

Thus Eq. 4-113 becomes

$$Y(x)=\frac{y_0}{2}\left[(\cosh\lambda x+\cos\lambda x)+\frac{(1-\cosh\lambda L)}{\sinh\lambda L}\sinh\lambda x + \frac{(1-\cos\lambda L)}{\sin\lambda L}\sin\lambda x\right] \tag{4-119}$$

and the solution given by Eq. 4-106 yields

$$y(t,x)=\frac{y_0}{2}\left[(\cosh\lambda x+\cos\lambda x)+\frac{(1-\cosh\lambda L)}{\sinh\lambda L}\sinh\lambda x + \frac{(1-\cos\lambda L)}{\sin\lambda L}\sin\lambda x\right]\sin\omega_f t \tag{4-120}$$

The displacements $y_b(t,x)$ of the member relative to the motion of the end supports is

$$y_b(t,x)=y(t,x)-y_s(t)=[Y(x)-y_o]\sin\omega_f t$$

or

$$y_b(t,x)=\frac{y_0}{2}\left[(\cosh\lambda x+\cos\lambda x)+\frac{(1-\cosh\lambda L)}{\sinh\lambda L}\sinh\lambda x + \frac{(1-\cos\lambda L)}{\sin\lambda L}\sin\lambda x-1\right]\sin\omega_f t \tag{4-121}$$

The solution of other problems involving support motion can be carried out in a similar manner. It should be pointed out, however, that approximate solutions would be more appropriate for the more complicated problems.

4-7 DIFFERENTIAL EQUATION OF MOTION FOR THIN PLATES

The derivation of the differential equation of motion for thin plates is based on the assumptions that the thickness h of the plate is small compared to other dimensions of the plate and that plane cross sections before strain remain plane after the plate is strained. In addition, it is further assumed that the middle surface of the plate is not strained, the external load is applied normal to the surface of the plate, and the influence of shear and rotating inertia is neglected.

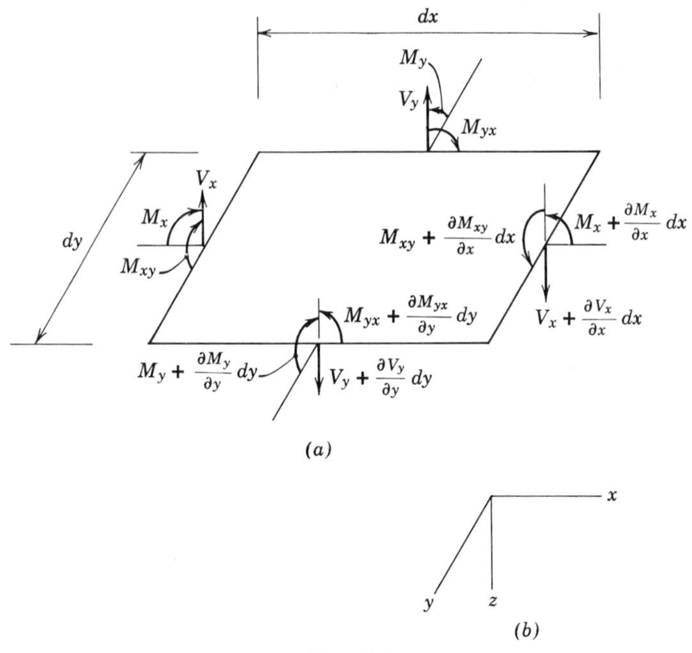

Fig. 4-4

By considering an element of the plate of sides dx, dy, and thickness h, its free-body diagram is shown in Fig. 4-4*a*. In this figure, V_x and V_y are the dynamic shear forces per unit of length parallel to the y and x axes, respectively. M_x and M_y are the moments and M_{xy} is the twisting moment. The load intensity per unit surface area of the plate is q and ρ is the volume density. The positive directions of the x, y, and z axes are shown in Fig. 4-4*b*. By summing up the forces in the z direction and equating them to zero, we obtain

$$\frac{\partial V_x}{\partial x} + \frac{\partial V_y}{\partial y} + q + \rho h \ddot{w} = 0 \tag{4-122}$$

The last term in this equation is the inertia force, and w is the deflection of the plate in the z direction.

By equating to zero the sum of the moments about the x and y axes and neglecting higher-order terms, we have the following equations:

$$\frac{\partial M_{xy}}{\partial x} - \frac{\partial M_y}{\partial y} + V_y = 0 \tag{4-123}$$

$$\frac{\partial M_{xy}}{\partial y} + \frac{\partial M_x}{\partial x} - V_x = 0 \tag{4-124}$$

With $M_{yx} = M_{xy}$, Eqs. 4-122, 4-123, and 4-124 yield the expression

$$\frac{\partial^2 M_x}{\partial x^2} + \frac{\partial^2 M_y}{\partial y^2} - 2\frac{\partial^2 M_{xy}}{\partial x \partial y} = q - \rho h \ddot{w} \tag{4-125}$$

From the theory of plates,[15]

$$M_x = -D\left(\frac{\partial^2 w}{\partial x^2} + \nu \frac{\partial^2 w}{\partial y^2}\right) \tag{4-126}$$

$$M_y = -D\left(\frac{\partial^2 w}{\partial y^2} + \nu \frac{\partial^2 w}{\partial x^2}\right) \tag{4-127}$$

$$M_{xy} = -M_{yx} = D(1-\nu)\frac{\partial^2 w}{\partial x \partial y} \tag{4-128}$$

where

$$D = \frac{Eh^3}{12(1-\nu^2)} \tag{4-129}$$

and ν is the Poisson ratio. Thus, by introducing Eqs. 4-126, 4-127, and 4-128 into q. 4-125, the following differential equation of motion for thin plates is obtained:

$$D\left(\frac{\partial^4 w}{\partial x^4} + 2\frac{\partial^4 w}{\partial x^2 \partial y^2} + \frac{\partial^4 w}{\partial y^4}\right) + \rho h \ddot{w} = q \tag{4-130}$$

or

$$D\,\nabla^4 w + \rho h \ddot{w} = q \tag{4-131}$$

where

$$\nabla^4 = \partial^4/\partial x^4 + 2\partial^4/\partial x^2 \partial y^2 + \partial^4/\partial y^4$$

For free vibration $q = 0$ and the differential equation becomes

$$D\,\nabla^4 w + \rho h \ddot{w} = 0 \tag{4-132}$$

Equations 4-131 and 4-132 can be used to determine the dynamic response and free vibration, respectively, of thin plates of various boundary

[15]See References 17 and 81.

conditions. It should be pointed out, however, that direct solution of the differential equation is usually difficult and procedures such as those discussed in Chapters 5, 9, and 10 are often employed.

PROBLEMS

4-1 For the uniform single-span beams shown in Fig. P4-1 determine the natural frequencies of vibration and the corresponding mode shapes

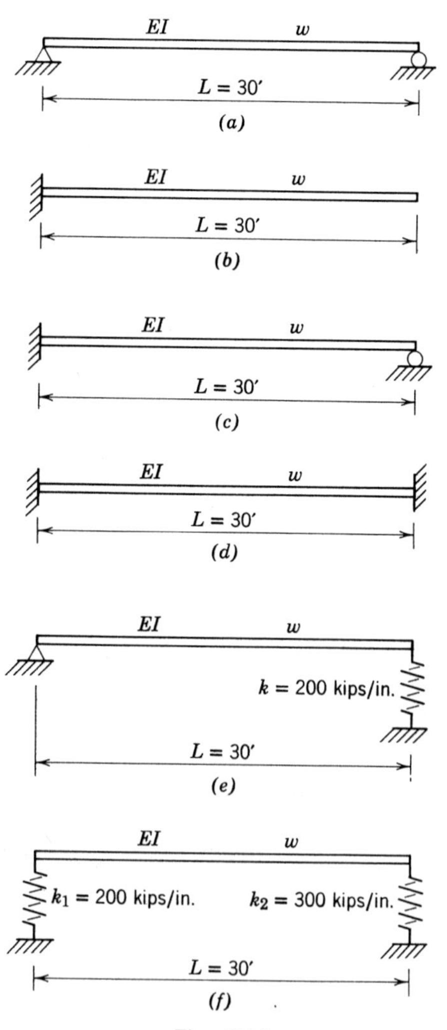

Fig. P4-1

by using Eq. 4-1 and appropriate boundary conditions. The stiffness EI is constant and equal to 30×10^6 kip-in.^{2} and the weight w is equal to 0.6 kips/ft. Neglect damping.

4-2 By using Eq. 4-1 and appropriate boundary conditions, determine the natural frequencies and the corresponding mode shapes for the uniform beams shown below. The stiffness EI is 45×10^6 kip-in.^{2} and the weight w is 0.6 kips/ft. Neglect damping. (Fig. P4-2.)

4-3 By using the appropriate three-moment equations, determine the natural frequencies of vibration and the corresponding normal modes

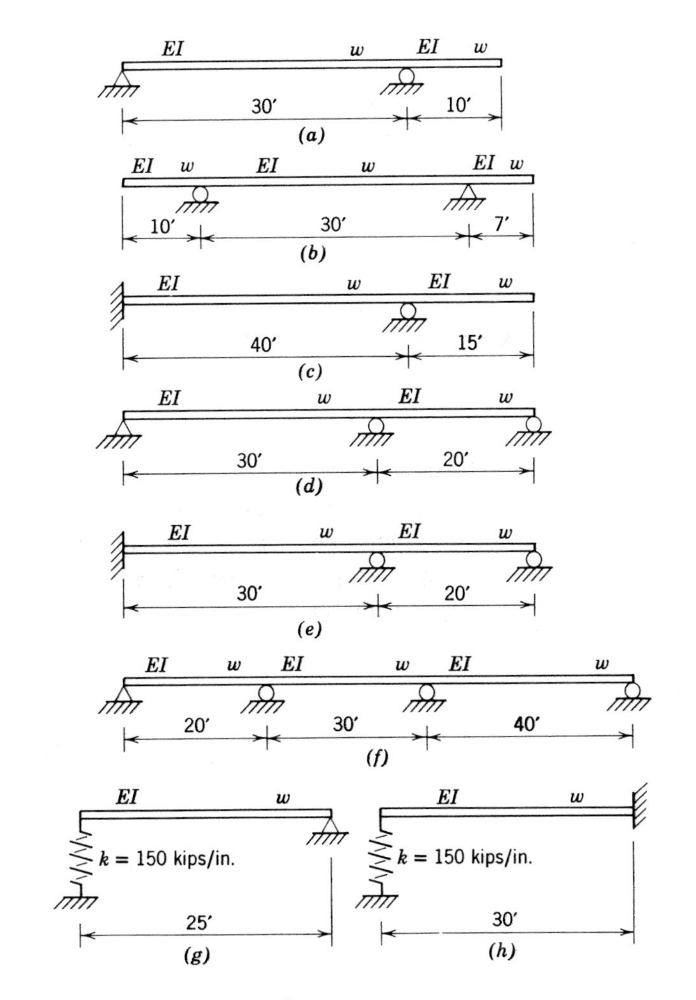

Fig. P4-2

of the two-span continuous beams in Problem 4-2. EI is 60×10^6 kip-in.^{2} and $w = 0.8$ kips/ft. Neglect damping.

4-4 Repeat Problem 4-3 by using Eq. 4-62.

4-5 The uniform single-span beams in Problem 4-1 are assumed to be subjected to a suddenly applied constant force q of infinite duration and uniformly distributed throughout their length. If q is 300 lb/in., determine in each case the expression for the dynamic vertical displacement by solving the appropriate differential equation. The stiffness EI and weight w are as in Problem 4-1.

4-6 Repeat Problem 4-5 when $q(t,x) = q \cos \omega_f t$, where q is uniformly distributed throughout the length of each beam. Its value is 300 lb/in. and $\omega_f = 10\pi$.

4-7 Repeat Problem 4-5 by assuming that the single-span uniform beams are loaded with a suddenly applied constant force F = 10 kips at the center length of each beam and of infinite duration.

4-8 Repeat Problem 4-7 when the concentrated force $F(t)$ at the center length of each beam is $F(t) = F \sin \omega_f t$, where $F = 10$ kips and $\omega_f = 10\pi$.

4-9 The supports of the uniform simply supported beam in Problem 4-1*a* are subjected to a vertical harmonic motion $y_s(t) = y_0 \sin \omega_f t$, where $y_0 = 2$ in. and $\omega_f = 4\pi$. Determine the expression for the displacement $y_b(t,x)$ of the member relative to the motion of the end supports. The stiffness EI is 30×10^6 kip-in.^{2} and $w = 0.6$ kips/ft.

4-10 Repeat Problem 4-9 for $y_s(t) = y_0 \cos \omega_f t$, where $y_0 = 3$ in. and $\omega_f = 2\pi$, 3π, 4π, 5π, and 6π.

4-11 By using the results of Problem 4-9, determine the maximum vertical displacement and the maximum bending stress at midspan.

4-12 By using the results of Problem 4-10, determine the maximum vertical displacement and the maximum bending stress at midspan.

5

MODAL ANALYSIS

5-1 INTRODUCTION

A very powerful method that can be used to determine the dynamic response of complicated structural dynamic problems is known as the method of modal analysis. Basically it can be considered to be an energy method, because the modal equations are derived by using Lagrange's equation discussed in Section 1-4.

In this chapter the modal equations for spring-mass systems, beams, frames, and plates are derived and are applied to solve various types of structural dynamics problems. In addition, an approximate procedure is also discussed in order to simplify the application of this method to particular types of problems.

The method of Stodola and iteration procedure, as well as the dynamic response of frames with flexible girders, are also discussed.

5-2 MODAL EQUATIONS FOR SPRING-MASS SYSTEMS

The dynamic analysis of complicated multiple degree of freedom systems can become convenient by using the method of modal analysis. The modal equations required in the application of this method are derived by utilizing Lagrange's equation given by Eq. 1-37. Theoretically, the application of this method is limited to linearly elastic systems and to cases where the dynamic forces acting on a structure have the same time variation.

In modal analysis, the response of a multidegree system is computed individually for each normal mode, and the total response is obtained by superimposing the responses of the individual modes. The modal equations, as they are derived in this section, permit each mode to be represented by an equivalent one degree of freedom system whose dynamic response is obtained independently by using the methods discussed in Chapter 2.

The development of the modal equations is initiated by considering a general system consisting of r masses, j springs, and M normal modes. If $\dot{a}_{ip}$ is the velocity component of mass i in the p mode, the total kinetic energy T of the system at a time t is

$$T = \sum_{i=1}^{r} \tfrac{1}{2} m_i \left(\sum_{P=1}^{M} \dot{a}_{ip} \right)^2 \tag{5-1}$$

The total strain energy U stored in the springs is

$$U = \sum_{n=1}^{j} \tfrac{1}{2} k_n \left(\sum_{P=1}^{M} \delta_{np} \right)^2 \tag{5-2}$$

where δ_{np} is the relative displacement of the ends of the spring n in the p mode and k_n is the constant of the nth spring.

Equations 5-1 and 5-2 can be written as

$$T = \sum_{i=1}^{r} \tfrac{1}{2} m_i \sum_{P=1}^{M} \dot{a}_{ip}^2 \tag{5-3}$$

and

$$U = \sum_{n=1}^{j} \tfrac{1}{2}k_n \sum_{P=1}^{M} \delta_{np}^2 \tag{5-4}$$

This is permissible, because according to the orthogonality properties of normal modes the cross products contained in the squared series in Eqs. 5-1 and 5-2 are zero.

The work W_e of the external forces $F_i(t)$ is

$$W_e = \sum_{i=1}^{r} F_i(t) \sum_{P=1}^{M} a_{ip} \tag{5-5}$$

By selecting arbitrarily a modal displacement Y_p, preferably the displacement of one of the masses, all the displacements a_{ip} of the masses can be expressed in proportion to the modal displacement Y_p. On this basis, the quantities a_{ip}, $\dot{a}_{ip}$, and δ_{np} are expressed as follows:

$$a_{ip} = Y_p \beta_{ip} \qquad \dot{a}_{ip} = \dot{Y}_p \beta_{ip} \qquad \delta_{np} = Y_p \beta_{\delta np} \tag{5-6}$$

where β_{ip} and $\beta_{\delta np}$ define the characteristic shape of the mode and are given by the expressions

$$\beta_{ip} = \frac{a_{ip}}{Y_p} = \frac{\dot{a}_{ip}}{\dot{Y}_p} \qquad \beta_{\delta np} = \frac{\delta_{np}}{Y_p} \tag{5-7}$$

Equations 5-3, 5-4, and 5-5 can now be written as

$$T = \sum_{i=1}^{r} \tfrac{1}{2}m_i \sum_{P=1}^{M} \dot{Y}_p^{\,2} \beta_{ip}^{\,2} \tag{5-8}$$

$$U = \sum_{n=1}^{j} \tfrac{1}{2}k_n \sum_{P=1}^{M} Y_p^{\,2} \beta_{\delta np}^{\,2} \tag{5-9}$$

$$W_e = \sum_{i=1}^{r} F_i(t) \sum_{P=1}^{M} Y_p \beta_{ip} \tag{5-10}$$

By selecting the modal displacement Y_p as the generalized coordinate q_i, the following equations are written:

$$\frac{d}{dt}\left(\frac{\partial T}{\partial \dot{Y}_p}\right) = \ddot{Y}_p \sum_{i=1}^{r} m_i \beta_{ip}^{\,2}$$

$$\frac{\partial U}{\partial Y_p} = Y_p \sum_{n=1}^{j} k_n \beta_{\delta n p}^2$$

$$\frac{\partial W_e}{\partial Y_p} = \sum_{i=1}^{r} F_i(t) \beta_{ip}$$

Utilizing the three expressions and Lagrange's equation given by Eq. 1-37, we have

$$\ddot{Y}_p \sum_{i=1}^{r} m_i \beta_{ip}^2 + Y_p \sum_{n=1}^{j} k_n \beta_{\delta n p}^2 = \sum_{i=1}^{r} F_i(t) \beta_{ip} \qquad (5\text{-}11)$$

Equation 5-11 is known as the modal equation of motion. It can be written in a somewhat different form by making the following substitutions:

$$m_e = \sum_{i=1}^{r} m_i \beta_{ip}^2 \qquad k_e = \sum_{n=1}^{j} k_n \beta_{\delta n p}^2 \qquad F_e(t) = \sum_{i=1}^{r} F_i(t) \beta_{ip} \qquad (5\text{-}12)$$

where m_e, k_e, and $F_e(t)$ are the equivalent mass, spring constant, and force, respectively, of an equivalent one-degree system. With this in mind, Eq. 5-11 is written as

$$m_e \ddot{Y}_p + k_e Y_p = F_e(t) \qquad (5\text{-}13)$$

Equation 5-13 represents the equivalent one-degree system corresponding to the p mode of vibration. The kinetic energy, internal strain energy, and the work done by all external forces are equal to the same quantities of the complete system vibrating in the p mode alone. Thus each normal mode can be analyzed independently as a one-degree system, and the dynamic response corresponding to a mode can be determined by using the methods discussed in Chapter 2.

Equation 5-13 can be also written as

$$\ddot{Y}_p + \omega_p^2 Y_p = \frac{F_e(t)}{m_e} \qquad (5\text{-}14)$$

or, by using Eq. 5-12,

$$\ddot{Y}_p + \omega_p^2 Y_p = g(t) \frac{\sum_{i=1}^{r} F_i \beta_{ip}}{\sum_{i=1}^{r} m_i \beta_{ip}^2} \qquad (5\text{-}15)$$

where $g(t)$ is the time variation of the force. That is, $F_i(t) = g(t)F_i$.

The two summations in Eq. 5-15 are readily computed for a given mode and loading. Thus the first step in applying this equation is to determine the required natural frequencies ω of the system and the corresponding modes of vibration. When these quantities are known, the dynamic response of the system in each mode is computed by using Eq. 5-15. It should be noted, however, that the time function $g(t)$ should be the same for all acting forces. This limitation can be removed if the modal equations are solved by using numerical methods. If the amplitudes of a mode are normalized so that

$$\sum_{i=1}^{r} m_i \beta_{ip}^2 = 1 \tag{5-16}$$

then Eq. 5-15 becomes

$$\ddot{Y}_p + \omega_p^2 Y_p = g(t) \sum_{i=1}^{r} F_i \beta_{ip} \tag{5-17}$$

In each mode, the modal static deflection is given by the expression

$$Y_{p_{st}} = \frac{F_e}{k_e} = \frac{F_e}{\omega_p^2 m_e}$$

or, by using Eq. 5-12,

$$Y_{p_{st}} = \frac{\sum_{i=1}^{r} F_i \beta_{ip}}{\omega_p^2 \sum_{i=1}^{r} m_i \beta_{ip}^2} \tag{5-18}$$

Therefore,

$$Y_p(t) = Y_{p_{st}} (\Gamma)_p \tag{5-19}$$

and

$$Y_{p_{max}} = Y_{p_{st}} (\Gamma)_{p_{max}} \tag{5-20}$$

Here, the magnification factor $(\Gamma)_p$ depends only on the time function $g(t)$ and the natural frequency ω_p. Thus the solutions and graphs of Chapter 2 can be used for the analysis of multiple degree of freedom systems. By superimposing all modes, the total deflection $y_i(t)$ of the mass m_i is obtained from the relation:

$$y_i(t) = \sum_{P=1}^{M} Y_{p_{st}} \beta_{ip}(\Gamma)_p \tag{5-21}$$

For just one mode, say p, this displacement is

$$y_{ip}(t) = Y_{p_{st}} \beta_{ip}(\Gamma)_p \tag{5-22}$$

In the following section, the method of modal analysis is used to determine the dynamic response of idealized frames.

5-3 IDEALIZED FRAMES OR BUILDINGS

The method of modal analysis is used here to determine the dynamic response of the two-story rigid frame in Fig. 3-5*b*. The idealized system and the variation of the dynamic force $F_1(t)$ are shown in Figs. 3-5*c* and 5*d*, respectively. The two natural frequencies and the two periods of the undamped vibration are determined in Section 3-4, and they are as follows:

$$\omega_1 = 5.10 \text{ rps} \qquad \omega_2 = 20.00 \text{ rps}$$

$$f_1 = 0.81 \text{ Hz} \qquad f_2 = 3.19 \text{ Hz}$$

$$\tau_1 = 1.232 \text{ sec} \qquad \tau_2 = 0.314 \text{ sec}$$

The amplitude relationships defining the shape of the two modes are given by Eqs. 1-102 and 1-103. Thus, if X_1 and X_2 are defined as the amplitudes of W_1 and W_2, respectively, the amplitude relationship of the first mode is

$$X_2^{(1)} = C_1 X_1^{(1)} \tag{5-23}$$

where

$$C_1 = \frac{k_2}{k_2 - m_2\omega_1^2} = \frac{32.10}{32.10 - (0.137)(5.10)^2} = 1.125$$

Thus

$$X_2^{(1)} = 1.125 X_1^{(1)} \tag{5-24}$$

For the second mode,

$$X_2^{(2)} = C_2 X_1^{(2)} \tag{5-25}$$

where

$$C_2 = \frac{k_2}{k_2 - m_2\omega_2^2} = \frac{32.10}{32.10 - (0.137)(20.00)^2}$$

$$= -1.415$$

Therefore,

$$X_2^{(2)} = -1.415X_1^{(2)} \tag{5-26}$$

Assuming $X_1 = 1.00$ in Eqs. 5-24 and 5-26, the normal mode shapes of the frame are as shown in Fig. 5-1. That is,

First mode	*Second mode*
$X_1 = 1.000$	$X_1 = \ \ 1.000$
$X_2 = 1.125$	$X_2 = -1.415$

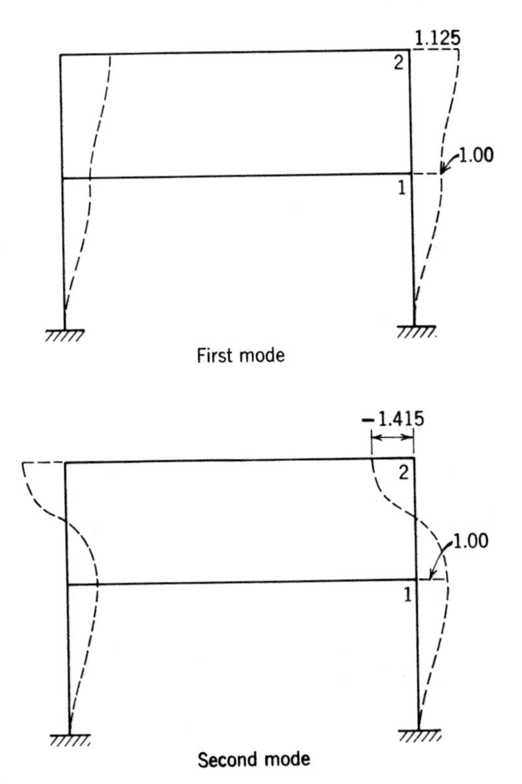

Fig. 5-1

The modal static deflections for the two modes are obtained by using Eq. 5-18. That is,

$$Y_{1\,\mathrm{st}} = \frac{\sum_{i=1}^{2} F_i \beta_{i1}}{\omega_1^2 \sum_{i=1}^{2} m_i \beta_{i1}^2}$$

$$= \frac{(3.0)(1.0) + (2.10)(1.125)}{(5.1)^2[(0.217)(1.0)^2 + (0.137)(1.125)^2]} = 0.527 \text{ in.}$$

$$Y_{2\,\mathrm{st}} = \frac{\sum_{i=1}^{2} F_i \beta_{i2}}{\omega_2^2 \sum_{i=1}^{2} m_i \beta_{i2}^2}$$

$$= \frac{(3.0)(1.0) + (2.1)(-1.415)}{(20.0)^2[(0.217)(1.0)^2 + (0.137)(-1.415)^2]} = 0.000153 \text{ in.}$$

The maximum magnification factor for each mode is obtained from the graph in Fig. 2-5*a*. Thus for the first mode

$$\frac{t_d}{\tau_1} = \frac{0.3}{1.232} = 0.243 \qquad (\Gamma)_{1\,\max} = 0.72$$

and for the second mode

$$\frac{t_d}{\tau_2} = \frac{0.3}{0.314} = 0.955 \qquad (\Gamma)_{2\,\max} = 1.56$$

The maximum first and second mode deflections of point 2, Fig. 5-1, are designated as x_{21} and x_{22}, respectively. By utilizing Eq. 5-22,

$$x_{21} = Y_{1\,\mathrm{st}} \beta_{21} (\Gamma)_{1\,\max}$$

$$= (0.527)(1.125)(0.72) = 0.427 \text{ in.}$$

$$x_{22} = Y_{2\,\mathrm{st}} \beta_{22} (\Gamma)_{2\,\max}$$

$$= (0.000153)(-1.415)(1.56) = -0.000338 \text{ in.}$$

Superposition of these two modal displacements yields the maximum horizontal deflection $x_{2\,\max}$ of point 2 of the frame. An upper limit of $x_{2\,\max}$

can be obtained by adding numerically the two modal displacements. That is,

$$x_{2\,\mathrm{max}} = x_{21} + x_{22} = 0.427 + 0.000338$$

$$= 0.4273 \text{ in.}$$

The numerical sum of the modal responses is a conservative estimate of the deflection. For practical problems, however, the response of some modes usually predominates over the response of others, and the numerical sum is a reasonable estimate of the total response. The results above, for example, show that the major contribution to the response of the frame comes from the first mode. In fact, the response of the second mode can be considered to be negligible. For structural problems, the differences between the values of the natural frequencies are usually large so that the first mode often predominates over the response of the others. Thus a reasonable estimate of the total response can be obtained by using only the first few modes. In fact, the response of the first mode is often sufficient for many practical problems.

The maximum value of x_2 that was obtained by using the acceleration impulse extrapolation method is shown in Table 3-3 and it is equal to 0.4308 in. The maximum deflection $x_{1\,\mathrm{max}}$ can be obtained in a similar manner.

A more sophisticated approach regarding the computation of $x_{2\,\mathrm{max}}$ is to maximize Eq. 5-21. This procedure is usually complicated, but, if needed, it can be carried out graphically. For the problem above, for example, the magnification factors for the time intervals $t < t_d$ and $t > t_d$ are given in Table 2-1. The summation in Eq. 5-21 consists of two modal components. Each component can be individually plotted with respect to time, and from these two plots the time of maximum response can be approximately evaluated by inspection. From Eq. 5-21,

$$x_2(t) = Y_{1\,\mathrm{st}}\beta_{21}(\Gamma)_1 + Y_{2\,\mathrm{st}}\beta_{22}(\Gamma)_2$$

$$= (0.527)(1.125)(\Gamma)_1 + (-0.000152)(-1.415)(\Gamma)_2$$

or

$$x_2(t) = 0.593(\Gamma)_1 + 0.000217(\Gamma)_2 \tag{5-27}$$

The expression for $(\Gamma)_1$ can be obtained from Table 2-1 with $\omega = \omega_1 = 5.10$ rps. In a similar manner $(\Gamma)_2$ can be determined. Each term of Eq. 5-27 is then separately plotted with respect to time, and from these graphs the time of maximum response is approximately evaluated.

The shear forces, moments, and stresses are evaluated by utilizing the relative displacements of the ends of the springs. These displacements can be found in a manner similar to the one used for $x_{2\max}$. For example, by considering the upper story, we find the characteristic amplitude $\beta_{\delta 2p}$ for each mode p to be

$$\beta_{\delta 21} = \beta_{21} - \beta_{11} = 1.125 - 1.00 = 0.125$$

$$\beta_{\delta 22} = \beta_{22} - \beta_{12} = -1.415 - 1.00 = -2.415$$

The maximum relative story displacement in each mode is obtained from the expression

$$\delta_{2p_{\max}} = Y_{p_{st}} \beta_{\delta 2p} (\Gamma)_{p_{\max}} \tag{5-28}$$

Thus, for each mode,

$$\delta_{21\max} = (0.527)(0.125)(0.72)$$

$$= 0.0472 \text{ in.}$$

$$\delta_{22\max} = (0.000153)(-2.415)(1.56)$$

$$= -0.00057 \text{ in.}$$

Again, the contribution of the first mode predominates. The numerical sum of the two displacements yields the total maximum relative displacement $\delta_{2\max}$ of point 2. That is,

$$\delta_{2\max} = \delta_{21\max} + \delta_{22\max}$$

$$= 0.0472 + 0.0006 = 0.0478 \text{ in.}$$

The acceleration impulse extrapolation method, Table 3-3, gives

$$\delta_{2\max} = x_{2\max} - x_{1\max}$$

$$= 0.4308 - 0.3838 = 0.0470 \text{ in.}$$

The results of both methods are nearly identical. The numerical sum of the modal components was used for $\delta_{2\max}$. If the loads $F_i(t)$ had not diminished to zero, the algebraic sum of these components would be more appropriate.

The maximum shear force $V_{2\max}$ at the top of each column of the second story is

$$V_{2\max} = \frac{k_2 \delta_2}{2} = \frac{(32.10)(0.0470)}{2} = 0.75 \text{ kips}$$

The maximum moment $M_{2\max}$ in the same column is

$$M_{2\,max} = \frac{V_2 L_2}{2} = \frac{(0.75)(12)(12)}{2} = 54.0 \text{ kip-in.}$$

and the maximum stress $\sigma_{2\,max}$ is

$$\sigma_{2\,max} = \frac{M_{2\,max}}{S} = \frac{54.0}{26.4} = 2.05 \text{ ksi}$$

The stresses at the tops of the columns of the first story can be obtained in a similar manner. In fact, these values are much higher.

By using the numerical sum of the modal components, one is always on the conservative side. In cases where the response of the first mode does not predominate, it might be required to maximize Eq. 5-21 as discussed earlier.

5-4 MODAL EQUATIONS FOR INFINITE DEGREE OF FREEDOM SYSTEMS

In Section 5-2, the modal equations for spring-mass systems were derived by making use of Lagrange's equation. In this section the method of modal analysis is extended to include systems with infinite degrees of freedom. Thus, by following the procedure of Section 5-2, an equation analogous to Eq. 5-15 is written as

$$\ddot{Y}_p + \omega_p^2 Y_p = g(t) \frac{\int_0^L q_1(x)\beta_p(x)\,dx}{m\int_0^L \beta_p^2(x)\,dx} \tag{5-29}$$

Equation 5-29 is the modal equation for a p mode and applies to any beam span that is acted upon by distributed dynamic loads. See for example Fig. 1-4a in Section 1-3. In the expression above, $q(t,x) = g(t)q_1(x)$ is the distributed dynamic load with $q_1(x)$ representing its distribution along the length of the span and $g(t)$ its variation with time. The length of the span is L, Y_p is the modal amplitude, m is its uniform mass per unit of length, $\beta_p(x)$ is the mode shape expression of the p mode, and ω_p is the corresponding natural frequency. The derivation of Eq. 5-29 is based on the assumption that there is no damping and that EI is constant.

If the external loading on the beam consists of concentrated time-varying forces $F_1(t), F_2(t), \ldots, F_r(t)$, the numerator on the right-hand side of Eq. 5-29 becomes a summation involving one term for each concentrated load. Thus, if the time function $g(t)$ is the same for all concentrated loads, Eq. 5-29 becomes

$$\ddot{Y}_p + \omega_p^2 Y_p = g(t) \frac{\sum_{i=1}^{r} F_i \beta_p(x_i)}{m \int_0^L \beta_p^2(x)\, dx} \tag{5-30}$$

where $\beta_p(x_i)$ is the amplitude of the p mode under the force F_i.

If the beam in Fig. 5-2 is subjected to a constant force F_1 moving along the length of the beam with a constant velocity v, the modal equation for the p mode is

$$\ddot{Y}_p + \omega_p^2 Y_p = \frac{F_1 \beta_p(vt)}{m \int_0^L \beta_p^2(x)\, dx} \tag{5-31}$$

Here the time t is zero when x is zero. This means that t is measured from the time at which F_1 is at the left support of the beam and starting to move to the right. On this basis, the distance a in Fig. 5-2 is equal to vt.

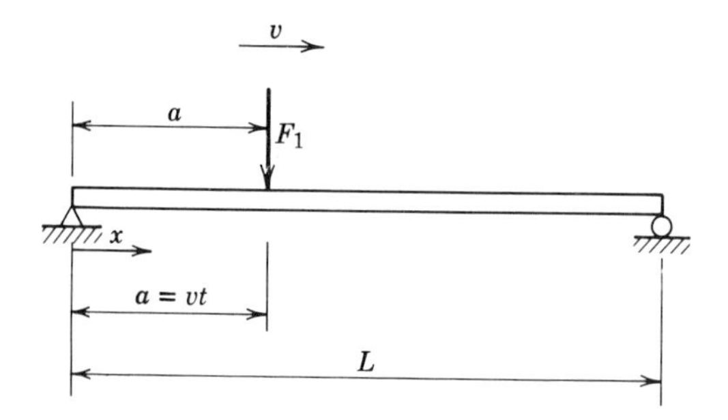

Fig. 5-2

The solution of these equations requires the computation of the mode shapes of the elastic system. This subject is discussed in Chapter 4. It should be noted, however, that such computations become very tedious for the more complicated problems, and the approximate methods discussed later in this book would be more appropriate.

The modal equations, as in Section 5-2, represent equivalent one-degree systems that can be solved to determine the dynamic response in each mode. The procedures discussed in Chapter 2 can be used for this purpose. Thus, for the case of distributed dynamic loads, expressions analgous to those in Section 5-2 are written as follows:

$$Y_{p_{st}} = \frac{\int_0^L q_1(x)\beta_p(x)\,dx}{\omega_p^2 m \int_0^L \beta_p^2(x)\,dx} \tag{5-32}$$

$$Y_p(t) = (\Gamma)_p Y_{p_{st}} \tag{5-33}$$

$$y(t,x) = \sum_{p=1}^{\infty} Y_p(t)\beta_p(x) \tag{5-34}$$

The summation in Eq. 5-34 represents superposition of the responses of the infinite modes. In practice, however, a reasonable solution can be obtained by using only the first few modes of the system.

If the external dynamic loads are concentrated, the expressions become

$$Y_{p_{st}} = \frac{\sum_{i=1}^{r} F_i\,\beta_p(x_i)}{\omega_p^2 m \int_0^L \beta_p^2(x)\,dx} \tag{5-35}$$

$$Y_p(t) = (\Gamma)_p Y_{p_{st}} \tag{5-36}$$

$$y(t,x) = \sum_{p=1}^{\infty} Y_p(t)\beta_p(x) \tag{5-37}$$

Applications of these equations are given in the next two sections.

5-5 DYNAMIC RESPONSE OF BEAMS

The modal equations of the preceding section are applied here to determine the dynamic response of simple beam problems. As a first application, let it be assumed that it is required to determine the dynamic response of the uniform simply supported beam loaded as shown in Fig. 4-3*a*. The mode shapes of the member are given by Eq. 1-151:

$$\beta_p(x) = \sin\frac{p\pi x}{L} \qquad p = 1, 2, 3, \ldots$$

Thus Eq. 5-32 yields

$$Y_{p_{st}} = \frac{\int_0^L q \sin(p\pi x/L)\,dx}{\omega_p^2 m \int_0^L \sin^2(p\pi x/L)\,dx}$$

$$= -\frac{2q}{p\pi\omega_p^2 m}(\cos p\pi - 1)$$

This equation shows that when $p=2, 4, 6, \ldots$, the static deflection $Y_{p_{st}}$ is zero. Therefore, the only modes that count for dynamic response are the odd ones. Thus

$$Y_{p_{st}} = \frac{4q}{p\pi\omega_p^2 m} \qquad p=1, 3, 5, \ldots \tag{5-38}$$

The magnification factor $(\Gamma)_p$ for the p mode can be determined[16] from Eq. 2-9 by setting $y_0 = \dot{y}_0 = 0$ and $F/k = y_{st}$. This factor, defined as the ratio $y(t)/y_{st}$, is

$$(\Gamma)_p = \frac{1}{1-(\omega_f/\omega_p)^2}\left(\sin\omega_f t - \frac{\omega_f}{\omega_p}\sin\omega_p t\right) \qquad p=1, 3, 5, \ldots \tag{5-39}$$

where ω_f is the frequency of the force and ω_p are the natural frequencies of the simply supported beam. Substituting Eqs. 5-38 and 5-39 into Eq. 5-33, the modal response of the p mode is

$$Y_p(t) = \frac{4q}{p\pi\omega_p^2 m[1-(\omega_f/\omega_p)^2]}\left(\sin\omega_f t - \frac{\omega_f}{\omega_p}\sin\omega_p t\right) \tag{5-40}$$

Thus, by Eq. 5-34, the total response of the beam is

$$y(t,x) = \frac{4q}{m\pi}\sum^{P}\frac{1}{p\omega_p^2[1-(\omega_f/\omega_p)^2]}\left(\sin\omega_f t - \frac{\omega_f}{\omega_p}\sin\omega_p t\right)\sin\frac{p\pi x}{L}$$

$$p=1, 3, 5, \ldots \tag{5-41}$$

This equation shows that when the frequency ω_f is equal to one of the odd-numbered natural frequencies of the beam, its deflections $y(t,x)$ become infinitely large. It is interesting to note that ω_p^2 appears in the denominator of Eq. 5-41. This indicates that the contribution of the higher modes to the total response of the system is rather small. Thus an accurate solution could be obtained by considering the responses of the first few odd-numbered modes of the member. When $y(t,x)$ is known, the bending moments and

[16] The magnification factor is also given in Table 2-1.

shear forces at any t and x can be determined by applying Eqs. 4-80 and 4-81, respectively. The bending and shear stresses can be computed from known formulas of strength of materials.

As a second example, let it be assumed that the uniform simply supported beam in Fig. 4-3a is acted upon by two concentrated dynamic forces $F_1 \sin \omega_f t$ and $F_2 \sin \omega_f t$ located at $x = L/4$ and $x = L/2$, respectively. The mode shapes of the beam are given by Eq. 1-151. Hence Eq. 5-35 yields

$$Y_{p\text{st}} = \frac{F_1 \sin(p\pi/4) + F_2 \sin(p\pi/2)}{\omega_p^2 m \int_0^L \sin^2(p\pi x/L)\, dx}$$

$$= \frac{2[F_1 \sin(p\pi/4) + F_2 \sin(p\pi/2)]}{\omega_p^2 mL} \qquad (5\text{-}42)$$

Here, the first term in the numerator becomes equal to zero when $p = 4, 8, 12, \ldots$, and the second term becomes zero when $p = 2, 4, 6, \ldots$.

The magnification factor $(\Gamma)_p$ is given by Eq. 5-39 with $p = 1, 2, 3, \ldots$, because only certain terms of Eq. 5-42 go to zero for certain values of p. By using Eqs. 5-39 and 5-42, Eq. 5-36, yields

$$Y_p(t) = \frac{2[F_1 \sin(p\pi/4) + F_2 \sin(p\pi/2)]}{\omega_p^2 mL[1 - (\omega_f/\omega_p)^2]} \left(\sin \omega_f t - \frac{\omega_f}{\omega_p} \sin \omega_p t \right) \qquad (5\text{-}43)$$

Thus, by Eq. 5-37, the total response of the member is

$$y(t,x) = \frac{2}{mL} \sum^{P} \frac{[F_1 \sin(p\pi/4) + F_2 \sin(p\pi/2)]}{\omega_p^2 [1 - (\omega_f/\omega_p)^2]} \times \left(\sin \omega_f t - \frac{\omega_f}{\omega_p} \sin \omega_p t \right) \sin \frac{p\pi x}{L} \qquad p = 1, 2, 3, \ldots \qquad (5\text{-}44)$$

It should be noted again that when $\omega_f = \omega_p$, the deflections $y(t,x)$ become infinite. Since ω_p^2 is in the denominator of Eq. 5-44, the contribution of the higher modes are rather small, and an accurate solution can be obtained by considering only the first few modes.

5-6 MOVING LOADS

The method of modal analysis can be used to determine the dynamic response of beams acted upon by moving loads. The case to be considered here is a constant force F moving with constant velocity along the length of the beam in Fig. 5-2. At $t=0$ the force F is assumed to be at the left support and moving to the right. The mode shapes of the member are given by the expression

$$\beta_p(x)=\sin\frac{p\pi x}{L} \qquad p=1,2,3,\ldots \tag{5-45}$$

Thus, by Eq. 5-31,

$$\ddot{Y}_p+\omega_p^2 Y_p=\frac{F\sin(p\pi vt/L)}{m\int_0^L \sin^2(p\pi x/L)\,dx}$$

or

$$\ddot{Y}_p+\omega_p^2 Y_p=\frac{2F}{mL}\sin(p\pi vt/L) \tag{5-46}$$

The above equation shows that the time function is $\sin\Omega_p t$, where Ω_p is taken equal to $p\pi v/L$. Thus the magnification factor $(\Gamma)_p$ is given by Eq. 5-39 with $\omega_f=\Omega_p$ and $p=1, 2, 3,\ldots$. The modal static deflection $Y_{p\text{st}}$ is,

$$Y_{p\text{st}}=\frac{2F}{\omega_p^2 mL} \tag{5-47}$$

and the modal response $Y_p(t)$ is

$$Y_p(t)=(\Gamma)_p Y_{p\text{st}}$$

$$=\frac{2F}{mL}\frac{1}{\omega_p^2[1-(\Omega_p/\omega_p)^2]}\left(\sin\Omega_p t-\frac{\Omega_p}{\omega_p}\sin\omega_p t\right) \tag{5-48}$$

The total response is given by Eq. 5-37. Thus, by using Eqs. 5-37, 5-45, and 5-48,

$$y(t,x)=\frac{2F}{mL}\sum^{P}\frac{1}{\omega_p^2[1-(\Omega_p/\omega_p)^2]}\left(\sin\Omega_p t-\frac{\Omega_p}{\omega_p}\sin\omega_p t\right)\sin\frac{p\pi x}{L}$$

$$p=1,2,3,\ldots \tag{5-49}$$

Again, the contribution of the higher modes is rather small, and the displacements $y(t,x)$ become infinite when Ω_p becomes equal to ω_p. In practice, the phenomenon of resonance is unlikely to occur, because a very high velocity v is needed for Ω_p to obtain the value of ω_p.

The solution of other problems can be carried out in a similar manner.

5-7 MODAL EQUATION FOR SIMPLY SUPPORTED THIN PLATES

The modal equation for a thin rectangular plate simply supported on all edges and acted upon by a uniformly distributed dynamic load $q(t)$ is derived in this section. The thickness h and mass m of the plate are uniform and the x, y, z axes are taken as shown in Fig. 5-3.

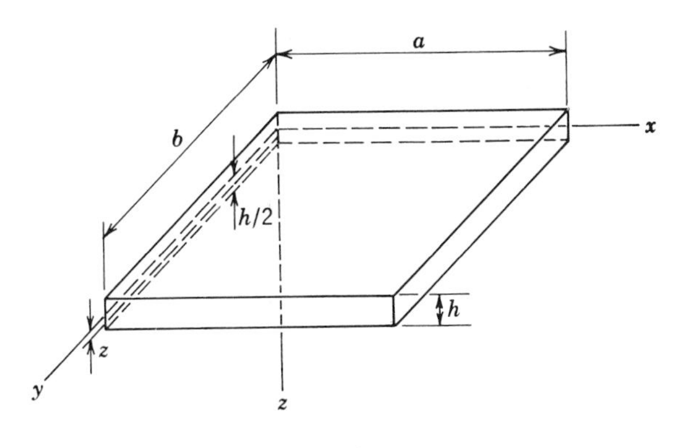

Fig. 5-3

By following the theory of thin plates, the dynamic deflection w of the plate in the z direction can be taken as

$$w = \sum_{n=1}^{\infty} \sum_{p=1}^{\infty} A_{np} \sin \frac{n\pi x}{a} \sin \frac{p\pi y}{b} \tag{5-50}$$

This equation satisfies the boundary conditions of zero displacements and zero moments at all four edges. The symbol A_{np} characterizes the modal displacement at the center of the plate.

The modal equation for a mode of vibration is obtained by using Lagrange's equation derived in Section 1-4. By neglecting damping and noting that $\partial T / \partial A_{np} = 0$, this equation can be written as

$$\frac{d}{dt}\left(\frac{\partial T}{\partial \dot{A}_{np}} \right) + \frac{\partial U}{\partial A_{np}} = \frac{\partial W_e}{\partial A_{np}} \tag{5-51}$$

Each possible combination of the integer values of n and p in Eq. 5-50 characterizes a modal shape. Here, the general shape

$$w = A_{np} \sin \frac{n\pi x}{a} \sin \frac{p\pi y}{b} \tag{5-52}$$

is considered.

The kinetic energy dT of any element of the plate is equal to $m\dot{w}^2\, dx\, dy/2$, where m is the mass per unit area of the plate. Thus the total kinetic energy T is

$$T = \frac{m}{2}\int_0^a \int_0^b \left(\dot{A}_{np} \sin\frac{n\pi x}{a} \sin\frac{p\pi y}{b}\right) dx\, dy$$

$$= \frac{mab}{8}\dot{A}_{np}$$

and

$$\frac{d}{dt}\left(\frac{\partial T}{\partial \dot{A}_{np}}\right) = \frac{mab}{4}\ddot{A}_{np} \qquad (5\text{-}53)$$

From the theory of plates, the total strain energy U is given by the expression

$$U = \frac{Eh^3}{24(1-\nu^2)}\int_0^a \int_0^b \left[\left(\frac{\partial^2 w}{\partial x^2}\right) + \left(\frac{\partial^2 w}{\partial y^2}\right) + 2\nu \frac{\partial^2 w}{\partial x^2}\frac{\partial^2 w}{\partial y^2} + 2(1-\nu)\left(\frac{\partial^2 w}{\partial x \partial y}\right)^2\right] dx\, dy$$

where E is the modulus of elasticity and ν is the Poisson ratio. By substituting Eq. 5-52 into the equation above and performing the required integrations, we have

$$U = \frac{Eh^3 ab\pi^4 A_{np}{}^2}{96(1-\nu^2)}\left[\frac{n^2}{a^2} + \frac{p^2}{b^2}\right]^2$$

Thus

$$\frac{\partial U}{\partial A_{np}} = \frac{Eh^3 ab\pi^4}{48(1-\nu^2)} A_{np}\left[\frac{n^2}{a^2} + \frac{p^2}{b^2}\right]^2 \qquad (5\text{-}54)$$

The work W_e by the external uniformly distributed dynamic load $q(t)$ is

$$W_e = q(t)\int_0^a \int_0^b A_{np} \sin\frac{n\pi x}{a} \sin\frac{p\pi y}{b}\, dx\, dy$$

$$= q(t)\frac{ab}{np\pi^2} A_{np}[\cos n\pi - 1][\cos p\pi - 1]$$

and

$$\frac{\partial W_e}{\partial A_{np}} = q(t)\frac{ab}{np\pi^2}[\cos n\pi - 1][\cos p\pi - 1] \tag{5-55}$$

By substituting Eqs. 5-53, 5-54, and 5-55 into Eq. 5-51, we obtain

$$\frac{abm}{4}\ddot{A}_{np} + \frac{Eh^3ab\pi^4}{48(1-\nu^2)}\left[\frac{n^2}{a^2}+\frac{p^2}{b^2}\right]A_{np}$$
$$= q(t)\frac{ab}{np\pi^2}[\cos n\pi - 1][\cos p\pi - 1]$$

or

$$\ddot{A}_{np} + \frac{Eh^3\pi^4}{12(1-\nu^2)m}\left[\frac{n^2}{a^2}+\frac{p^2}{b^2}\right]^2 A_{np}$$
$$= q(t)\frac{4}{npm\pi^2}[\cos n\pi - 1][\cos p\pi - 1] \tag{5-56}$$

This expression is the modal equation for a mode of vibration of the plate and can be used to determine the dynamic response of the plate corresponding to this mode. From it one can easily note that the natural frequencies ω_i of the plate are given by the expression

$$\omega_i = \pi^2\left[\frac{n^2}{a^2}+\frac{p^2}{b^2}\right]\left[\frac{Eh^3}{12(1-\nu^2)m}\right]^{\frac{1}{2}} \tag{5-57}$$

or

$$\omega_i = \pi^2\left[\frac{n^2}{a^2}+\frac{p^2}{b^2}\right]\sqrt{D/m} \tag{5-58}$$

where

$$D = \frac{Eh^3}{12(1-\nu^2)} \tag{5-59}$$

The various combinations of the integer values of n and p in Eq. 5-58 will yield the natural frequencies ω_1, ω_2, $\omega_3, \ldots$ of the plate. For the first mode $n = p = 1$. Thus Eqs. 5-56 and 5-58 yield

$$\ddot{A}_{11} + \omega_1^2 A_{11} = q(t)\frac{16}{m\pi^2} \tag{5-60}$$

and

$$\omega_1 = \pi^2 \left[\frac{1}{a^2} + \frac{1}{b^2} \right] \sqrt{D/m} \tag{5-61}$$

5-8 DYNAMIC RESPONSE OF SIMPLY SUPPORTED PLATES

The modal equation for rectangular simply supported plates is derived in the preceding section and is given by Eq. 5-56. The uniformly distributed dynamic load $q(t)$ acting on the plate can be written as $q(t) = qg(t)$ where $g(t)$ is its time variation. For a mode of vibration, Eq. 5-56 shows that the modal static displacement $(A_{np})_{st}$ is

$$(A_{np})_{st} = \frac{\dfrac{4q[\cos n\pi - 1][\cos p\pi - 1]}{npm\pi^2}}{\dfrac{Eh^3\pi^4}{12(1-\nu^2)m}\left[\dfrac{n^2}{a^2} + \dfrac{p^2}{b^2}\right]^2}$$

$$= \frac{48q[\cos n\pi - 1][\cos p\pi - 1](1-\nu^2)}{Eh^3 np\pi^6 \left[\dfrac{n^2}{a^2} + \dfrac{p^2}{b^2}\right]^2} \tag{5-62}$$

The dynamic modal displacement A_{np} is

$$A_{np} = (A_{np})_{st}\Gamma_{np} \tag{5-63}$$

where Γ_{np} is the magnification factor for the given time function in the np mode. For example, the static displacement $(A_{11})_{st}$ for the first mode is

$$(A_{11})_{st} = \frac{192q(1-\nu^2)}{Eh^3\pi^6\left(\dfrac{1}{a^2} + \dfrac{1}{b^2}\right)^2} \tag{5-64}$$

and the dynamic displacement A_{11} for the same mode is

$$A_{11} = (A_{11})_{st}\Gamma_{11} \tag{5-65}$$

Equations 5-64 and 5-65 are derived from Eqs. 5-62 and 5-63, respectively, by setting $n = p = 1$.

The dynamic displacements w of the plate corresponding to a mode np can be found from Eq. 5-52. Thus for a mode np,

$$w = A_{np} \sin \frac{n\pi x}{a} \sin \frac{p\pi y}{b}$$

$$= (A_{np})_{st} \Gamma_{np} \sin \frac{n\pi x}{a} \sin \frac{p\pi y}{b} \qquad (5\text{-}66)$$

The total response can be obtained by superimposing the responses of all modes. Usually the first few modes are sufficient for practical purposes.

With known displacements, the bending stresses σ_x and σ_y in the x and y directions of the plate, respectively, can be obtained from known formulas of the theory of plates. That is,

$$\sigma_x = -\frac{Ez}{1-\nu^2}\left(\frac{\partial^2 w}{\partial x^2} + \nu \frac{\partial^2 w}{\partial y^2}\right) \qquad (5\text{-}67)$$

$$\sigma_y = -\frac{Ez}{1-\nu^2}\left(\frac{\partial^2 w}{\partial y^2} + \nu \frac{\partial^2 w}{\partial x^2}\right) \qquad (5\text{-}68)$$

where z is the distance from the median plane of the plate shown in Fig. 5-3. The maximum stress will occur at the location $z = h/2$.

As an illustration, let it be assumed that it is required to determine the maximum dynamic bending stress in a flat rectangular simply supported steel plate whose sides a and b are 90 and 60 in., respectively. The thickness h is 2.0 in., the Poisson ratio ν is 0.25, and the modulus of elasticity E is 30×10^6 psi. The mass m per square inch of area of the plate is the weight of this area divided by the acceleration of gravity g, and is equal to 0.000124 lb-sec^2/in.^{3}. The plate is subjected to a uniform dynamic pressure $p(t)$ of 60.0 psi that is suddenly applied at time $t = 0$ and removed suddenly at $t = t_d = 0.03$ sec. It is required to determine the dynamic response of the plate by considering only the first mode of vibration.

The maximum displacement w of the plate is at the center. By Eq. 5-64,

$$(A_{11})_{st} = \frac{(192)(60)(1-0.25^2)}{(30)(10^6)(2^3)\pi^6(1/90^2 + 1/60^2)^2}$$

$$= 0.291 \text{ in.}$$

By Eq. 5-61, the frequency ω_1 of the first mode is

$$\omega_1 = \pi^2\left[\frac{1}{90^2} + \frac{1}{60^2}\right]\left[\frac{(30)(10^6)(8)}{12(1-0.25^2)(0.000124)}\right]^{\frac{1}{2}}$$

$$= 166\pi^2$$

and the period of vibration τ_1 is

$$\tau_1 = \frac{2\pi}{\omega_1} = \frac{1}{83\pi} = 0.00384 \text{ sec}$$

The maximum magnification factor $(\Gamma_{11})_{\max}$ can be obtained from Fig. 2-6*a*. Thus with

$$\frac{t_d}{\tau_1} = \frac{0.03}{0.00384} = 7.8$$

Fig. 2-6*a* yields

$$(\Gamma_{11})_{\max} = 2.0$$

Therefore, the maximum dynamic deflection at the center of the plate is

$$w_{\max} = (A_{11})_{\max} = (A_{11})_{\text{st}}(\Gamma_{11})_{\max}$$
$$= (0.291)(2.0) = 0.582 \text{ in.}$$

The maximum dynamic bending stresses can be determined from Eqs. 5-67 and 5-68. Thus with $z = h/2$, $x = a/2$, $y = b/2$, and

$$\frac{\partial^2 w}{\partial x^2} = -(A_{11})_{\max} \frac{\pi^2}{a^2}$$

$$\frac{\partial^2 w}{\partial y^2} = -(A_{11})_{\max} \frac{\pi^2}{b^2}$$

the maximum stresses $(\sigma_x)_{\max}$ and $(\sigma_y)_{\max}$ are as follows:

$$(\sigma_x)_{\max} = (A_{11})_{\max} \frac{Eh\pi^2}{2(1-\nu^2)} \left(\frac{1}{a^2} + \frac{\nu}{b^2} \right)$$
$$= 35{,}200 \text{ psi}$$

$$(\sigma_y)_{\max} = (A_{11})_{\max} \frac{Eh\pi^2}{2(1-\nu^2)} \left(\frac{1}{b^2} + \frac{\nu}{a^2} \right)$$
$$= 56{,}500 \text{ psi}$$

These values show that the maximum bending stress is 56,500 psi, and it is acting in the y direction at the center of the plate. Other plate problems or slabs can be solved in a similar manner.

5-9 DYNAMIC RESPONSE OF LUMPED PARAMETER SYSTEMS

The dynamic response of beamlike structures can be determined with reasonable accuracy by lumping the continuous mass and dynamic load at discrete points along the length of the member. In this manner, approximate methods of analysis can be used to determine the natural frequencies and mode shapes that are required in modal analysis in order to solve the more complex problems.

Consider for example the beam in Fig. 5-4*a* and let it be assumed that it is acted upon by the distributed dynamic force $q(t, x) = q(x)g(t)$ as shown. In order to be somewhat general, both the beam stiffness and load function $q(x)$ are assumed to be variable along its length. A reasonable solution to this problem can be obtained by dividing the length of the member into a sufficient number of segments ΔL as shown in Fig. 5-4*b*. The weight of the

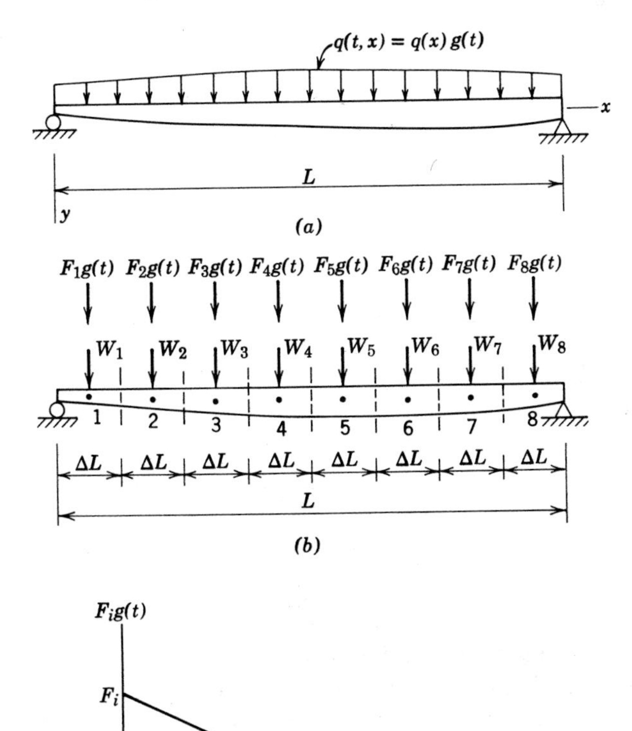

Fig. 5-4

member corresponding to each segment as well as the distributed force $q(x)$ within ΔL are concentrated at the center of each segment as shown. The lumped weights are designated as W_1, W_2, $W_3, \ldots, W_8$, and the concentrated dynamic forces are designated as $F_1(t)$, $F_2(t)$, $F_3(t), \ldots, F_8(t)$. Under this condition, the beam is assumed to be weightless between weight concentration points, but its elastic properties are retained. In this manner, the degrees of freedom of the member are equal to the number of the concentrated weights. The time variation $g(t)$ of the concentrated dynamic forces is the same as that of $g(t,x)$.

The required natural frequencies of vibration and the corresponding mode shapes can be determined by using either the method of Stodola and the iteration procedure discussed in Sections 5-10, 5-11, and 5-12 or the methods discussed in Chapters 6 and 7.

When the required natural frequencies and mode shapes are computed, the dynamic response of the beam in a p mode can be determined by using the modal equation

$$\ddot{Y}_p + \omega_p^2 Y_p = g(t) \frac{\sum_{i=1}^{r} F_i \beta_{ip}}{\sum_{i=1}^{r} m_i \beta_{ip}^2} \tag{5-69}$$

In this equation, ω_p is the natural frequency of the p mode, F_i is the concentrated force at point i, $g(t)$ is the time variation of F_i, m_i is the weight W_i at i divided by the acceleration of gravity, and β_{ip} is the mode amplitude of m_i in the p mode.

The modal static displacement $Y_{p_{st}}$ is

$$Y_{p_{st}} = \frac{\sum_{i=1}^{r} F_i \beta_{ip}}{\omega_p^2 \sum_{i=1}^{r} m_i \beta_{ip}^2} \tag{5-70}$$

The maximum magnification factor $(\Gamma_p)_{max}$ in the p mode is that of a single-degree spring-mass system acted upon by the time-varying force $F_i g(t)$, and it can be obtained as discussed earlier in this chapter. For example, if the time variation $g(t)$ is as shown in Fig. 5-4*c*, $(\Gamma)_{max}$ can be computed from the appropriate expression in Table 2-1 with $\omega = \omega_p$ or by using the graph in Fig. 2-5*a*.

The maximum dynamic amplitude $(y_{ip})_{max}$ of a point i by considering only the p mode is

$$(y_{ip})_{max} = Y_{p_{st}} \beta_{ip} (\Gamma_p)_{max} \tag{5-71}$$

The total response $(y_i)_{max}$ of point i can be determined by superimposing the responses of all modes. Usually the first few modes are sufficient. Thus

$$(y_i)_{max} = \sum_{p=1}^{n} Y_{p_{st}} \beta_{ip} (\Gamma_p)_{max} \tag{5-72}$$

where n is the number of modes considered.

The maximum dynamic bending moment M_{ip} at a section i in the p mode can be obtained by determining the static moment at section i produced by the F_i loads, Fig. 5-4*b*, and multiplying it by $(\Gamma_p)_{max}$. Superposition of the responses of the considered modes will yield the total maximum bending moment at section i. With known moments, the dynamic bending stresses can be computed in the usual way.

A better approximation of the bending moment in a p mode would be to determine the static moment at section i due to the distributed load $q(x)$, Fig. 5-4*a* and to multiply it by $(\Gamma_p)_{max}$. Then superposition of the considered modal responses will yield the maximum bending moment at section i.

The accuracy of this method depends on the number of ΔL segments used in Fig. 5-4*b*. Usually, 5 to 10 segments will yield reasonable results.

If the member in Fig. 5-4*a* is acted upon by concentrated dynamic forces instead of the distributed force $q(t, x)$, the same procedure as above can be used. The modal equation is the same, except that the summation in the numerator, Eqs. 5-69 and 5-70 includes only the points i where the concentrated dynamic loads are acting.

5-10 STODOLA'S METHOD AND ITERATION PROCEDURE

It was pointed out in the preceding section that the dynamic response of beamlike structures can be determined with reasonable accuracy by lumping the continuous mass and distributed dynamic load at discrete points along the length of the member. A convenient method for determining the natural frequencies and mode shapes that are required to analyze these structures for dynamic response is the method of Stodola combined with an iteration procedure. This method can be applied to members of either uniform or variable EI, as well as spring-mass systems. In this section the application of this method is limited to beam spans of uniform[17] EI.

The method of Stodola is predicated on the idea that, at a principal mode

[17] Members of variable EI are discussed in Chapter 7.

of vibration of constant angular frequency ω, the vibrating system is acted upon by inertia forces $-m_i\ddot{y}_i = m_i\omega^2 y_i$, where y_i is the dynamic amplitude of mass m_i at position i. For each mass concentration point i, an equation of motion is written which expresses the amplitudes y_i in terms of flexibility coefficients a_{ij} and the inertia loads $m_i\omega^2 y_i$ causing the amplitudes y_i. This means that if the inertia forces of the mth mode are applied statically to the structure, the resulting deflection curve is the mth mode shape.

Consider the member in Fig. 5-5 and let it be assumed that it vibrates at a mode of vibration of frequency ω. The weight of the beam is assumed to be lumped at the points shown in the figure. If the beam, in addition to its weight, supports other weights securely attached to the member, they should be included in the analysis in the same way as the lumped weights. The equations of motion under the weight concentration points $1, 2, \ldots, n$, can be written by following the procedure used in Section 1-13. Thus, by using Eq. 1-157 with the P_i loads replaced by the inertia loads $m_i\omega^2 y_i$, the resulting set of equations of motion is

$$
\begin{aligned}
y_1 &= a_{11}(m_1\omega^2 y_1) + a_{12}(m_2\omega^2 y_2) + \cdots + a_{1n}(m_n\omega^2 y_n) \\
y_2 &= a_{21}(m_1\omega^2 y_1) + a_{22}(m_2\omega^2 y_2) + \cdots + a_{2n}(m_n\omega^2 y_n) \\
&\cdots\cdots\cdots\cdots\cdots\cdots\cdots\cdots\cdots \\
y_n &= a_{n1}(m_1\omega^2 y_1) + a_{n2}(m_2\omega^2 y_2) + \cdots + a_{nn}(m_n\omega^2 y_n)
\end{aligned}
\tag{5-73}
$$

In matrix form, the equations are written as

$$
\begin{bmatrix} y_1 \\ y_2 \\ \vdots \\ y_n \end{bmatrix} = \omega^2 \begin{bmatrix} a_{11}m_1 & a_{12}m_2 & \cdots & a_{1n}m_n \\ a_{21}m_1 & a_{22}m_2 & \cdots & a_{2n}m_n \\ \cdots & \cdots & \cdots & \cdots \\ a_{n1}m_1 & a_{n2}m_2 & \cdots & a_{nn}m_n \end{bmatrix} \begin{bmatrix} y_1 \\ y_2 \\ \vdots \\ y_n \end{bmatrix} \tag{5-74}
$$

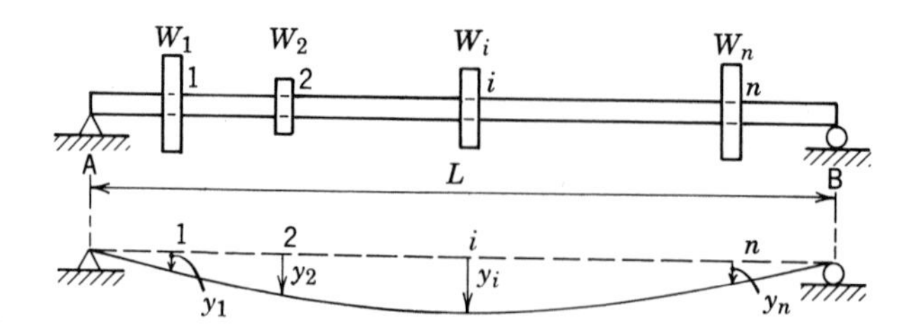

Fig. 5-5

or

$$\{y\} = \omega^2 [M] \{y\} \tag{5-75}$$

Returning to Eq. 5-73, dividing each term by ω^2, and setting $1/\omega^2 = \lambda$, this system of equations can be rewritten as

$$\begin{aligned}
(a_{11}m_1 - \lambda)\, y_1 + a_{12}m_2y_2 + \cdots + a_{1n}m_ny_n &= 0 \\
a_{21}m_1y_1 + (a_{22}m_2 - \lambda)\, y_2 + \cdots + a_{2n}m_ny_n &= 0 \\
\cdots\cdots\cdots\cdots\cdots\cdots\cdots\cdots \\
a_{n1}m_1y_1 + a_{n2}m_2y_2 + \cdots + (a_{nn}m_n - \lambda)\, y_n &= 0
\end{aligned} \tag{5-76}$$

For a nontrivial solution, the determinant of the coefficients in Eq. 5-76 must be zero. That is

$$\begin{vmatrix}
(a_{11}m_1 - \lambda) & a_{12}m_2 & \cdots & a_{1n}m_n \\
a_{21}m_1 & (a_{22}m_2 - \lambda) & \cdots & a_{2n}m_n \\
\cdots & \cdots & \cdots & \cdots \\
a_{n1}m_1 & a_{n2}m_2 & \cdots & (a_{nn}m_n - \lambda)
\end{vmatrix} = 0 \tag{5-77}$$

This determinant is known as the frequency determinant which upon expansion yields a frequency equation of the form

$$\lambda^n + c_1\lambda^{n-1} + c_2\lambda^{n-2} + \cdots + c_{n-1}\lambda + c_n = 0 \tag{5-78}$$

Since $\lambda = 1/\omega^2$, the n roots $\lambda_1, \lambda_2, \ldots, \lambda_n$ of the equation lead to the computation of the n frequencies $\omega_1, \omega_2, \ldots, \omega_n$ of the beam. For each value of λ, the corresponding mode shapes are computed from Eq. 5-76. Each frequency ω depends only on the shape of the corresponding mode. Therefore, the modes are usually normalized by setting one of the amplitudes $y_1, y_2, \ldots, y_n$ equal to unity.

For n greater than 3 the expansion of the determinant in Eq. 5-77 and consequently the solution of Eq. 5-78 become very laborious and require the use of an electronic digital computer. The solution may be simplified by using an iteration procedure. In this manner, a slide rule or a desk calculator will be sufficient for the solution of many complicated problems. If preferred, an electronic computer can be easily used for the iteration procedure, however.

In this section, the iteration procedure is applied to Eq. 5-74 to determine the fundamental frequency of vibration and the corresponding mode shape. Then, in Section 5-12, the higher frequencies of vibration and mode shapes are determined in a similar manner by making use of the orthogonality properties of the normal modes and the appropriate sets of equations of motion.

The iteration procedure is initiated by assuming arbitrarily that $y_1 = y_2 = \cdots = y_n = 1.0$ in the right-hand column of Eq. 5-74 and by performing the matrix multiplication. If preferred, other values for $y_1, y_2, \ldots, y_n$ can be used. The result of this multiplication is a column matrix that is normalized by taking as a common factor any one of its elements, preferably the smallest one. The elements of the normalized column matrix represent a better approximation of the amplitudes $y_1, y_2, \ldots, y_n$ of the fundamental mode. The procedure is repeated until the amplitudes $y_1, y_2, \ldots, y_n$ of the last trial are approximately equal to the corresponding ones obtained in the preceding trial. These amplitudes characterize the shape of the fundamental mode. The fundamental frequency ω is obtained by substituting the computed values of $y_1, y_2, \ldots, y_n$ into any one of the expressions in Eq. 5-73 and solving for ω.

The iteration procedure is rapidly convergent. Usually three to five repetitions will yield an excellent approximation of the fundamental frequency and mode shape. The mechanics of the method are best explained by an example.

Let it be assumed that it is required to determine the fundamental frequency of vibration and the corresponding mode shape of the simply supported beam in Fig. 5-6*a*. The beam supports four weights attached securely at the indicated locations. Its weight is assumed to be negligible and its constant stiffness EI is equal to 90×10^6 kip-in.^{2} The weight of the member, if required, can be taken into consideration by lumping it at discrete points along its length, as discussed in Section 5-9.

The flexibility coefficients a_{ij} are determined by applying unit loads at points 1, 2, 3, and 4 as shown in Figs. 5-6*b*, 5-6*c*, 5-6*d*, and 5-6*e*, respectively, and using known methods of strength of materials. They are as follows:

$$a_{11} = 1.43 \times 10^{-4} \qquad a_{13} = 4.13 \times 10^{-4}$$
$$a_{21} = 3.14 \times 10^{-4} \qquad a_{23} = 10.36 \times 10^{-4}$$
$$a_{31} = 4.13 \times 10^{-4} \qquad a_{33} = 15.35 \times 10^{-4}$$
$$a_{41} = 3.84 \times 10^{-4} \qquad a_{43} = 15.36 \times 10^{-4}$$

$$a_{12} = 3.14 \times 10^{-4} \qquad a_{14} = 3.84 \times 10^{-4}$$
$$a_{22} = 7.60 \times 10^{-4} \qquad a_{24} = 9.81 \times 10^{-4}$$
$$a_{32} = 10.36 \times 10^{-4} \qquad a_{34} = 15.36 \times 10^{-4}$$
$$a_{42} = 9.81 \times 10^{-4} \qquad a_{44} = 15.75 \times 10^{-4}$$

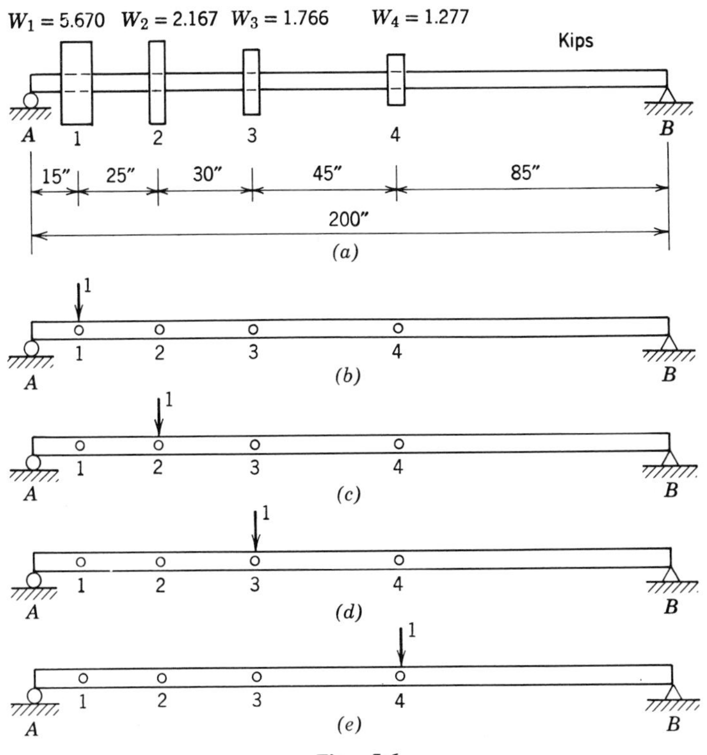

Fig. 5-6

It should be noted that $a_{ij} = a_{ji}$.

By utilizing Eq. 5-74,

$$\begin{bmatrix} y_1 \\ y_2 \\ y_3 \\ y_4 \end{bmatrix} = \frac{\omega^2}{10^4 g} \begin{bmatrix} (5.670)(1.43) & (2.167)(3.14) & (1.766)(4.13) & (1.277)(3.84) \\ (5.670)(3.14) & (2.167)(7.60) & (1.766)(10.36) & (1.277)(9.81) \\ (5.670)(4.13) & (2.167)(10.36) & (1.766)(15.35) & (1.277)(15.36) \\ (5.670)(3.84) & (2.167)(9.81) & (1.766)(15.36) & (1.277)(15.75) \end{bmatrix} \begin{bmatrix} y_1 \\ y_2 \\ y_3 \\ y_4 \end{bmatrix}$$

or

$$\begin{bmatrix} y_1 \\ y_2 \\ y_3 \\ y_4 \end{bmatrix} = \frac{4.91\omega^2}{10^4 g} \begin{bmatrix} 1.65 & 1.39 & 1.49 & 1.00 \\ 3.62 & 3.35 & 3.73 & 2.55 \\ 4.77 & 4.58 & 5.52 & 4.00 \\ 4.43 & 4.33 & 5.52 & 4.09 \end{bmatrix} \begin{bmatrix} y_1 \\ y_2 \\ y_3 \\ y_4 \end{bmatrix} \qquad (5\text{-}79)$$

where g is the acceleration of gravity.

The iteration procedure for Eq. 5-79 is initiated by assuming a shape $y_1 = y_2 = y_3 = y_4 = 1.00$. By using these values in the right-hand column of Eq. 5-79 and carrying out the matrix multiplication, we have

$$\begin{bmatrix} y_1 \\ y_2 \\ y_3 \\ y_4 \end{bmatrix} = \frac{4.91\omega^2}{10^4 g} \begin{bmatrix} 1.65 + 1.39 + 1.49 + 1.00 \\ 3.62 + 3.35 + 3.73 + 2.55 \\ 4.77 + 4.58 + 5.52 + 4.00 \\ 4.43 + 4.33 + 5.52 + 4.09 \end{bmatrix}$$

$$= \frac{4.91\omega^2}{10^4 g} \begin{bmatrix} 5.53 \\ 13.25 \\ 18.87 \\ 18.37 \end{bmatrix}$$

By selecting 5.53 as a common factor of the column matrix,

$$\begin{bmatrix} y_1 \\ y_2 \\ y_3 \\ y_4 \end{bmatrix} = \frac{(4.91)(5.53)\omega^2}{10^4 g} \begin{bmatrix} 1.00 \\ 2.39 \\ 3.41 \\ 3.32 \end{bmatrix}$$

The amplitudes $y_1=1.00$, $y_2=2.39$, $y_3=3.41$, and $y_4=3.32$ represent a better approximation of the fundamental mode shape. The procedure is now repeated by using these values in the right-hand column of Eq. 5-79 and performing again the matrix multiplication. That is,

$$\begin{bmatrix} y_1 \\ y_2 \\ y_3 \\ y_4 \end{bmatrix} = \frac{4.91\omega^2}{10^4 g} \begin{bmatrix} 1.65 & 1.39 & 1.49 & 1.00 \\ 3.62 & 3.35 & 3.73 & 2.55 \\ 4.77 & 4.58 & 5.52 & 4.00 \\ 4.43 & 4.33 & 5.52 & 4.09 \end{bmatrix} \begin{bmatrix} 1.00 \\ 2.39 \\ 3.41 \\ 3.32 \end{bmatrix}$$

$$= \frac{4.91\omega^2}{10^4 g} \begin{bmatrix} 1.65+\ 3.32+\ 5.08+\ 3.32 \\ 3.62+\ 8.00+12.70+\ 8.46 \\ 4.77+10.95+18.80+13.30 \\ 4.43+10.35+18.80+13.60 \end{bmatrix}$$

$$= \frac{4.91\omega^2}{10^4 g} \begin{bmatrix} 13.37 \\ 32.78 \\ 47.82 \\ 47.18 \end{bmatrix}$$

By selecting 13.37 as a common factor,

$$\begin{bmatrix} y_1 \\ y_2 \\ y_3 \\ y_4 \end{bmatrix} = \frac{(4.91)(13.37)\omega^2}{10^4 g} \begin{bmatrix} 1.00 \\ 2.45 \\ 3.58 \\ 3.53 \end{bmatrix}$$

One more repetition by using $y_1 = 1.00$, $y_2 = 2.45$, $y_3 = 3.58$, and $y_4 = 3.53$ in Eq. 5-79 yields

$$\begin{bmatrix} y_1 \\ y_2 \\ y_3 \\ y_4 \end{bmatrix} = \frac{4.91\omega^2}{10^4 g} \begin{bmatrix} 13.92 \\ 34.18 \\ 49.86 \\ 49.18 \end{bmatrix} = \frac{(4.91)(13.92)\omega^2}{10^4 g} \begin{bmatrix} 1.00 \\ 2.45 \\ 3.58 \\ 3.53 \end{bmatrix} \qquad (5\text{-}80)$$

It should be noted that the amplitudes of the last repetition are the same as the ones of the previous trial. Thus the correct shape of the fundamental mode is characterized by the magnitudes $y_1 = 1.00$, $y_2 = 2.45$, $y_3 = 3.58$, and $y_4 = 3.53$. The fundamental frequency ω is determined by using one of the expressions of Eq. 5-80. The expression for y_1 for example, with $y_1 = 1.00$, yields

$$1.00 = \frac{(4.91)(13.92)\omega^2}{10^4 g}(1.00)$$

or, by solving for ω,

$$\omega = \left[\frac{(10^4)(386)}{(4.91)(13.92)} \right]^{\frac{1}{2}} = 237.5 \text{ rps}$$

In hertz, the fundamental frequency f is

$$f = \frac{\omega}{2\pi} = \frac{237.5}{2\pi} = 37.80 \text{ Hz}$$

The solution of other problems can be obtained in a similar manner.

5-11 ITERATION PROCEDURE USING STIFFNESS COEFFICIENTS

The equations of motion for points $1, 2, \ldots, n$ in Fig. 5-5 can be also written in terms of stiffness coefficients k_{ij} by using the procedure described in Section 1-14. Thus, by using Eq. 1-162 with the P_i loads replaced by the inertia forces $m_i\omega^2 y_i$, the following equations are written:

$$
\begin{aligned}
m_1\omega^2 y_1 &= k_{11}y_1 + k_{12}y_2 + \cdots + k_{1n}y_n \\
m_2\omega^2 y_2 &= k_{21}y_1 + k_{22}y_2 + \cdots + k_{2n}y_n \\
&\cdots\cdots\cdots\cdots\cdots\cdots \\
m_n\omega^2 y_n &= k_{n1}y_1 + k_{n2}y_2 + \cdots + k_{nn}y_n
\end{aligned}
\tag{5-81}
$$

In matrix form, the equations are written as

$$
\begin{bmatrix} y_1 \\ y_2 \\ \vdots \\ y_n \end{bmatrix} = \frac{1}{\omega^2}
\begin{bmatrix}
\dfrac{k_{11}}{m_1} & \dfrac{k_{12}}{m_1} & \cdots & \dfrac{k_{1n}}{m_1} \\
\dfrac{k_{21}}{m_2} & \dfrac{k_{22}}{m_2} & \cdots & \dfrac{k_{2n}}{m_2} \\
\cdots & \cdots & \cdots & \cdots \\
\dfrac{k_{n1}}{m_n} & \dfrac{k_{n2}}{m_n} & \cdots & \dfrac{k_{nn}}{m_n}
\end{bmatrix}
\begin{bmatrix} y_1 \\ y_2 \\ \vdots \\ y_n \end{bmatrix}
\tag{5-82}
$$

or

$$
\{y\} = \frac{1}{\omega^2}[N]\{y\} \tag{5-83}
$$

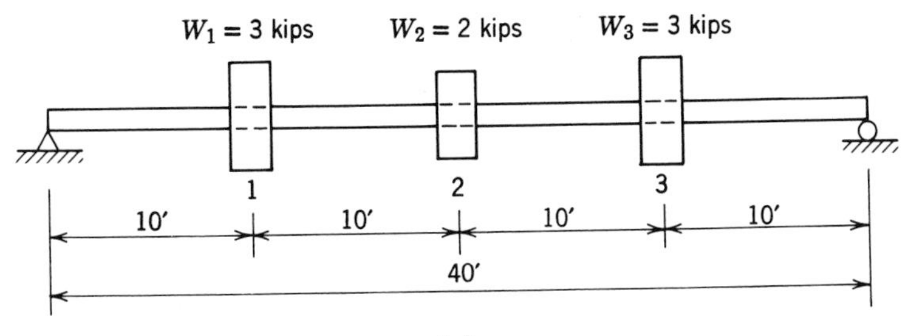

(a)

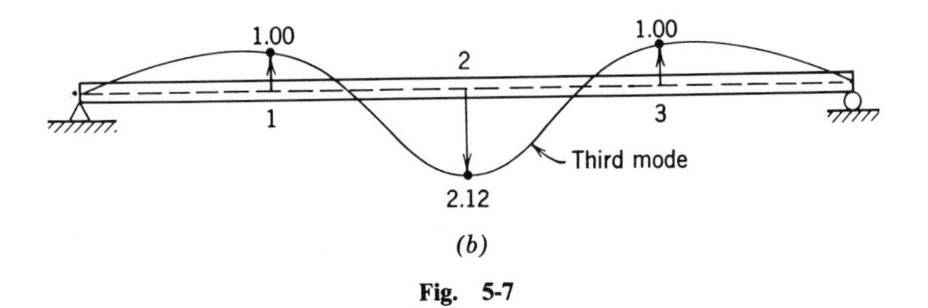

(b)

Fig. 5-7

The iteration procedure discussed in Section 5-10 can be again used here to solve Eq. 5-82. However, in this case the iteration of this equation will converge to the highest natural frequency and mode shape of the system. The lower frequencies and modes can be determined by using the procedure discussed in Section 5-12. The following example illustrates the application of the iteration procedure to determine the highest natural frequency and mode for a beam.

Consider the simply supported beam of negligible weight which supports three attached weights as shown in Fig. 5-7*a*. The stiffness EI of the member is constant and equal to 30×10^6 kip-in.^{2}. Its highest natural frequency and mode shape will be computed by using the iteration procedure to solve Eq. 5-82.

The stiffness coefficients for points 1, 2, and 3 are calculated in Section 1-15:

$$k_{11} = 171.0 \text{ kips/in.} \qquad k_{12} = -162.2 \text{ kips/in..}$$

$$k_{21} = -162.2 \text{ kips/in.} \qquad k_{22} = 235.0 \text{ kips/in.}$$

$$k_{31} = 66.6 \text{ kips/in.} \qquad k_{32} = -162.2 \text{ kips/in.}$$

$$k_{13} = 66.6 \text{ kips/in.}$$

$$k_{23} = -162.2 \text{ kips/in.}$$

$$k_{33} = 171.0 \text{ kips/in.}$$

Equation 5-82 yields

$$\begin{bmatrix} y_1 \\ y_2 \\ y_3 \end{bmatrix} = \frac{g}{\omega^2}\begin{bmatrix} 57.0 & -54.1 & 22.1 \\ -81.1 & 117.5 & -81.1 \\ 22.2 & -54.1 & 57.0 \end{bmatrix}\begin{bmatrix} y_1 \\ y_2 \\ y_3 \end{bmatrix} \tag{5-84}$$

The iteration procedure is initiated by assuming a trial shape $y_1 = y_3 = 1.00$ and $y_2 = -1.00$. By substituting these values into Eq. 5-84 and carrying out the matrix multiplication,

$$\begin{bmatrix} y_1 \\ y_2 \\ y_3 \end{bmatrix} = \frac{g}{\omega^2} \begin{bmatrix} 57.0 + 54.1 + 22.2 \\ -81.1 - 117.5 - 81.1 \\ 22.2 + 54.1 + 57.0 \end{bmatrix}$$

$$= \frac{g}{\omega^2} \begin{bmatrix} 133.3 \\ -297.7 \\ 133.3 \end{bmatrix} = \frac{133.3\,g}{\omega^2} \begin{bmatrix} 1.00 \\ -2.09 \\ 1.00 \end{bmatrix}$$

The amplitudes $y_1 = y_3 = 1.00$ and $y_2 = -2.09$ give a better approximation of the shape of the highest mode. By using these values in Eq. 5-84 and repeating the procedure,

$$\begin{bmatrix} y_1 \\ y_2 \\ y_3 \end{bmatrix} = \frac{g}{\omega^2} \begin{bmatrix} 192.2 \\ -408.2 \\ 192.2 \end{bmatrix} = \frac{192.2\,g}{\omega^2} \begin{bmatrix} 1.00 \\ -2.12 \\ 1.00 \end{bmatrix}$$

By using $y_1 = y_3 = 1.00$ and $y_2 = -2.12$ in Eq. 5-84,

$$\begin{bmatrix} y_1 \\ y_2 \\ y_3 \end{bmatrix} = \frac{g}{\omega^2} \begin{bmatrix} 193.7 \\ -412.2 \\ 193.7 \end{bmatrix} = \frac{193.7\,g}{\omega^2} \begin{bmatrix} 1.00 \\ -2.12 \\ 1.00 \end{bmatrix} \qquad (5\text{-}85)$$

These amplitudes are the same as those obtained in the previous trial. Thus the correct shape of the highest mode is characterized by the magnitudes $y_1 = y_3 = 1.00$ and $y_2 = -2.12$.

Since the beam is assumed to be of negligible weight and there are only three concentrated weights attached to the beam, the system has three degrees of freedom and thus three natural frequencies of vibration. The highest frequency, designated as ω_3, is determined by using one of the expressions in Eq. 5-85. The expression for y_1, with $y_1 = 1.00$, yields

$$1.00 = \frac{193.7\,g}{\omega_3^2}(1.00)$$

or

$$\omega_3 = \sqrt{(193.7)(386)} = 273.5 \text{ rps}$$

In hertz,

$$f_3 = \frac{\omega_3}{2\pi} = 43.5 \text{ Hz}$$

The mode shape corresponding to this frequency is shown in Fig. 5-7*b*. Other problems can be solved in a similar manner.

It should be pointed out, however, that the analysis of a structural system for dynamic response is usually sufficient by considering only the first few modes. Therefore, stiffness coefficients are usually computed for cases, such as frames, where flexibility coefficients are more difficult to determine. If the inverse of the stiffness matrix is used, the iteration procedure will yield first the fundamental frequency of vibration and the corresponding mode.

5-12 HIGHER FREQUENCIES OF VIBRATION AND MODE SHAPES

In this section, the method of Stodola is used to determine the higher frequencies of vibration and the corresponding mode shapes. The procedure is illustrated by a numerical example.

Consider the beam in Fig. 5-8*a* that supports three concentrated weights located as shown. For simplicity, the beam is assumed to have negligible weight with constant stiffness $EI = 90 \times 10^6$ kip-in.^{2}. It is required to compute its three natural frequencies of vibration and the corresponding mode shapes.

The flexibility coefficients with unit loads at points 1, 2, and 3 are computed by using handbook formulas:

$$a_{11} = 134.0 \times 10^{-4} \qquad a_{12} = 77.5 \times 10^{-4} \qquad a_{13} = 43.5 \times 10^{-4}$$

$$a_{21} = 77.5 \times 10^{-4} \qquad a_{22} = 119.7 \times 10^{-4} \qquad a_{23} = 75.3 \times 10^{-4}$$

$$a_{31} = 43.5 \times 10^{-4} \qquad a_{32} = 75.3 \times 10^{-4} \qquad a_{33} = 67.5 \times 10^{-4}$$

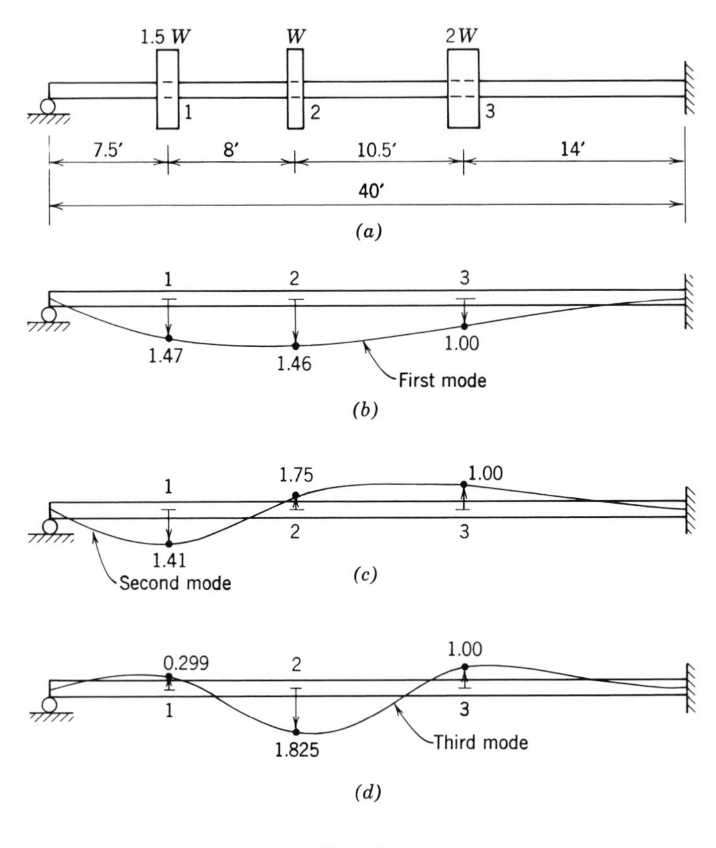

Fig. 5-8

By using Eq. 5-73, the equations of motion are written as

$$
\begin{aligned}
C^{(n)}y_1^{(n)} &= 205.5\,y_1^{(n)} + 77.5\,y_2^{(n)} + 87.0\,y_3^{(n)} \\
C^{(n)}y_2^{(n)} &= 116.3\,y_1^{(n)} + 119.7\,y_2^{(n)} + 150.6\,y_3^{(n)} \\
C^{(n)}y_3^{(n)} &= 65.3\,y_1^{(n)} + 75.3\,y_1^{(n)} + 135.0\,y_3^{(n)}
\end{aligned} \tag{5-86}
$$

where

$$
C = \frac{g(10)^4}{\omega^2 W} \tag{5-87}
$$

The superscript denotes mode number.

First Mode

The expressions in Eq. 5-86 are valid for all modes of vibration. For the fundamental mode, these equations are written in matrix form as

$$\begin{bmatrix} y_1 \\ y_2 \\ y_3 \end{bmatrix}^{(1)} = \frac{1}{C^{(1)}} \begin{bmatrix} 205.5 & 77.5 & 87.0 \\ 116.3 & 119.7 & 150.6 \\ 65.3 & 75.3 & 135.0 \end{bmatrix} \begin{bmatrix} y_1 \\ y_2 \\ y_3 \end{bmatrix} \tag{5-88}$$

where the superscript (1) denotes first mode. By applying the iteration procedure, the fundamental frequency ω_1 and the corresponding mode amplitudes are

$$\omega_1 = 106.5\sqrt{1/W} \qquad \begin{aligned} y_1^{(1)} &= 1.47 \\ y_2^{(1)} &= 1.46 \\ y_3^{(1)} &= 1.00 \end{aligned} \tag{5-89}$$

The shape of this mode is shown in Fig. 5-8*b*.

Second Mode

The second mode is determined by utilizing the orthogonality properties given in Eq. 1-156. By using the amplitudes given in Eq. 5-89, an orthogonality condition valid for $n=2$ and $n=3$ is written as

$$(1.50)(1.47)y_1^{(n)} + (1.00)(1.46)y_2^{(n)} + (2.00)(1.00)y_3^{(n)} = 0$$

or

$$1.10y_1^{(n)} + 0.73y_2^{(n)} + y_3^{(n)} = 0 \tag{5-90}$$

Thus, for the second mode, the equation yields

$$y_3^{(2)} = 1.10y_1^{(2)} - 0.73y_2^{(2)} \tag{5-91}$$

Equation 5-91 is used to eliminate y_3 from the first two expressions in Eq. 5-86. This yields

$$\begin{aligned} C^{(2)}y_1^{(2)} &= 109.8y_1^{(2)} + 14.0y_2^{(2)} \\ C^{(2)}y_2^{(2)} &= -49.4y_1^{(2)} + 9.7y_2^{(2)} \end{aligned} \tag{5-92}$$

The set of two equations is valid for the second and third modes. In matrix notation they are written as

$$\begin{bmatrix} y_1 \\ y_2 \end{bmatrix}^{(2)} = \frac{1}{C^{(2)}} \begin{bmatrix} 109.8 & 14.0 \\ -49.4 & 9.7 \end{bmatrix} \begin{bmatrix} y_1 \\ y_2 \end{bmatrix}^{(2)} \tag{5-93}$$

The iteration procedure for Eq. 5-93 is initiated by assuming $y_1 = 1.00$ and $y_2 = -1.00$. If the procedure is repeated four times, Eq. 5-93 converges to

$$\begin{bmatrix} y_1 \\ y_2 \end{bmatrix}^{(2)} = \frac{102.6}{C^{(2)}} \begin{bmatrix} 1.88 \\ -1.00 \end{bmatrix} \tag{5-94}$$

Thus the second frequency ω_2 is

$$\omega_2 = \sqrt{g(10^4)/102.6W} = 194.0\sqrt{1/W}$$

and $y_1^{(2)} = 1.88$, $y_2^{(2)} = -1.00$. Thus Eq. 5-91 yields

$$y_3^{(2)} = (1.10)(1.88) - (0.73)(-1.00) = -1.335$$

By making $y_3^{(2)} = 1.00$, the shape of the second mode is given by the amplitudes

$$\begin{aligned} y_1^{(2)} &= 1.41 \\ y_2^{(2)} &= -0.75 \\ y_3^{(2)} &= -1.00 \end{aligned} \tag{5-95}$$

The shape of this mode is shown in Fig. 5-8*c*.

Third Mode

By using Eq. 1-156 and the amplitudes given in Eq. 5-95, the orthogonality relation for the third mode yields

$$y_3^{(3)} = 1.06\,y_1^{(3)} - 0.375\,y_2^{(3)} \tag{5-96}$$

By Eqs. 5-91 and 5-96,

$$1.06\,y_1{}^{(3)} - 0.375\,y_2{}^{(3)} = -1.10\,y_1{}^{(3)} - 0.73\,y_2{}^{(3)}$$

or

$$2.16\,y_1{}^{(3)} = -0.355\,y_2{}^{(3)}$$

If $y_2^{(3)}$ is taken equal to 1.00, then $y_1^{(3)} = -0.164$, and Eq. 5-96 yields $y_3{}^{(3)} = -0.549$. Thus, by making $y_3^{(3)} = -1.00$, the shape of the third mode is given by

$$\begin{aligned} y_1{}^{(3)} &= -0.299 \\ y_2{}^{(3)} &= +1.825 \\ y_3{}^{(3)} &= -1.000 \end{aligned} \qquad (5\text{-}97)$$

which is shown in Fig. 5-8d. The frequency ω_3 of this mode is obtained by using one of the expressions in Eq. 5-92. This yields

$$\omega_3 = \sqrt{g(10^4)/24.1W} = 401.0\sqrt{1/W}$$

It should be pointed out, however, that the accuracy of the higher frequencies and modes depends on the accuracy of the fundamental frequency and mode shape. The method discussed here is approximate and some small error is introduced in the computation of the fundamental mode. This error becomes progressively larger as the higher modes are computed. Eventually, reliable accuracy can only be obtained for a certain number of the lower modes. In the problem above, for example, the accuracy of ω_3 and its corresponding mode is questionable.

In modal analysis, the first few modes are usually sufficient to determine the dynamic response of a system with reasonable accuracy. Therefore, the limitations described above are not very serious for this method.

5-13 VIBRATION OF BRIDGES

Approximate methods of analysis are often applied to determine the dynamic response of highway bridges. The required natural frequencies of vibration and the corresponding mode shapes can be determined by using the method of Stodola and iteration procedure. These quantities can be calculated with reasonable accuracy by considering one of its interior girders together with the associated weight of the bridge deck that is supported by the girder. A usual procedure is to consider 100% of the associated weight if there is composite action between bridge deck and girders, and 50% if such composite action does not take place.

To illustrate the procedure, let it be assumed that the three-span continuous beam in Fig. 5-9*a* is an interior girder of a highway bridge. The indicated load of 900 lb/lin. ft represents the weight of the girder together with the associated weight of the bridge deck that is supported by the girder. For example, if the girders are spaced at equal distances b, the weight of the bridge deck that is supported by an interior girder is usually taken as the one

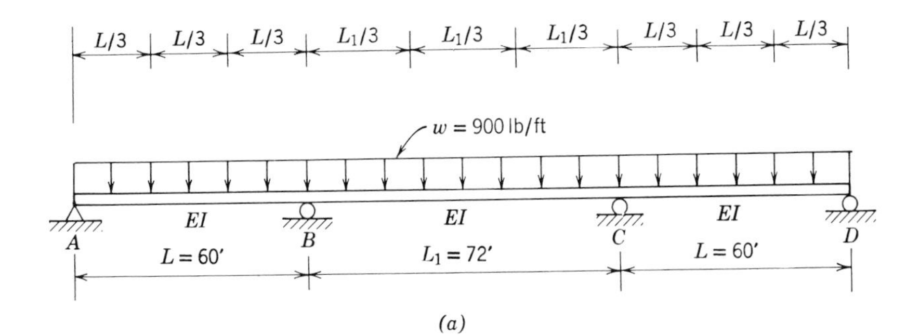

(*a*)

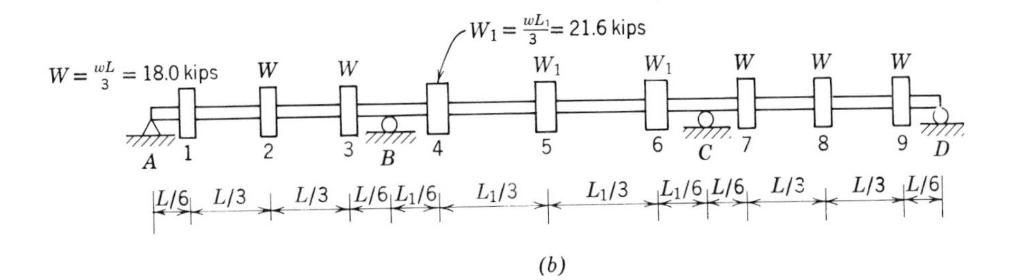

(*b*)

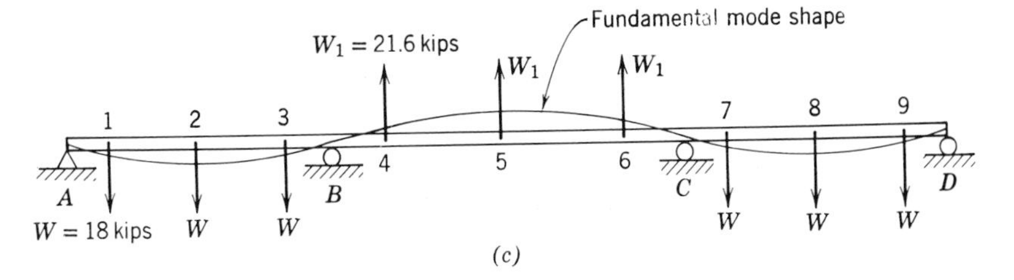

(*c*)

Fig. 5-9

included in the two strips of width $b/2$ on each side of the girder. The stiffness EI is assumed to be constant and equal to 30×10^8 kip-in.^{2}.

In applying the method of Stodola, the weight of the girder and deck is lumped at points 1, 2, 3,...,9, as shown in Fig. 5-9*b*. The beam is then assumed to be weightless between weight concentration points, but its elastic properties are retained. The natural frequencies of vibration and the corresponding mode shapes can be determined as explained in Sections 5-10 and 5-12. The fundamental frequency and mode are first determined by using the procedure discussed in Section 5-10. Then the higher frequencies and mode shapes are computed as discussed in Section 5-12.

It should be pointed out, however, that the equations of motion that are valid for all modes of vibration should be written by assuming all weights to act in the direction of the fundamental mode as shown in Fig. 5-9*c*. It is important to note this fact, because during vibration the forces are acting in the direction of the dynamic deflection produced by these forces. To illustrate the procedure, the fundamental frequency and mode shape of the girder are determined.

The flexibility coefficients are computed by applying unit loads at points 1, 2, 3,...,9 in the usual way. They are shown in Table 5-1. By substituting the appropriate values for the a's and m's, Eq. 5-74 yields

$$\begin{bmatrix} y_1 \\ y_2 \\ y_3 \\ y_4 \\ y_5 \\ y_6 \\ y_7 \\ y_8 \\ y_9 \end{bmatrix} = \frac{1.96\omega^2}{10^3 g} \begin{bmatrix} 18.35 & 26.65 & 8.98 & 8.17 & 10.60 & 3.62 & 1.57 & 2.31 & 1.00 \\ 25.65 & 51.00 & 8.43 & 19.95 & 24.50 & 8.43 & 3.65 & 5.35 & 2.31 \\ 8.98 & 8.43 & 14.30 & 12.90 & 17.20 & 5.73 & 2.45 & 3.65 & 1.57 \\ 6.80 & 16.63 & 10.75 & 20.70 & 29.80 & 10.96 & 4.77 & 7.00 & 3.01 \\ 8.82 & 20.50 & 14.35 & 29.80 & 78.20 & 29.80 & 14.35 & 20.50 & 8.82 \\ 3.01 & 7.00 & 4.77 & 10.96 & 29.80 & 20.70 & 10.75 & 16.63 & 6.80 \\ 1.57 & 3.65 & 2.45 & 5.73 & 17.20 & 12.90 & 14.30 & 8.43 & 8.98 \\ 2.31 & 5.35 & 3.65 & 8.43 & 24.50 & 19.95 & 8.43 & 51.00 & 25.65 \\ 1.00 & 2.31 & 1.57 & 3.62 & 10.60 & 8.17 & 8.98 & 26.65 & 18.35 \end{bmatrix} \begin{bmatrix} y_1 \\ y_2 \\ y_3 \\ y_4 \\ y_5 \\ y_6 \\ y_7 \\ y_8 \\ y_9 \end{bmatrix} \tag{5-98}$$

The iteration procedure may be initiated by assuming $y_1 = y_2 = \cdots = y_9 = 1.00$ in the right-hand column of Eq. 5-98 and carrying out the matrix multiplication as in Section 5-10. By applying this procedure and repeating it about five times, the matrix equation converges to the following values:

TABLE 5-1

Point of Unit Load	Deflection at Point								
	1	2	3	4	5	6	7	8	9
1	0.001995	0.002790	0.000977	−0.000741	−0.000962	−0.000328	0.000171	0.000251	0.000109
2	0.002790	0.005550	0.000918	−0.001810	−0.002230	−0.000762	0.000397	0.000582	0.000251
3	0.000977	0.000918	0.001555	−0.001170	−0.001560	−0.000520	0.000267	0.000397	0.000171
4	−0.000741	−0.001810	−0.001170	0.001875	0.002705	0.000995	−0.000520	−0.000762	−0.000328
5	−0.000962	−0.002230	−0.001560	0.002705	0.007080	0.002705	−0.001560	−0.002230	−0.000962
6	−0.000328	−0.000762	−0.000520	0.000995	0.002705	0.001875	−0.001170	−0.001810	−0.000741
7	0.000171	0.000397	0.000267	−0.000520	−0.001560	−0.001170	0.001555	0.000918	0.000977
8	0.000251	0.000582	0.000397	−0.000762	−0.002230	−0.001810	0.000918	0.005550	0.002790
9	0.000109	0.000252	0.000171	−0.000328	−0.000962	−0.000741	0.000977	0.002790	0.001995

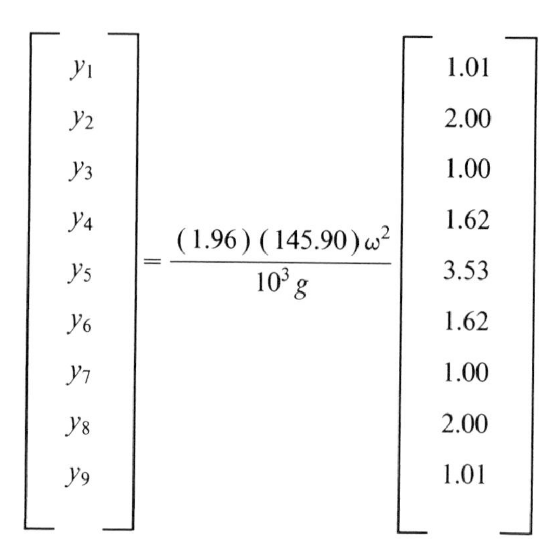

$$\begin{bmatrix} y_1 \\ y_2 \\ y_3 \\ y_4 \\ y_5 \\ y_6 \\ y_7 \\ y_8 \\ y_9 \end{bmatrix} = \frac{(1.96)(145.90)\omega^2}{10^3 g} \begin{bmatrix} 1.01 \\ 2.00 \\ 1.00 \\ 1.62 \\ 3.53 \\ 1.62 \\ 1.00 \\ 2.00 \\ 1.01 \end{bmatrix}$$

The fundamental frequency is obtained from the equation

$$1.00 = \frac{(1.96)(145.90)\omega^2}{10^3 g}(1.00)$$

where g is the acceleration of gravity. This yields

$$\omega = 36.80 \text{ rps}$$

$$f = \frac{\omega}{2\pi} = 5.87 \text{ Hz}$$

The mode shape is characterized by the amplitudes

$$y_1 = 1.01 \qquad y_6 = 1.62$$
$$y_2 = 2.00 \qquad y_7 = 1.00$$
$$y_3 = 1.00 \qquad y_8 = 2.00$$
$$y_4 = 1.62 \qquad y_9 = 1.01$$
$$y_5 = 3.53$$

The higher frequencies and mode shapes can be determined by using the procedure discussed in Section 5-12. Since the method is approximate, only the first four modes would be of reasonable accuracy.

It should be pointed out, however, that the structural flexibility of a bridge depends on its susceptibility to vibration. If the design satisfies static load criteria, this does not suggest that vibration and other dynamic criteria are automatically satisfied. This is particularly true for long-span bridges. In such cases, the structural integrity and safety of a bridge can be secured only if vibration and dynamic response criteria are incorporated in its design. The usual safety and impact factors may not be sufficient. Analytical and experimental studies made by the author and others at the Michigan State Highway Department suggest that the natural fundamental frequency of vibration for bridge girders should not be less than 5.5 Hz.

5-14 DYNAMIC RESPONSE OF FRAMES WITH FLEXIBLE GIRDERS

In Chapter 3 the dynamic response of two-story and multistory rigid frames was determined by assuming that the girders were infinitely rigid. In this section the flexibility of the girders is taken into account. The procedure is illustrated by a numerical example.

Consider the two-story rigid frame building in Fig. 3-5*a*, and let it be assumed that it is required to determine its dynamic response under the action of externally applied dynamic forces. The elastic properties of this building are uniform throughout its length, and its dynamic response is computed by considering only the interior frame in Fig. 3-5*b* together with the associated wall and floor areas. The time variation of the dynamic forces is shown in Fig. 3-5*d*. The steel sections of the girders and columns are 21WF 62 and 10WF 25, respectively.

The stiffness coefficients k_{ij} are determined by applying unit deflections at each floor level and computing the resulting holding forces. The procedure is discussed in Section 1-15, and the k_{ij} coefficients for this problem are determined by applying the method of moment distribution. The values of these coefficients, as well as the column end moments, are shown in Fig. 5-10. The mass of the frame, together with the associated wall and floor areas, is lumped at the floor levels as discussed in Section 3-4. They are as follows:

$$m_1 = 0.217 \frac{\text{kip-sec}^2}{\text{in.}}$$

$$m_2 = 0.137 \frac{\text{kip-sec}^2}{\text{in.}}$$

If x_1 and x_2 are the horizontal displacements of masses m_1 and m_2, respectively, the equations of motion in terms of stiffness coefficients are written as

$$m_1\ddot{x}_1+k_{11}x_1+k_{12}x_2=F_1(t)$$
$$m_2\ddot{x}_2+k_{21}x_1+k_{22}x_2=0.7F_1(t) \tag{5-99}$$

The two natural frequencies of the frame can be determined from the expressions

$$m_1\ddot{x}_1+k_{11}x_1+k_{12}x_2=0$$
$$m_2\ddot{x}_2+k_{21}x_1+k_{22}x_2=0 \tag{5-100}$$

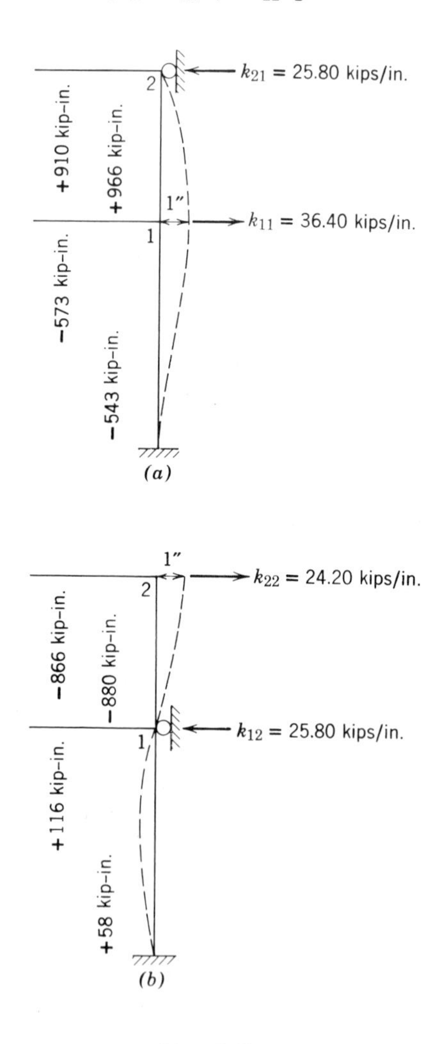

Fig. 5-10

Thus, by expressing x_1 and x_2 as

$$x_1 = X_1 \sin \omega t \quad x_2 = X_2 \sin \omega t$$

Eq. 5-100 yields

$$\begin{aligned} m_1\omega^2 X_1 &= k_{11}X_1 + k_{12}X_2 \\ m_2\omega^2 X_2 &= k_{21}X_1 + k_{22}X_2 \end{aligned} \tag{5-101}$$

On substitution of the appropriate values of the k's and m's, the expressions in Eq. 5-101 yield

$$\begin{aligned} 0.217\omega^2 X_1 &= 36.40X_1 - 25.80X_2 \\ 0.137\omega^2 X_2 &= -25.80X_1 + 24.20X_2 \end{aligned} \tag{5-102}$$

These equations are valid for both modes of vibration of the frame.

The two natural frequencies and the corresponding mode shapes are determined by using the method of Stodola and iteration procedure. Since stiffness coefficients are used, the second mode is determined first. Thus, in matrix notation, the expressions in Eq. 5-102 are written as

$$\begin{bmatrix} X_1 \\ X_2 \end{bmatrix}^{(2)} = \frac{1}{\omega^2} \begin{bmatrix} 167.5 & -119.0 \\ -188.5 & 177.0 \end{bmatrix} \begin{bmatrix} X_1 \\ X_2 \end{bmatrix} \tag{5-103}$$

The iteration procedure yields

$$\omega_2 = 17.90 \text{ rps}$$

for the second frequency ω_2 of the frame. The corresponding mode shape is given by the amplitudes,

$$X_1^{(2)} = 1.00$$

$$X_2^{(2)} = -1.30$$

The fundamental frequency ω_1 is determined by applying the orthogonality properties of the two modes and is found to be

$$\omega_1 = 4.65 \text{ rps}$$

The corresponding mode shape is

$$X_1^{(1)} = 1.00$$

$$X_2^{(1)} = 1.21$$

The dynamic response of the frame is determined by using the method of modal analysis. By Eq. 5-18, the modal static displacements $Y_{1\,st}$ and $Y_{2\,st}$ for the two modes are as follows:

$$Y_{1\,st} = \frac{(3.00)(1.00) + (2.10)(1.21)}{(4.65)^2[(0.217)(1.00)^2 + (0.137)(1.21)^2]} = 0.613 \text{ in.}$$

$$Y_{2\,st} = \frac{(3.00)(1.00) + (2.10)(-1.30)}{(17.9)^2[(0.217)(1.00)^2 + (0.137)(-1.30)^2]} = 0.00182 \text{ in.}$$

The maximum magnification factor for each mode is obtained from Fig. 2-5*a*. Thus

$$(\Gamma)_{1\,max} = 0.62$$

$$(\Gamma)_{2\,max} = 1.45$$

The horizontal displacements x_1 and x_2 at the tops of the columns of the first and second story, respectively, are computed by adding numerically the displacement responses of the individual modes. That is,

$$x_1 = x_{11} + x_{12}$$
$$x_2 = x_{21} + x_{22} \qquad (5\text{-}104)$$

where x_{11} and x_{12} are the responses of modes 1 and 2, respectively, for the first story, and x_{21} and x_{22} are those of the second story. It should be pointed out, however, that the numerical sum of the responses is a conservative estimate of the deflections x_1 and x_2.

By applying Eq. 5-21 for $p = 1$ and $p = 2$, separately,

$$x_{11} = (0.613)(1.00)(0.62) = 0.380 \text{ in.}$$

$$x_{12} = (0.00182)(1.00)(1.45) = 0.00264 \text{ in.}$$

and

$$x_{21} = (0.613)(1.21)(0.62) = 0.460 \text{ in.}$$

$$x_{22} = (0.00182)(-1.30)(1.45) = -0.00343 \text{ in.}$$

Thus, from Eq. 5-104,

$$x_1 = 0.380 + 0.00264 = 0.3826 \text{ in.}$$

$$x_2 = 0.460 + 0.00343 = 0.4634 \text{ in.}$$

These results are somewhat higher compared to those obtained in Section 3-4, where the girders were assumed to be infinitely rigid.

The bending moments at the top and bottom of each column of the frame are determined by multiplying the moments due to unit deflections in Fig. 5-10 by the actual deflections. For example, the moment responses of modes 1 and 2 at the top of a column of the second story are as follows:

First mode

$$(0.380)(910) + (0.460)(-866) = -52.0 \text{ kip-in.}$$

Second mode

$$(0.00264)(910) + (-0.00343)(-866) = 5.4 \text{ kip-in.}$$

The numerical sum of these moments is 57.4 kip-in. In section 3-4 the value of this moment was found to be equal to 54.0 kip-in. The stress σ_2 at the top of this column is

$$\sigma_2 = \frac{M_2}{S} = \frac{57.4}{26.4} = 2.17 \text{ ksi}$$

where S is the section modulus of the column section.

In a similar manner, the moment responses at the top of a column of the first story are obtained. They are as follows:

First mode

$$(0.380)(-573) + (0.460)(116) = -164.7 \text{ kip-in.}$$

Second mode

$$(0.00264)(-573) + (-0.00343)(116) = -1.9 \text{ kip-in.}$$

The numerical sum of these moments is 166.6 kip-in. and the stress σ_1 is

$$\sigma_1 = \frac{M_1}{S} = \frac{166.6}{26.4} = 6.32 \text{ ksi}$$

These results show that the maximum stress occurs at the tops of the first story columns.

The natural frequencies and the mode shapes of the frame could be also obtained by using flexibility coefficients instead of stiffness coefficients. In fact, this would be advisable, because for many practical problems the fundamental mode alone yields sufficiently accurate results. In the frame above, for example, the contribution of the second mode is small and could be neglected.

When the stiffness coefficients are computed, the flexibility coefficients can be found, because the inverse of the stiffness matrix yields the flexibility matrix. For the problem above, the stiffness matrix $[N]$ is

$$[N]=\begin{bmatrix} k_{11} & k_{12} \\ k_{21} & k_{22} \end{bmatrix}=\begin{bmatrix} 36.4 & -25.8 \\ -25.8 & 24.2 \end{bmatrix}$$

$$[N^{-1}]=[M]=\begin{bmatrix} a_{11} & a_{12} \\ a_{21} & a_{22} \end{bmatrix}=\begin{bmatrix} 0.1160 & 0.1236 \\ 0.1236 & 0.1727 \end{bmatrix}$$

The a_{ij} flexibility coefficients correspond to unit loads applied individually at points 1 and 2 and computing in each case the horizontal displacements of points 1 and 2. By using flexibility coefficients, the Stodola method and iteration procedure will yield first the fundamental frequency of vibration and the corresponding mode shape.

PROBLEMS

5-1 If the dynamic force $F_1(t)$ acting on the two-story building in Fig. 3-5*b* is a force of 5 kips applied suddenly at $t=0$ and then released suddenly at $t=0.3$ sec, determine the maximum horizontal displacements at the girder levels and the maximum bending stresses in the columns. Assume that the girders of the frame are rigid and apply the method of modal analysis.

5-2 Repeat Problem 5-1 for a dynamic force $F_1(t)$ of 5 kips applied suddenly to the frame at $t=0$ and of infinite duration.

5-3 Repeat Problem 5-1 when $F_1(t)$ is a force of 7 kips applied suddenly to the frame at $t=0$ and decreasing linearly to zero at $t=0.5$ sec.

5-4 For the steel frames of Problem 3-27, determine the maximum horizontal displacements at the girder levels and the maximum bending stresses in the columns by utilizing the method of modal analysis. The dynamic force $F_1(t)$ is a suddenly applied constant force of 5 kips at $t=0$ and of infinite duration. Assume that the girders are rigid.

5-5 Repeat Problem 5-4 when the force $F_1(t)$ is a suddenly applied constant force of 5 kips at $t=0$ and is suddenly removed at $t=0.2$ sec.

5-6 Repeat Problem 5-4 when the force $F_1(t)$ is a suddenly applied force of 10 kips at $t=0$ and decreasing linearly to zero at $t=0.2$ sec.

5-7 Repeat Problem 4-5 by using the method of modal analysis.

5-8 Repeat Problem 4-6 by using the method of modal analysis.

5-9 Repeat Problem 4-7 by using the method of modal analysis.

5-10 The uniform single span beams in Problem 4-1 are assumed to be loaded by three concentrated dynamic forces $F_1(t)$, $F_2(t)$, and $F_3(t)$ acting at $L/4$, $L/2$, and $3L/4$, respectively. If $F_1(t)=F_1=10$ kips, $F_2(t)=F_2=20$ kips, $F_3(t)=F_3=15$ kips, and all three forces are suddenly applied at $t=0$ and suddenly removed at $t=0.3$ sec, determine the maximum vertical displacement at the center length of each beam by applying the method of modal analysis. The stiffness EI and the weight w are as in Problem 4-1. Neglect damping.

5-11 Repeat Problem 5-10 by computing the maximum bending moment and maximum bending stress in each beam case. Assume that each beam is made of steel with a rectangular cross section 6 in. wide and 14 in. deep and modulus of elasticity E equal to 30×10^6 psi.

5-12 The uniform simply supported beam in Fig. 5-2 is subjected to a constant force $F=10$ kips moving with a constant velocity $v=0.5$ mi/min. Determine the expression for the dynamic displacement $y(t, x)$ of the beam by superimposing the response of the first three modes. At $t=0$ the force F is at the left support. The stiffness EI is 30×10^6 kip-in.^{2}, $L=50$ ft, and its uniform weight w is 0.6 kips/ft.

5-13 Repeat Problem 5-12 for $F=20$ kips and $L=30$, 40, 50, and 60 ft.

5-14 For the beam in Problem 5-12, determine the maximum vertical displacement and the maximum bending stress at midspan.

5-15 For the beam in Problem 5-13, determine in each case the maximum vertical displacement and maximum bending stress at midspan and compare results.

5-16 The sides a and b of a flat rectangular simply supported steel plate are 120 and 70 in., respectively. The thickness h is 2 in., the Poisson ratio ν is 0.25, and $E=30\times10^6$ psi. The plate is subjected to a uniform dynamic pressure of 40 psi applied suddenly at $t=0$ that is of infinite duration. Determine the maximum displacement and the maximum bending stress at the center of the plate by considering only the response of the first mode.

5-17 Repeat Problem 5-16 for a uniform dynamic pressure of 40 psi applied suddenly at $t=0$ and removed suddenly at $t=0.03$ sec.

5-18 Repeat Problem 5-16 for a uniform dynamic pressure of 70 psi applied suddenly at $t=0$ and decreased linearly to zero at $t=0.08$ sec.

5-19 Repeat Problem 5-16 by considering the responses of the first three modes and superimposing results.

5-20 The uniform beams in Problem 4-2 are of constant stiffness $EI=45\times10^6$ kip-in.^{2} and weight $w=0.6$ kips/ft. By lumping the weight at discrete points along their length and applying the method of Stodola and iteration procedure, determine in each case the fundamental natural frequency and the corresponding mode of vibration. Use flexibility coefficients and neglect damping.

5-21 If the entire length of each of the uniform beams in Problem 4-2 is acted upon by a uniformly distributed force of 5 kips applied suddenly at $t=0$ and of infinite duration, determine the maximum vertical displacement at the center of each span. Use the natural fundamental frequency and the corresponding mode determined in Problem 5-20 and apply the appropriate equations of modal analysis.

5-22 Repeat Problem 5-20 by computing the first three natural frequencies of vibration and the corresponding mode shapes.

5-23 Repeat Problem 5-21 by superimposing the responses of the first three modes.

5-24 Repeat Problem 5-21 by using a uniformly distributed force of 10 kips applied suddenly at $t=0$ and decreased linearly to zero at $t=0.15$ sec.

5-25 If the dynamic force $F_1(t)$ acting on the two-story frame in Fig. 3.5*b* is a force of 5 kips applied suddenly at $t=0$ and then released suddenly at $t=0.3$ sec, determine the maximum horizontal displacements at the girder levels and the maximum bending stresses in the columns by considering the flexibility of the steel girders. All girders have a 21W 62 steel section.

5-26 Repeat Problem 5-25 when the dynamic force $F_1(t)$ is a force of 5 kips applied suddenly at $t=0$ sec that is of infinite duration.

5-27 By applying the method of modal analysis and considering the flexibility of the girders, determine the maximum horizontal displacements at the girder levels of the steel frames in Problem 3-27 and the maximum bending stresses in the columns. Assume that $F_1(t)$ is a suddenly applied force of 6 kips at $t=0$ sec and released suddenly at $t=0.4$ sec.

6

METHODS OF VIBRATION

6-1 INTRODUCTION

In this chapter methods for the computation of natural frequencies and mode shapes of beamlike structures and spring-mass systems are discussed. These methods can be used to determine the vibration response of structural systems, as well as their dynamic response due to time-varying forces if they are combined with the method of modal analysis.

The well-known methods of Rayleigh and Myklestad are discussed in the first two Sections. The remaining sections, except Section 6-9, are devoted to the concept of transfer matrices with applications to various structural problems. The method of Stodola and iteration procedure is treated in Chapter 5. Transfer matrices can be easily handled by a digital computer, because they involve basic operations of matrix algebra.

In the last section of this chapter the concept of the dynamic hinge is presented. By this concept, the fundamental frequency of complicated beam problems can be obtained by using only a portion of a beam span.

6-2 RAYLEIGH'S METHOD

In 1877 Lord Rayleigh proposed a method by which natural frequencies and mode shapes of a freely vibrating elastic system can be determined by equating its maximum potential energy to its maximum kinetic energy at an extreme configuration of the system. Initially, the application of this method was somewhat limited to the computation of the fundamental frequency and mode of relatively simple problems. Later, with modern developments in the fields of science and technology, the method was extended to include the computation of higher frequencies and mode shapes of more complicated problems.

The discussion is initiated by considering the uniform simply supported beam in Fig. 6-1*a*. The dynamic elastic line of the fundamental mode is shown in the same figure. If w is the weight of the member per unit of length, $w\,dx/g$ the mass dm of the element dx at location x, ω the fundamental natural frequency of vibration, η the dynamic displacement at x during the vertical motion of the beam, and y the amplitude of the fundamental frequency ω, then at distance x from support A the element $w\,dx/g$ of the member vibrates at the natural frequency ω of the member.

The beam vibrates harmonically, hence the displacement η can be represented by any one of the expressions

$$\eta = y\cos\omega t \qquad \eta = y\sin\omega t \tag{6-1}$$

Thus, at the frequency ω, the amplitude of the element $w\,dx/g$ is y, its maximum velocity $\dot{\eta}_{\max}$ is $y\omega$, and its maximum kinetic energy dT is

$$dT = \frac{1}{2}dm\dot{\eta}_{\max}^2 = \frac{1}{2}\left(\frac{w\,dx}{g}\right)(\omega y)^2 = \frac{w}{2g}\omega^2 y^2\,dx \tag{6-2}$$

By considering all the elements of the member, the maximum kinetic energy

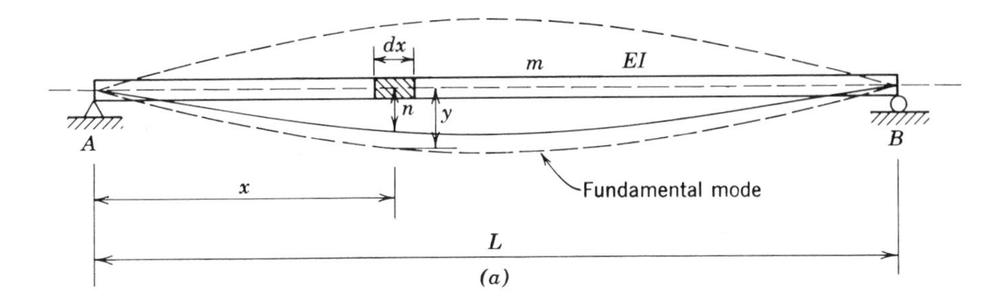

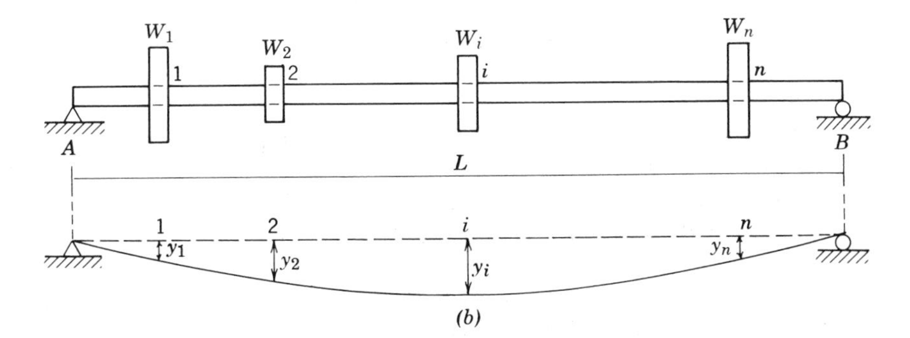

Fig. 6-1

T of the beam is

$$T = \frac{w}{2g}\,\omega^2 \int_0^L y^2\,dx \tag{6-3}$$

For the fundamental mode, an extreme configuration occurs when $\eta = y$. Thus, from strength of materials, the maximum potential energy dU of the element is given by the expression

$$dU = M\frac{d\theta}{2} = \frac{EI}{2}\left(\frac{d^2y}{dx^2}\right)^2 dx \tag{6-4}$$

By considering all the elements of the beam, we obtain its maximum potential energy U:

$$U = \frac{EI}{2}\int_0^L \left(\frac{d^2y}{dx^2}\right)^2 dx \tag{6-5}$$

By equating T and U and solving for ω^2,

$$\omega^2 = \frac{EIg}{w} \frac{\int_0^L (d^2y/dx^2)^2 dx}{\int_0^L y^2 dx} \tag{6-6}$$

The fundamental frequency ω of the beam can be determined from the equation above, provided that the expression for the dynamic elastic line $y(x)$ is known. If EI and w are variable, Eq. 6-6 should be written as

$$\omega^2 = g \frac{\int_0^L EI (d^2y/dx^2)^2 dx}{\int_0^L wy^2 dx} \tag{6-7}$$

The expression for y is usually unknown, and an approximate shape for the dynamic elastic line $y(x)$ is often used. The accuracy of the solution will then depend on how closely $y(x)$ approximates the true shape of the fundamental mode. For many practical problems, a reasonable solution is often obtained by assuming that $y(x)$ is the static elastic line produced by the dead weight w of the member. For example, the static elastic line of a uniformly simply supported beam due to its dead weight w is

$$y(x) = \frac{w}{2EI}(x^4 - 2Lx^3 + L^3x) \tag{6-8}$$

By using this expression in Eq. 6-6, the resulting fundamental frequency ω of the member is determined:

$$\omega = \frac{9.88}{L^2}\sqrt{EI/m}$$

This is very close to the exact solution found in Section 1-11.

If the beam is fixed at both ends, the static elastic line $y(x)$ due to w is given by

$$y(x) = \frac{w}{24EI}(xL - x^2)^2 \tag{6-9}$$

If this expression is substituted into Eq. 6-6, the resulting frequency ω is about 6% higher than the true one.

If in addition to its weight the beam supports the weights $W_1, W_2, \ldots, W_i, \ldots, W_n$ causing the dynamic deflections $y_1, y_2, \ldots, y_i, \ldots, y_n$, respectively, under the weight concentration points, Fig. 6-1*b*, the total maximum kinetic energy of the member is

$$T = \frac{w}{2g}\omega^2 \int_0^L y^2\,dx + \sum_{i=1}^{n} \frac{W_i}{g}\omega^2 y_i^2 \tag{6-10}$$

The expression for the total maximum potential energy, however, is of the same form as Eq. 6-5, but with curvature d^2y/dx^2 that takes into consideration the effects of the concentrated weights. Again, by equating T and U, the expression for the fundamental frequency ω can be obtained provided that y and y_i are known.

If the static elastic line is used, the work done by gravity on the weights in Fig. 6-1*b* is

$$\tfrac{1}{2}\sum_{i=1}^{n} W_i y_i$$

The work done by gravity on the distributed weight wL of the beam is obtained by integrating $wy\,dx$ over the length of the member. Therefore, for any beam, the general expression for the fundamental frequency ω is

$$\omega^2 = g\frac{\int_0^L wy\,dx + \sum_{i=1}^{n} W_i y_i}{\int_0^L wy^2\,dx + \sum_{i=1}^{n} W_i y_i^2} \tag{6-11}$$

Equation 6-11 can be used for any type of weight distribution and for any number of concentrated weights. If the distributed weight of the beam is lumped at discrete points along its length as discussed in Section 5-9, Eq. 6-11 becomes

$$\omega^2 = g\frac{\sum_{i=1}^{n} W_i y_i}{\sum_{i=1}^{n} W_i y_i^2} \tag{6-12}$$

where the summations in this equation include the lumped weights, as well as the additional weights supported by the beam. Equations 6-11 and 6-12, if desired, can take into account shear displacements and elastic displacements of its supports.

By selecting an arbitrary displacement Y, preferably that of a weight W_i, we have

$$y_i = Y\beta_i \qquad \beta_i = \frac{y_i}{Y} \tag{6-13}$$

where β_i characterizes the shape of the mode. Thus, Eq. 6-12 becomes

$$\omega^2 = g\frac{\sum_{i=1}^{n} W_i\beta_i}{Y\sum_{i=1}^{n} W_i\beta_i^2} \tag{6-14}$$

or, with $W_i = m_i g$,

$$\omega^2 = g\frac{\sum_{i=1}^{n} m_i\beta_i}{Y\sum_{i=1}^{n} m_i\beta_i^2} \tag{6-15}$$

Equations 6-12, 6-14, and 6-15 are all similar, and any one of these expressions can be used to determine a first approximation of the fundamental frequency of structural systems.

A better approximation of the fundamental frequency and mode is obtained by applying a repetitive procedure. A first approximation is obtained by using, for example, Eq. 6-14. If the value of the computed frequency is designated as ω_1, the inertia forces W_i' of the corresponding mode are determined from the expression

$$W_i' = \frac{W_i}{g}\omega_1^2 y_i \tag{6-16}$$

or, by Eq. 6-13,

$$W_i' = \frac{Y\omega_1^2}{g} W_i\beta_i$$

Here the quantity $Y\omega_1^2/g$ is constant and can be omitted, because the frequency depends only on the shape of the mode. Thus W_i' is taken as

$$W_i' = W_i\beta_i \tag{6-17}$$

where β_i are the mode amplitudes of the preceding trial. With W_1' applied to the member as static loads, the new displacements y_i' are determined by using known methods of strength of materials. The new value ω_2 of the frequency is computed from the expression

$$\omega_2^{\,2} = g\frac{\sum_{i=1}^{n} W_i{}'y_i'}{\sum_{i=1}^{n} W_i(y_i')^2} \qquad (6\text{-}18)$$

The procedure is rapidly convergent, and it usually requires two to four repetitions for an accurate value of the frequency and mode. Again, a modal displacement Y' can be selected so that

$$y_i' = Y'\beta_i' \qquad \text{and} \qquad \beta_i' = \frac{y_i'}{Y'} \qquad (6\text{-}19)$$

where β_i' provides a better approximation of the mode shape. On this basis, Eq. 6-18 is written as

$$\omega_2^{\,2} = g\frac{\sum_{i=1}^{n} W_i{}'\beta_i'}{Y'\sum_{i=1}^{n} W_i(\beta_i')^2} \qquad (6\text{-}20)$$

As an application, let it be assumed that it is required to determine the fundamental frequency and mode of the bridge girder in Fig. 5-9*a*, Section 5-13. The weight of the member is lumped at the indicated points and is applied as static load as shown in Fig. 5-9*c*. The resulting deflection shape that approximates the shape of the fundamental mode would also be as shown. By applying methods of strength of materials the deflections y_i are as follows:

$y_1 = 0.1572$ in. $\quad y_6 = -0.2210$ in.

$y_2 = 0.2930$ in. $\quad y_7 = 0.1475$ in.

$y_3 = 0.1475$ in. $\quad y_8 = 0.2930$ in.

$y_4 = -0.2210$ in. $\quad y_9 = 0.1572$ in.

$y_5 = -0.4420$ in.

By using these values in Eq. 6-12, we obtain a first approximation of the fundamental frequency:

$$\omega_1 = 37.6 \text{ rps}$$

$$f_1 = \frac{\omega_1}{2\pi} = 5.98 \text{ Hz}$$

Selecting y_3 as the modal displacement Y and applying Eq. 6-13, the values β_i characterizing the shape of the mode are:

$$\beta_1 = 1.065 \qquad \beta_6 = -1.500$$

$$\beta_2 = 1.985 \qquad \beta_7 = 1.000$$

$$\beta_3 = 1.000 \qquad \beta_8 = 1.985$$

$$\beta_4 = -1.500 \qquad \beta_9 = 1.065$$

$$\beta_5 = -2.990$$

If better accuracy is desired, Eq. 6-18 or 6-20 should be used to repeat the procedure. By Eq. 6-17, the inertia loads $W_i{}'$ are

$$W_1{}' = W_1\,\beta_1 = 19.20 \text{ kips} \qquad W_6{}' = W_6\,\beta_6 = -32.40 \text{ kips}$$

$$W_2{}' = W_2\,\beta_2 = 35.80 \text{ kips} \qquad W_7{}' = W_7\,\beta_7 = 18.00 \text{ kips}$$

$$W_3{}' = W_3\,\beta_3 = 18.00 \text{ kips} \qquad W_8{}' = W_8\,\beta_8 = 35.80 \text{ kips}$$

$$W_4{}' = W_4\,\beta_4 = -32.40 \text{ kips} \qquad W_9{}' = W_9\,\beta_9 = 19.20 \text{ kips}$$

$$W_5{}' = W_5\,\beta_5 = -64.50 \text{ kips}$$

The $W_i{}'$ loads are applied statically to the beam in Fig. 5-9*c*, and the new displacements y_i' are computed. By using these values in Eqs. 6-18 and 6-19, a better approximation of the frequency and mode shape, respectively, is obtained. The procedure can be repeated as many times as is required to obtain a desirable accuracy.

Higher Frequencies and Mode Shapes

The general procedure is to assume a shape for the mode to be determined and sweep out the parts related to the lower mode. Then, by applying Rayleigh's method as before, the procedure converges to the required mode. The methodology is illustrated by a numerical example.

Let it be assumed that it is required to determine the second higher frequency ω_2 of the beam in Fig. 5-8*a*. The fundamental mode is already computed in Section 5-12 and is shown in Fig. 5-8*b*. The mode amplitudes β_{pi} for $p=1$ and points $i=1, 2, 3$ are

$$\beta_{11}=1.47 \qquad \beta_{12}=1.46 \qquad \beta_{13}=1.00$$

The amplitudes β_{pi} of the second mode ($p=2$) are yet unknown and are designated as β_{21}, β_{22}, and β_{23}.

The first step is to assume a shape for the second mode. That is, assume

$$\beta_{21}=1.40 \qquad \beta_{22}=-0.70 \qquad \beta_{23}=-1.00$$

The components associated with the fundamental mode shape are $b_1\,\beta_{1i}$, where b_1 is given by the expression

$$b_1=\frac{\sum_{i=1}^{3} m_i\beta_{2i}\beta_{1i}}{\sum_{i=1}^{3} m_i\beta_{1i}^2} \tag{6-21}$$

The mode amplitude $(\beta_{2i})_r$ of any mass m_i that does not include the component effects of the fundamental mode is given by the expression

$$(\beta_{2i})_r=\beta_{2i}-b_1\,\beta_{1i} \tag{6-22}$$

From Eq. 6-21

$$b_1=\frac{1.02}{7.38}=0.1355$$

Thus, by Eq. 6-22,

$$(\beta_{21})_1=\ \ 1.40-(0.1355)(1.47)=\ \ 1.201$$

$$(\beta_{22})_2=-0.70-(0.1355)(1.46)=-0.898$$

$$(\beta_{23})_3=-1.00-(0.1355)(1.00)=-1.136$$

The corrected values of the assumed shape of the second mode are swept of the fundamental mode. By the method of Lord Rayleigh, applied as before, the procedure will converge to the frequency ω_2 of the second mode. The procedure is initiated by computing first the inertia forces. From Eq. 6-17, with $W_i=m_i g$,

$$W_i'=W_i\,\beta_i=m_i g\beta_i=Q_i g \tag{6-23}$$

where

$$Q_i = m_i \beta_i \qquad (6\text{-}24)$$

can be taken as the inertia force in the same way as Eq. 6-17. Thus, by substituting Eq. 6-23 into Eq. 6-18,

$$\omega^2 = g \frac{\sum_{i=1}^{n} Q_i \, \beta_i'}{Y' \sum_{i=1}^{n} m_i (\beta_i')^2} \qquad (6\text{-}25)$$

By using Eq. 6-24, the inertia forces Q_i for this problem are

$$Q_1 = m_1(\beta_{21})_1 = \frac{W}{g}(1.50)(1.201) = 1.80\frac{W}{g}$$

$$Q_2 = m_2(\beta_{22})_2 = -0.898\frac{W}{g}$$

$$Q_3 = m_3(\beta_{23})_3 = -2.272\frac{W}{g}$$

By applying the inertia forces as static loads on the beam in Fig. 5-8*a*, the deflections y_i' at points $i = 1, 2, 3$ are determined. With y_3' as modal displacement Y', the new mode amplitudes β_{21}', β_{22}', and β_{23}' are computed. The frequency ω_2 of the second mode is then determined by applying Eq. 6-25. This yields

$$\omega_2 = 181.5\sqrt{1/W}$$

If better accuracy is desired, the procedure should be repeated by using the amplitudes β_{21}', β_{22}', and β_{23}' as the assumed shape of the second mode and proceeding as above.

If there are more than one mode preceding the one to be determined and if their shapes are known, Eq. 6-21 should be applied for each mode. Thus a more general expression in place of Eq. 6-21 is written as

$$b_k = \frac{\sum_{i=1}^{n} m_i \beta_{pi} \beta_{ki}}{\sum_{i=1}^{n} m_i \beta_{ki}^2} \qquad k = 1, 2, 3, \ldots \qquad (6\text{-}26)$$

In this equation p is the mode to be determined, k represents the number of modes preceding the p mode, and n is the number of masses. Under these conditions, Eq. 6-22 becomes

$$(\beta_{pi})_r = \beta_{pi} - \sum_{k=1}^{p-1} b_k \beta_{ki} \tag{6-27}$$

The required frequency is usually obtained with good accuracy with only a few repetitions, but more repetitions are needed for the mode shape.

6-3 MYKLESTAD METHOD FOR FREE FLEXURAL VIBRATIONS

A general method for the computation of natural frequencies and mode shapes of beams of uniform and variable stiffness, was developed independently by N. O. Myklestad[18] and M. A. Prohl[19]. Later, W. T. Thomson[20] rearranged the equations for systematic tabular calculations and extended the method to more general problems. In this section the application of this method is limited to uncoupled bending vibrations.

The Myklestad method is basically an extension of the Holzer method for torsional vibrations. It is, however, more complicated because for flexural vibration four elastic parameters must be computed for each length segment of the beam while only one is needed for torsional vibration problems.

In this method, the member is divided into an arbitrary number of weightless segments, and the total weight of the beam is represented by point masses at the juncture points of the segments. Within each segment the stiffness properties are assumed constant and equal to the mean value of the segment. If the beam vibrates at a natural frequency ω, the free-body diagram of a typical segment is as shown in Fig. 6-2b. The length of the segment extending between stations $j-1$ and j is l_j and its uniform stiffness is $(EI)_j$. The reversed direction of the inertia force $m_{j-1}\ddot{y}_{j-1}$ is shown in this figure as $m_{j-1}\omega^2 y_{j-1}$.

By applying to the segment in Fig. 6-2b the equilibrium conditions, the following two equations are written:

$$V_j = V_{j-1} + m_{j-1}\omega^2 y_{j-1} \tag{6-28}$$

$$M_j = M_{j-1} + V_{j-1} l_j + m_{j-1}\omega^2 y_{j-1} l_j \tag{6-29}$$

In addition, for the l_j segment the expressions for the scope θ_j and

[18] See Reference 33.

[19] See Reference 34.

[20] See Reference 35.

deflection y_j are written as follows:

$$\theta_j = \theta_{j-1} + \frac{l_j^2}{2(EI)_j}(V_{j-1} + m_{j-1}\omega^2 y_{j-1}) + \frac{l_j}{(EI)_j} M_{j-1}$$

$$= \theta_{j-1} + \frac{l_j^2}{2(EI)_j} V_j + \frac{l_j}{(EI)_j} M_{j-1} \tag{6-30}$$

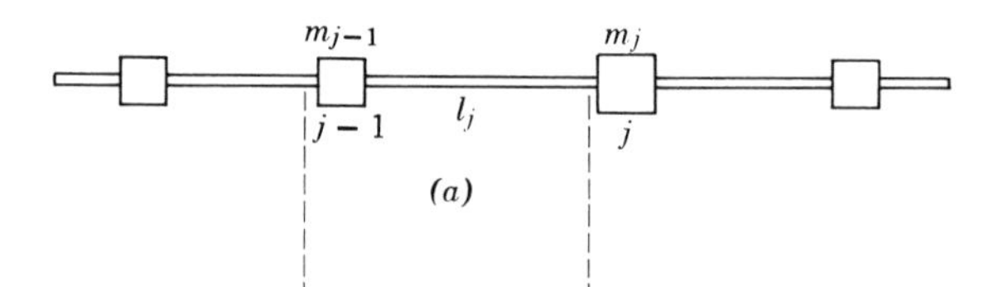

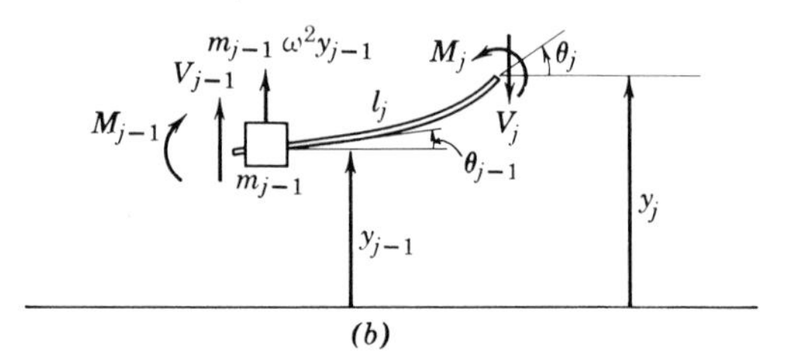

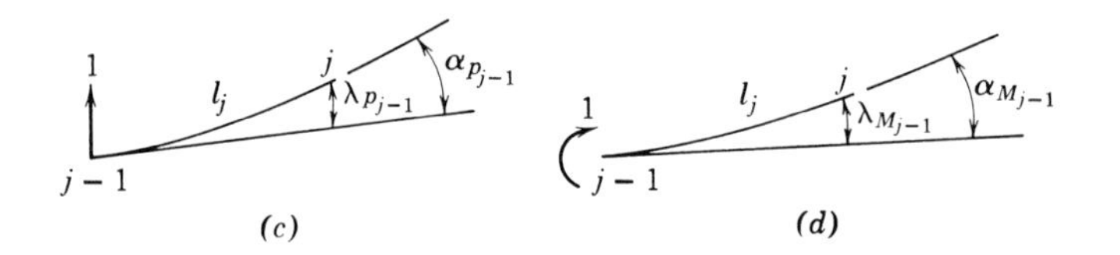

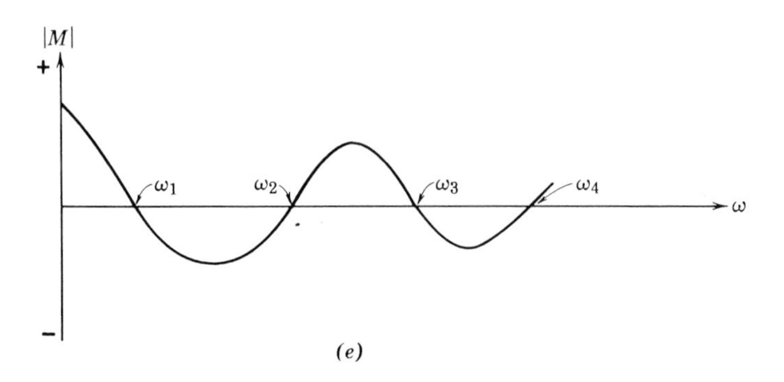

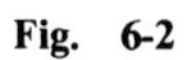

Fig. 6-2

$$y_j = y_{j-1} + \theta_{j-1} l_j + \frac{l_j^3}{6(EI)_j}(V_{j-1} + m_{j-1}\omega^2 y_{j-1}) + \frac{l_j^2}{2(EI)_j} M_{j-1}$$

$$= y_{j-1} + \theta_{j-1} l_j + \frac{l_j^3}{6(EI)_j} V_j + \frac{l_j^2}{2(EI)_j} M_{j-1} \qquad (6\text{-}31)$$

In Eq. 6-30, the coefficient $\alpha_{P_{j-1}} = l_j^2/2(EI)_j$ is the slope of the beam at station j relative to the tangent drawn at station $j-1$, due to a unit force at $j-1$, Fig. 6-2*c*, and $\alpha_{M_{j-1}} = l_j/(EI)_j$ is the slope at station j relative to a tangent at $j-1$, due to a unit moment at $j-1$, Fig. 6-2*d*. In Eq. 6-31, the coefficient $\lambda_{P_{j-1}} = l_j^3/6(EI)_j$ is the vertical distance from station j to the tangent at station $j-1$, Fig. 6-2*c*, due to a unit load at $j-1$, and $\lambda_{M_{j-1}} = l_j^2/2(EI)_j$ is the vertical distance from station j to the tangent at station $j-1$, Fig. 6-2*d*, due to a unit moment at $j-1$. The moment area method was used to compute these coefficients. Equations 6-30 and 6-31 can be also determined by using Eqs. 6-28 and 6-29 and the known expressions from strength of materials relating y, θ, M, and V.

If the number of beam segments l_j is n, that is, $j = 1, 2, \ldots, n$, the concentrated masses $m_0, m_1, m_2, \ldots, m_n$ are located at stations $0, 1, 2, \ldots, n$, respectively. For each station, Eqs. 6-28 through 6-31 can be used to compute the quantities y, θ, M, and V. The procedure is initiated by starting at the first mass concentration point, that is, station zero, where two boundary conditions are known. By inserting these two conditions into Eqs. 6-28 through 6-31, two out of the four quantities y_0, θ_0, M_0, and V_0 can be defined. In addition, at the same station, the normalizing value of unity can be assigned, if desired, to the third quantity. Thus at station zero the unknown quantities will be either one or two if the third quantity is not yet normalized. At present, let it be considered that the unknown quantities are two and assume a value of ω. Under these conditions, the quantities y_1, θ_1, M_1, and V_1 at the second station can be determined from the equations above in terms of numerical values and the two unknowns at station zero.

By proceeding from station to station successively, the four quantities at the last station, that is station n, are expressed in terms of numerical values and the two unknowns from station zero. The two boundary conditions of the last station provide the criteria for the computation of the natural frequencies. These criteria are better illustrated by using an actual beam case.

Consider, for example, a simply supported beam whose left end is the station zero and the right end is the last station n. The boundary conditions at station zero are $y_0 = M_0 = 0$. At station 1, Eqs. 6-28, 6-29, 6-30, and 6-31 yield

$$
\begin{aligned}
V_1 &= V_0 \\
M_1 &= l_1 V_0 \\
\theta_1 &= \theta_0 + \frac{l_1^{\,2}}{2(EI)_1} V_0 \\
y_1 &= l_1\,\theta_0 + \frac{l_1^{\,3}}{6(EI)_1} V_0
\end{aligned}
\tag{6-32}
$$

By using the results in Eq. 6-32 and applying again Eqs. 6-28 through 6-31 for station 2,

$$
\begin{aligned}
V_2 &= m_1\omega^2 l_1\,\theta_0 + \left(1 + \frac{m_1\omega^2 l_1^{\,3}}{6(EI)_1}\right) V_0 \\
M_2 &= m_1\omega^2 l_1 l_2\,\theta_0 + \left(l_1 + l_2 + \frac{m_1\omega^2 l_2 l_1^{\,3}}{6(EI)_1}\right) V_0 \\
\theta_2 &= \left(1 + \frac{m_1\omega^2 l_2^2 l_1}{2(EI)_2}\right)\theta_0 \\
&\quad + \left(\frac{l_1^{\,2}}{2(EI)_1} + \frac{l_1 l_2}{(EI)_2} + \frac{l_2^{\,2}}{2(EI)_2} + \frac{m_1\omega^2 l_1^{\,3} l_2^{\,2}}{12(EI)_1(EI)_2}\right) V_0 \\
y_2 &= \left(l_1 + l_2 + \frac{m_1\omega^2 l_1 l_2^{\,3}}{6(EI)_2}\right)\theta_0 \\
&\quad + \left(\frac{l_1^{\,3}}{6(EI)_1} + \frac{l_1^{\,2} l_2}{2(EI)_1} + \frac{l_1 l_2^{\,2}}{2(EI)_2} + \frac{l_2^{\,3}}{6(EI)_2} + \frac{m_1\omega^2 l_1^{\,3} l_2^{\,3}}{36(EI)_1(EI)_2}\right) V_0
\end{aligned}
\tag{6-33}
$$

In a similar manner, the computations can be carried out until the nth station is reached. It is easily observed that the quantities V, M, θ, and y at each station are linear combinations of the unknown quantities V_0 and θ_0 of station 1. Thus, at station n, the following expressions can be written for V_n, M_n, θ_n, and y_n:

$$
\begin{aligned}
V_n &= A_1\,\theta_0 + A_2 V_0 \\
M_n &= B_1\,\theta_0 + B_2 V_0 \\
\theta_n &= C_1\,\theta_0 + C_2 V_0 \\
y_n &= D_1\,\theta_0 + D_2 V_0
\end{aligned}
\tag{6-34}
$$

where the constants A_1, A_2, B_1, B_2, and so on, involve quantities such as those of the coefficients of θ_0 and V_0 in Eq. 6-33. Thus, for an assumed frequency ω, these constants can take numerical values.

At station n the boundary conditions of the simply supported beam are $y_n = M_n = 0$, and Eq. 6-34 yields

$$\begin{aligned} B_1\,\theta_0 + B_2 V_0 &= 0 \\ D_1\,\theta_0 + D_2 V_0 &= 0 \end{aligned} \tag{6-35}$$

A nontrivial solution is obtained only if the determinant of the coefficients of θ_0 and V_0 in Eq. 6-35 is zero. This yields

$$|M| = \begin{vmatrix} B_1 & B_2 \\ D_1 & D_2 \end{vmatrix} = 0 \tag{6-36}$$

The determinant in Eq. 6-36 is known as the frequency determinant. The values of ω that satisfy Eq. 6-36 are the natural frequencies of the beam. A convenient way to determine these frequencies would be to plot the values of the determinant $|M|$ against assumed values of ω as shown in Fig. 6-2*e*. The zero intersects are the natural frequencies $\omega_1, \omega_2, \ldots$, of the beam. For each value of ω, the corresponding mode shape is determined by substituting this value into Eqs. 6-28 through 6-31 and computing the value of y at each station by following the procedure used to write Eqs. 6-32 and 6-33. In performing these computations, the normalized value of unity can be used for V_0, while θ_0, since ω is known, can be determined by using one of the expressions in Eq. 6-35. Thus numerical values of y are obtained at all stations, which define the mode shape of the beam.

Equations 6-28 through 6-31 can be also expressed in matrix notation that yields a relationship between y, θ, M, and V at station j and the same quantities at station $j-1$. This relationship is

$$\begin{bmatrix} y \\ \theta \\ M \\ V \end{bmatrix}_j = \begin{bmatrix} \left(\dfrac{m_{j-1}\omega^2 l_j^{\,3}}{6(EI)_j} + 1\right) & l_j & \dfrac{l_j^{\,2}}{2(EI)_j} & \dfrac{l_j^{\,3}}{6(EI)_j} \\ \dfrac{m_{j-1}\omega^2 l_j^{\,2}}{2(EI)_j} & 1 & \dfrac{l_j}{(EI)_j} & \dfrac{l_j^{\,2}}{2(EI)_j} \\ m_{j-1}\omega^2 l_j & 0 & 1 & l_j \\ m_{j-1}\omega^2 & 0 & 0 & 1 \end{bmatrix} \begin{bmatrix} y \\ \theta \\ M \\ V \end{bmatrix}_{j-1} \tag{6-37}$$

or

$$\{v\}_j = [K]_{j-1}\{v\}_{j-1} \tag{6-38}$$

The column vectors $\{v\}_j$ and $\{v\}_{j-1}$ at stations j and $j-1$, respectively, are called the state vectors, and they involve the quantities y, θ, M, and V at these two stations. The relationship between y, θ, M, and V at stations j and $j-1$ is obtained by making use of the square matrix $[K]_{j-1}$ which is called the transfer matrix at station $j-1$.

The matrix equation introduces the concept of transfer matrices which provides a convenient way to analyze structural systems for both static and dynamic response. The sections that follow are devoted to the construction of transfer matrices for various types of structural systems and the way in which such transfer matrices are used to solve structural[21] problems.

6-4 TRANSFER MATRICES FOR SPRING-MASS SYSTEMS

The construction of transfer matrices for spring-mass systems is discussed in this section.

Consider the system in Fig. 6-3*a* consisting of the masses m_{j-1}, m_j, and m_{j+1} connected by the linear massless springs of constants k_{j-1}, k_j, k_{j+1}, and k_{j+2}. The masses are permitted to move only in the horizontal direction, and their displacements are denoted by the symbols x_{j-1}, x_j, and x_{j+1} as shown.

If the system is vibrating freely at a circular frequency ω, the free-body diagrams in Fig. 6-3*b* depict the forces and displacements on each side of spring k_j and mass m_j. The letter L is used to denote left-hand side and R to denote right-hand side.

For the spring in Fig. 6-3*b*, the following two equations are written:

$$F_{j-1}^R = F_j^L \tag{6-39}$$

$$x_j^L = x_{j-1}^R + \frac{F_{j-1}^R}{k_j} \tag{6-40}$$

In matrix form, the two equations are written as

$$\begin{bmatrix} x \\ F \end{bmatrix}_j^L = \begin{bmatrix} 1 & \dfrac{1}{k_j} \\ 0 & 1 \end{bmatrix} \begin{bmatrix} x \\ F \end{bmatrix}_{j-1}^R \tag{6-41}$$

[21]An extensive treatment of this subject can be found in the works of Pestel and Leckie, Reference 15, and in the works of Argyris, References 36 through 39.

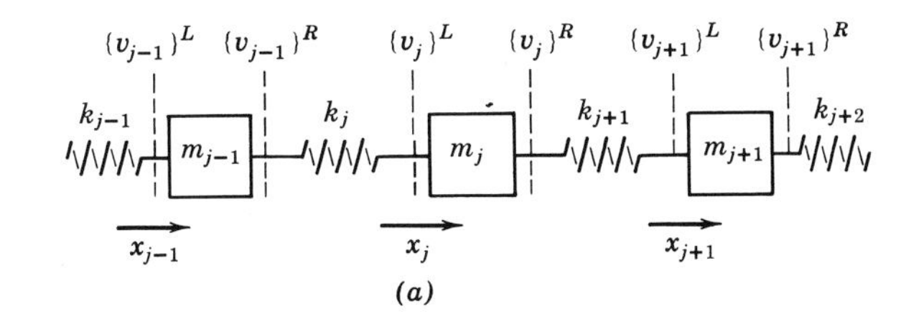

(a)

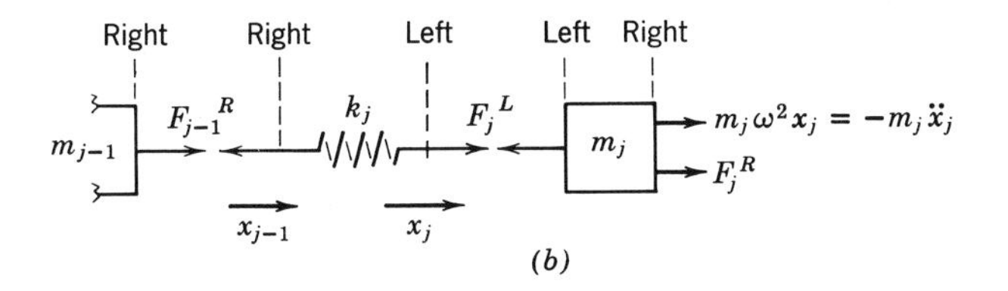

(b)

Fig. 6-3

or

$$\{v_j\}^L = [F]\{v_{j-1}\}^R \tag{6-42}$$

In the equation above,

$$\{v_j\}^L = \begin{bmatrix} x \\ F \end{bmatrix}_j^L \qquad \{v_{j-1}\}^R = \begin{bmatrix} x \\ F \end{bmatrix}_{j-1}^R \tag{6-43}$$

are the state vectors involving the displacements x_j and x_{j-1} at stations j and $j-1$, respectively, and the internal forces F_j and F_{j-1} at the same stations. The matrix

$$[F] = \begin{bmatrix} 1 & \dfrac{1}{k_j} \\ 0 & 1 \end{bmatrix} \tag{6-44}$$

is the field transfer matrix that expresses $\{v_j\}^L$ in terms of $\{v_{j-1}\}^R$.

For the mass m_j in Fig. 6-3b, the set of two equations is

$$x_j^R = x_j^L \tag{6-45}$$

$$F_j^R = -m_j\omega^2 x_j + F_j^L \tag{6-46}$$

In matrix form they are written as

$$\begin{bmatrix} x \\ F \end{bmatrix}_j^R = \begin{bmatrix} 1 & 0 \\ -m_j\omega^2 & 1 \end{bmatrix} \begin{bmatrix} x \\ F \end{bmatrix}_j^L \tag{6-47}$$

or

$$\{v_j\}^R = [P]\{v_j\}^L \tag{6-48}$$

In this expression $\{v_j\}^R$ and $\{v_j\}^L$ are the state vectors to the right and left, respectively, of the mass m_j, and

$$P = \begin{bmatrix} 1 & 0 \\ -m_j\omega^2 & 1 \end{bmatrix} \tag{6-49}$$

is the point transfer matrix expressing $\{v_j\}^R$ in terms of $\{v_j\}^L$.

By substituting Eq. 6-42 into Eq. 6-48,

$$\{v_j\}^R = [P][F]\{v_{j-1}\}^R \tag{6-50}$$

Equation 6-50 expresses $\{v_j\}^R$ in terms of $\{v_{j-1}\}^R$ in Fig. 6-3a. By carrying out the indicated matrix multiplication,

$$\begin{bmatrix} x \\ F \end{bmatrix}_j^R = \begin{bmatrix} 1 & \dfrac{1}{k_j} \\ -m_j\omega^2 & \left(1 - \dfrac{m_j\omega^2}{k_j}\right) \end{bmatrix} \begin{bmatrix} x \\ F \end{bmatrix}_{j-1}^R \tag{6-51}$$

$$\{v_n\}^R = [P_n][F_n][P_{n-1}][F_{n-1}]\cdots[P_1][F_1]\{v_0\}^R \tag{6-52}$$

or

$$\{v_n\}^R = [M]\{v_0\}^R \tag{6-53}$$

where

$$[M] = [P_n][F_n][P_{n-1}][F_{n-1}]\cdots[P_1][F_1] \tag{6-54}$$

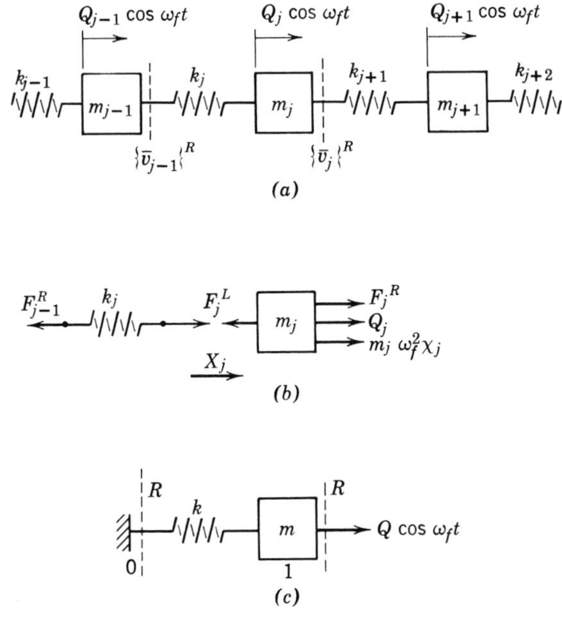

Fig. 6-4

Equation 6-53 expresses the state vector $\{v_n\}^R$ at the nth station in term of the state vector $\{v_0\}^R$ of the first station.

If the spring-mass system is also subjected to dynamic forces as shown in Fig. 6-4*a*, the construction of transfer matrices can be carried out in a similar manner. From Fig. 6-4*b*, the two equations for the mass m_i are

$$X_j^R = X_j^L \tag{6-55}$$

$$F_j^R = F_j^L - m_j\omega_f^2 X_j^L - Q_j \tag{6-56}$$

where X_j and Q_j are the maximum values of the displacement and force, respectively, at station j.

Equations 6-55 and 6-56 are expressed in matrix form as

$$\begin{bmatrix} X \\ F \\ \hline 1 \end{bmatrix}_j^R = \left[\begin{array}{cc|c} 1 & 0 & 0 \\ -m_j\omega_f^2 & 1 & -Q_j \\ \hline 0 & 0 & 1 \end{array}\right] \begin{bmatrix} X \\ F \\ \hline 1 \end{bmatrix}_j^L \tag{6-57}$$

or

$$\{\bar{v}_j\}^R = [\bar{P}]\{\bar{v}_j\}^L \tag{6-58}$$

Above, $\{\bar{v}_j\}^R$ and $\{\bar{v}_j\}^L$ are the extended state vectors and $[\bar{P}]$ is the extended transfer point matrix.

By using the free-body diagram of the spring k_j in Fig. 6-4*b*, the two equations are written as

$$F_j^L = F_{j-1}^R \tag{6-59}$$

$$X_j^L = X_{j-1}^R + \frac{F_{j-1}^R}{k_j} \tag{6-60}$$

In matrix form, the two expressions yield

$$\begin{bmatrix} X \\ F \\ \hline 1 \end{bmatrix}_j^L = \left[\begin{array}{cc|c} 1 & \dfrac{1}{k_j} & 0 \\ 0 & 1 & 0 \\ \hline 0 & 0 & 1 \end{array}\right] \begin{bmatrix} X \\ F \\ \hline 1 \end{bmatrix}_{j-1}^R \tag{6-61}$$

or

$$\{\bar{v}_j\}^L = [\bar{F}]\{\bar{v}_{j-1}\}^R \tag{6-62}$$

where $\{\bar{v}_j\}^L$ and $\{\bar{v}_{j-1}\}^R$ are the extended state vectors, and $[\bar{F}]$ is the extended field matrix.

By substituting Eq. 6-62 into Eq. 6-58,

$$\{\bar{v}_j\}^R = [\bar{P}][\bar{F}]\{\bar{v}_{j-1}\}^R \tag{6-63}$$

which expresses $\{\bar{v}_j\}^R$ in terms of $\{\bar{v}_{j-1}\}^R$ in Fig. 6-4*a*. By carrying out the indicated matrix multiplication, Eq. 6-63 yields

$$\begin{bmatrix} X \\ F \\ \hline 1 \end{bmatrix}_j^R = \left[\begin{array}{cc|c} 1 & \dfrac{1}{k_j} & 0 \\ -m_j\omega_f^2 & \left(1 - \dfrac{m_j\omega_f^2}{k_j}\right) & -Q_j \\ \hline 0 & 0 & 1 \end{array}\right] \begin{bmatrix} X \\ F \\ \hline 1 \end{bmatrix}_{j-1}^R \tag{6-64}$$

In the next section the transfer matrices constructed above are used to determine the vibration response of spring-mass systems.

6-5 VIBRATION OF SPRING-MASS SYSTEMS BY TRANSFER MATRICES

Consider the spring-mass system in Fig. 6-5*a* that is constrained to move in the horizontal direction only, and let it be assumed that it is required to determine the natural frequencies and mode shapes of its undamped free vibration. From Eq. 6-52, the expression relating $\{v_0\}^R$ and $\{v_2\}^R$ is

$$\{v_2\}^R = [P_2][F_2][P_1][F_1]\{v_0\}^R \tag{6-65}$$

where $[P_1]$ and $[P_2]$ are the point transfer matrices for stations 1 and 2, respectively, and $[F_1]$ and $[F_2]$ are the field transfer matrices for the springs.

By utilizing Eqs. 6-44 and 6-49 and carrying out the indicated matrix

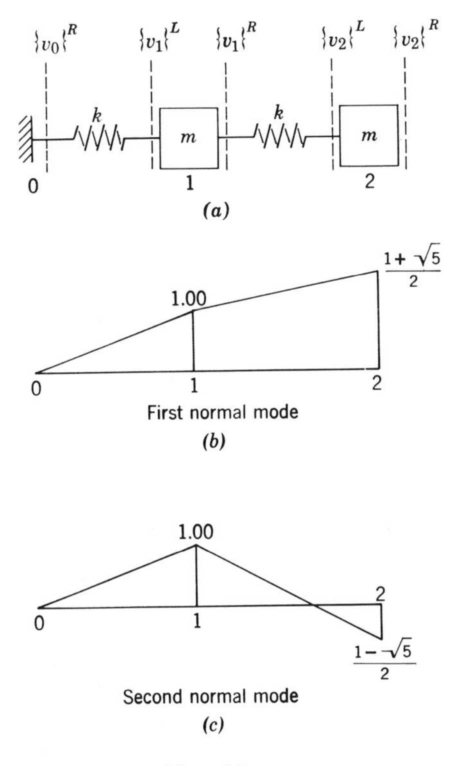

Fig. 6-5

multiplications, Eq. 6-65 yields

$$\begin{bmatrix} x \\ 0 \end{bmatrix}_2^R = \begin{bmatrix} & \frac{1}{k}\left(2-\frac{m\omega^2}{k}\right) \\ & \left(\frac{m^2\omega^4}{k^2}-\frac{3m\omega^2}{k}+1\right) \end{bmatrix} \begin{bmatrix} 0 \\ F \end{bmatrix}_0^R \tag{6-66}$$

It should be noted that the displacement x_0^R and spring force F_2^R are zero. In Eq. 6-66, the computations in the first column of the matrix are omitted because they are not required in the solution.

The second row of the matrix in Eq. 6-66 yields the frequency equation

$$\frac{m^2\omega^4}{k^2}-\frac{3m\omega^2}{k}+1=0$$

or

$$\omega^4-\frac{3K}{m}\omega^2+\frac{k^2}{m^2}=0 \tag{6-67}$$

Thus

$$\omega_1^2=\frac{k}{m}\frac{3-\sqrt{5}}{2} \qquad \omega_2^2=\frac{k}{m}\frac{3+\sqrt{5}}{2} \tag{6-68}$$

The two positive values ω_1 and ω_2 are the two natural frequencies of the system.

With known ω_1^2 and ω_2^2, the corresponding mode shapes can be determined. For example, the state vectors at stations 1 and 2 in terms of the state vector at station zero, individually, are as follows:

$$\{v_1\}^L=[F_1]\{v_0\}^R=\begin{bmatrix} & \frac{1}{k} \\ & 1 \end{bmatrix}\begin{bmatrix} 0 \\ F \end{bmatrix}_0^R$$

$$\{v_1\}^R=[P_1][F_1]\{v_0\}^R=\begin{bmatrix} & \frac{1}{k} \\ & \left(1-\frac{m\omega^2}{k}\right) \end{bmatrix}\begin{bmatrix} 0 \\ F \end{bmatrix}_0^R$$

$$\{v_2\}^L = [F_2][P_1][F_1]\{v_0\}^R = \begin{bmatrix} \begin{vmatrix} \frac{1}{k}\left(2 - \frac{m\omega^2}{k}\right) \\ \left(1 - \frac{m\omega^2}{k}\right) \end{vmatrix} \end{bmatrix} \begin{bmatrix} 0 \\ F \end{bmatrix}_0^R$$

$$\{v_2\}^R = [P_2][F_2][P_1][F_1]\{v_0\}^R = \begin{bmatrix} \begin{vmatrix} \frac{1}{k}\left(2 - \frac{m\omega^2}{k}\right) \\ \left(\frac{m^2\omega^4}{k^2} - \frac{3m\omega^2}{k} + 1\right) \end{vmatrix} \end{bmatrix} \begin{bmatrix} 0 \\ F \end{bmatrix}_0^R$$

By using the values given by Eq. 6-68 and assuming that $F_0 = 1$, the relations yield the following results: For $\omega^2 = \omega_1^2$,

$$\{v_1\}^L = \begin{bmatrix} 0 & \frac{1}{k} \\ 0 & 1 \end{bmatrix} \quad \{v_1\}^R = \begin{bmatrix} 0 & \frac{1}{k} \\ 0 & \left(\frac{-1+\sqrt{5}}{2}\right) \end{bmatrix}$$

$$\{v_2\}^L = \begin{bmatrix} 0 & \frac{1+\sqrt{5}}{2k} \\ 0 & \left(\frac{-1+\sqrt{5}}{2}\right) \end{bmatrix} \quad \{v_2\}^R = \begin{bmatrix} 0 & \frac{1+\sqrt{5}}{2k} \\ 0 & 0 \end{bmatrix}$$

For $\omega^2 = \omega_2^2$,

$$\{v_1\}^L = \begin{bmatrix} 0 & \frac{1}{k} \\ 0 & 1 \end{bmatrix} \quad \{v_1\}^R = \begin{bmatrix} 0 & \frac{1}{k} \\ 0 & -\left(\frac{1+\sqrt{5}}{2}\right) \end{bmatrix}$$

$$\{v_2\}^L = \begin{bmatrix} 0 & \left(\frac{1-\sqrt{5}}{2k}\right) \\ 0 & -\left(\frac{1+\sqrt{5}}{2}\right) \end{bmatrix} \quad \{v_2\}^R = \begin{bmatrix} 0 & \left(\frac{1-\sqrt{5}}{2k}\right) \\ 0 & 0 \end{bmatrix}$$

The first row in each of these matrix expressions gives the amplitude at the specified station. The amplitude at station 0 is zero. By setting arbitrarily the amplitude x_1 at station 1 as equal to unity, the mode shape corresponding to the frequency ω_1 is given by the amplitudes

$$x_0 = 0 \qquad x_1 = 1 \qquad x_2 = \frac{1+\sqrt{5}}{2}$$

and the one corresponding to ω_2 is

$$x_0 = 0 \qquad x_1 = 1 \qquad x_2 = \frac{1-\sqrt{5}}{2}$$

The mode shapes are illustrated in Figs. 6-5*b* and 6-5*c*.

As a second example, the steady-state response of the system in Fig. 6-4*c* is determined. By utilizing the boundary conditions at the ends of the system and applying Eq. 6-64,

$$\begin{bmatrix} X \\ 0 \\ \hline 1 \end{bmatrix}_1^R = \left[\begin{array}{cc|c} 1 & \dfrac{1}{k} & 0 \\ -m\omega_f^2 & \left(1 - \dfrac{m\omega_f^2}{k}\right) & -Q \\ \hline 0 & 0 & 1 \end{array}\right] \begin{bmatrix} 0 \\ F \\ \hline 1 \end{bmatrix}_0^R$$

The matrix equation yields the expressions

$$X_1 = \frac{F_0}{k}$$

$$F_0\left(1 - \frac{m\omega_f^2}{k}\right) - Q = 0$$

The simultaneous solution of the two equations gives

$$Q = kX_1 - m\omega_f^2 X_1$$

or

$$X_1 = \frac{Q}{k - m\omega_f^2} = \frac{Q}{k}\,\frac{1}{1-(\omega_f/\omega)^2}$$

This is the same as Eq. 2-6 in Section 2-2, representing the maximum amplitude of m during motion. Its time variation can be assumed to be the same as that of the harmonic force $Q\cos\omega_f t$. Other problems can be solved in a similar manner.

6-6 TRANSFER MATRICES FOR FLEXURAL SYSTEMS

The vibration response of beamlike systems of either uniform or variable EI and mass can be determined by using field and point transfer matrices. The mass of such systems is usually lumped at discrete points along their length as discussed in Section 6-3. The elements of the state vector $\{v_j\}$ at a station j are the displacements y_j and θ_j, the internal moment M_j, and the internal shear force V_j.

Consider for example the system in Fig. 6-6*a* that is assumed to vibrate freely in the transverse direction. At a mode of vibration the free-body diagram of the lumped mass m_j is shown in Fig. 6-6*b*. By utilizing the continuity conditions for moment, slope, and deflection as well as vertical equilibrium, the following four equations are written:

$$\begin{aligned} y_j^R &= y_j^L \qquad & M_j^R &= M_j^L \\ \theta_j^R &= \theta_j^L \qquad & V_j^R &= V_j^L - m_j\omega^2 y_j^L \end{aligned} \tag{6-69}$$

In matrix form the four equations are

$$\begin{bmatrix} -y \\ \theta \\ M \\ V \end{bmatrix}_j^R = \begin{bmatrix} 1 & 0 & 0 & 0 \\ 0 & 1 & 0 & 0 \\ 0 & 0 & 1 & 0 \\ m_j\omega^2 & 0 & 0 & 1 \end{bmatrix} \begin{bmatrix} -y \\ \theta \\ M \\ V \end{bmatrix}_j^R \tag{6-70}$$

or

$$\{v_j\}^R = [P]\{v_j\}^L \tag{6-71}$$

where $\{v_j\}^R$ and $\{v_j\}^L$ are the state vectors at station j and $[P]$ is the point transfer matrix. The negative sign for y is used in Eq. 6-70 so that all the elements of $[P]$ are shown as positive.

The free-body diagram of the massless element l_j is shown in Fig. 6-6*b* and its deformed configuration in Fig. 6-6*c*. By applying vertical equilibrium and also moment equilibrium about station $j-1$, the following two

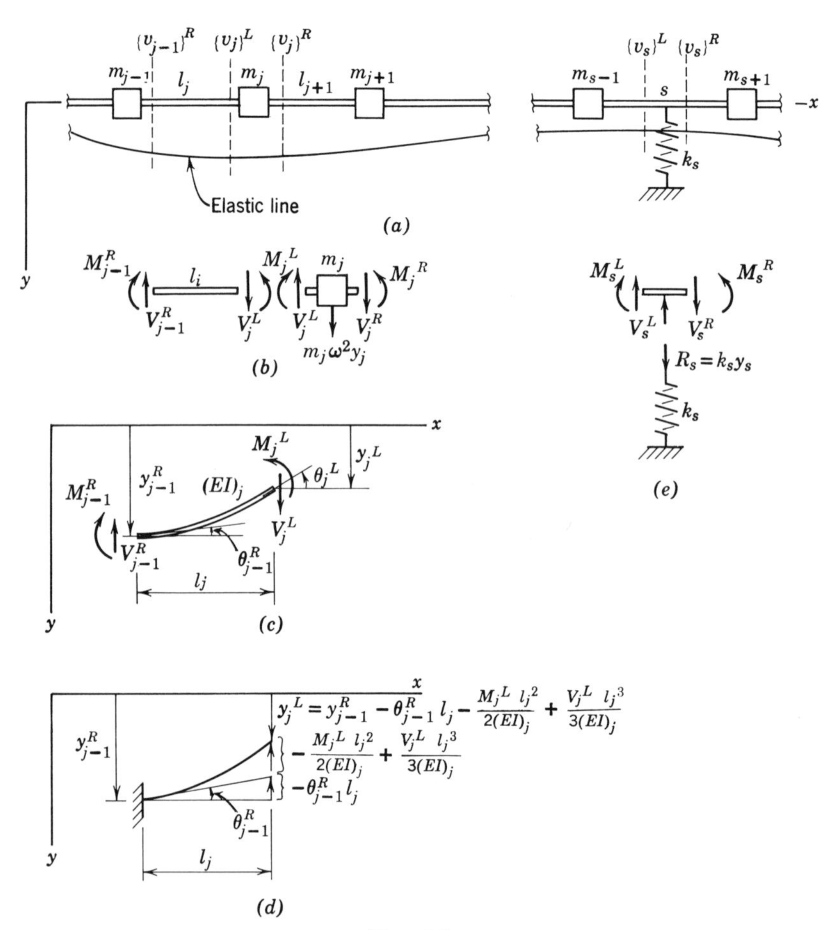

Fig. 6-6

equations are obtained:

$$V_j^L = V_{j-1}^R \tag{6-72}$$

$$M_j^L = M_{j-1}^R + V_j^L l_j \tag{6-73}$$

In addition, the expressions for y_j^L and θ_j^L are written as

$$y_j^L = y_{j-1}^R - \theta_{j-1}{}^R l_j + \left[-\frac{M_j^L l_j{}^2}{2(EI)_j} + \frac{V_j^L l_j{}^3}{2(EI)_j} \right] \tag{6-74}$$

$$\theta_j^L=\theta_{j-1}^R+\left[\frac{M_j^L l_j}{(EI)_j}-\frac{V_j^L l_j^{\,2}}{2(EI)_j}\right] \tag{6-75}$$

The bracketed terms are obtained by assuming l_j to be a cantilever beam fixed at $j-1$ and acted upon by the moment M_j^L and shear force V_j^L at j. See for example Fig. 6-6*d*. By utilizing Eqs. 6-72 and 6-73, Eqs. 6-74 and 6-75 are written as

$$-y_j^L=-y_j^R+l_j\,\theta_{j-1}^R+\frac{l_j^{\,2}}{2(EI)_j}M_{j-1}^R+\frac{l_j^{\,3}}{6(EI)_j}V_{j-1}^R \tag{6-76}$$

$$\theta_j^L=\theta_{j-1}^R+\frac{l_j}{(EI)_j}M_{j-1}^R+\frac{l_j^{\,2}}{2(EI)_j}V_{j-1}^R \tag{6-77}$$

In matrix form, Eqs. 6-72, 6-73, 6-76, and 6-77 yield

$$\begin{bmatrix}-y\\ \theta\\ M\\ V\end{bmatrix}_j^L=\begin{bmatrix}1 & l_j & \dfrac{l_j^{\,2}}{2(EI)_j} & \dfrac{l_j^{\,3}}{6(EI)_j}\\ 0 & 1 & \dfrac{l_j}{(EI)_j} & \dfrac{l_j^{\,2}}{2(EI)_j}\\ 0 & 0 & 1 & l_j\\ 0 & 0 & 0 & 1\end{bmatrix}\begin{bmatrix}-y\\ \theta\\ M\\ V\end{bmatrix}_{j-1}^R \tag{6-78}$$

or

$$\{v_j\}^L=[F]\{v_{j-1}\}^R \tag{6-79}$$

where $[F]$ is the field transfer matrix relating the state vectors $\{v_j\}^L$ and $\{v_{j-1}\}^R$ in Fig. 6-6*a*.

By substituting Eq. 6-79 into Eq. 6-71, a matrix expressing $\{v_j\}^R$ in terms of $\{v_{j-1}\}^R$ is obtained. That is,

$$\{v_j\}^R=[P][F]\{v_{j-1}\}^R \tag{6-80}$$

Carrying out the indicated matrix multiplication, Eq. 6-80 yields

$$\begin{bmatrix} -y \\ \theta \\ M \\ V \end{bmatrix}_j^R = \begin{bmatrix} 1 & l_j & \dfrac{l_j^{\,2}}{2(EI)_j} & \dfrac{l_j^{\,3}}{6(EI)_j} \\ 0 & 1 & \dfrac{l_j}{(EI)_j} & \dfrac{l_j^{\,2}}{2(EI)_j} \\ 0 & 0 & 1 & 1 \\ m_j\omega^2 & m_j\omega^2 l_j & \dfrac{m_j\omega^2 l_j^{\,2}}{2(EI)_j} & \left(1+\dfrac{m_j\omega^2 l_j^{\,3}}{6(EI)_j}\right) \end{bmatrix} \begin{bmatrix} -y \\ \theta \\ M \\ V \end{bmatrix}_{j-1}^R \tag{6-81}$$

If a member is supported elastically at a point s by a linear spring of constant k_s, Fig. 6-6a, a point transfer matrix relating the state vectors $\{v_s\}^L$ and $\{v_s\}^R$ is easily obtained. The four equations relating the elements of $\{v_s\}^L$ and $\{v_s\}^R$, Figs. 6-6a and 6-6e, are

$$\begin{aligned} -y_s^R &= -y_s^L & M_s^R &= M_s^L \\ \theta_s^R &= \theta_s^L & V_s^R &= k_s y_s + V_s^L \end{aligned} \tag{6-82}$$

In matrix form the four equations are written as

$$\begin{bmatrix} -y \\ \theta \\ M \\ V \end{bmatrix}_s^R = \begin{bmatrix} 1 & 0 & 0 & 0 \\ 0 & 1 & 0 & 0 \\ 0 & 0 & 1 & 0 \\ -k_s & 0 & 0 & 1 \end{bmatrix} \begin{bmatrix} -y \\ \theta \\ M \\ V \end{bmatrix}_s^L \tag{6-83}$$

or

$$\{v_s\}^R = [P]\{v_s\}^L \tag{6-84}$$

where $[P]$ is the point transfer matrix relating $\{v_s\}^R$ and $\{v_s\}^L$.

These transfer matrices can take a more convenient form for computational purposes by making the substitutions

$$\begin{aligned} y^* &= \frac{y}{l} & M^* &= \frac{Ml}{EI} \\ \theta^* &= \theta & V^* &= \frac{Vl^2}{EI} \end{aligned} \tag{6-85}$$

and designating as $a(EI)$ and bl the variations in stiffness EI and segments l, respectively, if such variations exist. On this basis, Eq. 6-78 is written as

$$\begin{bmatrix} -y^* \\ \theta^* \\ M^* \\ V^* \end{bmatrix}_j^L = \begin{bmatrix} 1 & b & \dfrac{b^2}{2a} & \dfrac{b^3}{6a} \\ 0 & 1 & \dfrac{b}{a} & \dfrac{b^2}{2a} \\ 0 & 0 & 1 & b \\ 0 & 0 & 0 & 1 \end{bmatrix} \begin{bmatrix} -y^* \\ \theta^* \\ M^* \\ V^* \end{bmatrix}_{j-1}^R \tag{6-86}$$

or

$$\{v_j^*\}^L = [F_j^*]\{v_{j-1}^*\}^R \tag{6-87}$$

To conform with the new notation, the point transfer matrix given by Eq. 6-70 is written as

$$\begin{bmatrix} -y^* \\ \theta^* \\ M^* \\ V^* \end{bmatrix}_j^R = \begin{bmatrix} 1 & 0 & 0 & 0 \\ 0 & 1 & 0 & 0 \\ 0 & 0 & 1 & 0 \\ \dfrac{\omega^2 l_j^3 m_j}{(EI)_j} & 0 & 0 & 1 \end{bmatrix} \begin{bmatrix} -y^* \\ \theta^* \\ M^* \\ V^* \end{bmatrix}_j^L \tag{6-88}$$

or

$$\{v_j^*\}^R = [P_j^*]\{v_j^*\}^L \tag{6-89}$$

For the same reason, Eq. 6-83 is rearranged as

$$\begin{bmatrix} -y^* \\ \theta^* \\ M^* \\ V^* \end{bmatrix}_s^R = \begin{bmatrix} 1 & 0 & 0 & 0 \\ 0 & 1 & 0 & 0 \\ 0 & 0 & 1 & 0 \\ -\dfrac{k_s l_s^3}{(EI)_s} & 0 & 0 & 1 \end{bmatrix} \begin{bmatrix} -y^* \\ \theta^* \\ M^* \\ V^* \end{bmatrix}_s^L \tag{6-90}$$

or

$$\{v_s^*\}^R = [P_s^*]\{v_s^*\}^L \tag{6-91}$$

The changes made in these equations are self-explanatory.

6-7 FLEXURAL VIBRATIONS BY TRANSFER MATRICES

The transfer matrices derived in the preceding section are used here to solve a beam problem. It should be pointed out, however, that the computations for the more complicated problems become tedious, and the use of an electronic digital computer is suggested. Matrix multiplication programs are usually available.

Consider the weightless beam in Fig. 6-7 supporting the heavy masses m and $2m$ at the indicated locations. The beam vibrates freely in the transverse direction, and it is required to determine its two natural frequencies of vibration and the corresponding mode shapes.

The relationship between the state vectors $\{v_0\}^R$ and $\{v_3\}^R$ in Fig. 6-7 is obtained by eliminating the intermediate state vectors as in Sections 6-4 and 6-5. This yields

$$\{v_3^*\}^R = [P_3^*][F_3^*][P_2^*][F_2^*][P_1^*][F_1^*]\{v_0^*\}^R$$

$$= [M^*]\{v_0^*\}^R \tag{6-92}$$

where

$$[M^*] = [P_3^*][F_3^*][P_2^*][F_2^*][P_1^*][F_1^*] \tag{6-93}$$

The product $[M^*]$ is determined by using the modified matrix equations (6-86), (6-88), and (6-90), as needed, and applying the multiplication scheme explained in the appendix. This is shown in Table 6-1. The first and third columns are omitted because they are not needed in the frequency computations.

The boundary conditions at station 0 are $y_0{}^{*R} = M_0^{*R} = 0$, and those on the right-hand side of station 3 are $M_3^{*R} = V_3^{*R} = 0$. By taking into consideration these boundary conditions, the required natural frequencies

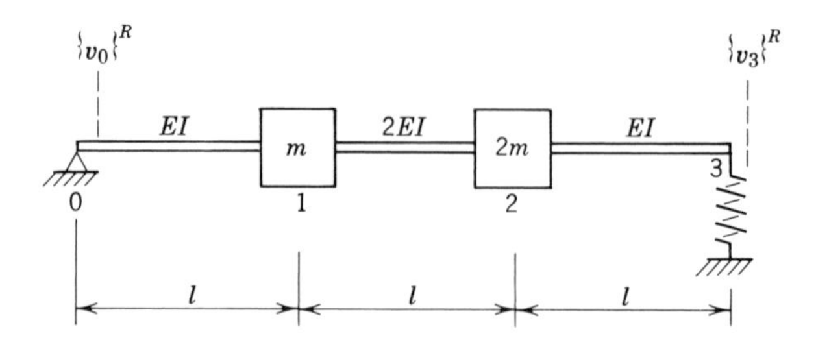

Fig. 6-7

should be determined by considering the last two rows of the product [M *], Table 6-1, and Eq. 6-92. This yields

$$\left[\frac{\omega^4 l^6 m^2}{6(EI)^2}+\frac{6\omega^2 l^3 m}{EI}\right]\theta_0{}^{*R}+\left[\frac{\omega^4 l^6 m^2}{36(EI)^2}+\frac{7\omega^2 l^3 m}{3EI}+3\right]V_0^{*R}=0 \qquad (6\text{-}94)$$

$$\left[\frac{\omega^4 l^6 m^2}{36(EI)^2}\left(6-\frac{kl^3}{EI}\right)+\frac{5\omega^2 l^3 m}{3EI}\left(3-\frac{kl^3}{EI}\right)-\frac{3kl^3}{EI}\right]\theta_0{}^{*R}$$

$$+\left[\frac{\omega^4 l^6 m^2}{216(EI)^2}\left(6-\frac{kl^3}{EI}\right)+\frac{\omega^2 l^3 m}{6EI}\left(13-\frac{3kl^3}{EI}\right)-\frac{41kl^3}{12EI}+1\right]V_0^{*R}=0$$

(6-95)

or, by using Eq. 6-85,

$$\left[\frac{\omega^4 l^4 m^2}{6EI}+6\omega^2 lm\right]\theta_0^R+\left[\frac{\omega^4 l^6 m^2}{36(EI)^2}+\frac{7\omega^2 l^3 m}{3EI}+3\right]V_0^R=0$$

$$\left[\frac{\omega^4 l^4 m^2}{36EI}\left(6-\frac{kl^3}{EI}\right)+\frac{5}{3}\omega^2 lm\left(3-\frac{kl^3}{EI}\right)-3kl\right]\theta_0^R$$

$$+\left[\frac{\omega^4 l^6 m^2}{216(EI)^2}\left(6-\frac{kl^3}{EI}\right)+\frac{\omega^2 l^3 m}{6EI}\left(13-\frac{3kl^3}{EI}\right)-\frac{41kl^3}{12EI}+1\right]V_0^R=0$$

For a nontrivial solution, the determinant of the coefficients of θ_0^R and V_0^R must be zero. That is,

$$\begin{vmatrix} \left[\frac{\omega^4 l^4 m^2}{6EI}+6\omega^2 lm\right] & \left[\frac{\omega^4 l^6 m^2}{36(EI)^2}+\frac{7\omega^2 l^3 m}{3EI}+3\right] \\ \left[\frac{\omega^4 l^4 m^2}{36(EI)}\left(6-\frac{kl^3}{EI}\right)+\frac{5}{3}\omega^2 lm\left(3-\frac{kl^3}{EI}\right)-3kl\right] & \left[\frac{\omega^4 l^6 m^2}{216(EI)^2}\left(6-\frac{kl^3}{EI}\right)+\frac{\omega^2 l^3 m}{6EI}\left(13-\frac{3kl^3}{EI}\right)-\frac{41kl^3}{12EI}+1\right] \end{vmatrix}=0$$

TABLE 6-1

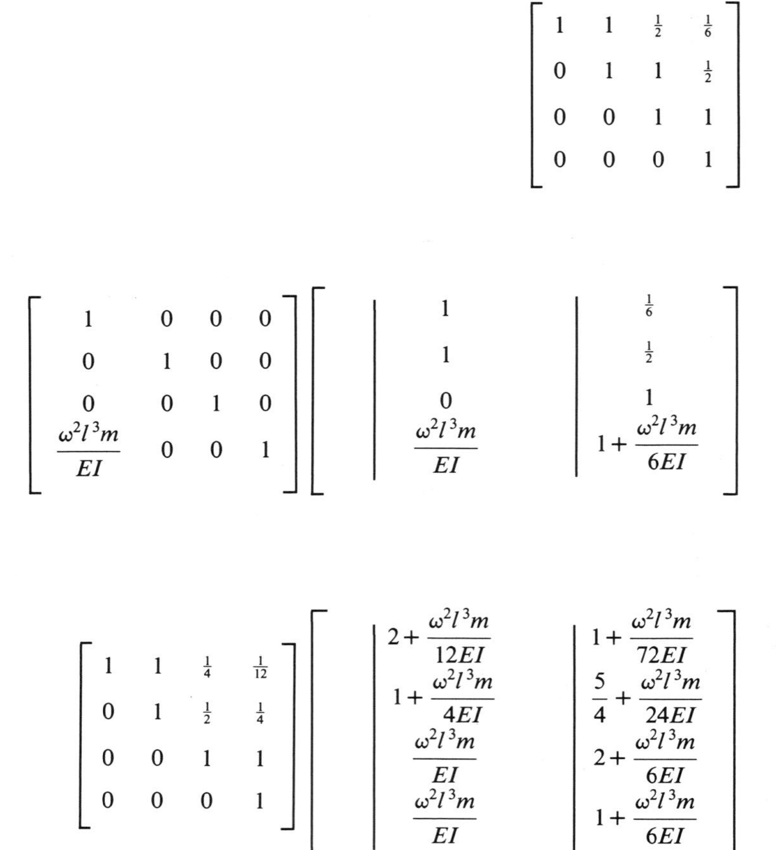

$$\begin{bmatrix} 1 & 1 & \frac{1}{2} & \frac{1}{6} \\ 0 & 1 & 1 & \frac{1}{2} \\ 0 & 0 & 1 & 1 \\ 0 & 0 & 0 & 1 \end{bmatrix}$$

$$\begin{bmatrix} 1 & 0 & 0 & 0 \\ 0 & 1 & 0 & 0 \\ 0 & 0 & 1 & 0 \\ \frac{\omega^2 l^3 m}{EI} & 0 & 0 & 1 \end{bmatrix} \left[\begin{array}{c|c|c} & 1 & \frac{1}{6} \\ & 1 & \frac{1}{2} \\ & 0 & 1 \\ & \frac{\omega^2 l^3 m}{EI} & 1 + \frac{\omega^2 l^3 m}{6EI} \end{array} \right]$$

$$\begin{bmatrix} 1 & 1 & \frac{1}{4} & \frac{1}{12} \\ 0 & 1 & \frac{1}{2} & \frac{1}{4} \\ 0 & 0 & 1 & 1 \\ 0 & 0 & 0 & 1 \end{bmatrix} \left[\begin{array}{c|c|c} & 2 + \frac{\omega^2 l^3 m}{12EI} & 1 + \frac{\omega^2 l^3 m}{72EI} \\ & 1 + \frac{\omega^2 l^3 m}{4EI} & \frac{5}{4} + \frac{\omega^2 l^3 m}{24EI} \\ & \frac{\omega^2 l^3 m}{EI} & 2 + \frac{\omega^2 l^3 m}{6EI} \\ & \frac{\omega^2 l^3 m}{EI} & 1 + \frac{\omega^2 l^3 m}{6EI} \end{array} \right]$$

$$
\begin{bmatrix} 1 & 0 & 0 & 0 \\ 0 & 1 & 0 & 0 \\ 0 & 0 & 1 & 0 \\ \dfrac{2\omega^2 l^3 m}{EI} & 0 & 0 & 1 \end{bmatrix}
\left[\begin{array}{|c|c}
 & 2+\dfrac{\omega^2 l^3 m}{12EI} & 1+\dfrac{\omega^2 l^3 m}{72EI} \\
 & 1+\dfrac{\omega^2 l^3 m}{4EI} & \dfrac{5}{4}+\dfrac{\omega^2 l^3 m}{24EI} \\
 & \dfrac{\omega^2 l^3 m}{EI} & 2+\dfrac{\omega^2 l^3 m}{6EI} \\
 & \dfrac{\omega^4 l^6 m^2}{6(EI)^2}+\dfrac{5\omega^2 l^3 m}{EI} & \dfrac{\omega^4 l^6 m^2}{36(EI)^2}+\dfrac{13\omega^2 l^3 m}{6EI}+1
\end{array}\right]
$$

$$
\begin{bmatrix} 1 & 1 & \frac{1}{2} & \frac{1}{6} \\ 0 & 1 & 1 & \frac{1}{2} \\ 0 & 0 & 1 & 1 \\ 0 & 0 & 0 & 1 \end{bmatrix}
\left[\begin{array}{|c|c}
 & \dfrac{\omega^4 l^6 m^2}{36(EI)^2}+\dfrac{5\omega^2 l^3 m}{3EI}+3 & \dfrac{\omega^4 l^6 m^2}{216(EI)^2}+\dfrac{\omega^2 l^3 m}{2EI}+\dfrac{41}{12} \\
 & \dfrac{\omega^4 l^6 m^2}{12(EI)^2}+\dfrac{15\omega^2 l^3 m}{4EI}+1 & \dfrac{\omega^4 l^6 m^2}{72(EI)^2}+\dfrac{31\omega^2 l^3 m}{24EI}+\dfrac{15}{4} \\
 & \dfrac{\omega^4 l^6 m^2}{6(EI)^2}+\dfrac{6\omega^2 l^3 m}{EI} & \dfrac{\omega^4 l^6 m^2}{36(EI)^2}+\dfrac{7\omega^2 l^3 m}{3EI}+3 \\
 & \dfrac{\omega^4 l^6 m^2}{6(EI)^2}+\dfrac{5\omega^2 l^3 m}{EI} & \dfrac{\omega^4 l^6 m^2}{36(EI)^2}+\dfrac{13\omega^2 l^3 m}{6EI}+1
\end{array}\right]
$$

$$
\begin{bmatrix} 1 & 0 & 0 & 0 \\ 0 & 1 & 0 & 0 \\ 0 & 0 & 1 & 0 \\ -\dfrac{kl^3}{EI} & 0 & 0 & 1 \end{bmatrix}
\left[\begin{array}{|c|c}
 & \dfrac{\omega^4 l^6 m^2}{36(EI)^2}+\dfrac{5\omega^2 l^3 m}{3EI}+3 & \dfrac{\omega^4 l^6 m^2}{216(EI)^2}+\dfrac{\omega^2 l^3 m}{2EI}+\dfrac{41}{12} \\
 & \dfrac{\omega^4 l^6 m^2}{12(EI)^2}+\dfrac{15\omega^2 l^3 m}{4EI}+1 & \dfrac{\omega^4 l^6 m^2}{72(EI)^2}+\dfrac{31\omega^2 l^3 m}{24EI}+\dfrac{15}{4} \\
 & \dfrac{\omega^4 l^6 m^2}{6(EI)^2}+\dfrac{6\omega^2 l^3 m}{EI} & \dfrac{\omega^4 l^6 m^2}{36(EI)^2}+\dfrac{7\omega^2 l^3 m}{3EI}+3 \\
 & \left[\dfrac{\omega^4 l^6 m^2}{36(EI)^2}\left(6-\dfrac{kl^3}{EI}\right)-\dfrac{3kl^3}{EI} + \dfrac{5\omega^2 l^3 m}{3EI}\left(3-\dfrac{kl^3}{EI}\right)\right] & \left[\dfrac{\omega^4 l^6 m^2}{216(EI)^2}\left(6-\dfrac{kl^3}{EI}\right)-\dfrac{41kl^3}{12EI} + \dfrac{\omega^2 l^3 m}{6EI}\left(13-\dfrac{3kl^3}{EI}\right)+1\right]
\end{array}\right] = [M^*]
$$

This is the frequency determinant that by expansion yields the characteristic equation

$$\omega^4\left[\frac{35kl^6m^2}{72(EI)^2}+\frac{l^3m^2}{EI}\right]-\omega^2\left[\frac{17kl^3m}{2EI}+9m\right]+9k=0 \qquad (6\text{-}96)$$

By assuming $l=200$ in. $EI=30\times10^6$ kip-in.^{2}, $m=5.18\times10^{-3}$ kip-sec^2/in., and $k=300$ kips/in., Eq. 6-96 yields

$$\omega^4-12{,}800\omega^2+9.68(10)^6=0$$

The two natural frequencies are

$$\omega_1=28.15\text{ rps} \qquad \omega_2=109.50\text{ rps}$$

In hertz,

$$f_1=\frac{\omega_1}{2\pi}=4.51\text{ Hz} \qquad f_2=\frac{\omega_2}{2\pi}=17.40\text{ Hz}$$

With ω_1 and ω_2 calculated, the mode shapes can be computed by using the results in Table 6-1 and proceeding as in Section 6-5. That is, the state vectors at stations 1, 2, and 3 are expressed in terms of the state vector at station 0 in the usual way. Since $y_0{}^{*R}=M_0^{*R}=0$, the unknowns at station 0 are $\theta_0{}^{*R}$ and V_0^{*R} and their relationship can be evaluated. In this manner, numerical values for the elements of the state vectors at stations 1, 2, and 3 are obtained. The numerical values of the deflections y for $\omega=\omega_1$ and $\omega=\omega_2$, characterize the shapes of the two modes

Other problems can be solved in a similar manner.

6-8 TRANSFER MATRICES FOR CONTINUOUS BEAMS

Natural frequencies of vibration and mode shapes of continuous beams of either uniform or variable stiffness EI can be determined by using the concept of transfer matrices. If the intermediate supports of a continuous beam are elastic springs, the procedure used in preceding sections is directly applicable. If, however, the intermediate supports are rigid, additional discontinuities are introduced into the problem which must be taken into account.

Consider, for example, the continuous beam in Fig. 6-8a. In this case, the unknown discontinuity is the reaction Q_j at station j. At this station the condition corresponding to this discontinuity is that the vertical deflection of the beam is zero. The procedure for computing the natural frequencies of this beam can be initiated by dividing the member into a number of segments as before, and lumping the mass of the beam at the juncture points of these segments. Let it be assumed that this is done and that the transfer matrices $[B]$ and $[C]$ relating the state vectors $\{v_0\}^R$ with $\{v_j\}^L$ and $\{v_j\}^R$ with $\{v_n\}^L$, respectively, are known.

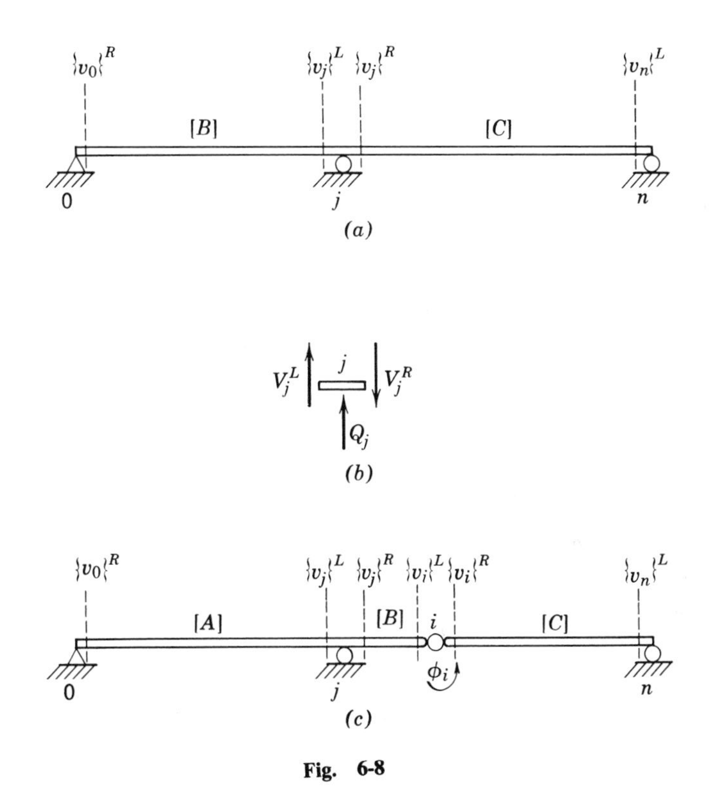

Fig. 6-8

At station zero, the boundary conditions are $y_0 = M_0 = 0$. Thus the unknown quantities at this station are θ_0 and V_0. By starting from station zero with unknowns θ_0 and V_0 and proceeding along the length of the beam, the unknown discontinuity Q_j should be introduced. The expression relating the state vectors $\{v_0\}^R$ and $\{v_j\}^L$ is

$$\{v_j\}^L = [B]\{v_0\}^R \tag{6-97}$$

and is used in the usual way. Since $y_0 = M_0 = 0$, the first and the third columns of matrix $[B]$ can be omitted. Thus Eq. 6-97 can be written as

$$\{v_j\}^L = \begin{bmatrix} b_{12} & b_{14} \\ b_{22} & b_{24} \\ b_{32} & b_{34} \\ b_{42} & b_{44} \end{bmatrix} \begin{bmatrix} \theta \\ V \end{bmatrix}_0^R \tag{6-98}$$

Here, only the elements of the second and fourth columns of $[B]$ are shown.

At station j, by using the free-body diagram in Fig. 6-8b, the following expression is written:

$$V_j^R = V_j^L + Q_j \tag{6-99}$$

where Q_j is the reaction at station j that causes a discontinuity in the shear force. Thus the vector containing all the unknowns is $\{\theta_0 \; V_0 \; Q_j\}$.

A relation between the state vectors $\{v_0\}^R$ and $\{v_j\}^R$ in Fig. 6-8a can be written as

$$\begin{bmatrix} b_{12} & b_{14} & 0 \\ b_{22} & b_{24} & 0 \\ b_{32} & b_{34} & 0 \\ b_{42} & b_{44} & 1 \end{bmatrix} \begin{bmatrix} \theta_0 \\ V_0 \\ Q_j \end{bmatrix} = \begin{bmatrix} y \\ \theta \\ M \\ V \end{bmatrix}_j^R \tag{6-100}$$

In this equation, the reaction Q_j is taken into consideration by attaching to the fourth row of the transfer matrix $[B]$ a quantity equal to unity. This setup permits one to transfer directly from station zero to the right of station j. At station j, the deflection y_j^R is zero. Thus, from Eq. 6-100,

$$b_{12}\,\theta_0 + b_{14}V_0 = 0 \tag{6-101}$$

In order to transfer just to the left of the station n, it is only necessary to multiply the enlarged matrix $[B]$ by the transfer matrix $[C]$. That is,

$$[D]=[C][B]=\begin{bmatrix} c_{11} & c_{12} & c_{13} & c_{14} \\ c_{21} & c_{22} & c_{23} & c_{24} \\ c_{31} & c_{32} & c_{33} & c_{34} \\ c_{41} & c_{42} & c_{43} & c_{44} \end{bmatrix}\begin{bmatrix} b_{12} & b_{14} & 0 \\ b_{22} & b_{24} & 0 \\ b_{32} & b_{34} & 0 \\ b_{42} & b_{44} & 1 \end{bmatrix}$$

or

$$[D]=[C][B]=\begin{bmatrix} d_{11} & d_{12} & d_{13} \\ d_{21} & d_{22} & d_{23} \\ d_{31} & d_{32} & d_{33} \\ d_{41} & d_{42} & d_{43} \end{bmatrix} \tag{6-102}$$

This completes the procedure. Thus the relation between the state vectors $\{v_0\}^R$ and $\{v_n\}^L$ in Fig. 6-8*a* is

$$\begin{bmatrix} d_{11} & d_{12} & d_{13} \\ d_{21} & d_{22} & d_{23} \\ d_{31} & d_{32} & d_{33} \\ d_{41} & d_{42} & d_{43} \end{bmatrix}\begin{bmatrix} \theta_0 \\ V_0 \\ Q_j \end{bmatrix}=\begin{bmatrix} y \\ \theta \\ M \\ V \end{bmatrix}_n^L \tag{6-103}$$

At station n, the boundary conditions suggest that $y_n^L=M_n^L=0$. Thus, from Eq. 6-103, the following two expressions are written:

$$d_{11}\,\theta_0+d_{12}V_0+d_{13}Q_j=0 \tag{6-104}$$

$$d_{31}\,\theta_0+d_{32}V_0+d_{33}Q_j=0 \tag{6-105}$$

Equations 6-101, 6-104, and 6-105 form a system of three homogeneous equations involving the unknowns θ_0, V_0, and Q_j. These equations are written again as

$$\begin{aligned} b_{12}\,\theta_0 + b_{14}V_0 &= 0 \\ d_{11}\,\theta_0 + d_{12}V_0 + d_{13}Q_j &= 0 \\ d_{31}\,\theta_0 + d_{32}V_0 + d_{33}Q_j &= 0 \end{aligned} \tag{6-105}$$

For a solution other than the trivial one, the determinant of the coefficients in Eq. 6-105 should be zero. That is,

$$\begin{vmatrix} b_{12} & b_{14} & 0 \\ d_{11} & d_{12} & d_{13} \\ d_{31} & d_{32} & d_{33} \end{vmatrix} = 0 \tag{6-106}$$

The elements of this determinant involve the natural frequencies ω of the beam. Thus the values of ω that satisfy Eq. 6-106 are the natural frequencies of the continuous beam. With known values of ω, the corresponding mode shapes can be found in the usual way.

It should be pointed out that Eq. 6-106 involves the solution of a third-order determinant, while only a second-order determinant is the usual case in beam problems. The reason for this change is the discontinuity in the shear introduced by the reaction Q_j at station j. If an additional discontinuity is introduced, a fourth-order determinant is obtained. Thus each discontinuity increases the order of the determinant by one.

Discontinuities in a continuous beam problem can be also introduced by the presence of internal hinges such as the one at station i in Fig. 6-8*c*. Let it be assumed that such hinges are frictionless. In this case, the unknown discontinuities in Fig. 6-8*c* are the reaction Q_j at station j and the slope change ϕ_i at station i. The corresponding conditions are those of zero deflection at j and zero bending moment at i. The natural frequencies of this beam can be determined by proceeding as in the preceding problem. For convenience, let it be assumed that the transfer matrices $[A]$, $[B]$, and $[C]$, Fig. 6-8*c*, between stations zero and j, j and i, and i and n, respectively, are already computed.

At station zero, the boundary conditions are $y_0 = M_0 = 0$. Thus the relation between the state vectors $\{v_0\}^R$ and $\{v_j\}^L$ is

$$\{v_j\}^L = \begin{bmatrix} a_{12} & a_{14} \\ a_{22} & a_{24} \\ a_{32} & a_{34} \\ a_{42} & a_{44} \end{bmatrix} \begin{bmatrix} \theta \\ V \end{bmatrix}_0^R \tag{6-107}$$

At station j, the discontinuity due to the reaction Q_j can be introduced as in the preceding problem, and a relation between the state vectors $\{v_0\}^R$ and $\{v_j\}^R$ is as follows:

$$\begin{bmatrix} a_{12} & a_{14} & 0 & 0 \\ a_{22} & a_{24} & 0 & 0 \\ a_{32} & a_{34} & 0 & 0 \\ a_{42} & a_{44} & 1 & 0 \end{bmatrix} \begin{bmatrix} \theta_0 \\ V_0 \\ Q_j \\ \phi_i \end{bmatrix} = \begin{bmatrix} y \\ \theta \\ M \\ V \end{bmatrix}_j^R \tag{6-108}$$

Here the vector containing the four unknowns is $\{\theta_0 \, V_0 \, Q_j \, \phi_i\}$, and the reaction Q_j is taken into account by the quantity of unity in the fourth row of the transfer matrix $[A]$.

From Eq. 6-108, since y_j^R is zero, the following expression is written:

$$a_{12}\theta_0 + a_{14}V_0 = 0 \tag{6-109}$$

In order to transfer just to the left of station i, the enlarged matrix $[A]$ is multiplied by the transfer matrix $[B]$. That is,

$$[D] = [B][A] = \begin{bmatrix} b_{11} & b_{12} & b_{13} & b_{14} \\ b_{21} & b_{22} & b_{23} & b_{24} \\ b_{31} & b_{32} & b_{33} & b_{34} \\ b_{41} & b_{42} & b_{43} & b_{44} \end{bmatrix} \begin{bmatrix} a_{12} & a_{14} & 0 & 0 \\ a_{22} & a_{24} & 0 & 0 \\ a_{32} & a_{34} & 0 & 0 \\ a_{42} & a_{44} & 1 & 0 \end{bmatrix}$$

or

$$[D]=\begin{bmatrix} d_{11} & d_{12} & d_{13} & 0 \\ d_{21} & d_{22} & d_{23} & 0 \\ d_{31} & d_{32} & d_{33} & 0 \\ d_{41} & d_{42} & d_{43} & 0 \end{bmatrix} \tag{6-110}$$

At station i, because of the presence of the frictionless hinge, there is a discontinuity in the slope. At this station, the following equation is written:

$$\theta_i^R=\theta_i^L+\phi_i \tag{6-111}$$

where ϕ_i is the slope change. The relation between the state vectors $\{v_0\}^R$ and $\{v_i\}^R$ is

$$\begin{bmatrix} d_{11} & d_{12} & d_{13} & 0 \\ d_{21} & d_{22} & d_{23} & 1 \\ d_{31} & d_{32} & d_{33} & 0 \\ d_{41} & d_{42} & d_{43} & 0 \end{bmatrix}\begin{bmatrix} \theta_0 \\ V_0 \\ Q_j \\ \phi_i \end{bmatrix}=\begin{bmatrix} y \\ \theta \\ M \\ V \end{bmatrix}_i^R \tag{6-112}$$

where the slope discontinuity ϕ_i is introduced by placing a quantity of unity in the second row of the square matrix of Eq. 6-112.

At station i, the bending moment M_i is zero. Thus, from Eq. 6-112, the following expression is written:

$$d_{31}\,\theta_0+d_{32}V_0+d_{33}Q_j=0 \tag{6-113}$$

which takes into account the condition of zero bending moment at station i.

The final step would be to transfer from the right side of station i to the left side of station n. This is easily accomplished by multiplying the square matrix $[D]$ in Eq. 6-112 by the transfer matrix $[C]$. That is,

$$[E]=[C][D]=\begin{bmatrix} c_{11} & c_{12} & c_{13} & c_{14} \\ c_{21} & c_{22} & c_{23} & c_{24} \\ c_{31} & c_{32} & c_{33} & c_{34} \\ c_{41} & c_{42} & c_{43} & c_{44} \end{bmatrix}\begin{bmatrix} d_{11} & d_{12} & d_{13} & 0 \\ d_{21} & d_{22} & d_{23} & 1 \\ d_{31} & d_{32} & d_{33} & 0 \\ d_{41} & d_{42} & d_{43} & 0 \end{bmatrix}$$

or

$$[E]=\begin{bmatrix} e_{11} & e_{12} & e_{13} & e_{14} \\ e_{21} & e_{22} & e_{23} & e_{24} \\ e_{31} & e_{32} & e_{33} & e_{34} \\ e_{41} & e_{42} & e_{43} & e_{44} \end{bmatrix} \tag{6-114}$$

Thus the relation between the state vectors $\{v_0\}^R$ and $\{v_n\}^L$ is

$$\begin{bmatrix} e_{11} & e_{12} & e_{13} & e_{14} \\ e_{21} & e_{22} & e_{23} & e_{24} \\ e_{31} & e_{32} & e_{33} & e_{34} \\ e_{41} & e_{42} & e_{43} & e_{44} \end{bmatrix} \begin{bmatrix} \theta_0 \\ V_0 \\ Q_j \\ \phi_i \end{bmatrix} = \begin{bmatrix} y \\ \theta \\ M \\ V \end{bmatrix}_n^L \tag{6-115}$$

At station n, the boundary conditions suggest that the deflection y_n and the bending moment M_n should be zero. Thus, from Eq. 6-115,

$$e_{11}\theta_0+e_{12}V_0+e_{13}Q_j+e_{14}\phi_i=0 \tag{6-116}$$

$$e_{31}\theta_0+e_{32}V_0+e_{33}Q_j+e_{34}\phi_i=0 \tag{6-117}$$

Equations 6-109, 6-113, 6-116, and 6-117 form a system of four homogeneous equations involving the unknowns θ_0, V_0, Q_j, and ϕ_i. For a nontrivial solution, the determinant of their coefficients should be zero. That is,

$$\begin{vmatrix} a_{12} & a_{14} & 0 & 0 \\ d_{31} & d_{32} & d_{33} & 0 \\ e_{11} & e_{12} & e_{13} & e_{14} \\ e_{31} & e_{32} & e_{33} & e_{34} \end{vmatrix}=0 \tag{6-118}$$

The elements of this frequency determinant involve the natural frequencies ω of the beam in Fig. 6-8c. The values of ω that satisfy Eq. 6-118 are the natural frequencies of the beam.

6-9 THE DYNAMIC HINGE CONCEPT[22]

For many structural systems and machine elements, it is often required to determine just the fundamental frequency of free vibration to verify that vibration design criteria are met. Highway bridge girders and machine shafts are examples of such types of problems. A reasonable solution regarding the fundamental frequency can be obtained relatively easily by applying the concept of the dynamic hinge as discussed in this section.

The concept of the dynamic hinge is predicated on the assumption that the elastic line of the pseudostatic system and the dynamic elastic line of the fundamental mode of free vibration are everywhere proportional. In practice, however, this assumption is only approximately satisfied, but for many practical problems it yields satisfactory results. The pseudostatic system is defined as the static system whose gravitational weight is assumed to act in the direction of the dynamic elastic line of the fundamental mode.

Consider for example the n-span continuous beam in Fig. 6-9*a*. Each span of the member can be of either uniform or variable EI and mass. For convenience, although not necessary, the weight of the beam is lumped at discrete points along its length as discussed in Section 5-10. If the system vibrates in its fundamental mode, the lumped weights $W_1, W_2, \ldots, W_j$, are assumed to act in the direction of the dynamic elastic line of this mode as shown in Fig. 6-9*b*. This is defined as the pseudostatic system whose pseudostatic elastic line is shown in Fig. 6-9*c*. The dynamic elastic line of the fundamental mode is shown in Fig. 6-9*d*.

The dynamic hinge concept is predicated on the assumption that the shape of the pseudostatic elastic line in Fig. 6-9*c* is proportional to the shape of the fundamental mode in Fig. 6-9*d*. This assumption permits one to conclude that the inflation points at the fundamental mode configuration are at the same locations for both elastic lines. At these points, the pseudostatic moments and the dynamic moments are zero.

During free vibration, all elements of the member, as well as any portion of the member, vibrate with the same frequency ω. If a portion of the member is taken apart from the complete structure and its end boundary conditions are properly satisfied, this portion vibrates with the frequency ω of the complete system.

It was pointed out earlier in this chapter that the frequency is independent of the actual amplitudes of vibration, but is dependent on the shape of the mode. In addition, if the inertia forces of the fundamental mode are applied to the member as static loads, the static elastic line due to these loads will be the fundamental mode. On this basis, the assumption of proportionality

[22]For additional information the reader may refer to the author's work in References 25, 28, 78, and 79.

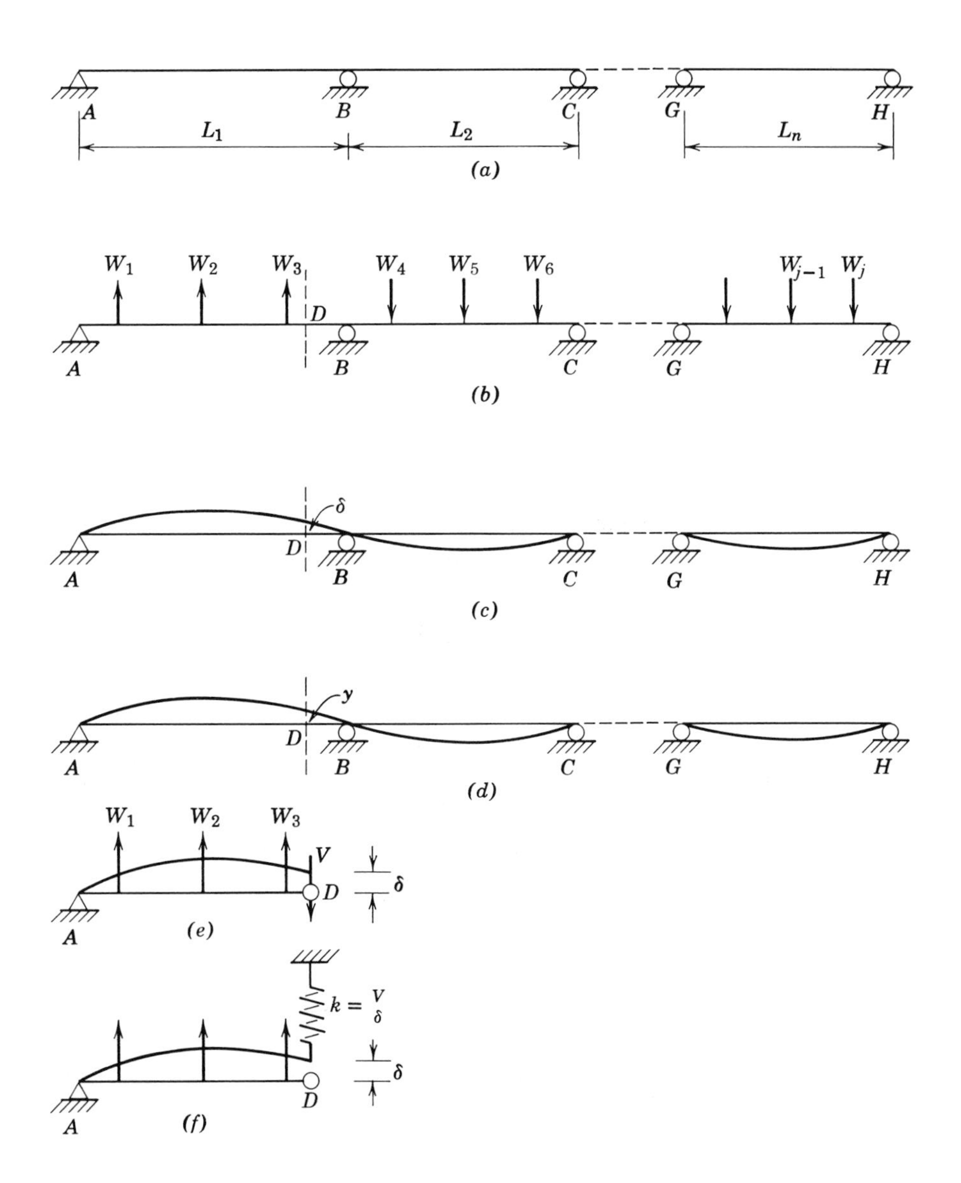

Fig. 6-9. (*a*) Static system. (*b*) Pseudostatic system. (*c*) Pseudostatic elastic line. (*d*) Dynamic elastic line. (*e*) Dynamically equivalent system.

between the elastic lines in Fig. 6-9*c* and 6-9*d* permits one to consider W_1, $W_2, \ldots, W_j$ as the inertia loads producing the proportional fundamental mode shape in Fig. 6-9*c*. The points of zero moment are determined by solving the pseudostatic system. Let it be assumed that a point is found and it is point D in Fig. 6-9*b*. At this point, the pseudostatic deflection δ in Fig. 6-9*c* is proportional to the amplitude y in Fig. 6-9*d*, and it can be found by using methods of strength of materials.

If portion AD is isolated as in Fig. 6-9*e*, the moment at D is zero, and the shear force V represents the influence of portion $DBC \ldots H$ on portion AD. The shear force V can be replaced by a linear spring of constant

$$k = \frac{V}{\delta} \tag{6-119}$$

as in Fig. 6-9*f*. The system in Fig. 6-9*f* is the dynamically equivalent system and point D is the dynamic hinge. If the beam is assumed to be ideally hinged at this point, the shape of the mode does not alter because the moment is zero.

The fundamental frequency ω of the beam can be determined by using the dynamically equivalent system and applying Rayleigh's method. That is,

$$\omega^2 = g \frac{\sum_{i=1}^{n} W_i y_i}{\sum_{i=1}^{n} W_i y_i^2} \tag{6-120}$$

where y_i are the deflections under the weight concentration points in Fig. 6-9*f*. These deflections can be determined by using the system in Fig. 6-9*f* and applying known methods of strength of materials. The method of Stodola, if preferred, can be used to solve the dynamically equivalent system to determine ω.

The shear force V and deflection δ that are needed to determine k in Fig. 6-9*f* must be computed by using the complete pseudostatic system in Fig. 6-9*b*. This is not, however, a tremendous limitation, because V and δ can be determined with reasonable accuracy by using only the first three spans of the member. This is justified, because V and δ are not appreciably effected by the loading of the fourth span. This method is particularly advantageous for members consisting of many spans.

It should be noted that the dynamically equivalent system in Fig. 6-9*f* is statically determinate. In applying this concept, a statically determinate system should be obtained to avoid difficulties in satisfying the boundary conditions at the dynamic hinge points. If for example the end supports of the beam are fixed, there should be two points of zero moment at one of its spans. By using the portion between these two points, the dynamically equivalent system can be obtained in the same way as above. The following examples illustrate the application of this concept.

Consider the continuous beam in Fig. 6-10*a* whose uniform stiffness EI is 150×10^6 kip-in.^{2}. Let it be assumed that this member is a bridge girder and that it is required to determine the fundamental frequency of its free vibration. Its lumped weight as well as the weights supported by the beam are assumed to act in the direction of the fundamental mode as shown in Fig. 6-10*b*. This is the pseudostatic system whose elastic line is assumed to be proportional to the dynamic elastic line of the fundamental mode.

By applying methods of strength of materials, the point of zero moment is found to be located at point D in Fig. 6-10*b*. By solving the pseudostatic system, the shear force V_D and deflection δ_D, at D, are

$$V_D = 18.35 \text{ kips}$$

$$\delta_D = 0.07012 \text{ in.}$$

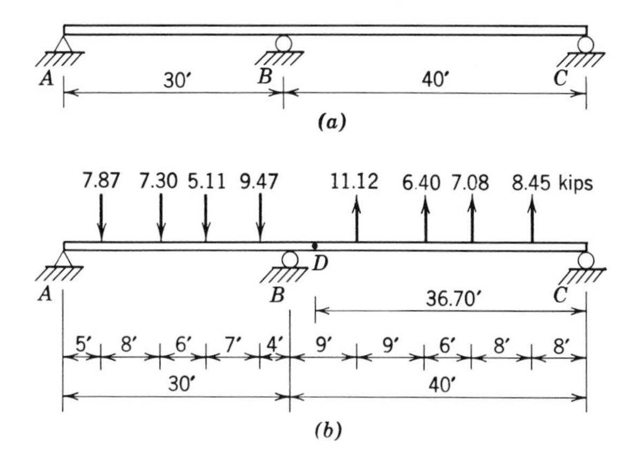

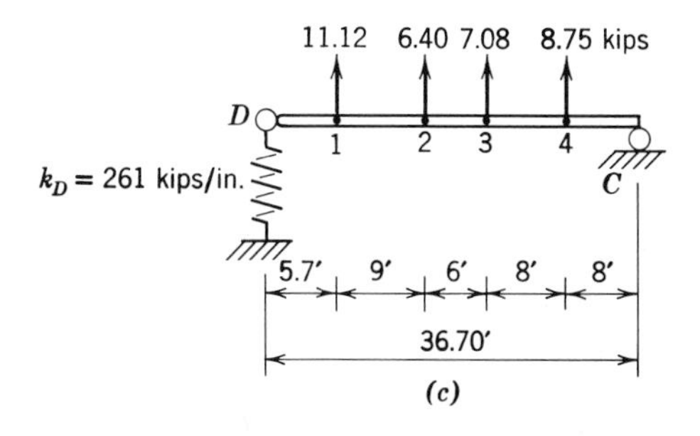

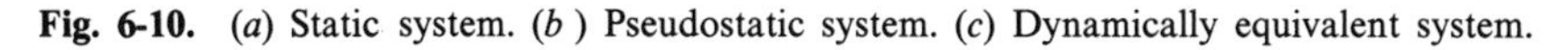

Fig. 6-10. (*a*) Static system. (*b*) Pseudostatic system. (*c*) Dynamically equivalent system.

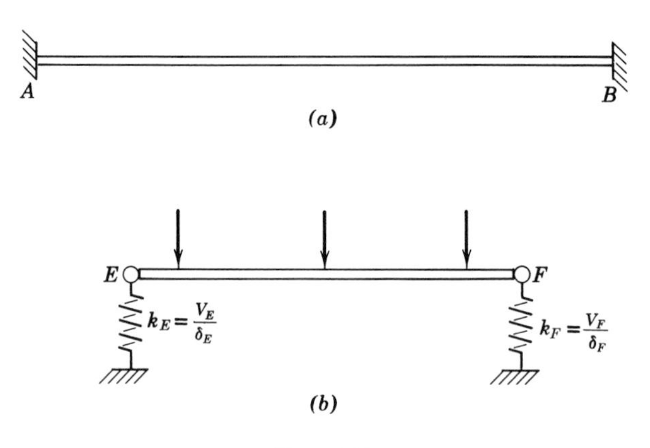

Fig. 6-11

Thus, from Eq. 6-119,

$$k_D = \frac{V_D}{\delta_D} = \frac{18.35}{0.07012} = 261.0 \text{ kips/in.}$$

The dynamically equivalent system is shown in Fig. 6-10*c*.

By solving this system, the deflections y_1, y_2, y_3, and y_4 at points 1, 2, 3, and 4, respectively, are

$$y_1 = 0.196 \text{ in.} \qquad y_3 = 0.301 \text{ in.}$$

$$y_2 = 0.306 \text{ in.} \qquad y_4 = 0.191 \text{ in.}$$

By Eq. 6-120,

$$\omega^2 = (386)\frac{7.98}{1.98} = 1560$$

or

$$\omega = 39.50 \text{ rps}$$

$$f = \frac{\omega}{2\pi} = 6.30 \text{ Hz}$$

The results obtained by using the complete system are in close agreement with the values above.

For the member in Fig. 6-11*a*, the dynamically equivalent system would be of the form shown in Fig. 6-11*b*. The shear forces and deflections at the points of zero moment E and F can be determined by sloving the pseudostatic system.

PROBLEMS

6-1 The uniform beams in Problem 4-1 are of constant stiffness $EI = 45 \times 10^6$ kip-in.^{2} and of uniform weight $w = 0.6$ kips/ft. By utilizing the static deflection curve produced by the weight w, determine the fundamental frequency of vibration by applying the method of Lord Rayleigh. Neglect damping.

6-2 Repeat Problem 6-1 by lumping the weight at discrete points along their length. Use flexibility coefficients and determine in each case

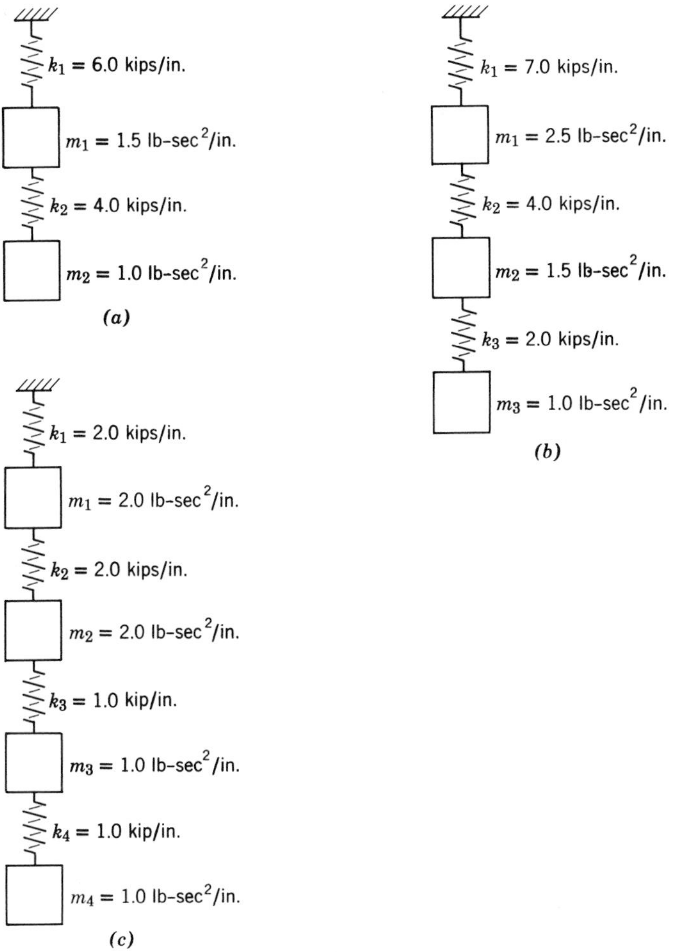

$k_1 = 6.0$ kips/in.
$m_1 = 1.5$ lb-sec^2/in.
$k_2 = 4.0$ kips/in.
$m_2 = 1.0$ lb-sec^2/in.
(a)

$k_1 = 7.0$ kips/in.
$m_1 = 2.5$ lb-sec^2/in.
$k_2 = 4.0$ kips/in.
$m_2 = 1.5$ lb-sec^2/in.
$k_3 = 2.0$ kips/in.
$m_3 = 1.0$ lb-sec^2/in.
(b)

$k_1 = 2.0$ kips/in.
$m_1 = 2.0$ lb-sec^2/in.
$k_2 = 2.0$ kips/in.
$m_2 = 2.0$ lb-sec^2/in.
$k_3 = 1.0$ kip/in.
$m_3 = 1.0$ lb-sec^2/in.
$k_4 = 1.0$ kip/in.
$m_4 = 1.0$ lb-sec^2/in.
(c)

Fig. P6-7

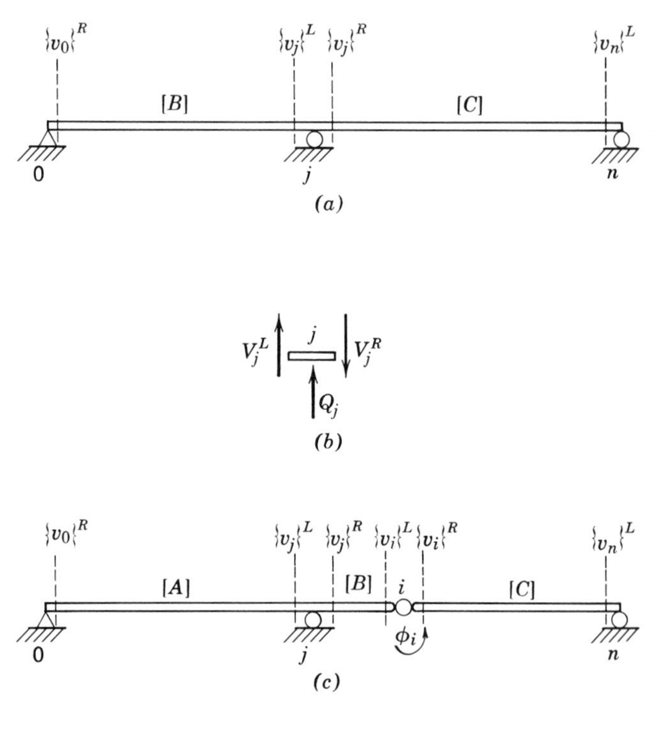

Fig. P6-8

the fundamental frequency and the corresponding mode shape. Repeat the procedure for as many times as necessary to obtain accurate solution.

6-3 Repeat Problem 5-20 by using the method of Lord Rayleigh.

6-4 Repeat Problem 6-2 and compute the first three natural frequencies of vibration and the corresponding mode shapes.

6-5 By applying the Myklestad method, determine the first two natural frequencies of vibration and the corresponding mode shapes of the

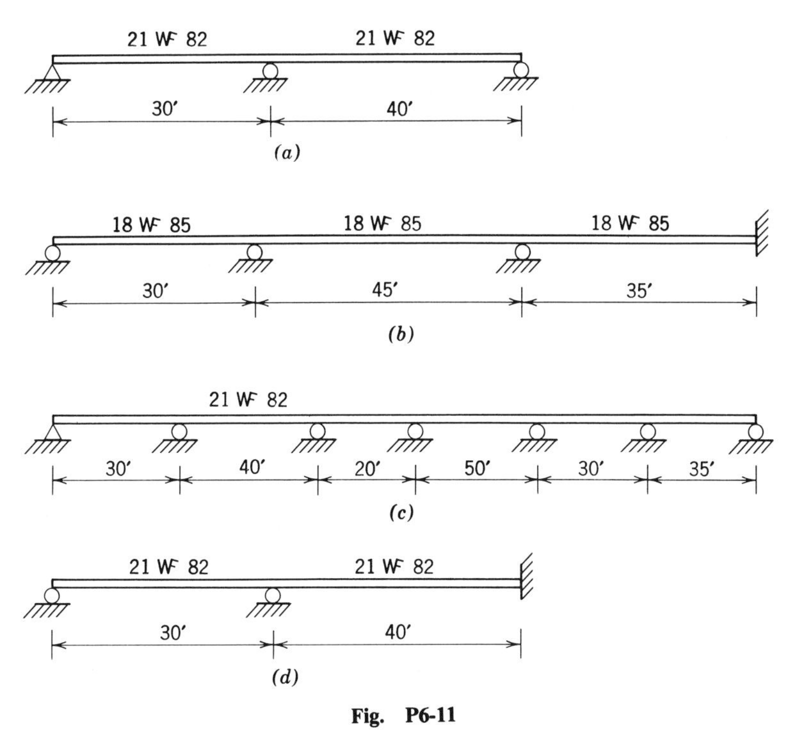

Fig. P6-11

beams shown in Problem 4-1. The stiffness EI is constant and equal to 30×10^6 kip-in.^{2} and the weight w is 0.6 kips/ft. Neglect damping.

6-6 The beams in Problem 6-5 are assumed to be loaded with a dynamic load of 5.0 kips which is uniform throughout the length of each beam. This load is applied suddenly at $t=0$ and released suddenly at $t=0.4$ sec. By using the results from Problem 6-5 and applying the method of modal analysis, determine the dynamic deflections and the dynamic bending moments at the center of each beam.

6-7 By using appropriate transfer matrices, determine the natural frequencies and the corresponding mode shapes of the free undamped spring-mass systems shown in Fig. P6-7.

6-8 The beams shown in Fig. P6-8 are assumed to be of negligible weight and they are supporting the attached weights as shown. By using appropriate transfer matrices, determine in each case the free undamped natural frequencies and the corresponding mode shapes.

6-9 By using appropriate transfer matrices, determine the first three natural frequencies and the corresponding mode shapes for each of the beams in Problem 4-1. The uniform stiffness EI is 30×10^6 kip-in.^{2} and $w=0.6$ kips/ft.

6-10 By using appropriate transfer matrices, determine the first three natural frequencies and the corresponding mode shapes for each of the beams in Problem 4-2. The uniform stiffness EI is 45×10^6 kip-in.^{2} and $w = 0.6$ kips/ft.

6-11 By applying the concept of the dynamic hinge, determine the fundamental frequency of vibration for the steel beams shown in Fig. P6-11. The modulus of elasticity E is 30×10^6 psi.

7

STRUCTURES WITH MEMBERS OF VARIABLE STIFFNESS

7-1 INTRODUCTION

The dynamic response of structures with members of variable stiffness is discussed in this chapter. This topic has received more attention during the past decade, because the use of members with variable stiffness is more common in modern structures. For example, the moment of inertia of the concrete girders of many highway bridges is variable.

The discussion in this chapter includes unique methods of analysis that simplify greatly the solution of complex problems. This method of approach can be used to determine static, dynamic, and vibration responses of structures composed of members with variable EI.

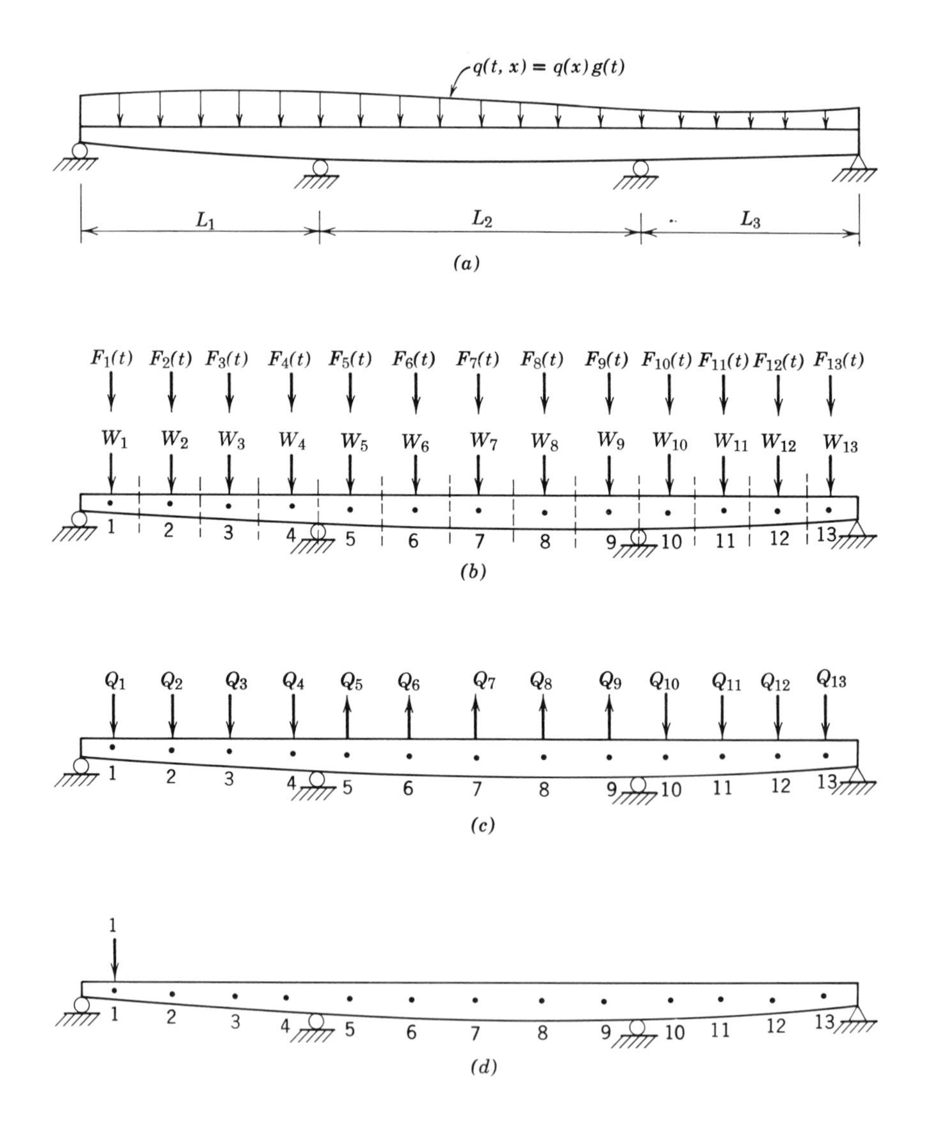

Fig. 7-1

7-2 DYNAMIC RESPONSE OF BEAMS

The dynamic response of beams with variable mass and stiffness can be determined by using the method of modal analysis as discussed in Chapter 5. A reasonable solution can be obtained by following the procedure discussed in Section 5-9.

Consider for example the three-span continuous beam in Fig. 7-1*a* that is subjected to a distributed dynamic force $q(t,x) = q(x)g(t)$ as shown. The function $q(x)$ represents the variation along the length of the beam and $g(t)$ is the time variation. By proceeding as in Section 5-9, the distributed dynamic force $q(t,x)$ and weight $w(x)$ can be concentrated at discrete points along the length of the member as shown in Fig. 7-1*b*. Here, the lumped weights are designated by $W_1, W_2, \ldots, W_{13}$ and the lumped forces by $F_1(t)$, $F_2(t), \ldots, F_{13}(t)$. Between weight concentration points the beam is assumed to be weightless, but its elastic properties are retained. The dynamic response of the member is then determined by using the lumped parameter system in Fig. 7-1*b* and applying the method of modal analysis as in Section 5-9.

It should be pointed out, however, that the application of the method of modal analysis requires the computation of the natural frequencies and the corresponding mode shapes of the beam. This can be accomplished by using either the method of Stodola and iteration procedure or any one of the methods discussed in Chapter 6. If the method of Stodola is to be applied, the loads are assumed to act in the direction shown in Fig. 7-1*c*, and the flexibility coefficients are determined by applying unit loads at points 1, $2, \ldots, 13$ in the usual way. See for example Fig. 7-1*d*.

The most difficult task in this problem is the computation of the flexibility coefficients. This difficulty rests primarily in the fact that the stiffness EI of the member is variable. The amount of labor required for such computations is greatly reduced if the method of the equivalent[23] systems is used. This method is treated in the sections that follow.

7-3 THEORY AND METHOD OF THE EQUIVALENT SYSTEMS[24]

The theory and method of the equivalent systems permit one to replace a member of variable stiffness $E_x I_x$ by a member of uniform stiffness $E_1 I_1$ by the application of a theorem of equivalent systems. The development of this

[23]This method is discussed in great detail in the author's work in Reference 24, 25, 27, 28, 30, and 79.

[24]For more information on the subject see author's work in References 24, 25, 27, 28, 30, and 79.

theory is predicated on the usual assumptions in the theory of small deflections and on the assumption that the line joining the centroids of adjacent cross sections is straight. If it is curved, this theory will apply with good accuracy if the ratio of the radius of curvature to the depth of any cross section of a member is large.

This method is general and applies to both statically determinate and statically indeterminate structural problems. In a given case, once an equivalent system is obtained, it may be used to compute rotations, deflections, natural frequencies, and mode shapes of vibration of the initial variable stiffness problem.

The existence of an equivalent system of uniform stiffness is based on the validity of the following theorem of the equivalent systems:

> A member of variable stiffness EI, can be replaced by any number of equivalent systems of uniform stiffness EI, where the deflection curve of each one of the equivalent systems is everywhere identical to the deflection curve of the original variable stiffness member.

This theorem can be proved by using concepts of strength of materials as follows:

The general differential equation of an elastic line is given by

$$\frac{d^2y}{dx^2} = -\frac{M_x}{E_x I_x} \tag{7-1}$$

By integrating twice,

$$y = \int \left[-\int \frac{M_x\,dx}{E_x I_x} \right] dx + C_1 \int dx + C_2. \tag{7-2}$$

where C_1 and C_2 are constants of integration and depend on the boundary conditions of a given problem.

The variable stiffness $E_x I_x$ of a member can be expressed as

$$E_x I_x = E_1 I_1 f(x) \tag{7-3}$$

where $E_1 I_1$ is an arbitrary reference value of the stiffness $E_x I_x$, and $f(x)$ is a function of x representing the variation of $E_x I_x$ with respect to reference value $E_1 I_1$. By substituting Eq. 7-3 into Eq. 7-2,

$$y = \frac{1}{E_1 I_1} \int \left[-\int \frac{M_x\,dx}{f(x)} \right] dx + C_1 \int dx + C_2 \tag{7-4}$$

For a beam of constant stiffness $E_1 I_1$ and with length and reference system of axes identical to the one represented by Eq. 7-4, its elastic line y_e is given by the expression

$$y_e = \frac{1}{E_1 I_1} \int \left[-\int M_e \, dx \right] dx + C_1' \int dx + C_2' \qquad (7\text{-}5)$$

Here M_e is the moment at any cross section x, and C_1' and C_2' are the constants of integration.

The elastic lines represented by Eqs. 7-4 and 7-5 are identical if

$$C_1 = C_1'$$

$$C_2 = C_2' \qquad (7\text{-}6)$$

and

$$\int \left[-\int \frac{M_x \, dx}{f(x)} \right] dx = \int \left[-\int M_e \, dx \right] dx \qquad (7\text{-}7)$$

The conditions in Eq. 7-6 are satisfied if the two members have the same length and boundary conditions. Equation 7-7 is satisfied if

$$M_e = \frac{M_x}{f(x)} \qquad (7\text{-}8)$$

This discussion proves that for a beam of variable stiffness $E_x I_x$ there exists an equivalent system of constant stiffness $E_1 I_1$ whose boundary conditions and length are the same as those of the variable stiffness member, but whose moment at any cross section x is given by Eq. 7-8. Thus the moment diagram of the equivalent system of constant stiffness $E_1 I_1$ can be determined from Eq. 7-8, provided that M_x and $f(x)$ can be computed. With known M_e, the shear force V_e and the applied load w_e of the equivalent system can be determined from the expressions

$$V_e = \frac{d}{dx}(M_e) = \frac{d}{dx}\left[\frac{M_x}{f(x)} \right] \qquad (7\text{-}9)$$

$$w_e = -\frac{d}{dx}(V_e) = -\frac{d^2}{dx^2}\left[\frac{M_x}{f(x)} \right] \qquad (7\text{-}10)$$

For a given variable stiffness problem, any number of equivalent systems can be obtained, because the reference stiffness value $E_1 I_1$ was selected arbitrarily. If $f(x) = 1$, then $E_x I_x = E_1 I_1$, and the stiffness of the equivalent system is identical to that of the original system.

The elastic line of the equivalent system is everywhere identical to that of the original variable stiffness member. Thus, by applying to the equivalent system known methods of strength of materials, the rotations and deflections of the initial system can be obtained.

Consider for example the beam in Fig. 7-2*a* whose moment of inertia variation is shown in Fig. 7-2*b*. The modulus of elasticity E is constant and the reference stiffness value EI_1 is selected as the one of the free end of the beam. Thus

$$I_x = \frac{(L + 3x)}{L} I_1 \tag{7-11}$$

and

$$f(x) = \frac{L + 3x}{L} \tag{7-12}$$

By applying simple statics, the moment M_x of the variable stiffness beam at any location x is

$$M_x = -\frac{wx^2}{2} \tag{7-13}$$

The moment M_e of the equivalent system of constant stiffness EI_1 is given by Eq. 7-8. Thus

$$M_e = \frac{M_x}{f(x)} = -\frac{wLx^2}{2L + 6x} \tag{7-14}$$

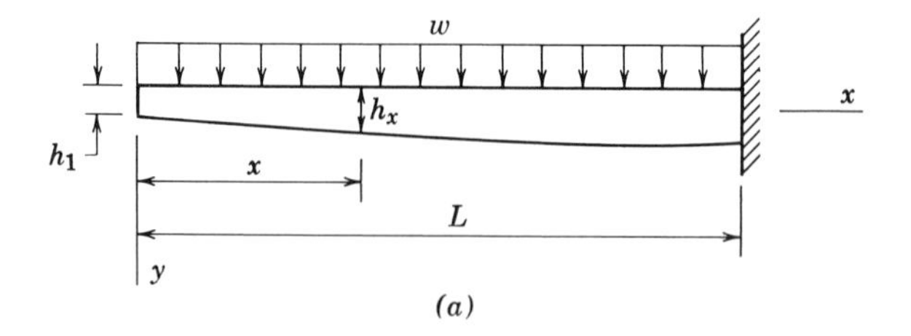

(*a*)

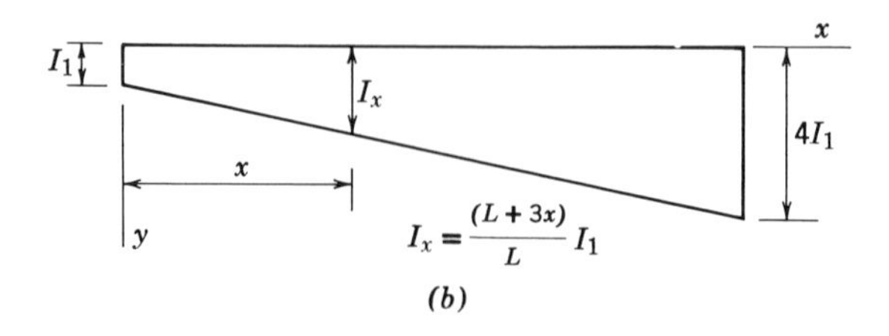

(*b*)

Fig. 7-2

By applying Eqs. 7-9 and 7-10,

$$V_e = \frac{d}{dx}(M_e) = -\frac{(4wL^2x + 6wLx^2)}{(2L + 6x)^2} \tag{7-15}$$

$$w_e = -\frac{d}{dx}(V_e) = \frac{8wL^3}{(2L + 6x)^3} \tag{7-16}$$

With $w = 1.00$ lb/in. and $L = 60$ in., the plots of Eqs. 7-14, 7-15, and 7-16 are shown in Figs. 7-3*c*, 7-3*d*, and 7-3*e*, respectively. Thus the equivalent

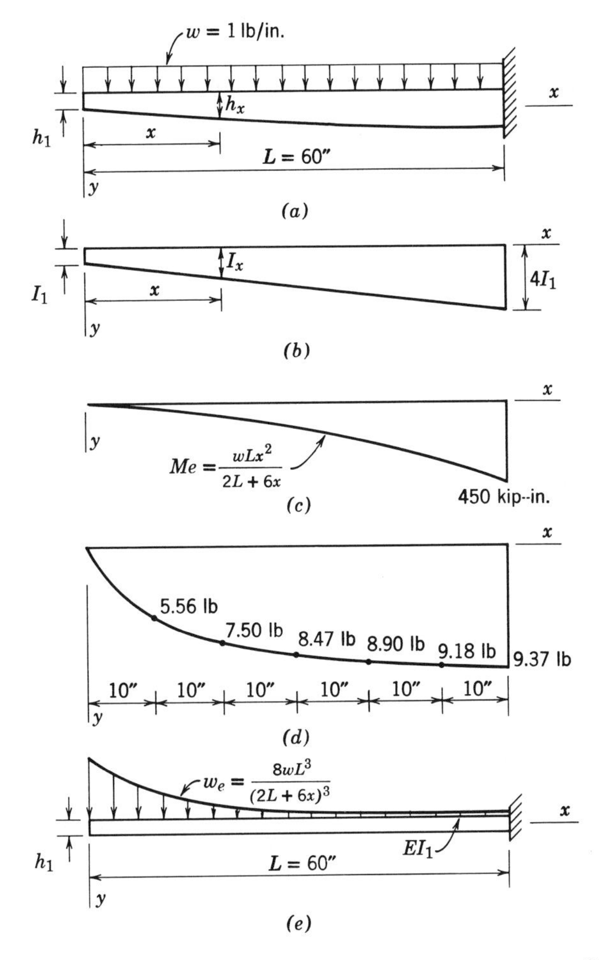

Fig. 7-3. (*a*) Original system. (*b*) Moment of inertia variation. (*c*) M_e moment diagram. (*d*) V_e shear force diagram. (*e*) Equivalent system.

system of uniform stiffness EI_1 is as shown in Fig. 7-3*e*, and its elastic line, length, and boundary conditions are the same as those of the variable stiffness beam in Fig. 7-3*a*.

The deflection at the free end can be determined by using the equivalent system and applying the conjugate beam method. The conjugate beam of the equivalent system is shown in Fig. 7-4*a*. By Fig. 7-4*b*,

$$y_A = M_A = \int_0^L \frac{M_e x}{EI_1}\,dx = \frac{wL}{2EI_1}\int_0^L \frac{x^3}{(L+3x)}\,dx$$

$$= \frac{wL}{2EI_1}\left[\int_0^L\left(\frac{x^2}{3} - \frac{Lx}{9} + \frac{L^2}{27}\right)dx - \frac{L^3}{27}\int_0^L \frac{3\,dx}{3(L+3x)}\right]$$

$$= \frac{wL}{2EI_1}\left[\frac{L^3}{9} - \frac{L^3}{18} + \frac{L^3}{27} - \frac{L^3}{(27)(3)}\log(L+3L) + \frac{L^3}{(27)(3)}\log(L)\right]$$

$$= \frac{wL}{54EI_1}\left[\frac{5L^3}{2} - \frac{L^3}{3}\log(4)\right] \tag{7-17}$$

Thus, with $w = 1.00$ lb/in., and $L = 60$ in., Eq. 7-17 yields

$$y_A = \frac{(489)(10)^3}{EI_1} \text{ (downward)} \tag{7-18}$$

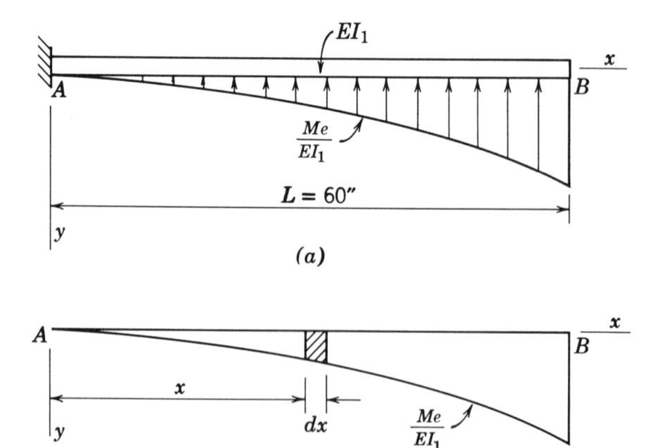

Fig. 7-4

Approximate Method of the Equivalent Systems

It should be pointed out, however, that the exact method of the equivalent systems as discussed above often becomes laborious for the more complicated problems. This difficulty is practically removed by introducing an accurate simplication in the derivation of the equivalent systems.

Consider for example the variable stiffness member in Fig. 7-5*a* and let it be assumed that the stiffness function $f(x) = E_x I_x / E_1 I_1$ is as shown in Fig. 7-5*b*. This graph is plotted by selecting a reference value $E_1 I_1$ and computing the values of $f(x) = E_x I_x / E_1 I_1$ at a sufficient number of cross sections along the length of the member. At the same cross sections, the corresponding values of the moment M_x of the member in Fig. 7-5*a* are computed and plotted as in Fig. 7-5*c*. By dividing the values of M_x by the corresponding values of $f(x)$, the moment diagram $M_e = M_x / f(x)$ of the equivalent system is shown plotted by the solid line in Fig. 7-5*d*.

The shape of M_e is now approximated with straight-line segments of length ΔS_n, $n = 1, 2, 3, \ldots$, as shown in Fig. 7-5*d*. The ΔS_n segments do not have to be of equal length, and their juncture points are usually located above or below the solid line curve, so that they approximately balance the areas added to or subtracted from the M_e diagram. This is shown in Fig. 7-5*d*. The number of such segments for an excellent approximation depends on the shape of the M_e diagram. Usually three to five segments judiciously selected are sufficient.

The equivalent shear force diagram is drawn as shown in Fig. 7-5*e*. For example, the shear force V_2 is equal to the difference in bending moment between sections at C and B, divided by the length ΔX_2. The equivalent system of uniform stiffness $E_1 I_1$ is determined as shown in Fig. 7-5*f* and is acted upon by three concentrated loads. The slope and deflection of any point on the elastic line of the variable stiffness beam in Fig. 7-5*a* are very accurately determined by using the equivalent system in Fig. 7-5*f* and applying known methods of strength of materials. For example, an approximation of the M_e diagram that yields two to four concentrated loads on the equivalent system, produces less than 1% error in the computed deflections.

If EI is uniform, then $f(x)$ is unity and the M_e diagram is identical to that of the original system. Its approximation with straight lines can be carried out in the same way as for the variable stiffness beams, and the equivalent system is loaded with simple concentrated loads.

This simplification makes the method easily applicable to any member of variable stiffness, whether or not the variation of its stiffness and loading can be expressed as continuous functions of x. The following example illustrates the application of this method.

Consider the cantilever beam in Fig. 7-3*a*, and let it be assumed that it is

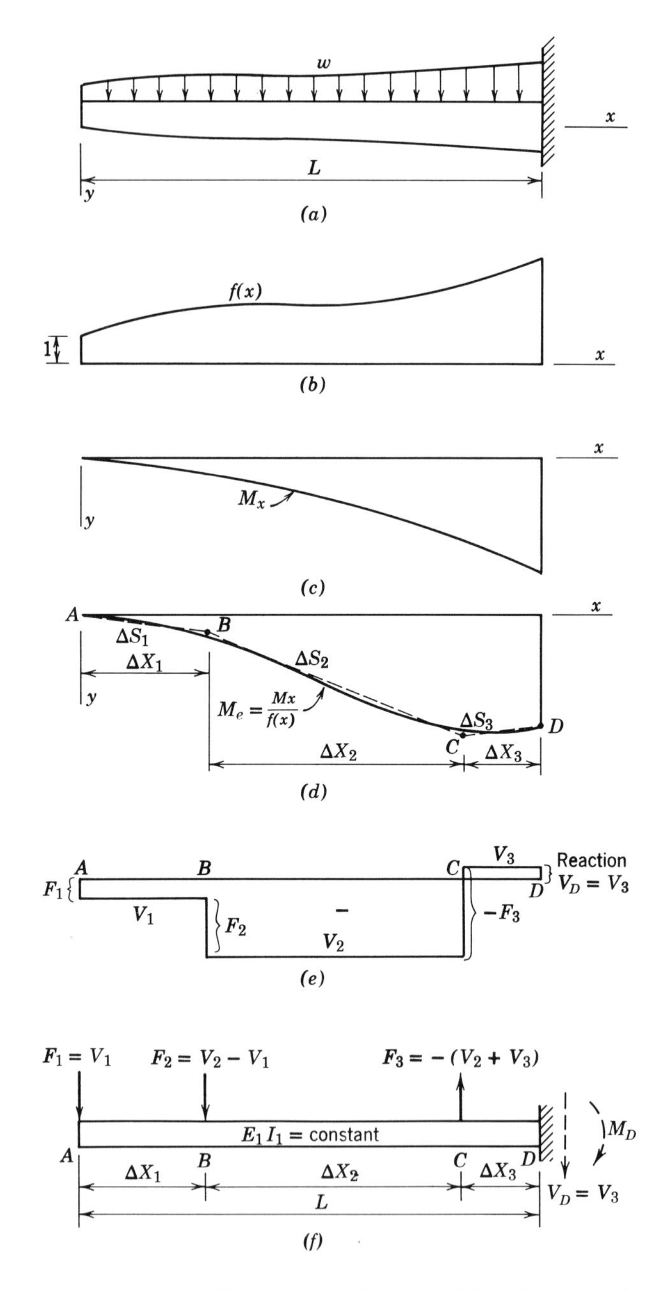

Fig. 7-5. (*a*) Actual system. (*b*) Stiffness function diagram. (*c*) Actual moment diagram. (*d*) Approximated M_e diagram. (*e*) Equivalent shear force diagram. (*f*) Equivalent system of constant stiffness.

required to determine an equivalent system of uniform stiffness EI_1 by applying the approximate method of the equivalent systems. In addition, by using the equivalent system and applying the area moment method, determine the deflection and rotation at the free end.

The M_e diagram is the same as the one in Fig. 7-3c, but is here plotted by

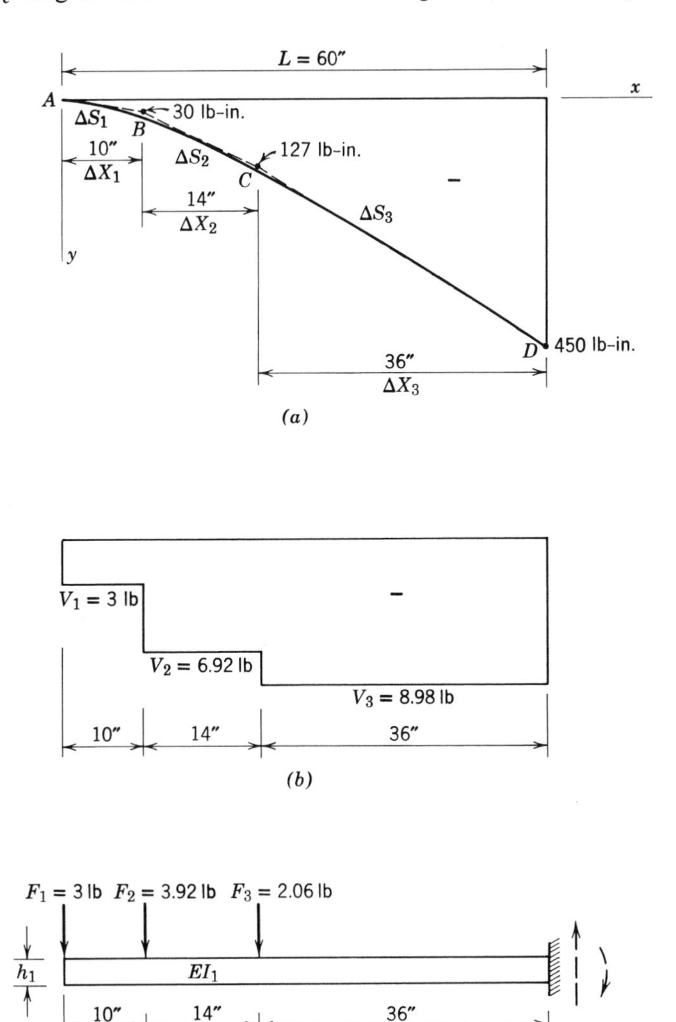

Fig. 7-6. (*a*) Approximated M_e diagram. (*b*) Equivalent shear force diagram. (*c*) Equivalent system.

considering a sufficient number of sections along the length of the member and dividing the computed values of M_x by the corresponding values of $f(x)$. The approximation of the M_e diagram with three straight-line segments ΔS_1, ΔS_2, and ΔS_3 is shown by the dashed lines in Fig. 7-6*a*. The equivalent shear force diagram appears in Fig. 7-6*b*, where

$$V_1 = \frac{\Delta M_B}{\Delta X_1} = -\frac{30}{10} = -3 \text{ lb}$$

$$V_2 = \frac{M_C - M_B}{\Delta X_2} = -\frac{127-30}{14} = -6.92 \text{ lb}$$

$$V_3 = \frac{M_D - M_C}{\Delta X_3} = -\frac{450-127}{36} = -8.98 \text{ lb}$$

The equivalent system of uniform stiffness EI_1 is shown in Fig. 7-6*c*, where $F_1 = V_1 = 3\text{ lb}, F_2 = V_2 - V_1 = 3.92\text{ lb}$, and $F_3 = V_3 - V_2 = 2.06$ lb. By applying the area moment method to the equivalent system in Fig. 7-6*c*,

$$y_A = \frac{(491.9)(10^3)}{EI_1} \quad \text{(downward)}$$

The error compared to the exact solution, Eq. 7-18, is

$$\text{error} = \frac{2900}{489{,}000} = \frac{1}{169} < 1\%$$

The slope θ_A at the free end is

$$\theta_A = -\frac{(11.63)(10^3)}{EI_1} \quad \text{(counterclockwise)}$$

The exact solution yields practically identical results.

Statically Indeterminate Beams

The method of the equivalent systems can be also used to solve statically indeterminate problems. Consider for example the beam in Fig. 7-7*a* and let it be assumed that it is required to determine an equivalent system of uniform stiffness. The procedure is initiated by selecting the reaction R_A and moment M_A at the end A as the redundants, Fig. 7-7*b*. The variation of the stiffness $f(x)$ and the moment diagram of the cantilever beam in Fig. 7-7*b*, by parts, are shown in Figs. 7-7*c* and 7-7*d*, respectively.

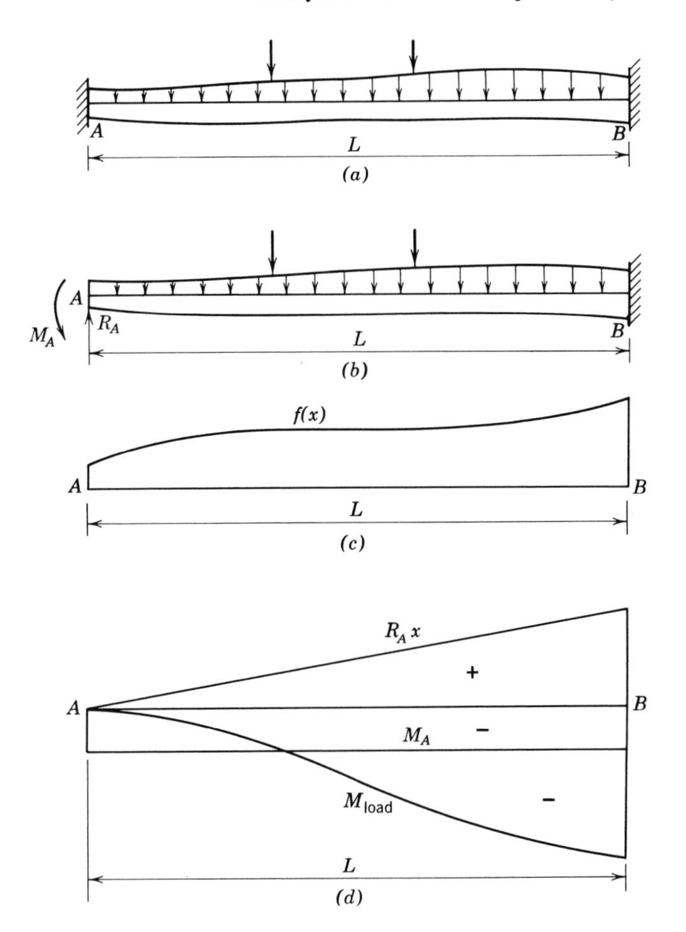

Fig. 7-7. (*a*) Original system. (*b*) Original system with M_A and R_A as redundants. (*c*) Stiffness function variation. (*d*) Actual moment diagram by parts.

For each moment diagram in Fig. 7-7*d*, an equivalent M_e diagram is determined in the usual way, because the cantilever beam is now statically determinate. These M_e diagrams, approximated by straight lines, are shown in Fig. 7-8*a*. By considering each approximated M_e diagram separately, the corresponding three equivalent systems of constant stiffness are determined as explained above; they are shown in Figs. 7-8*b*, 7-8*c*, and 7-8*d*. The λ's in these diagrams are numerical values obtained during derivation of the equivalent systems.

The redundants R_A and M_A are now computed by using the three equivalent systems and applying known methods of strength of materials.

For example, if the deflections at the end A in Figs. 7-8*b*, 7-8*c*, and 7-8*d* are designated, respectively, by δ_A', δ_A'', and δ_A''' and the rotations at the same end by θ_A', θ_A'', and θ_A''', then the boundary conditions at the end A of the initial system in Fig. 7-7*a* will be satisfied if

$$\delta_A' + \delta_A'' + \delta_A''' = 0 \tag{7-19}$$

$$\theta_A' + \theta_A'' + \theta_A''' = 0 \tag{7-20}$$

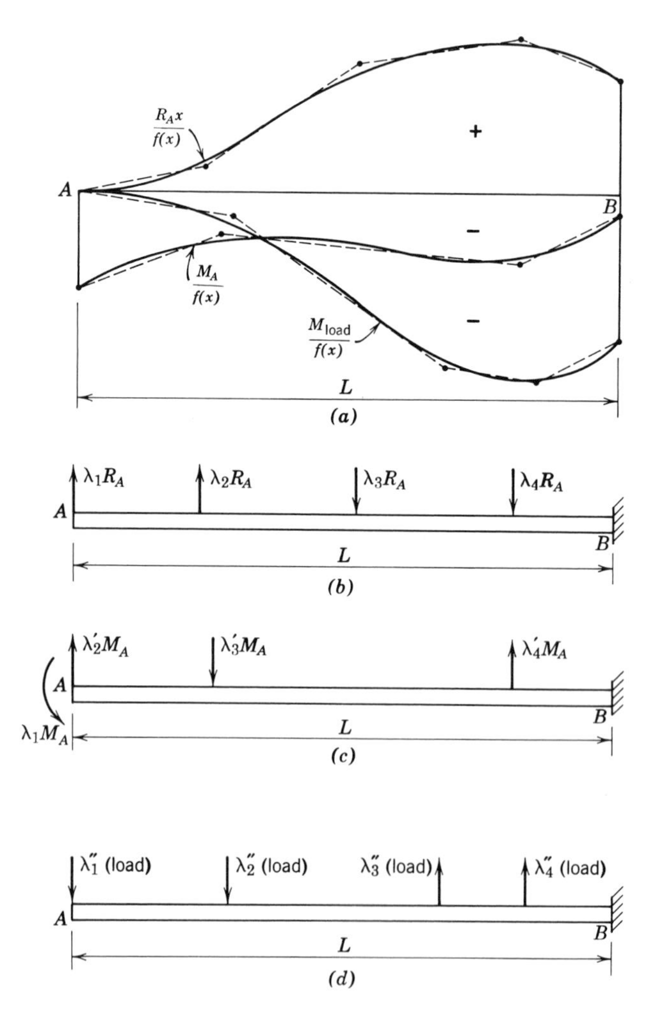

Fig. 7-8. (*a*) Approximated M_e diagrams. (*b*) Equivalent system for R_A. (*c*) Equivalent system for M_A. (*d*) Equivalent system for load.

Equations 7-19 and 7-20 are functions of R_A and M_A. Their simultaneous solution will yield the values of the reaction R_A and moment M_A at the fixed end A of the original system in Fig. 7-7*a*. With the redundant quantities determined, the values of the actual moments M_x at cross sections along the length of the member in Fig. 7-7*a* can be computed by applying simple statics. Thus, by using the approximate method of the equivalent systems as above, an equivalent system of constant stiffness is derived as shown in Fig. 7-9.

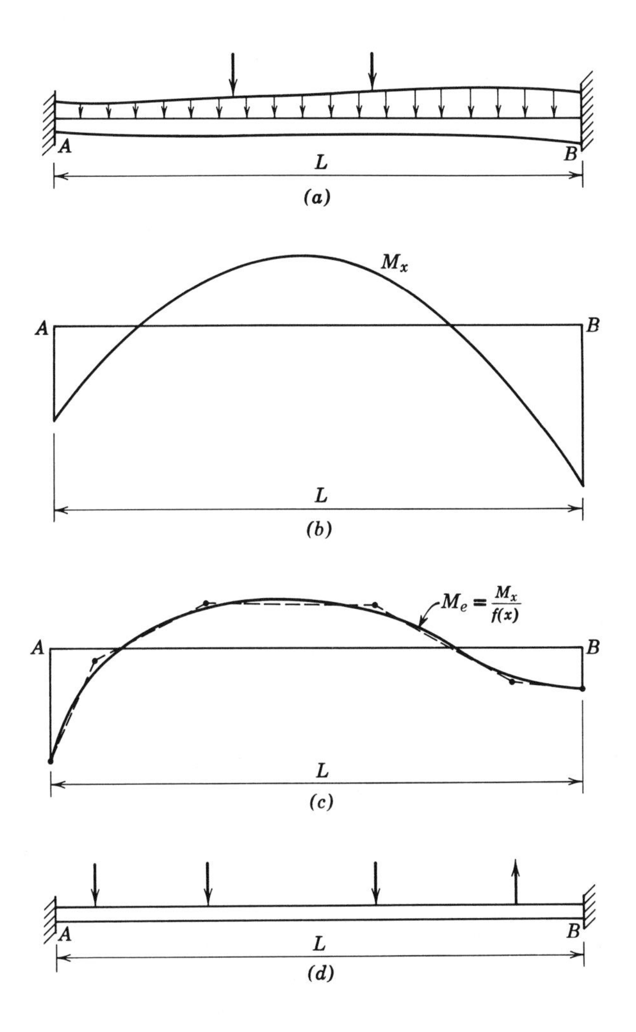

Fig. 7-9. (*a*) Original system. (*b*) Actual moment diagram. (*c*) M_e moment diagram. (*d*) Equivalent system.

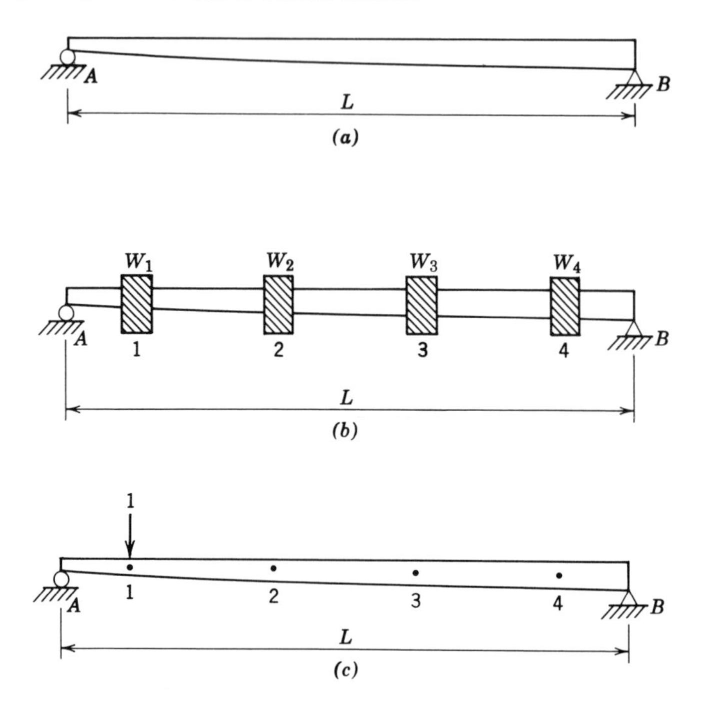

Fig. 7-10. (*a*) Initial system. (*b*) Lumped-mass system. (*c*) Calculation of flexibility coefficients by equivalent system method.

The equivalent system in Fig. 7-9*d* has the same length and boundary conditions and an almost identical elastic line as the variable stiffness member in Fig. 7-7*a*. Therefore, rotations and deflections of the variable stiffness member can be determined by solving the equivalent system. These values can be also determined, if preferred, by solving the three equivalent systems in Fig. 7-8 and superimposing the results.

The method of the equivalent systems provides a convenient way to solve problems in structural analysis, as well as to compute flexibility and stiffness coefficients that are needed in dynamic and vibration analysis of structures. If preferred, the computations can be easily handled by an electronic digital computer, since the required operations are readily adaptable to it.

7-4 NATURAL FREQUENCIES AND MODE SHAPES FOR BEAMS

The procedure for computing natural frequencies and mode shapes of variable stiffness members can be carried out in two ways. The first way consists of the usual arbitrary lumping of the weight, or mass, of the original

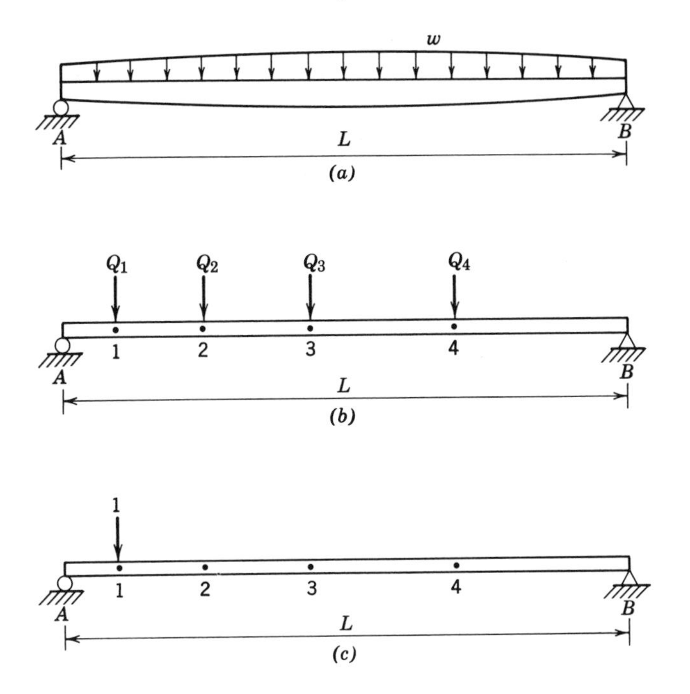

Fig. 7-11. (*a*) Original system. (*b*) Equivalent system. (*c*) Calculation of flexibility coefficients by using methods of strength of materials.

system in Fig. 7-10*a* at discrete points, as shown in Fig. 7-10*b*. The number of such lumped weights, or masses, should be at least twice the number of the natural frequencies to be determined. Then, by using the system in Fig. 7-10*b* and applying the method of Stodola and iteration procedure or one of the methods in Chapter 6, the natural frequencies and mode shapes of the member are determined.

In applying the method of Stodola, the required a_{ij} flexibility coefficients can be easily computed by using the method of the equivalent systems. For example, in Fig. 7-10*c*, a unit load is applied at point 1. The flexibility coefficients a_{11}, a_{21}, a_{31}, and a_{41} at points 1, 2, 3, and 4, respectively, due to the unit load, can be computed by solving the problem in Fig. 7-10*c* by the method of the equivalent systems. In a similar manner, the flexibility coefficients with unit loads at points 2, 3, and 4 can be obtained. When the flexibility coefficients are determined, the required natural frequencies and mode shapes are obtained by the method of Stodola as discussed in Chapter 5.

The second procedure uses an equivalent system which replaces the initial variable stiffness member. For example, in Fig. 7-11*a* the weight of the beam is assumed to be applied as distributed load w on the member as shown. If the member supports additional attached weights, these weights should be also considered as applied loads. Then, by applying the approximate method of the equivalent systems, an equivalent system of constant stiffness such as that in Fig. 7-11*b* is obtained.

The concentrated weights in this figure are equivalent weights that serve a definite purpose. Their magnitude and location on the equivalent system are determined in a way that compensates for the change in the stiffness, so that at the static equilibrium position the elastic lines of both systems are identical. Thus, under these conditions, it would be reasonable to conclude that the equivalent system in Fig. 7-11*b* will vibrate, at least for the first few modes, in a manner similar to that in Fig. 7-11*a*. Therefore, the required natural frequencies and mode shapes of the system in Fig. 7-11*a* can be determined by using the equivalent system in Fig. 7-11*b* and applying the method of Stodola and iteration procedure or one of the methods in Chapter 6.

In Table 7-1 an attempt is made to compute natural frequencies of vibration of a beam of various lengths by using several methods and to compare results. The loading on the beam that participates in its vibrational motion is as shown. For convenience, it is assumed to be uniformly distributed. The numerical computations were performed by using a digital computer. The methods used here are: *(a)* the method of Stodola and iteration procedure, by assuming that the stiffness EI between any two weight concentration points is constant and equal to the average value in the interval; *(b)* the method of Stodola and iteration procedure, but using the method of the equivalent systems as in Fig. 7-10*c* to determine the flexibility coefficients; *(c)* by using an equivalent system of constant stiffness as in Fig. 7-11*b* and applying the method of Stodola; and *(d)* the Myklestad method.

The results show that all methods yield reasonable results for the first three to five natural frequencies of vibration. All methods yield erroneous results for frequencies above the fifth one, even if the number of concentrated weights is larger than 20. This, of course, is not a tremendous limitation, because an accurate estimate for the dynamic response of structures is usually obtained by using only the first few modes.

In Table 7-2 a similar analysis is carried out by using the same methods to solve a uniform stiffness beam. In this case, however, the exact method is also included. The results are as shown in Table 7-2. All methods yield excellent results for the first five frequencies.

As an illustration, let it now be assumed that it is required to determine the fundamental frequency and the corresponding mode shape of the statically indeterminate beam in Fig. 7-12*a*. The modulus of elasticity E is constant and $EI_1 = 460 \times 10^6$ kip-in.^{2}, where I_1 is the moment of inertia of

TABLE 7-1

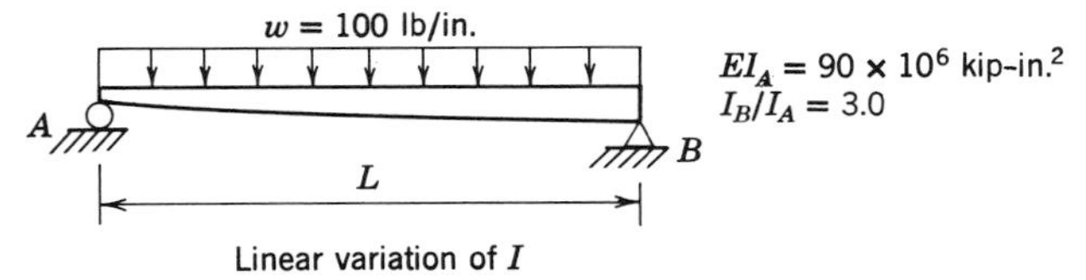

L (in.)	Method	10 Concentrated Masses ω (rps) ω_1	ω_2	ω_3	ω_4	ω_5	20 Concentrated Masses ω (rps) ω_1	ω_2	ω_3	ω_4	ω_5
	Stodola	807	3094	6233	7867	10404	807	3179	6811	7192	8323
	Equiv. systems*	807	3179	6823	8178	11103	807	3203	7101	7406	9027
100	Equiv. systems†	807	3202	7185	11866	18730	807	3200	7277	13229	18602
	Myklestad	807	3208	7199	12762	19828	807	3210	7208	12802	19998
	Stodola	202	774	1558	1967	2601	202	795	1703	1798	2081
	Equiv. systems*	202	795	1706	2045	2777	202	801	1775	1852	2258
200	Equiv. systems†	202	850	1896	2968	4886	202	850	1895	3312	4989
	Myklestad	202	802	1800	3191	4955	202	803	1802	3201	5002
	Stodola	90	344	693	874	1156	90	353	757	799	925
	Equiv. systems*	90	353	758	905	1234	90	356	789	823	1004
300	Equiv. systems†	90	378	843	1419	2217	90	357	842	1471	2208
	Myklestad	90	357	800	1418	2205	90	357	801	1423	2224

*Arbitrary mass lumping †Using an equivalent system

TABLE 7-2

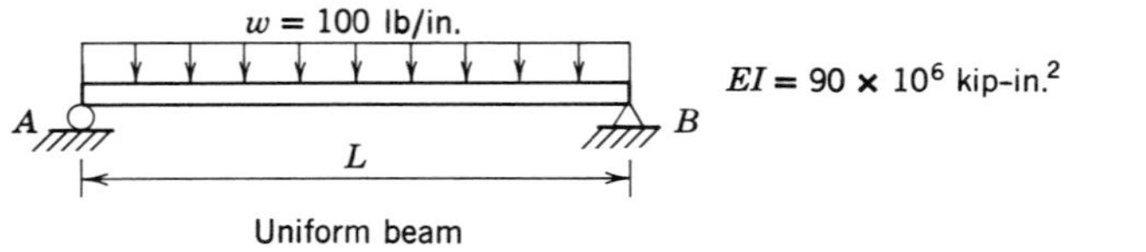

$EI = 90 \times 10^6$ kip-in.^{2}

L (in.)	Method	10 Concentrated Masses ω (rps) ω_1	ω_2	ω_3	ω_4	ω_5	20 Concentrated Masses ω (rps) ω_1	ω_2	ω_3	ω_4	ω_5
100	Stodola	582	2327	5232	9247	13092	582	2327	5235	9252	9300
	Equiv. systems*	582	2326	5232	9284	14273	582	2327	5235	9301	9465
	Equiv. systems†	582	2336	5269	9346	14308	582	2327	5220	9366	14555
	Myklestad	582	2328	5233	9284	14449	582	2327	5236	9306	14458
	Exact solution	582	2328	5238	9312	14550	582	2328	5238	9312	14550
200	Stodola	145	582	1308	2318	3532	145	582	1309	2325	2409
	Equiv. systems*	145	582	1308	2321	3610	145	582	1308	2327	2173
	Equiv. systems†	145	584	1317	2323	3641	145	582	1305	2293	3635
	Myklestad	146	582	1308	2321	3612	145	582	1309	2326	3644
	Exact solution	145	580	1305	2320	3625	145	580	1305	2320	3625
300	Stodola	65	258	581	1031	1596	65	259	582	1034	1080
	Equiv. systems*	65	258	581	1032	1607	65	259	582	1034	1084
	Equiv. systems†	65	260	585	1072	1663	65	259	580	1019	1630
	Myklestad	65	259	582	1032	1605	65	259	582	1034	1617
	Exact solution	65	260	585	1040	1625	65	260	585	1040	1625

*Arbitrary mass lumping †Using an equivalent system

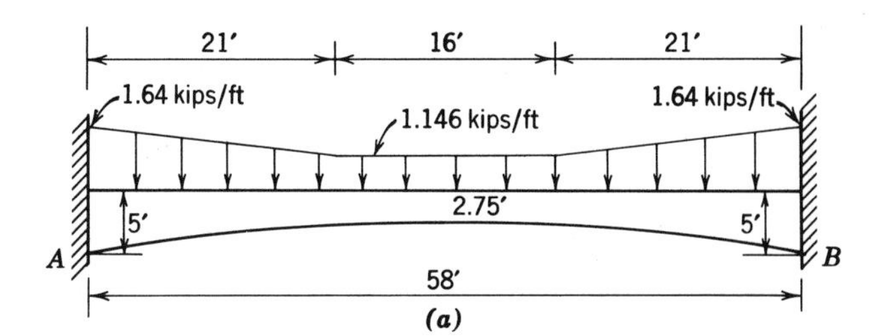

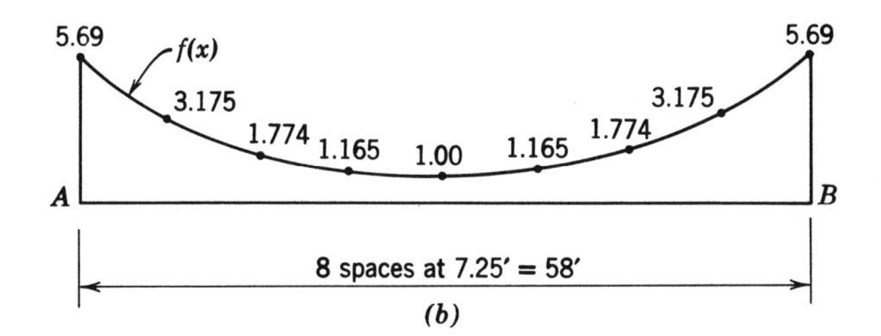

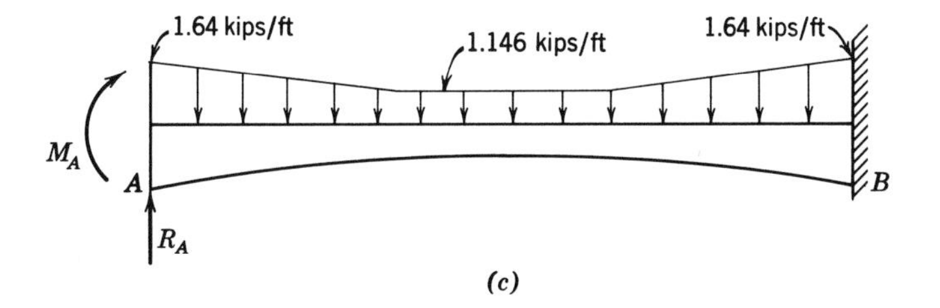

Fig. 7-12. (*a*) Original system. (*b*) Stiffness variation. (*c*) Original system with R_A and M_A as reduntants.

the beam at its center. The fundamental frequency and mode shape will be determined by using the equivalent system of constant stiffness EI_1 and applying the method of Stodola and iteration procedure.

The distributed weight of the member is somewhat approximated and it is assumed to act as shown in Fig. 7-12*a*. The variation *f(x)* of the stiffness, with EI_1 as reference value, is shown in Fig. 7-12*b*. By considering the reaction R_A and the moment M_A as the redundants, Fig. 7-12*c*, the moment diagrams due to R_A, M_A, and distributed loading are shown in Fig. 7-13. These are the actual moment diagrams of the cantilever beam in Fig. 7-12*c*. The ordinates of these diagrams are divided by the corresponding values of *f(x)* in Fig. 7-12*b* and their shapes are approximated with straight lines. The equivalent systems of uniform stiffness EI_1 for R_A, M_A, and distributed loading are determined as discussed in Section 7-3; they are as shown in Figs. 7-14*a*, 7-14*b*, and 7-14*c*.

The deflections δ_1, δ_2, and δ_3 at the end A in Figs. 7-14*a*, 7-14*b*, and 7-14*c*, respectively, and the corresponding rotations θ_1, θ_2, and θ_3 are determined by using the equivalent systems and applying the area moment method. They are as follows:

$$EI_1\,\delta_1 = 34{,}244R_A \qquad EI_1\,\theta_1 = 988.4R_A$$

$$EI_1\,\delta_2 = 986.5M_A \qquad EI_1\,\theta_2 = 33.94M_A$$

$$EI_1\,\delta_3 = -898{,}000 \qquad EI_1\,\theta_3 = -23{,}470$$

The boundary conditions at the end A of the original system in Fig. 7-12*a* are

$$\delta = \delta_1 + \delta_2 + \delta_3 = 0$$

$$\theta = \theta_1 + \theta_2 + \theta_3 = 0$$

or

$$34{,}244R_A + 986.5M_A - 988{,}000 = 0$$

$$988.4R_A + 33.94M_A - 23{,}470 = 0$$

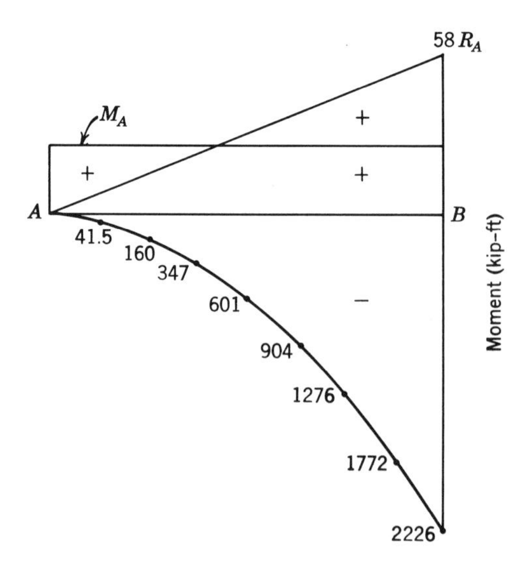

Fig. 7-13. Original moment diagram by parts.

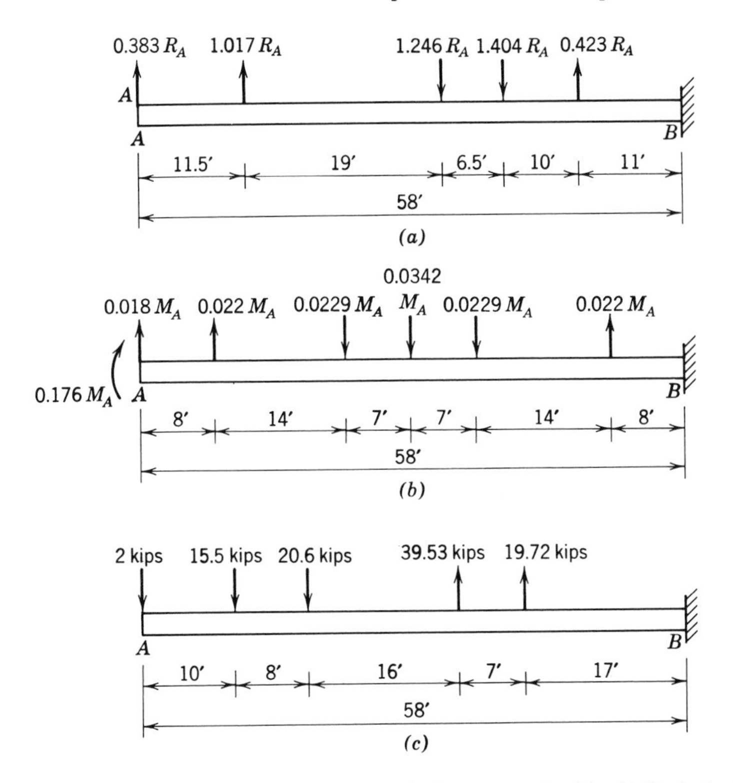

Fig. 7-14. (*a*) Equivalent system for R_A. (*b*) Equivalent system for M_A. (*c*) Equivalent system for loading.

Simultaneous solution of the two equations yields

$$R_A = 38.4 \text{ kips} \uparrow$$

$$M_A = 427.0 \text{ kip-ft} \circlearrowright$$

which are the actual reaction and moment at the end of A of the beam in Fig. 7-12*a*.

With R_A and M_A computed, the values of the moments at cross sections along the length of the member in Fig. 7-12*a* are obtained by applying simple statics. When these moments are divided by the corresponding values of $f(x)$, the equivalent moments $M_e = M_x / f(x)$ are determined; they are shown in Fig. 7-15*a*. In the same figure, the approximation of the shape of M_e with straight lines is also shown. The equivalent system of constant stiffness EI_1 is derived in the usual way, Fig. 7-15; it is shown in Fig. 7-15*c*. This system will be used to determine the fundamental frequency and the corresponding mode shape.

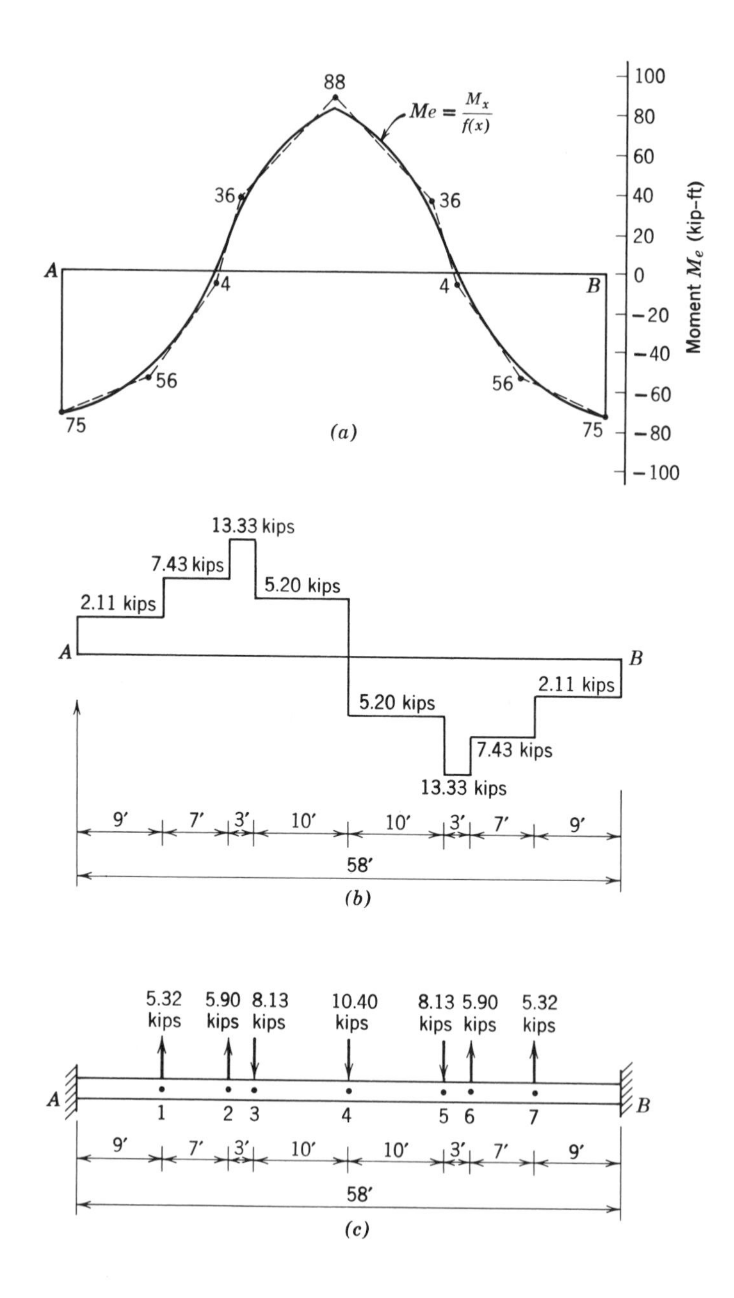

Fig. 7-15. (*a*) Approximated M_e diagram. (*b*) Shear force diagram of the equivalent system. (*c*) Equivalent system.

The concentrated equivalent weights in Fig. 7-15*c* are now assumed to act in the direction of the fundamental mode shape as stated earlier. With this in mind, the a_{ij} flexibility coefficients are determined in the usual way by applying unit loads at points 1, 2,...,7, Fig. 7-15*c*. They are as follows:

$a_{11}=5.52\times10^{-4}$ $a_{12}=9.02\times10^{-4}$ $a_{13}=9.61\times10^{-4}$

$a_{21}=9.02\times10^{-4}$ $a_{22}=19.40\times10^{-4}$ $a_{23}=22.10\times10^{-4}$

$a_{31}=9.61\times10^{-4}$ $a_{32}=22.10\times10^{-4}$ $a_{33}=26.15\times10^{-4}$

$a_{41}=8.73\times10^{-4}$ $a_{42}=22.05\times10^{-4}$ $a_{43}=27.60\times10^{-4}$

$a_{51}=5.52\times10^{-4}$ $a_{52}=13.34\times10^{-4}$ $a_{53}=17.63\times10^{-4}$

$a_{61}=4.00\times10^{-4}$ $a_{62}=10.60\times10^{-4}$ $a_{63}=13.34\times10^{-4}$

$a_{71}=1.50\times10^{-4}$ $a_{72}=4.00\times10^{-4}$ $a_{73}=5.22\times10^{-4}$

$a_{14}=8.73\times10^{-4}$ $a_{15}=5.22\times10^{-4}$

$a_{24}=22.05\times10^{-4}$ $a_{25}=13.34\times10^{-4}$

$a_{34}=27.60\times10^{-4}$ $a_{35}=17.63\times10^{-4}$

$a_{44}=38.20\times10^{-4}$ $a_{45}=27.60\times10^{-4}$

$a_{54}=27.60\times10^{-4}$ $a_{55}=26.15\times10^{-4}$

$a_{64}=22.05\times10^{-4}$ $a_{65}=22.10\times10^{-4}$

$a_{74}=8.73\times10^{-4}$ $a_{75}=9.61\times10^{-4}$

$a_{16}=4.00\times10^{-4}$ $a_{17}=1.50\times10^{-4}$

$a_{26}=10.60\times10^{-4}$ $a_{27}=4.00\times10^{-4}$

$a_{36}=13.34\times10^{-4}$ $a_{37}=5.22\times10^{-4}$

$a_{46}=22.05\times10^{-4}$ $a_{47}=8.73\times10^{-4}$

$a_{56}=22.10\times10^{-4}$ $a_{57}=9.61\times10^{-4}$

$a_{66}=19.40\times10^{-4}$ $a_{67}=9.02\times10^{-4}$

$a_{76}=9.02\times10^{-4}$ $a_{77}=5.52\times10^{-4}$

By applying Eq. 5-74,

$$\begin{bmatrix} y_1 \\ y_2 \\ y_3 \\ y_4 \\ y_5 \\ y_6 \\ y_7 \end{bmatrix} = \frac{7.98\omega^2}{10^4 g} \begin{bmatrix} 3.67 & 6.67 & 9.79 & 11.37 & 5.32 & 2.95 & 1.00 \\ 6.02 & 14.35 & 22.50 & 28.70 & 13.57 & 7.83 & 2.67 \\ 6.40 & 16.35 & 26.65 & 36.00 & 17.95 & 9.78 & 3.48 \\ 5.82 & 16.30 & 28.10 & 49.80 & 28.10 & 16.30 & 5.82 \\ 3.48 & 9.87 & 17.95 & 36.00 & 26.65 & 16.35 & 6.40 \\ 2.67 & 7.83 & 13.57 & 28.70 & 22.50 & 14.35 & 6.02 \\ 1.00 & 2.95 & 5.32 & 11.37 & 9.79 & 6.67 & 3.67 \end{bmatrix} \begin{bmatrix} y_1 \\ y_2 \\ y_3 \\ y_4 \\ y_5 \\ y_6 \\ y_7 \end{bmatrix}$$

By applying the iteration procedure, the matrix equation converges to

$$\begin{bmatrix} y_1 \\ y_2 \\ y_3 \\ y_4 \\ y_5 \\ y_6 \\ y_7 \end{bmatrix} = \frac{(7.98)(116.10)\omega^2}{10^4 g} \begin{bmatrix} 1.00 \\ 2.40 \\ 2.95 \\ 3.85 \\ 2.95 \\ 2.40 \\ 1.00 \end{bmatrix}$$

Three repetitions, starting with $y_1 = y_2 = \cdots = y_7 = 1.00$, were required for this purpose. The fundamental frequency ω is determined from the expression

$$\frac{(7.98)(116.10)\omega^2}{10^4 g} = 1$$

which yields

$$\omega = \left[\frac{10^4 (386)}{(7.98)(116.10)} \right]^{\frac{1}{2}} = 64.50 \text{ rps}$$

$$f = \frac{\omega}{2\pi} = \frac{64.50}{2\pi} = 10.25 \text{ Hz}$$

The mode shape is characterized by the magnitudes

$$y_1=1.00$$

$$y_2=2.40$$

$$y_3=2.95$$

$$y_4=3.85$$

$$y_5=2.95$$

$$y_6=2.40$$

$$y_7=1.00$$

7-5 DYNAMIC RESPONSE OF FRAMES WITH MEMBERS OF VARIABLE STIFFNESS

So far, the members of a frame have been assumed to be of uniform stiffness. In some cases, however, the stiffness of these members could vary along their length. Then, the analysis for dynamic response becomes simpler if the approximate method of the equivalent systems is used.

Consider for example the building frame in Fig. 7-16*a* and let it be assumed that the stiffness EI of its members is variable. If the stiffness of the girders in comparison to that of the columns is large, it would be reasonable to assume that the girders are infinitely rigid and to neglect the small rotations of the tops of the columns. On this basis, reasonable results would be obtained by analyzing the two-degree system in Fig. 7-16*b*. Here, the weights W_1 and W_2 represent the weight of the frame lumped at the girder levels in exactly the same manner as in Chapter 3. During motion, the displacements x_1 and x_2 in Fig. 7-16*b* will be equal to the horizontal displacement x_1 and x_2, respectively, of the frame in Fig. 7-16*a*, because the spring constants k_1 and k_2 are equal to the total stiffness of the columns at the girder levels.

The spring constant k_2, for example, is determined by considering each column of the second story of the frame as fixed at both ends and applying at its top a displacement δ equal to unity as shown in Fig. 7-16*c*. The problem to be solved is shown in Fig. 7-16*d*. The force $P=k$ and moment M produce at the top of the column a displacement $\delta=1$, and rotation θ equal to zero. The spring constant k_2 in Fig. 7-16*b* is the sum of the k 's of all columns of the second story of the frame. In a similar manner, the k 's at the top of the columns of the first story can be determined. Their sum will yield the spring constant k_1 in Fig. 7-16*b*.

The computation of the k's of the variable stiffness columns becomes convenient if the approximate method of the equivalent systems is used. The problem to be solved by this method is shown in Fig. 7-16*e*. The boundary conditions $\theta = 0$ and $\delta = 1$ are sufficient to determine k and M.

The dynamic displacements x_1 and x_2, as well as the dynamic shear forces, moments, and bending stresses, can be determined by using the

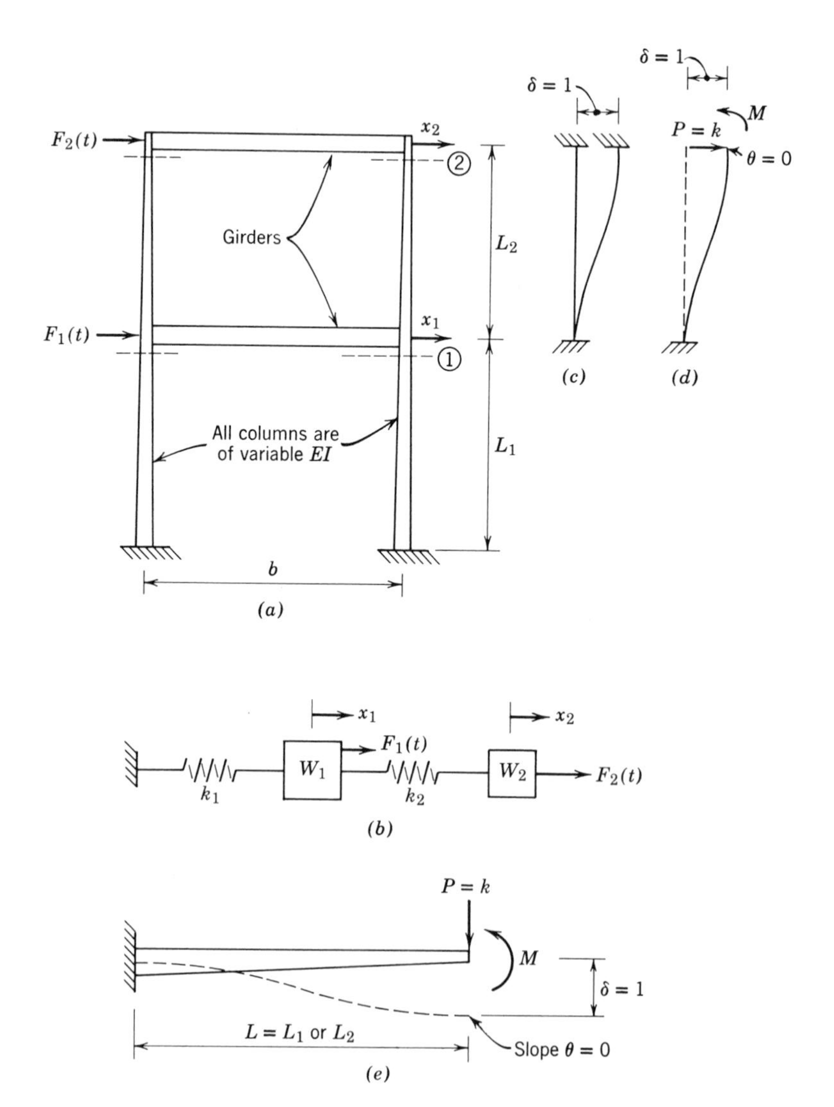

Fig. 7-16

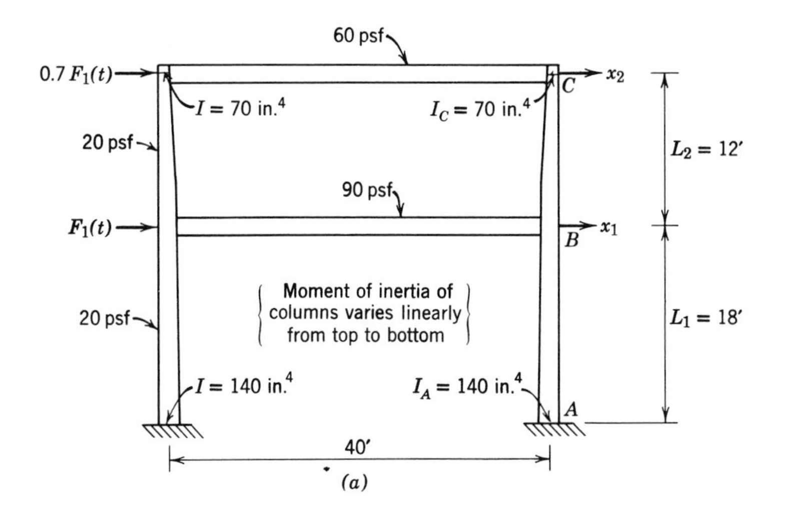
60 psf
0.7 $F_1(t)$
x_2
$I = 70$ in.^{4}
$I_C = 70$ in.^{4}
C
$L_2 = 12'$
20 psf
90 psf
$F_1(t)$
x_1
B
Moment of inertia of columns varies linearly from top to bottom
20 psf
$L_1 = 18'$
$I = 140$ in.^{4}
$I_A = 140$ in.^{4}
A
40′
(a)

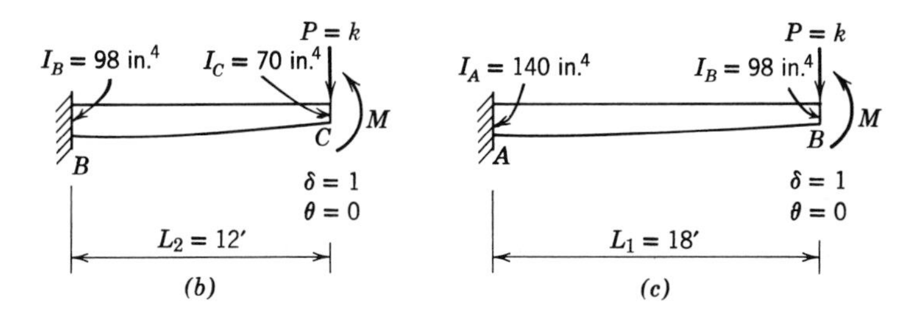
$P = k$
$I_B = 98$ in.^{4}
$I_C = 70$ in.^{4}
M
C
B
$\delta = 1$
$\theta = 0$
$L_2 = 12'$
(b)
$P = k$
$I_A = 140$ in.^{4}
$I_B = 98$ in.^{4}
M
B
A
$\delta = 1$
$\theta = 0$
$L_1 = 18'$
(c)

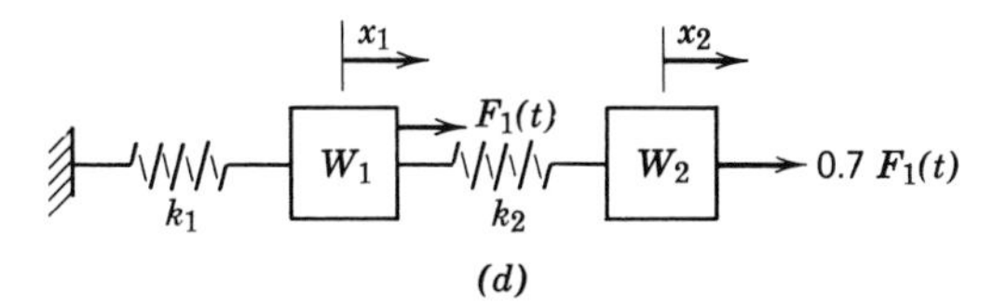
x_1
x_2
$F_1(t)$
W_1
W_2
0.7 $F_1(t)$
k_1
k_2
(d)

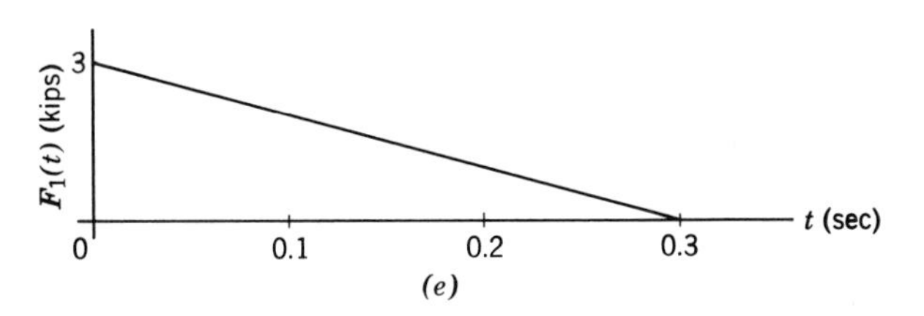
$F_1(t)$ (kips)
3
0
0.1
0.2
0.3
t (sec)
(e)

Fig. 7-17

spring-weight system of Fig. 7-16*b* and applying either the acceleration impulse extrapolation method in the usual way or the method of modal analysis. Other methods can be also used.

If it is desired to take into account the flexibility of the girders, the method discussed in Section 5-14 can be readily used. The required stiffness coefficients k_{ij} can be determined by following the procedure given in Section 1-15. The computation of these coefficients becomes easier if the approximate method of the equivalent systems is used. For example, if the columns are of variable stiffness EI, the computation of the fixed end moments due to a unit displacement, as well as the moment distribution factors, can be determined by applying the approximate method of the equivalent systems as discussed in the preceding sections. With these quantities computed, the moment distribution procedure and the evaluation of k_{ij} coefficients are carried out in the same way as in Section 1-15. The dynamic response is obtained as in Section 5-14.

As an illustration, let it be assumed that it is required to determine the dynamic response of the two-story rigid frame building in Fig. 3-5*a* whose material properties, loading, and so on, are made uniform throughout the length of the building. In this case, the five rigid frames forming the building are of the form shown in Fig. 7-17*a*. The girders are assumed to be infinitely rigid compared to the columns. The moment of inertia I of the columns is variable and varies linearly from the elevation C to the elevation A. At C, the moment of inertia I_C is 70 in.^{4}; it is 140 in.^{4} at A. The modulus of elasticity E is constant and equal to 30×10^6 psi. The entire building will be analyzed for dynamic response by considering only the interior frame in Fig. 7-17*a* together with the associated wall and floor areas.

The idealized two-degree system is shown in Fig. 7-17*d*. The spring constant k_1 is the sum of the k 's of the two columns of the first story of the frame. For each column, the k is determined by using the approximate method of the equivalent systems to solve the problem in Fig. 7-17*c*. In solving this problem, it should be noted that the boundary conditions to be satisfied at B are $\delta = 1$ and $\theta = 0$. The spring constant k_2 is the sum of the k 's of the columns of the second story. Each k is determined by solving the problem in Fig. 7-17*b*. The results are

$$k_1 = 8.32 \text{ kips/in.}$$

$$k_2 = 20.10 \text{ kips/in.}$$

The weight of the frame and associated wall and floor areas is lumped at the floor levels as in Chapter 3. The lumped weights W_1 and W_2 in Fig. 7-

17*d* are the same as in Fig. 3-5*c*;.

$$W_1 = 84.0 \text{ kips}$$

$$W_2 = 52.8 \text{ kips}$$

The corresponding masses m_1 and m_2 are

$$m_1 = \frac{W_1}{g} = \frac{84.0}{386.0} = 0.217 \text{ kip-sec}^2/\text{in.}$$

$$m_2 = \frac{W_2}{g} = \frac{52.8}{386.0} = 0.137 \text{ kip-sec}^2/\text{in.}$$

Applying Eq. 3-22, we have the two natural frequencies of vibration:

$$\omega_1 = 4.69 \text{ rps} \qquad f_1 = 0.747 \text{ Hz}$$

$$\omega_2 = 16.00 \text{ rps} \qquad f_2 = 2.545 \text{ Hz}$$

The smallest period of vibration τ_2 is

$$\tau_2 = \frac{1}{f_2} = 0.393 \text{ sec}$$

The dynamic displacements x_1 and x_2 can be determined by using the method of modal analysis, the acceleration impulse extrapolation method, or other suitable methods. At this time, the acceleration impulse extrapolation method will be used with a time interval Δt equal to 0.03 sec.

The time variation of the applied dynamic forces is shown in Fig. 7-17*e*. By applying Eqs. 3-23 and 3-24,

$$\ddot{x}_1 = 92.7(x_2 - x_1) - 38.3x_1 + 4.65F_1(t) \tag{7-21}$$

$$\ddot{x}_2 = -146.5\ (x_2 - x_1) + 5.10F_1(t) \tag{7-22}$$

The amplitude responses x_1 and x_2 for the time interval $0 \leqslant t \leqslant 0.3$ sec are shown in Table 7-3. At each time station, the accelerations $\ddot{x}_1$ and $\ddot{x}_2$ were determined from Eqs. 7-21 and 7-22, respectively. With $x_1$and x_2 known, the shear forces at the tops of the columns of the first and second story of the frame are determined in the usual way. For example, if V_B is the shear force at the top of a first story column, its value at $t = 0.30$ sec is

$$V_B = \frac{V_1}{2} = \frac{k_1 x_1}{2} = \frac{8.32}{2}(0.3311) = 1.385 \text{ kips}$$

TABLE 7-3

t (sec)	$92.7(x_2 - x_1)$	$38.3x_1$	$4.65F_1(t)$	$\ddot{x}_1$ Eq.7-21 (in./sec^2)	$\ddot{x}_1(\Delta t)^2$ (in.)	x_1 Eq.2-53 (in.)	$146.5(x_2 - x_1)$	$5.10F_1(t)$	$\ddot{x}_2$ Eq.7-22 (in./sec^2)	$\ddot{x}_2(\Delta t)^2$	x_2 Eq.2-53 (in.)
0	0	0	13.95	13.95	0.0126	0	0	15.30	15.30	0.0138	0
0.03	0.056	0.241	12.55	12.37	0.0111	0.0063*	0.088	13.75	13.66	0.0123	0.0069*
0.06	0.222	0.910	11.15	10.46	0.0094	0.0237	0.352	12.25	11.90	0.0107	0.0261
0.09	0.509	1.930	9.75	8.33	0.0075	0.0505	0.806	10.70	9.89	0.0089	0.0560
0.12	0.927	3.250	8.35	6.03	0.0054	0.0848	1.465	9.18	7.71	0.0069	0.0948
0.15	1.483	4.770	6.95	3.66	0.0033	0.1245	2.345	7.65	5.30	0.0048	0.1405
0.18	2.180	6.420	5.58	1.34	0.0012	0.1675	3.440	6.12	2.68	0.0024	0.1910
0.21	2.980	8.300	4.18	−1.14	−0.0010	0.2117	4.720	4.58	−0.14	−0.0001	0.2439
0.24	3.880	9.780	2.79	−3.11	−0.0028	0.2549	6.150	3.06	−3.09	−0.0028	0.2967
0.27	4.760	11.300	1.40	−5.14	−0.0046	0.2953	7.530	1.56	−5.97	−0.0054	0.3467
0.30						0.3311					0.3913

*Use Eq. 2-55 to determine this amplitude.

At the same time t, the shear force V_C at the top of a second story column is

$$V_C=\frac{V_2}{2}=\frac{k_2}{2}(x_2-x_1)=\frac{20.10}{2}(0.0602)=0.607 \text{ kips}$$

The moments at the ends of the columns are determined by using the appropriate problem in Figs. 7-17*b* and 7-17*c*. For example, if the moments at the ends of the second-story columns are required, then Fig. 7-17*b* is used. In this figure, P should be taken equal to $V_2/2$, or $k_2/2(x_2-x_1)$. With $\delta=(x_2-x_1)$ and $\theta=0$, the moment M in Fig. 7-17*b* is the moment M_C at the top of a second-story column in Fig. 7-17*a*. In other words, since M in Fig. 7-17*b* is derived for a deflection equal to unity, the moment M_C should be equal to $M(x_2-x_1)$. With M_C known, the moment at the other end of the column is obtained by simple statics. With known moments, the dynamic bending stresses can be evaluated by using formulas of strength of materials.

7-6 BEAMS ON ELASTIC SUPPORTS

If a member of variable stiffness is supported elastically by linear springs, its dynamic response due to external dynamic forces can be determined in the same way as for beams on rigid supports. Again, the method of the equivalent systems will facilitate the procedure. The purpose in this section is to discuss briefly how the method of the equivalent systems can be applied to beams on elastic supports.

Consider the variable stiffness beam in Fig. 7-18*a* that is elastically supported at the end A by a linear spring of constant k_A. The elastic line of the beam is the solid line A_1B in Fig. 7-18*b* and is attained by assuming that the member deflects first as a rigid body until A reaches the position A_1 and then as a simply supported beam on rigid supports. The rigid body displacement is represented by the dashed line in Fig. 7-18*b*. If X_A is the elastic reaction at A, then the vertical displacement δ_A of the spring is

$$\delta_A=\frac{X_A}{k_A} \tag{7-23}$$

At another point of distance x from the end A, the displacement y in Fig. 7-18*b* is

$$y=y_1+y_2 \tag{7-24}$$

The part y_1 is the rigid body translation and is given by the expression

$$y_1=\delta_A\frac{(L-x)}{L} \tag{7-25}$$

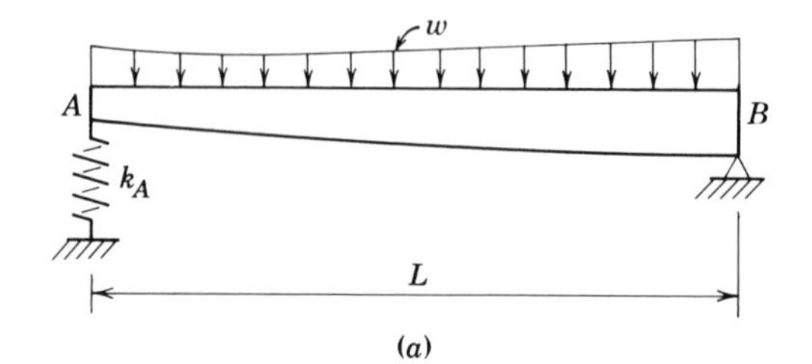

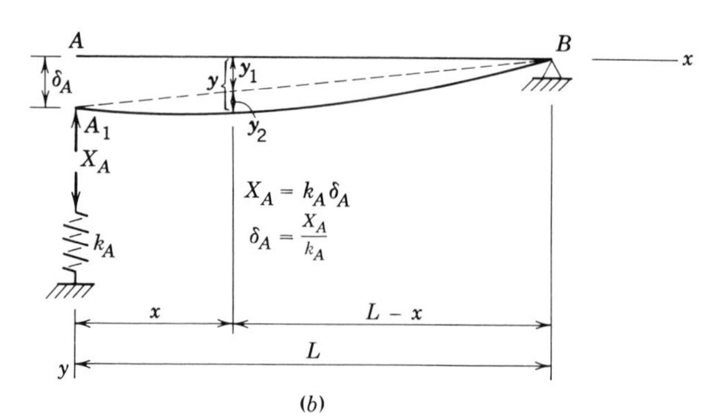

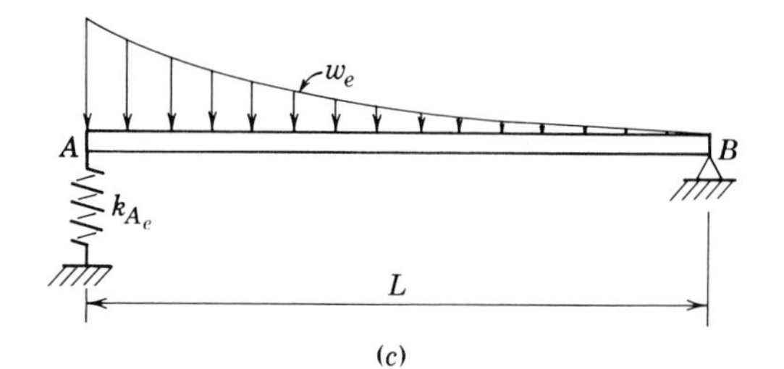

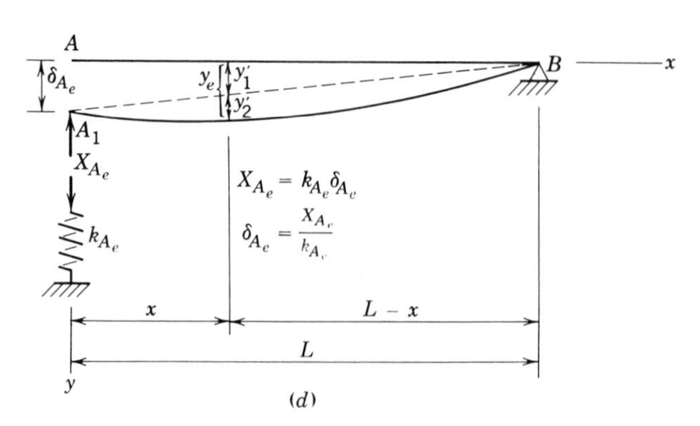

Fig. 7-18. (*a*) Original system. (*b*) Elastic line of the original system. (*c*) Equivalent system. (*d*) Elastic line of the equivalent system.

The portion y_2 in Eq. 7-24 represents the bending of the beam and is assumed to take place as soon as point A reaches the position A_1.

Based on the deformation configuration above and applying the exact method of the equivalent systems, an equivalent system of constant stiffness E_1I_1 is obtained. Let it be assumed that it is as the one shown in Fig. 7-18*c*. Its elastic line is the solid line A_1B in Fig. 7-18*d*. The deflection y_e at any distance x from the end A is

$$y_e = y_1' + y_2' \tag{7-26}$$

Equations 7-24 and 7-26 would be identical if

$$y_1 = y_1' \tag{7-27}$$

$$y_2 = y_2' \tag{7-28}$$

The condition in Eq. 7-28 is satisfied, because y_2' is the elastic line of the equivalent system in Fig. 7-18*c* produced by bending only. During bending, the ends of the beam were assumed to rest on rigid supports, and by virtue of the theory of the equivalent systems the elastic lines y_2 and y_2' should be everywhere identical.

The condition given by Eq. 7-27, however, would be satisfied if the support displacement δ_{A_e} in Fig. 7-18*d* were made equal to the support displacement δ_A in Fig. 7-18*b*. This can be accomplished by using an equivalent linear spring of constant k_{A_e} as shown in Fig. 7-18*c*. Thus, if the symbol X_{A_e} in Fig. 7-18*d* is used to denote the equivalent elastic reaction at the end A, then the displacement δ_{A_e} of the equivalent spring is

$$\delta_{A_e} = \frac{X_{A_e}}{k_{A_e}} \tag{7-29}$$

With $\delta_A = \delta_{A_e}$, Eqs. 7-23 and 7-29 yield

$$\frac{X_A}{k_A} = \frac{X_{A_e}}{k_{A_e}}$$

or

$$k_{A_e} = k_A \frac{X_{A_e}}{X_A} \tag{7-30}$$

Equation 7-30 yields the value of the constant k_{A_e} of the equivalent spring in Fig. 7-18*c*. In this manner, the static elastic line of the equivalent system in Fig. 7-18*c* will be everywhere identical to that of the original variable stiffness beam in Fig. 7-18*a*.

If the end B in Fig. 7-18*a* is also supported elastically by a linear spring of constant k_B, the constant k_{B_e} of the equivalent spring can be found in a

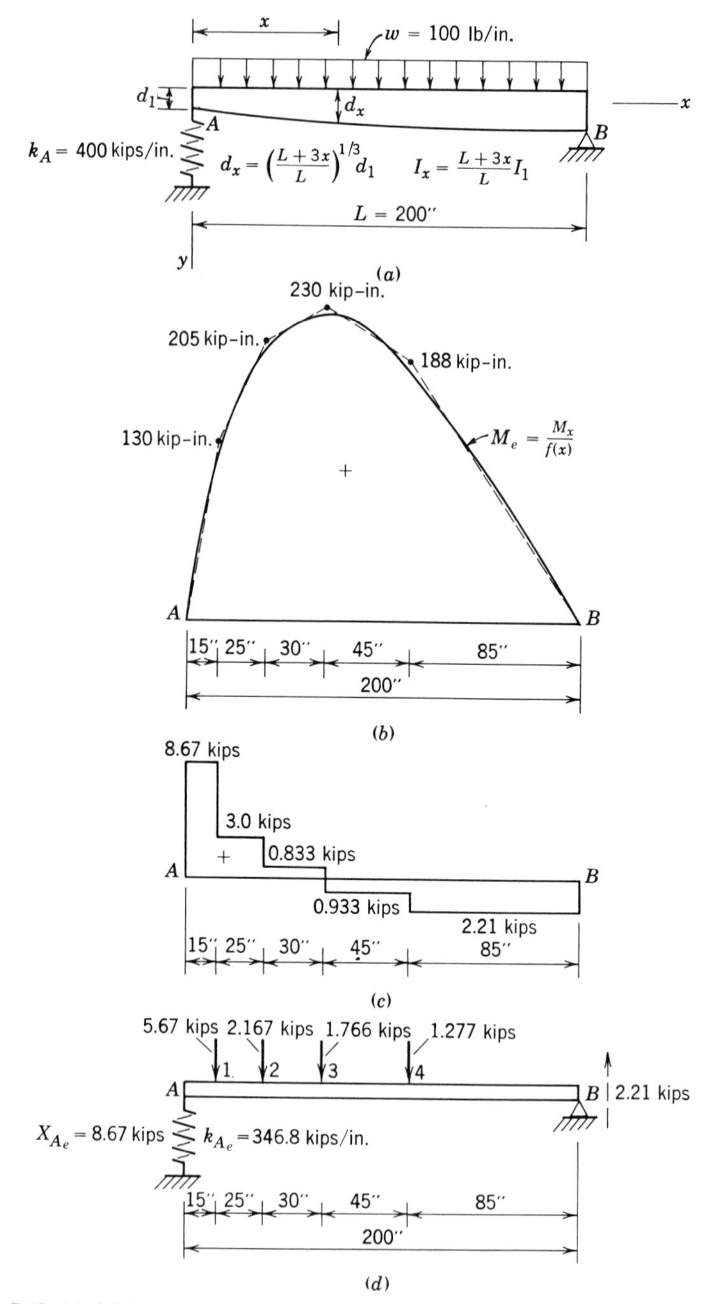

Fig. 7-19. (*a*) Original system. (*b*) Equivalent moment diagram. (*c*) Equivalent shear force diagram. (*d*) Equivalent system.

similar manner and is given by the expression

$$k_{B_e} = k_B \frac{X_{B_e}}{X_B} \tag{7-31}$$

In this equation, X_{B_e} is the elastic reaction at the end B of the equivalent system, and X_B is the one at the corresponding end of the original variable stiffness member.

It should be pointed out, however, that this solution is based on the exact method of the equivalent systems. In practical applications, the exact method becomes rather difficult for the more complicated problems, and the approximate method of the equivalent systems would be the more appropriate one to use.

As an illustration of the application of this method, consider the variable stiffness beam in Fig. 7-19*a* that is supported elastically at the end A. The modulus of elasticity E is constant and EI_1 is 90×10^6 kip-in.^{2}, where I_1 is the moment of inertia at the end A. The moment of inertia I_x varies linearly and is given by the expression

$$I_x = \frac{L + 3x}{L} I_1$$

The reactions at A and B, Fig. 7-19*a*, are 10 kips each. By proceeding in the usual manner, the values of I_x, M_x, and M_e at cross sections along the beam's length are as follows:

x (measured from A)	I_x	M_x (kip-in.)	$M_e = \dfrac{M_x}{f(x)}$ (kip-in.)
0	I_1	0	0
25 in.	$1.375I_1$	218.8	159.0
50 in.	$1.750I_1$	375.0	214.0
75 in.	$2.125I_1$	468.0	221.0
100 in.	$2.500I_1$	500.0	220.0
125 in.	$2.875I_1$	468.0	163.0
150 in.	$3.250I_1$	375.0	115.0
175 in.	$3.625I_1$	218.0	60.4
200 in.	$4.000I_1$	0	0

The equivalent moment diagram M_e approximated with straight lines, as well as the equivalent shear force diagram, are shown in Figs. 7-19*b* and 7-19*c*, respectively. The equivalent system of constant stiffness EI_1 is shown in Fig. 7-19*d*. It is loaded with concentrated loads, because the approximate method is used. In the same figure, the elastic reaction X_{A_e} is 8.67 kips. The constant k_{A_e} of the equivalent spring is determined from Eq. 7-30. That is,

$$k_{A_e} = k_A \frac{X_{A_e}}{X_A} = 400\left(\frac{8.67}{10}\right)$$

$$= 346.8 \text{ kips/in.}$$

The elastic line of the equivalent system in Fig. 7-19*d* is approximately the same as that of the variable stiffness beam in Fig. 7-19*a*. By using the equivalent system and applying the conjugate beam method, the deflection at 70 in. from end A was found to be 0.02542 in. The exact value is 0.02535 in. These two values are practically identical.

If a variable stiffness beam on elastic supports is statically indeterminate, the redundants are first determined by applying the method of the equivalent systems. Then, by proceeding as above, an equivalent system that is acted upon by concentrated loads is obtained. The methodology is similar to that applied to statically indeterminate problems in earlier sections.

PROBLEMS

7-1 Determine an exact equivalent system of uniform stiffness for each of the beams loaded as shown in Fig. P7-1. In each case the modulus of elasticity and the width b are constant. Neglect the weight of the members.

7-2 By using the exact equivalent systems derived in Problem 7-1, determine in each case the vertical displacements and rotations at points B and/or A.

7-3 If the variable stiffness beams in Problem 7-1 are made of steel of uniform width $b = 6$ in. and depth $h_1 = 12$ in., determine an equivalent system in each case by applying the approximate method of the equivalent systems. The modulus of elasticity E is 30×10^6 psi. Neglect the weight of the members.

7-4 By using the equivalent systems derived in Problem 7-3, determine in each case the vertical displacements and rotations of points B and/or A.

7-5 The variable stiffness steel beams loaded as shown in Fig. P7-5 are of rectangular cross section of uniform width $b = 8$ in. and depth $h_1 = 16$

in. Determine an equivalent system of uniform stiffness in each case by applying the approximate method of the equivalent systems. The modulus of elasticity E is 30×10^6 psi. Neglect the weight of the members.

7-6 By using the equivalent systems derived in Problem 7-5, determine in each case the vertical displacements and rotations at points A and C.

7-7 The steel variable stiffness beams shown in Fig. P7-7 have a rectangular cross section of uniform width $b = 9$ in. and depth $h_1 = 15$

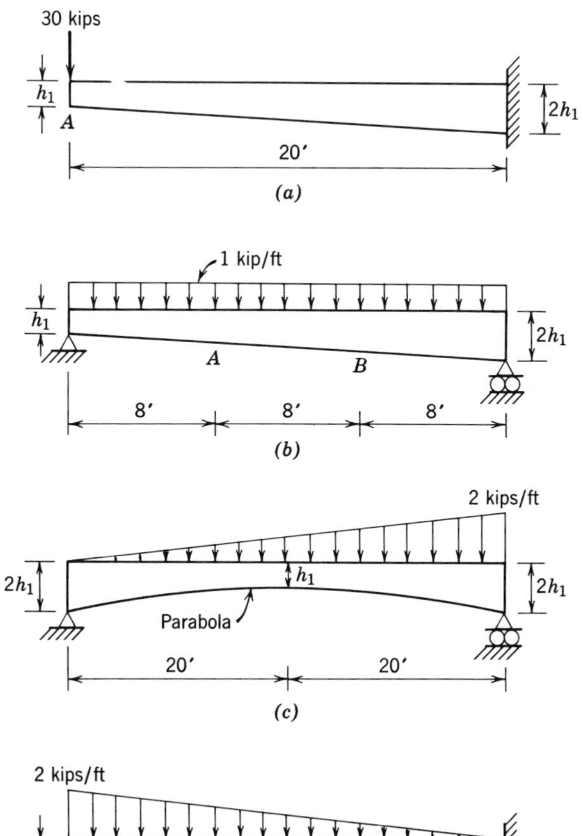

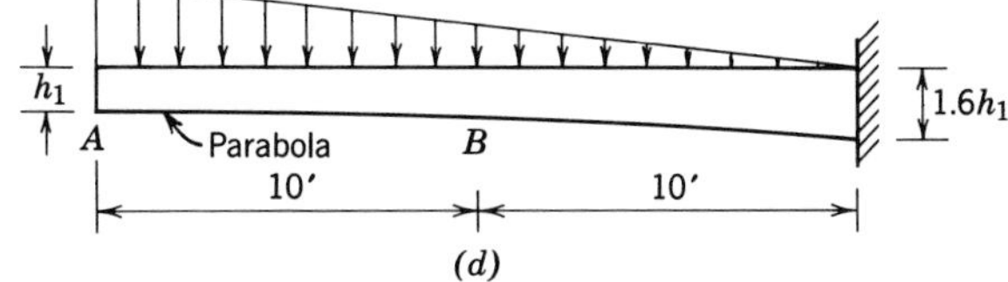

Fig. P7-1

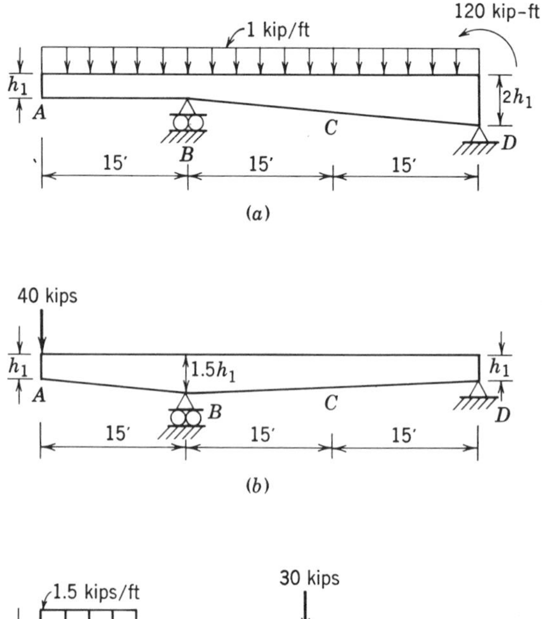

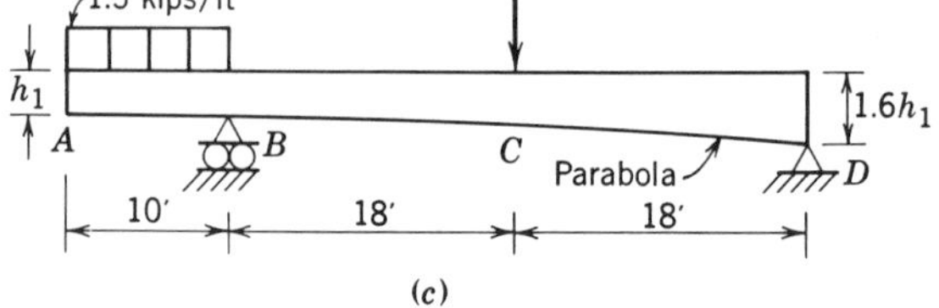

Fig. P7-5

in. By applying the approximate method of the equivalent systems and the method of Stodola and iteration procedure, determine in each case the first two natural frequencies of vibration and the corresponding normal modes. The modulus of elasticity E is 30×10^6 psi.

7-8 Repeat Problem 7-7 by using the approximate method of the equivalent systems and the method of Lord Rayleigh.

7-9 The entire length of the variable stiffness beams in Problem 7-7 is loaded with a uniformly distributed dynamic force of 5 kips applied suddenly at $t = 0$ and removed suddenly at $t = 0.8$ sec. By applying the method of modal analysis and using the method of the equivalent systems and Stodola to compute the first two natural frequencies and the corresponding mode shapes, determine in each case the maximum dynamic vertical displacement at midspan.

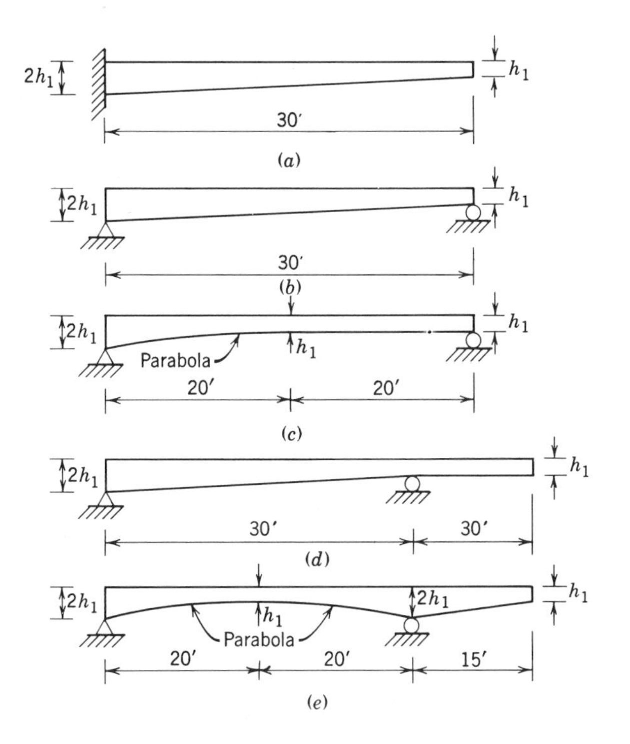

Fig. P7-7

7-10 If the beams of Problem 7-7 are loaded as indicated in Problem 7-9, determine in each case the maximum dynamic bending moment and the maximum dynamic bending stress at midspan.

7-11 Repeat Problem 7-9 when one concentrated dynamic force is applied at midspan and another concentrated dynamic force is applied at the free end (for the cases where there is a free end). Both forces have a value of 7 kips at $t = 0$ that decreases linearly to zero at $t = 0.6$ sec.

7-12 The steel variable stiffness beams shown in Fig. P7-12 have a rectangular cross section of uniform width $b = 10$ in. and depth $h_1 = 16$ in. By applying the approximate method of the equivalent systems and the method of Stodola and iteration procedure, determine in each case the first two natural frequencies of vibration and the corresponding normal modes. The modulus of elasticity E is 30×10^6 psi.

7-13 Repeat Problem 7-12 by using the approximate method of the equivalent systems and the method of Lord Rayleigh.

7-14 The entire length of the variable stiffness beams in Problem 7-12 is loaded with a uniformly distributed dynamic force of 6 kips applied suddenly at $t=0$ and removed suddenly at $t=0.6$ sec. By applying the method of modal analysis and using the methods of the equivalent systems and Stodola to compute the first two natural frequencies and normal mode shapes, determine in each case the maximum vertical displacement at the center of each span.

7-15 If the beams of Problem 7-12 are loaded as indicated in Problem 7-14, determine in each case the maximum dynamic bending moment and the maximum dynamic bending stress at the center of each span.

7-16 Repeat Problem 7-14 for the case where the uniformly distributed dynamic force of 6 kips is applied suddenly at $t=0$ and decreased linearly to zero at $t=0.6$ sec.

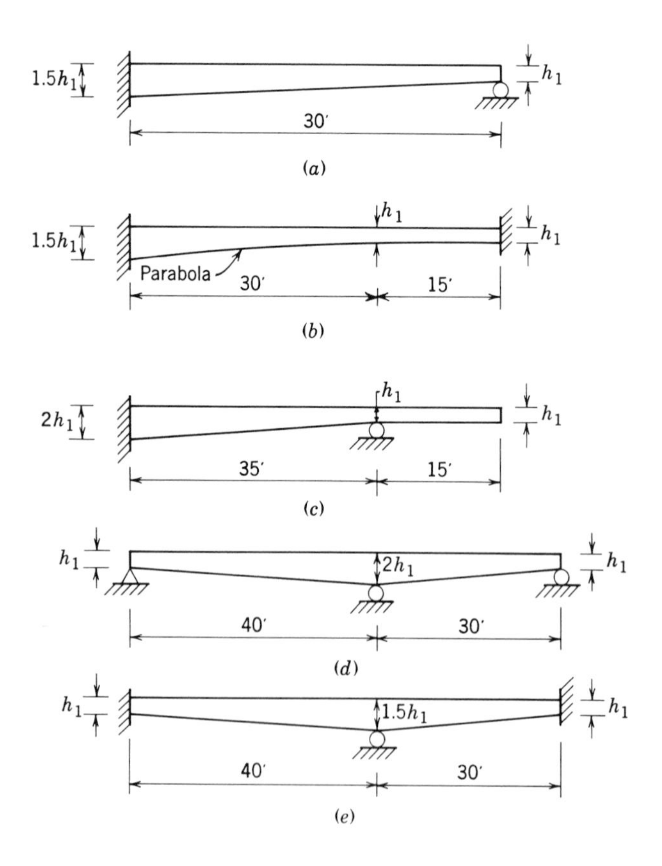

Fig. P7-12

7-17 The variable stiffness beams shown in Fig. P7-17 are made of steel and have a rectangular cross section of uniform width $b = 8$ in. and depth $h_1 = 14$ in. By applying the approximate method of the equivalent systems and the method of Stodola and iteration procedure, determine in each case the fundamental natural frequency of vibration and the corresponding mode shapes. The modulus of elasticity E is 30×10^6 psi.

7-18 Repeat Problem 7-17 by using the approximate method of the equivalent systems and the method of Lord Rayleigh.

7-19 The entire length of the variable stiffness beams in Problem 7-17 is loaded with a uniformly distributed dynamic force of 4 kips applied suddenly at $t = 0$ and of infinite duration. By applying the method of modal analysis and using the method of the equivalent systems and Stodola to compute the fundamental natural frequency and mode shape, determine in each case the maximum dynamic vertical displacement at midspan. Also determine the maximum bending stress for each case at midspan.

7-20 The girders of the steel frames shown in Fig. P7-20 are assumed to be infinitely rigid compared to the columns. The dynamic force $F_1(t)$ is a suddenly applied force of 6 kips at $t = 0$ and lasting for an infinite period of time. By using the approximate method of the equivalent

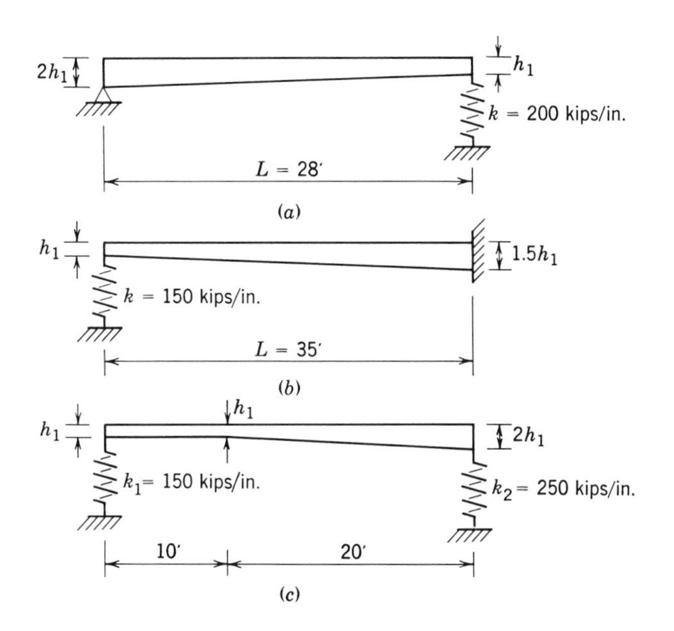

Fig. P7-17

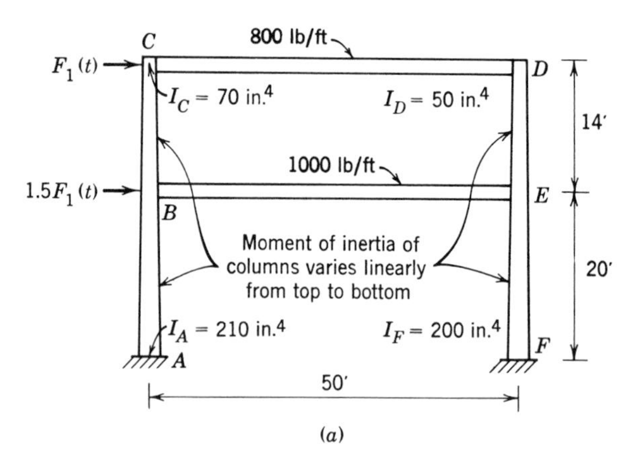

(a)

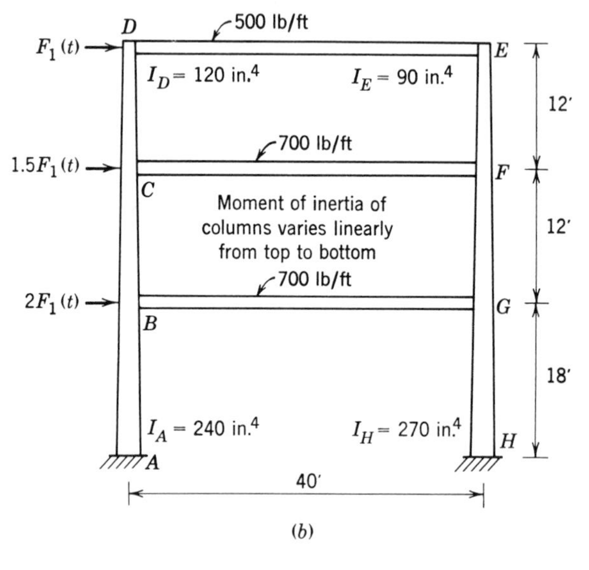

(b)

Fig. P7-20

systems and the acceleration impulse extrapolation method, determine the maximum horizontal displacements at the girder levels and the maximum bending stresses in the columns. The modulus of elasticity E is 30×10^6 psi. Neglect the weight of the columns.

7-21 Repeat Problem 7-20 by using the approximate method of the equivalent systems and the method of modal analysis.

7-22 Repeat Problem 7-20 for the case where the dynamic force $F_1(t)$ is a suddenly applied force of 6 kips at $t = 0$ and decreasing linearly to zero at $t = 0.4$ sec.

8

FOURIER AND LAPLACE TRANSFORMS

8-1 INTRODUCTION

The solution of structural dynamics problems involving periodic and nonperiodic excitations could be efficiently obtained by using methods of transformed calculus. The discussion in this chapter is concentrated primarily on Fourier series, Fourier transforms, and Laplace transforms, with applications to spring-mass systems and structures that can be idealized as spring-mass systems. This method of approach, together with the convolution theorem discussed in the last section of this chapter, prepares also some of the ground work for the analysis of structures subjected to random excitations as discussed in Chapter 12.

8-2 PERIODIC EXCITATIONS AND DISCRETE SPECTRA

If a forcing function $f(t)$ that excites an engineering system is defined in the time interval between zero and τ and satisfies the conditions that *(a)* $f(t)$ is periodic, *(b)* $f(t)$ has a finite number of discontinuities, and *(c)* the integral

$$\int_{-\tau/2}^{+\tau/2} |f(t)|\, dt \tag{8-1}$$

exists, then this function can be represented in the frequency domain by an infinite number of sinusoidal components. These components have frequencies that are harmonically related to the fundamental. The amplitude and relative phase of each of these frequency components can be expressed by a Fourier series[25] expansion of $f(t)$.

The Fourier series expansion that satisfies these conditions can have the form

$$f(t) = \frac{a_0}{\tau} + \frac{2}{\tau} \sum_{n=1}^{\infty} (a_n \cos n\omega_0 t + b_n \sin n\omega_0 t) \tag{8-2}$$

where

$$a_n = \int_{-\tau/2}^{+\tau/2} f(t) \cos n\omega_0 t\, dt \tag{8-3}$$

$$b_n = \int_{-\tau/2}^{+\tau/2} f(t) \sin n\omega_0 t\, dt \tag{8-4}$$

and

$$a_0 = \int_{-\tau/2}^{+\tau/2} f(t)\, dt \tag{8-5}$$

Here ω_0 is the fundamental frequency, τ is the period, and the relationship

[25] The Fourier series and its application to structural dynamics problems is also discussed in Section 2-10. The changes in notation introduced here are only for convenience in the discussion here.

between ω_0 and τ is given by

$$\omega_0 = \frac{2\pi}{\tau} \tag{8-6}$$

In exponential form, the Fourier series expansion of $f(t)$ is

$$f(t) = \sum_{n=-\infty}^{\infty} c_n e^{jn\omega_0 t} \tag{8-7}$$

where

$$c_n = \frac{1}{\tau} \int_{-\tau/2}^{+\tau/2} f(t)\, e^{-jn\omega_0 t}\, dt \tag{8-8}$$

In these expressions, j denotes the imaginary part of a complex number.
Equation 8-7 shows that a periodic function contains all frequencies, both negative and positive, that are harmonically related to the fundamental. The

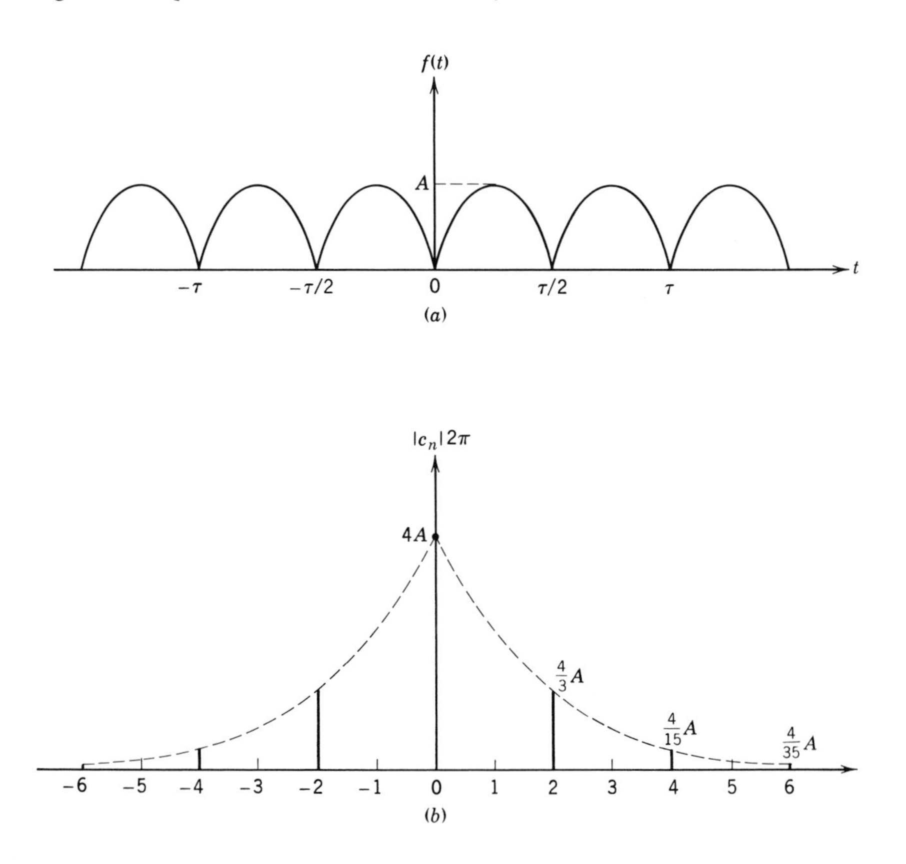

Fig. 8-1

magnitude of each frequency component is $|c_n|$, $n = 1, 2, 3, \ldots$, and they constitute the discrete spectra of the waveform.

As an illustration, consider the waveform in Fig. 8-1*a* and determine its Fourier series representation and its frequency spectrum. The solution is initiated by noting that the function $f(t)$ defining the waveform is given by the expression

$$f(t) = A\,|\sin \omega_0 t\,| \tag{8-9}$$

where A is as shown in Fig. 8-1*a* and ω_0 is given by Eq. 8-6.

By Eq. 8-8,

$$\tau c_n = \int_{-\tau/2}^{+\tau/2} A\,|\sin \omega_0 t\,| e^{-jn\omega_0 t}\,dt \tag{8-10}$$

By applying the Eulerian relation, Eq. 8-10 yields

$$\tau c_n = A \int_{-\tau/2}^{+\tau/2} |\sin \omega_0 t\,| \{ \cos n\omega_0 t - j \sin n\omega_0 t \}\,dt \tag{8-11}$$

or

$$\tau c_n = A \int_{-\tau/2}^{+\tau/2} |\sin \omega_0 t\,| \cos n\omega_0 t\,dt - jA \int_{-\tau/2}^{+\tau/2} |\sin \omega_0 t\,| \sin n\omega_0 t\,dt \tag{8-12}$$

It should be noted that $f(t)$ in Fig. 8-1*a* is symmetrical with respect to the vertical axis. Therefore, the second integral on the right-hand side of Eq. 8-12 can be proved to be equal to zero. Thus, by taking advantage of the symmetry involved, Eq. 8-12 is written as

$$\tau c_n = 2A \int_0^{\tau/2} \sin \omega_0 t \cos n\omega_0 t\,dt$$

or

$$\tau c_n = A \int_0^{\tau/2} \{ \sin (n+1)\omega_0 t - \sin (n-1)\omega_0 t \}\,dt \tag{8-13}$$

Integration of Eq. 8-13 yields

$$\tau c_n = -A \left[\frac{\cos (n+1)\pi}{(n+1)\omega_0} - \frac{\cos (n-1)\pi}{(n-1)\omega_0} - \frac{1}{(n+1)\omega_0} + \frac{1}{(n-1)\omega_0} \right] \tag{8-14}$$

For odd values of n, $n+1$ and $n-1$ in Eq. 8-14 are even. Thus, for odd values of n, c_n is zero. For even values of n, Eq. 8-14 yields

$$c_n = -\frac{A}{\tau\omega_0}\left(\frac{4}{n^2-1}\right) = -\frac{A}{2\pi}\left(\frac{4}{n^2-1}\right) \tag{8-15}$$

Therefore, the Fourier series expansion of $f(t)$ is

$$f(t) = \sum_{-\infty}^{\infty} c_n e^{jn\omega_0 t}$$

$$= \sum_{-\infty}^{\infty} -\frac{4A}{\omega_0\tau}\left(\frac{1}{n^2-1}\right)e^{jn\omega_0 t} \tag{8-16}$$

where even values of n should be used to evaluate the summation in Eq. 8-16.

The discrete frequency spectra is shown in Fig. 8-1*b*. The horizontal axis gives the values of n and the vertical axis the values $|c_n|2\pi$. The dashed line represents the envelope of the spectra which is shown to decrease as $1/n^2$.

8-3 NONPERIODIC EXCITATIONS

Mathematically, functions can be expressed by a Fourier series expansion. In physical situations, however, the forcing functions $f(t)$ are not always periodic. For example, forcing functions resulting from physical situations such as blast and earthquake are usually random, and the period of these functions cannot be observed. Even clocks run down or wear out, and also the period of the earth's rotation changes by an appreciable amount. For these types of problems a different approach should be followed in order to examine their frequency domain. This approach is known as the Fourier transform method and is derived from the Fourier series as follows.

To derive a frequency domain representation of nonperiodic functions, let τ in Eq. 8-8 approach infinity. The function $f(t)$ can then be considered to be nonperiodic. As τ approaches infinity ($\tau \rightarrow \infty$), all the Fourier coefficients c_n go to zero, provided that

$$\int_{t=-\infty}^{t=\infty} |f(t)|\,dt < \infty \tag{8-17}$$

where t is time. In addition, the fundamental frequency $\omega_0 = 2\pi/\tau$ goes to zero, hence each component of the series represented by Eq. 8-7 is practically equal to the adjacent ones. In other words, the smallness of the quantity

$$(n+1)\omega_0 - n\omega_0 = \frac{2\pi}{\tau} \tag{8-18}$$

is limited by the size of τ.

The difficulties of this procedure could be removed by introducing minor adjustments in Eqs. 8-7 and 8-8. That is, let it be defined that

$$\Delta\omega \equiv \omega_0 \equiv \frac{2\pi}{\tau} \tag{8-19}$$

and

$$F(n\omega_0) \equiv \tau c_n \tag{8-20}$$

yielding

$$\Delta\omega F(n\omega_0) = 2\pi c_n \tag{8-21}$$

where the symbol $\equiv$ denotes equal by definition.

By utilizing the definitions above, Eq. 8-8 yields

$$F(n\omega_0) = \lim_{\tau\to\infty} \int_{-\tau/2}^{+\tau/2} F(t) e^{-jn\omega_0 t}\, dt \tag{8-22}$$

or

$$F(\omega) = \int_{-\infty}^{\infty} f(t)\, e^{-j\omega t}\, dt \tag{8-23}$$

where ω is substituted for $n\omega_0$. In a similar manner, Eq. 8-7 yields

$$f(t) = \frac{1}{2\pi} \int_{-\infty}^{\infty} F(\omega)\, e^{j\omega t}\, d\omega \tag{8-24}$$

Equation 8-23 is known as the Fourier transform, while Eq. 8-24 is the inverse Fourier transform. Equation 8-23 is a synthesis of a given function of time which is composed of an infinite number of sines and cosines. In this equation, $F(\omega)$ is considered to represent the complex spectra density of $f(t)$. Equations 8-23 and 8-24 are known as the Fourier transform pair.

As an illustration, let it be assumed that it is required to determine the frequency spectra of the function shown in Fig. 8-2*a*. This spectra can be

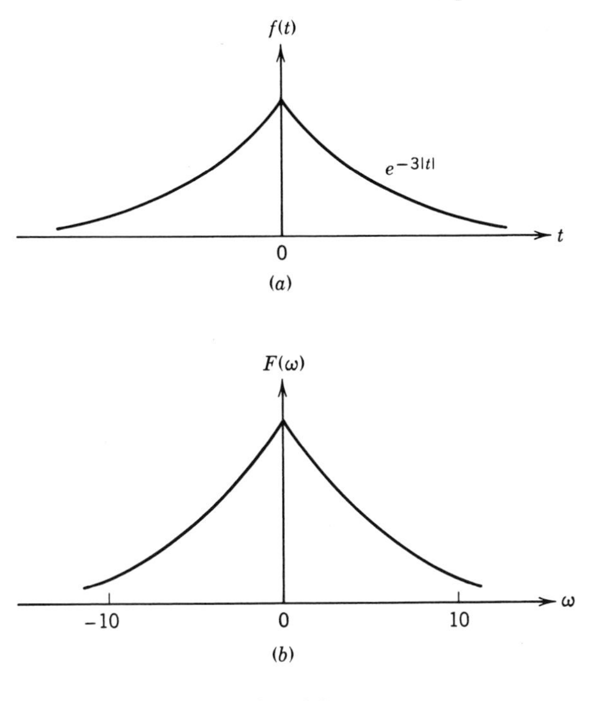

Fig. 8-2

obtained by using Eq. 8-23. Thus, from Eq. 8-23,

$$F(\omega) = \int_{-\infty}^{\infty} f(t)\, e^{-j\omega t} dt = \int_{-\infty}^{0} e^{3t} e^{-j\omega t} dt + \int_{0}^{\infty} e^{-3t} e^{-j\omega t} dt$$

$$= \frac{1}{3-j\omega} + \frac{1}{3+j\omega}$$

or

$$F(\omega) = \frac{6}{9+\omega^2} \tag{8-25}$$

Equation 8-25, if plotted, will yield the continuous frequency spectra shown in Fig. 8-2*b*.

So far, the Fourier series and Fourier transforms throughout the time domain have been considered. The forcing functions that are usually of interest to the engineer are the ones with zero values at $t < 0$ and not converging for $t > 0$. It is often possible to multiply such functions by a converging factor $e^{-\sigma t}$, where σ is a constant, and form a new function $\Phi(t)$

that converges. That is,

$$\Phi(t) = f(t)\, e^{-\sigma t} \tag{8-26}$$

Here σ is considered to be sufficiently large so that the integral

$$\int_{-\infty}^{\infty} \Phi(t)\, dt \tag{8-27}$$

is finite.

By applying to the function $\Phi(t)$ the Fourier transform given by Eq. 8-23, we have

$$F(\omega) = \int_{-\infty}^{\infty} \Phi(t)\, e^{-j\omega t}\, dt \tag{8-28}$$

Since $f(t)$ is zero for $t < 0$, Eq. 8-28 yields a Fourier transform of $\Phi(t)$ that is zero to the left of the origin. That is,

$$F(\omega) = \int_{-\infty}^{0} (0)\, e^{-j\omega t}\, dt + \int_{0}^{\infty} \Phi(t) e^{-j\omega t}\, dt$$

or

$$F(\omega) = \int_{0}^{\infty} \Phi(t)\, e^{-j\omega t}\, dt \tag{8-29}$$

By using Eq. 8-26 in Eq. 8-29, the transform becomes a function of both ω and σ instead of a function of ω alone. That is,

$$F(\omega,\sigma) = \int_{0}^{\infty} f(t) e^{-\sigma t} e^{-j\omega t}\, dt \tag{8-30}$$

Equation 8-30 can also be written as

$$F(\sigma + j\omega) = \int_{0}^{\infty} f(t)\, e^{-(\sigma + j\omega)t}\, dt \tag{8-31}$$

which is known as the direct Fourier integral transform of $f(t)$.

If in Eq. 8-31 the complex quantity $(\sigma + j\omega)$ is denoted by

$$s = \sigma + j\omega \tag{8-32}$$

then Eq. 8-31 yields

$$F(s) = \int_{0}^{\infty} f(t)\, e^{-st}\, dt \tag{8-33}$$

Equation 8-33 is the Laplace transform of $f(t)$. This shows that when σ is zero, Eq. 8-33 is identical to Eq. 8-23.

The Laplace transform of a time function transforms the function to the s plane. When $F(s)$ is known, the function $f(t)$ can be obtained from the expression

$$f(t) = -\frac{1}{2\pi j}\int_{\sigma-j\infty}^{\sigma+j\infty} F(s)\, e^{st}\, ds \tag{8-34}$$

with limits of integration $s=\sigma-j\infty$ when $\omega=-\infty$, and $s=\sigma+j\infty$ when $\omega=+\infty$. In the equation above, ds is equal to $jd\omega$.

Equations 8-33 and 8-34 constitute the Laplace transform pair. Table 8-1 shows Laplace transforms of certain functions, as well as derivatives, that will be used in later applications.

8-4 DYNAMIC RESPONSE OF SINGLE-DEGREE SPRING-MASS SYSTEMS

Consider the spring-mass system in Fig. 8-3 that is subjected to a dynamic force $F(t)$ and is under the influence of light viscous damping. The differential equation of motion is derived in Section 1-3:

$$m\ddot{y} + c\dot{y} + ky = F_0 f(t) \tag{8-35}$$

where $f(t)$ is the time function of the force and F_0 is its maximum value.

Equation 8-35 is rewritten as

$$\ddot{y} + \frac{c}{m}\dot{y} + \frac{k}{m}y = \frac{F_0}{m}f(t) \tag{8-36}$$

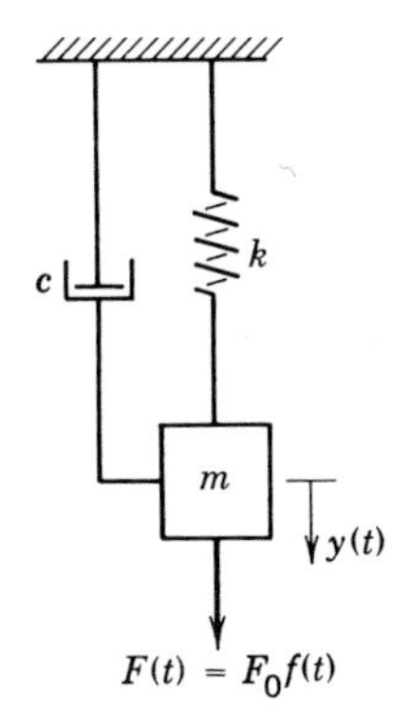

Fig. 8-3

TABLE 8-1 LAPLACE TRANSFORMS

Function $f(t)$	Laplace Transform $Lf(t) = L(s)$
1	$\frac{1}{s}$
a	$\frac{a}{s}$
$\delta(t)$	1
Ae^{-at}	$\frac{A}{s+a}$
$\frac{t^{a-1}}{(a-1)!}$	$\frac{1}{s^a}$
$\frac{e^{-a^2/4t}}{(\pi t)^{1/2}}$	$\frac{e^{-a\sqrt{s}}}{\sqrt{s}}$
$\frac{\sin \omega t}{\omega}$	$\frac{1}{s^2+\omega^2}$
$\cos \omega t$	$\frac{s}{s^2+\omega^2}$
$\frac{e^{-a^2/4t}}{2(\pi t)^{3/2}}$	$\frac{e^{-a\sqrt{s}}}{\pi a}$
$\frac{df(t)}{dt}$	$sf(s) - f(0)$
$\frac{d^n f(t)}{dt^n}$	$s^n F(s) - \sum_{k=1}^{n} s^{n-k} \frac{d^{k-1} f(0)}{dt^{k-1}}$

From Chapter 1, the natural undamped frequency ω of the system is

$$\omega = \sqrt{k/m} \tag{8-37}$$

and

$$\frac{c}{m} = 2\omega\zeta \tag{8-38}$$

where ζ is the damping ratio. Thus Eq. 8-36 yields

$$\ddot{y} + 2\omega\zeta\dot{y} + \omega^2 y = \frac{F_0}{m} f(t) \tag{8-39}$$

The Laplace transform of both sides of Eq. 8-39 yields

$$s^2 Y(s) - sY(0) - \dot{Y}(0) + 2\omega\zeta[sY(s) - Y(0)] + \omega^2 Y(s) = \frac{F_0}{m} F(s) \tag{8-40}$$

where $Y(0)$ and $\dot{Y}(0)$ are the displacement and velocity of the mass m at $t = 0$. If $Y(0) = \dot{Y}(0) = 0$, Eq. 8-40 becomes

$$Y(s)[s^2 + 2\omega\zeta s + \omega^2] = \frac{F_0}{m} F(s) \tag{8-41}$$

or, by solving for $Y(s)$,

$$Y(s) = \frac{(F_0/m)F(s)}{s^2 + 2\omega\zeta s + \omega^2} \tag{8-42}$$

Equation 8-42 relates the input $F_0F(s)$ to the output $Y(s)$ of the system through the characteristic impedance $Z(s)$, where

$$Z(s) = s^2 + 2\omega\zeta s + \omega^2 \tag{8-43}$$

On this basis, Eq. 8-42 is written as

$$Y(s) = \frac{(F_0/m)F(s)}{Z(s)} \tag{8-44}$$

Let it now be assumed that the time function $f(t)$ in Eq. 8-35 is

$$f(t) = \sin \omega_f t \tag{8-45}$$

where ω_f is the frequency of the force. The Laplace transform of Eq. 8-45 is

$$F(s) = L\{f(t)\} = \frac{\omega_f}{s^2+\omega_f^2} \tag{8-46}$$

By substituting Eq. 8-46 into Eq. 8-42 and making minor manipulations,

$$Y(s) = \frac{(F_0/k)\omega^2\omega_f}{(s^2+2\omega\zeta s+\omega^2)(s^2+\omega_f^2)} \tag{8-47}$$

or

$$Y(s) = \frac{(F_0/k)\omega^2\omega_f}{(s-s_1)(s-s_2)(s-s_3)(s-s_4)} \tag{8-48}$$

where s_1, s_2, s_3, and s_4 are the roots of the denominator in Eq. 8-47, when this denominator is set equal to zero. That is, the roots of

$$Z(s)(s^2+\omega_f^2) = 0 \tag{8-49}$$

By solving Eq. 8-49, the roots s_1, s_2, s_3, and s_4 are

$$\begin{aligned} s_1 &= -\zeta\omega + j\omega(1-\zeta^2)^{\frac{1}{2}} \qquad & s_3 &= -j\omega_f \\ s_2 &= -\zeta\omega - j\omega(1-\zeta^2)^{\frac{1}{2}} \qquad & s_4 &= j\omega_f \end{aligned} \tag{8-50}$$

The displacement $y(t)$ of the mass m can be obtained by taking the inverse Laplace transform of Eq. 8-48. But first, by using partial fractions expansion, Eq. 8-48 is written as

$$\begin{aligned} Y(s) &= \frac{(F_0/k)\omega^2\omega_f}{(s-s_1)(s-s_2)(s-s_3)(s-s_4)} \\ &= \frac{A}{s-s_1} + \frac{B}{s-s_2} + \frac{C}{s-s_3} + \frac{D}{s-s_4} \end{aligned} \tag{8-51}$$

where A, B, C, and D are constants to be determined. For example, the constant A can be determined by multiplying Eq. 8-51 by $(s-s_1)$ and evaluating A for $s=s_1$. This yields

$$A = \left. \frac{(F_0/k)\omega^2\omega_f}{[s+\zeta\omega+j\omega(1-\zeta^2)^{\frac{1}{2}}][s+j\omega_f][s-j\omega_f]} \right|_{s=s_1=-\zeta\omega+j\omega(1-\zeta^2)^{1/2}} \tag{8-52}$$

This equation, evaluated at $s=-\omega\zeta+j\omega(1-\zeta^2)^{1/2}$, yields

$$A=\frac{(F_0/k)\,\omega^2\omega_f}{2j\omega_d[\,(\omega\zeta-j\omega_d)^2+\omega_f^2\,]} \tag{8-53}$$

where

$$\omega_d=\omega(1-\zeta^2)^{\frac{1}{2}} \tag{8-54}$$

is the damped natural frequency of the system. In a similar manner,

$$B=\frac{(F_0/k)\,\omega^2\omega_f}{-2j\omega_d[\,(\omega\zeta+j\omega_d)^2+\omega_f^2\,]} \tag{8-55}$$

The constants C and D are complex conjugate numbers and their sum is also complex. Therefore, the sum of the last two terms on the right-hand side of Eq. 8-51 can be replaced by twice the real part of either one of the two terms. That is,

$$\frac{C}{s+j\omega_f}+\frac{D}{s-j\omega_f}=2\mathrm{Re}\frac{C}{s+j\omega_f} \tag{8-56}$$

where Re denotes real part. Thus, by using Eqs. 8-56 and 8-51,

$$\frac{(F_0/k)\,\omega^2\omega_f}{(s-s_1)\,(s-s_2)\,(s-s_3)\,(s-s_4)}=\frac{A}{s-s_1}+\frac{B}{s-s_2}+2\mathrm{Re}\frac{C}{s-s_3} \tag{8-57}$$

The constant C can be evaluated in the same way as A and B. That is, multiply both sides of the equation by $(s-s_3)$ and evaluate at $s=s_3$. This yields

$$C=2\mathrm{Re}\left.\frac{(F_0/k)\,\omega^2\omega_f}{[\,s+\zeta\omega-j\omega(1-\zeta^2)^{\frac{1}{2}}\,][\,s+\zeta\omega+j\omega(1-\zeta^2)^{\frac{1}{2}}\,][\,s-j\omega_f\,]}\right|_{s=-j\omega_f}$$

or

$$C=\mathrm{Re}\frac{(F_0/k)\,\omega^2}{-j[\,(\zeta\omega-j\omega_f)^2+\omega_d^2\,]} \tag{8-58}$$

By using Eqs. 8-53, 8-55, and 8-58 in Eq. 8-57 and then taking its inverse

Laplace transform, we have

$$y(t) = Ae^{-[\zeta\omega - j\omega(1-\zeta^2)^{1/2}]t} + Be^{-[\zeta\omega + j\omega(1-\zeta^2)^{1/2}]t} + Ce^{-j\omega_f t} \qquad (8\text{-}59)$$

where A, B, and C are as shown in Eqs. 8-53, 8-55, and 8-58, respectively. It may be shown that Eq. 8-59 is identical to Eq. 2-20 in Section 2-3.

The first two terms of Eq. 8-59 give the transient part of the response, while the last term gives the steady-state one due to the applied force $F(t)$.

8-5 DYNAMIC RESPONSE DUE TO A UNIT IMPULSE

Let it now be assumed that the force $F(t)$ acting on the spring-mass system in Fig. 8-3 is an impulse. That is,

$$F(t) = F_0\,\delta(t) \qquad (8\text{-}60)$$

where F_0 is the magnitude of the impulse at the time of occurrence and $\delta(t)$ is the Dirac delta function defined as

$$\delta(t) = 0 \qquad \text{for} \qquad t \neq 0$$

and

$$\int \delta(t)\,dt = 1 \qquad (8\text{-}61)$$

The dynamic response of the spring-mass system can be obtained by following the procedure used in the preceding section. That is, the function $F(s)$ in Eq. 8-42 is the Laplace transform of $\delta(t)$ which in Table 8-1 is shown to be equal to 1. Thus Eq. 8-42 yields

$$Y(s) = \frac{F_0/m}{s^2 + 2\omega\zeta s + \omega^2} = \frac{F_0/m}{Z(s)} \qquad (8\text{-}62)$$

where $1/Z(s)$ is defined as the transfer function $H(s)$ of the system. That is,

$$H(s) = \frac{1}{Z(s)} \qquad (8\text{-}63)$$

By using the method of partial fractions, Eq. 8-62 is written as

$$Y(s) = \frac{A}{(s-s_1)} + \frac{B}{(s-s_2)} \qquad (8\text{-}64)$$

where the constants A and B may be evaluated as discussed in Section 8-4:

$$A = \frac{(F_0 / k)\omega^2}{j\,2\omega(1-\zeta^2)^{\frac{1}{2}}} \tag{8-65}$$

$$B = \frac{(F_0 / k)\omega^2}{-j\,2\omega(1-\zeta^2)^{\frac{1}{2}}} \tag{8-66}$$

The dynamic response $y(t)$ of the spring-mass system can be obtained by taking the inverse Laplace transform of Eq. 8-64. That is,

$$y(t) = Ae^{-[\omega\zeta - j\omega(1-\zeta^2)^{1/2}]t} + Be^{-[\omega\zeta + j\omega(1-\zeta^2)^{1/2}]t} \tag{8-67}$$

where A and B are given by Eqs. 8-65 and 8-66, respectively.

By utilizing Eq. 8-63 and $s = j\omega$, Eq. 8-62 yields

$$Y(j\omega) = \frac{F_0}{m} H(j\omega) \tag{8-68}$$

The inverse Fourier transform of Eq. 8-68, by Eq. 8-24, is

$$y(t) = \frac{F_0}{2\pi m} \int_{-\infty}^{\infty} H(j\omega)\, e^{j\omega t} d\omega \tag{8-69}$$

Thus the output displacement $y(t)$ of the mass m due to an impulse is the inverse Fourier transform of $H(j\omega)$. If the impulse is a unit impulse, which implies that $F_0 = 1$, Eq. 8-69 is written as

$$y(t) = h(t) = \frac{1}{2\pi m} \int_{-\infty}^{\infty} H(j\omega)\, e^{j\omega t} d\omega \tag{8-70}$$

where $h(t)$ is used to denote the inverse Fourier transform of $H(j\omega)$.

If $h(t)$ is known, the transfer function $H(j\omega)$ is given by

$$H(j\omega) = m \int_{-\infty}^{\infty} h(t) e^{-j\omega t} dt \tag{8-71}$$

Again, Eqs. 8-70 and 8-71 constitute a Fourier transform pair. The use of this Fourier pair is shown later in this book.

8-6 DYNAMIC RESPONSE OF SYSTEMS WITH TWO OR MORE DEGREES OF FREEDOM

The method described in the preceding sections can be also used to determine the dynamic response of systems with two or more degrees of freedom. Consider, for example, the two-degree spring-mass system in Fig. 8-4. The dynamic forces F_1 and F_2 are suddenly applied to the masses m_1 and m_2, respectively, at $t=0$ and they are of infinite duration. In addition, the system is under the influence of viscous damping. The differential equations of motion for the masses can be derived in the usual way:

$$m_1\ddot{y}_1+k_1y_1-k_2(y_2-y_1)+c\dot{y}_1-c(\dot{y}_2-\dot{y}_1)=F_1 \qquad (8\text{-}72)$$

$$m_2\ddot{y}_2+k_2(y_2-y_1)+c(\dot{y}_2-\dot{y}_1)=F_2 \qquad (8\text{-}73)$$

The Laplace transforms of the two equations, by assuming the initial displacements and initial velocities to be zero, are

$$m_1s^2Y_1(s)+k_1Y_1(s)-k_2Y_2(s)+k_2Y_1(s)+csY_1(s)$$
$$-csY_2(s)+csY_1(s)=\frac{F_1}{s}$$

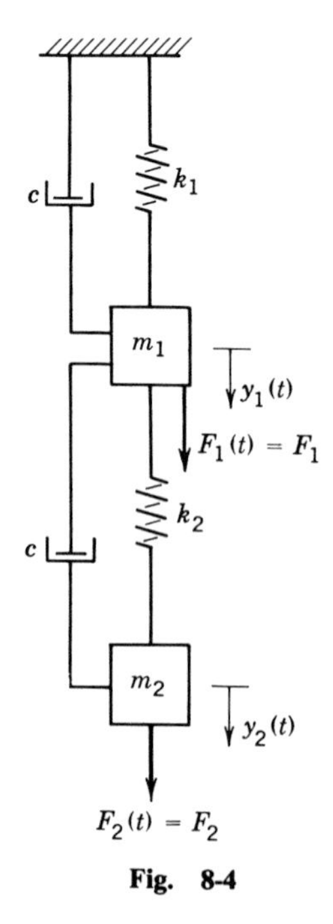

Fig. 8-4

$$m_2 s^2 Y_2(s) + k_2 Y_2(s) - k_2 Y_1(s) + csY_2(s) - csY_1(s) = \frac{F_2}{s}$$

The two equations, by rearranging terms, are written as

$$Y_1(s)[m_1 s^2 + k_1 + k_2 + 2cs] - Y_2(s)[cs + k_2] = \frac{F_1}{s} \tag{8-74}$$

$$-Y_1(s)[cs + k_2] + Y_2(s)[m_2 s^2 + k_2 + cs] = \frac{F_2}{s} \tag{8-75}$$

Applying Cramer's rule, the simultaneous solution of the two equations yields

$$Y_1(s) = \frac{\begin{vmatrix} \dfrac{F_1}{s} & -(cs + k_2) \\ \dfrac{F_2}{s} & (m_2 s^2 + k_2 + cs) \end{vmatrix}}{\begin{vmatrix} (m_1 s^2 + k_1 + k_2 + 2cs) & -(cs + k_2) \\ -(cs + k_2) & (m_2 s^2 + k_2 + cs) \end{vmatrix}} \tag{8-76}$$

$$Y_2(s) = \frac{\begin{vmatrix} (m_1 s^2 + k_1 + k_2 + 2cs) & \dfrac{F_1}{s} \\ -(cs + k_2) & \dfrac{F_2}{s} \end{vmatrix}}{\begin{vmatrix} (m_1 s^2 + k_1 + k_2 + 2cs) & -(cs + k_2) \\ -(cs + k_2) & (m_2 s^2 + k_2 + cs) \end{vmatrix}} \tag{8-77}$$

The determinant in the denominator on the right-hand side of Eq. 8-76 or Eq. 8-77 represents the impedance $Z(s)$ of the system. That is,

$$Z(s) = \begin{vmatrix} (m_1 s^2 + k_1 + k_2 + 2cs) & -(cs + k_2) \\ -(cs + k_2) & (m_2 s^2 + k_2 + cs) \end{vmatrix} \tag{8-78}$$

Thus Eqs. 8-76 and 8-77 can be written as

$$Y_1(s) = \frac{\Delta_1(s)}{Z(s)} \tag{8-79}$$

$$Y_2(s) = \frac{\Delta_2(s)}{Z(s)} \tag{8-80}$$

where

$$\Delta_1(s) = \begin{vmatrix} \frac{F_1}{s} & -(cs+k_2) \\ \frac{F_2}{s} & (m_2s^2+k_2+cs) \end{vmatrix} \tag{8-81}$$

$$\Delta_2(s) = \begin{vmatrix} (m_1s^2+k_1+k_2+2cs) & \frac{F_1}{s} \\ -(cs+k_2) & \frac{F_2}{s} \end{vmatrix} \tag{8-82}$$

Again, $1/Z(s)$ in the equations above represents the transfer function$H(s)$ of the system.

By expanding the determinants, Eqs. 8-76 and 8-77 yield

$$Y_1(s) = \frac{F_1m_2s^2+F_1k_2+F_1cs+F_2cs+F_2k_2}{s(m_1s^2+k_1+k_2+2cs)(m_2s^2+k_2+cs)-s(cs+k_2)^2} \tag{8-83}$$

$$y_2(s) = \frac{F_2m_1s^2+F_2k_1+F_2k_2+2F_2cs+F_1cs+F_1k_2}{s(m_1s^2+k_1+k_2+2cs)(m_2s^2+k_2+cs)-s(cs+k_2)^2} \tag{8-84}$$

The time displacements $y_1(t)$ and $y_2(t)$ of the masses m_1 and m_2, respectively, can be evaluated by using the method of partial fractions as in Section 8-4. For example, Eq. 8-83 can be written as

$$Y_1(s) = \frac{A_0}{s} + \frac{A_1}{(s-s_1)} + \frac{A_2}{(s-s_2)} + \frac{A_3}{(s-s_3)} + \frac{A_4}{(s-s_4)} \tag{8-85}$$

where A_0, A_1, A_2, A_3, and A_4 are constants that can be evaluated as shown in Section 8-4, and s_1, s_2, s_3, and s_4 are the roots of the denominator in Eq. 8-83 when it is set equal to zero. The inverse Laplace transform of Eq. 8-85 yields the expression for the displacement $y_1(t)$. That is,

$$y_1(t) = A_0 + A_1e^{s_1t} + A_2e^{s_2t} + A_3e^{s_3t} + A_4e^{s_4t} \tag{8-86}$$

In a similar manner, the expression for the displacement $y_2(t)$ of the mass m_2 can be found. That is,

$$y_2(t) = B_0 + B_1 e^{s_1 t} + B_2 e^{s_2 t} + B_3 e^{s_3 t} + B_4 e^{s_4 t} \tag{8-87}$$

where B_0, B_1, B_2, B_3, and B_4 are constants to be evaluated as in Section 8-4, and s_1, s_2, s_3, and s_4 are the roots of the denominator in Eq. 8-84 when this denominator is set equal to zero.

8-7 CONVOLUTION

An important mathematical operation that relates input to output in the time or frequency domain is the convolution technique. Consider for example two functions $x(t)$ and $h(t)$ that are well behaved within their respective time domain. The operation of convolution for $x(t)$ and $h(t)$ is defined by the result $y(t)$, where

$$y(t) = \int_{-\infty}^{\infty} x(t-\rho)h(\rho)\,d\rho \tag{8-88}$$

Here ρ represents a dummy variable of integration. In shorthand notation, the operation of convolution is written in the equivalent form

$$y(t) = x(t) * h(t) \tag{8-89}$$

The Fourier transform pairs of $x(t)$ and $h(t)$ can be determined by using Eqs. 8-23 and 8-24. The Fourier transform of Eq. 8-88 is denoted by $Y(\omega)$:

$$Y(\omega) = \int_{-\infty}^{\infty} h(\rho)\,d\rho \int_{-\infty}^{\infty} e^{-j\omega t} x(t-\rho)\,dt \tag{8-90}$$

If the variables in Eq. 8-90 are changed so that

$$u = t - \rho \qquad t = \infty \qquad u = \infty$$

$$du = dt \qquad t = -\infty \qquad u = -\infty$$

then $Y(\omega)$ becomes

$$Y(\omega) = \int_{-\infty}^{\infty} h(\rho)e^{-j\omega\rho}\,d\rho \int_{-\infty}^{\infty} x(u)e^{-j\omega u}\,du \tag{8-91}$$

Identification of terms indicates that Eq. 8-91 can be written as

$$Y(\omega) = H(\omega)X(\omega) \tag{8-92}$$

where

$$H(\omega)=\int_{-\infty}^{\infty} h(\rho)\, e^{-j\omega\rho}\, d\rho$$

$$X(\omega)=\int_{-\infty}^{\infty} x(u)\, e^{-j\omega u}\, du$$

The result in Eq. 8-92 indicates that convolution in the time domain is equivalent to multiplication in the frequency domain. In a similar manner, it may be shown that convolution in the frequency domain is equivalent to multiplication in the time domain.

If the frequency response $H(\omega)$ as well as the input[26] function $x(t)$ of a system are known, the output $y(t)$ can be determined by taking the inverse transform of Eq. 8-92. That is

$$y(t)=\frac{1}{2\pi}\int_{-\infty}^{\infty} H(\omega)X(\omega)\, e^{j\omega t}\, d\omega \tag{8-93}$$

With respect to the dummy variable ρ, Eq. 8-93 yields

$$y(t)=\int_{-\infty}^{\infty} h(\rho)x(t-\rho)\, d\rho \tag{8-94}$$

or

$$y(t)=\int_{-\infty}^{\infty} x(\rho)h(t-\rho)\, d\rho \tag{8-95}$$

The integrals in Eqs. 8-94 and 8-95 are the convolution integrals and provide a direct method of computing the output $y(t)$ due to a given input. For example, the output $y(t)$ can be the displacement $y(t)$ of the spring-mass system in Fig. 8-3 due to a unit impulse. In this system, the function $x(t)$ above is represented by $x(t)=F(t)=F_0\,\delta(t)$.

The convolution theorem has found many applications in engineering. It is used in Chapter 12 to develop the theory of random analysis.

PROBLEMS

8-1 Determine the Fourier series expansion of the waveform shown in Fig. P8-1 and plot the discrete frequency spectra.

8-2 Repeat Problem 8-1 for the case where the function $f(t)$ is equal to e^{-t}.

[26] The frequency domain of $x(t)$ is represented by $X(\omega)$ in Eq. 8-92.

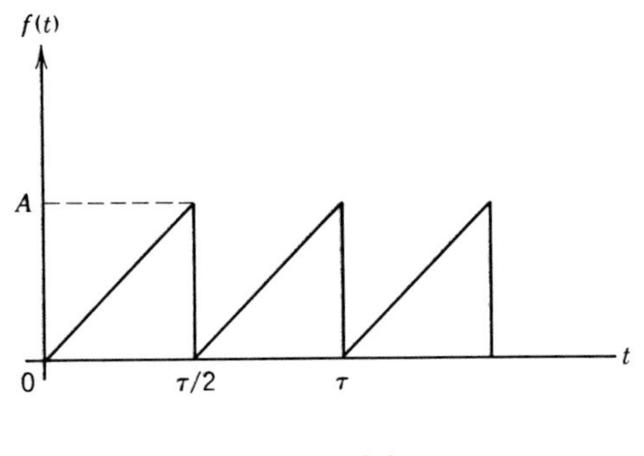

Fig. P8-1

8-3 Determine the frequency spectra for the waveform shown in Fig. P8-3.

8-4 Determine the frequency spectra for the function

$$f(t)=\begin{cases} 0 & t<0 \\ e^{-t/\tau}\sin\omega_0 t & t>0 \end{cases}$$

8-5 For the coupled pendulums shown in Fig. P8-5, determine the displacements x_1 and x_2. The initial displacements at $t=0$ are $x_1(0)=x_2(0)=0$, and the initial velocities at $t=0$ are $\dot{x}_1(0)=b$ and $\dot{x}_2(0)=0$.

8-6 The response of the spring-mass system in Fig. 8-3, where $F(t)=F\sin\omega_f t$, is given by Eq. 8-59. With initial conditions $y(0)=\dot{y}(0)=0$ at $t=0$, show that Eqs. 8-59 and 2-20 are identical.

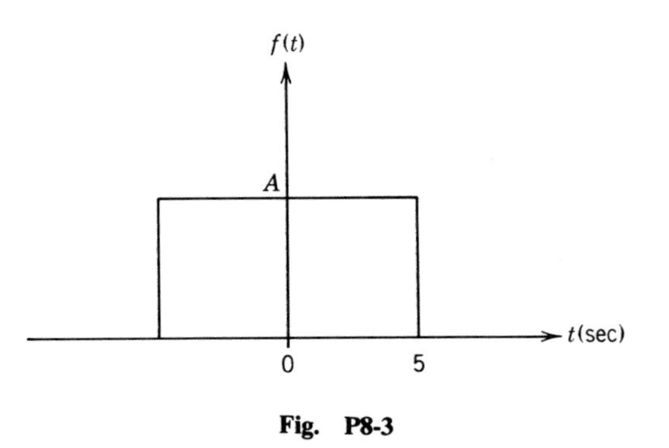

Fig. P8-3

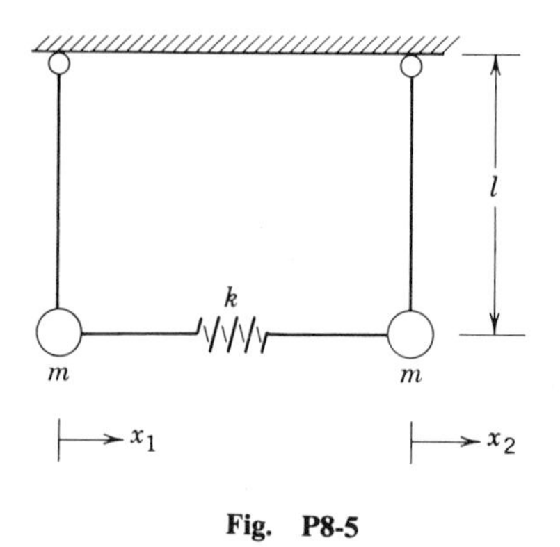

Fig. P8-5

8-7 If the initial conditions for the spring-mass system in Fig. 8-3 are $y(0) = y_0$ and $\dot{y}(0) = \dot{y}_0$ at $t = 0$, determine the transfer function, the impedance and the response $y(t)$ of the system. Separate $y(t)$ into transient and steady-state response terms. The forcing function $F(t)$ is equal to $F \sin \omega_f t$.

8-8 Repeat Problem 8-7 with initial conditions $y(0) = \dot{y}(0) = 0$ at $t = 0$ and $F(t) = F \cos \omega_f t$.

8-9 Repeat Problem 8-7 with initial conditions $y(0) = y_0$, $\dot{y}(0) = \dot{y}_0$ at $t = 0$ and $F(t) = F \cos \omega_f t$.

8-10 The response of the spring-mass system in Fig. 8-4 is given by Eqs. 8-86 and 8-87. Determine the constants $A_0, A_1, A_2, A_3, A_4, B_0, B_1, B_2, B_3$, and B_4.

8-11 Determine the response of the spring-mass system in Fig. 8-4 when $F_1(t) = F_1 \sin \omega_f t$ and $F_2(t) = F_2 \sin \omega_f t$. The initial displacements and initial velocities at $t = 0$ are zero.

8-12 Repeat Problem 8-11 for $F_1(t) = F_1 \cos \omega_f t$ and$F_2(t) = F_2 \cos \omega_f t$.

9

VARIATIONAL APPROACH

9-1 INTRODUCTION

Variational problems are gaining in popularity today and many applications can be found in engineering fields such as controls, dynamics, and structural analysis. In fact, an application of the calculus of variation could be also attributed to the ancient Greeks in solving the problem of determining a closed curve having a length l and encircling an area A that is maximal. This curve can be proved to be the circumference of a circle.

The development of the calculus of variation was actually begun in 1696, and it became independent after the discoveries of L. Euler (1707–1783), whom many claim to be the founder of this field.

The purpose of the calculus of variation is to find methods by which maxima or minima of functionals can be determined. Functionals are variable values that depend on a variable, or variables, running through a set of functions. The methods for determining extrema, that is solving a given variational problem, are similar to those used to determine maxima and minima of ordinary functions. This is easily observed by comparing the principles associated with functionals and those of ordinary functions.

The development of the field of the calculus of variation was initiated and influenced by three basic problems: *(a)* the brachistochrone problem suggested by Johann Bernoulli in 1696, requiring the determination of the path of quickest descent; *(b)* the problem of geodesies dealing with the determination of the line of minimum length lying on a given surface $\Phi(x, y, z) = 0$ and joining two given points of this surface; and *(c)* the isoperimetric problem of determining a closed curve of length l encircling an area A that is maximal. The ancient Greeks indicated that such a curve should be the circumference of a circle.

The second problem was solved by Johann Bernoulli in 1697, and a general method of solution was given by J. Lagrange and L. Euler. The third problem, somewhat modified, was further treated in detail by L. Euler.

The sections that follow introduce the calculus of variation and its applications to structural dynamics problems.

9-2 VARIATIONAL PROPERTIES

Some of the important properties of functionals and their variation are discussed in this section. A functional is represented by the symbol Ω and is written as

$$\Omega = \Omega(y(x)) \tag{9-1}$$

where $y(x)$ is the independent function. That is, for a given function $y(x)$, from a certain class of functions, there corresponds a certain value of Ω. The quantity $\delta y = y(x) - y_1(x)$, representing the difference of two functions, is defined as the variation of a functional. If small variations of $y(x)$ lead always to small variations of the functional $\Omega(y(x))$, then the functional $\Omega(y(x))$ is said to be continuous. This, of course, leads to the question of what variations of $y(x)$ may be considered small. This question can be answered by defining the order of closeness of two functions $y = y(x)$ and $y = y_1(x)$.

The functions $y = y(x)$ and $y = y_1(x)$ are close in the sense of closeness of order zero, if the absolute values of their difference $y(x) - y_l(x)$ are small for all values of x defining these two functions. If the absolute values of their differences $y(x) - y_1(x)$, as well as $y'(x) - y_1'(x)$, are small, then the two functions are said to be in the sense of closeness of order 1. If the absolute values of the differences of their nth derivative are small, then the two functions are close in the sense of closeness of order n. If they are close in the sense of closeness of order n, they are also close in the sense of any order less than n.

For functionals of the form $\Omega(y(x))$, the variation may be defined as the derivative of the functional $\Omega(y(x) + \alpha\delta y)$ with respect to α at $\alpha = 0$, where α is a variable. Thus the variation of a functional $\Omega(y(x))$ may be expressed as

$$\frac{\partial}{\partial \alpha}\Omega(y(x) + \alpha\delta y)|_{\alpha=0} \tag{9-2}$$

At $y = y_0(x)$ the functional $\Omega(y(x))$ is maximum if

$$\Omega(y(x)) - \Omega(y_0(x)) = \Delta\Omega < 0, \tag{9-3}$$

that is, when the values of the functional $\Omega(y(x))$ for arbitrary curves neighboring $y = y_0(x)$ are not greater than $\Omega(y_0(x))$. If $\Delta\Omega = 0$ only when $y(x) = y_0(x)$, then the functional $\Omega(y(x))$ takes an absolute maximum along $y = y_0(x)$. When the functional has a minimum value along $y = y_0(x)$, then $\Delta\Omega > 0$ for all curves sufficiently close to $y = y_0(x)$.

Hence, without proof, the following theorem applies:

If the variation of a functional $\Omega(y(x))$ exists and Ω takes on a maximum or a minimum along $y = y_0(x)$, then $\Delta\Omega = 0$ along $y = y_0(x)$.

Thus the variation vanishes along the curves that make the functional a maximum or a minimum. The order of closeness should be taken into consideration when speaking about extrema of functionals. A maximum or a minimum is called a *weak* maximum or minimum if the class of neighboring curves are in the sense of closeness of order 1. In investigating necessary conditions for an extremum, it is not required to make a distinction between *strong* and *weak* maximum or minimum. This distinction becomes a requirement when dealing with sufficiency conditions for an extremum.

The discussion above applies equally well to functionals involving several independent functions such as

$$\Omega(y_1(x), y_2(x), \ldots, y_n(x))$$

or to functionals consisting of one or more functions having more than one independent variable, that is, to a functional of the form

$$\Omega(y(x_1, x_2, \ldots, x_n))$$

or

$$\Omega(y_1(x_1, x_2, \ldots, x_n), y_2(x_1, x_2, \ldots, x_n), \ldots, y_n(x_1, x_2, \ldots, x_n))$$

9-3 NECESSARY CONDITIONS FOR AN EXTREMUM

The necessary conditions for a functional to have an extremum are represented by the Euler equation which was derived by Euler in 1744. The discussion is initiated by considering the family of problems where the boundaries A and B, Fig. 9-1, of all admissible curves are fixed. That is, $y(x_0) = y_0$ and $y(x_1) = y_1$.

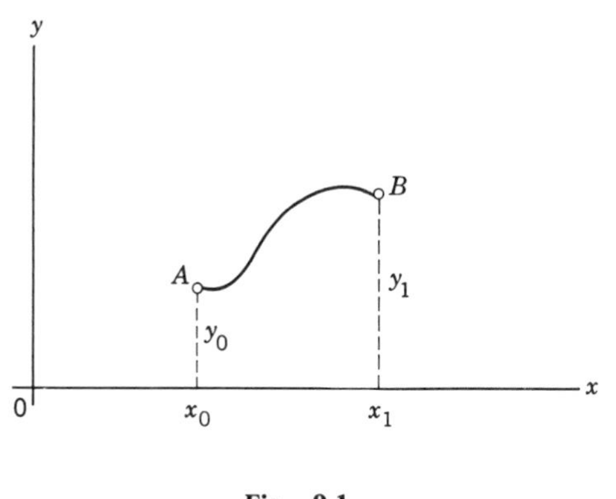

Fig. 9-1

The simplest form of a functional in this family of problems can be written as

$$\Omega(y(x)) = \int_{x_0}^{x_1} F(x, y, y')\, dx \tag{9-4}$$

where y and y' are functions of x, and y' denoting the derivative with respect to x. The necessary condition for this functional to have an extremum is

$$F_y = \frac{d}{dx} F_{y'} = 0 \tag{9-5}$$

or, explicitly,

$$F_y - F_{xy'} - F_{yy'}y' - F_{y'y'}y'' = 0 \tag{9-6}$$

where F_y and $F_{y'}$ in Eq. 9-5 denote, respectively, partial derivatives of $F(x, y, y')$ with respect to y and y'.

Equation 9-5 or 9-6 is known as the Euler equation. Only the integral curves $y = y(x, C_1, C_2)$, called extremals, where C_1 and C_2 are constants of integration depending on the boundary conditions of the problem, can make the functional given by Eq. 9-4 have an extremum. The derivation of the Euler equation is omitted here, but it can be found in the literature[27]. The proof can be carried out by utilizing the fundamental theorem discussed in the preceding section. That is, equating to zero the variation of the functional represented by Eq. 9-4.

It should be pointed out, however, that the Euler equation provides only the necessary condition for a functional to have an extremum. That is, if an extremum exists, it must satisfy the Euler equation, because only extremals can cause a functional to have an extremum. On the other hand, the Euler equation provides no guarantee that such an extremum is really an extremum. It cannot even indicate if this extremum is a maximum or a minimum. These answers can be obtained only from the sufficiency conditions. For many engineering problems it is not necessary to refer to sufficiency conditions, because the answer may be obtained from physical interpretations. Sufficiency conditions are very useful in research work.

The Euler equations for the functional

$$\Omega(y_1(x), y_2(x)) = \int_{x_0}^{x_1} F(x, y_1, y_2, y_1', y_2')\, dx \tag{9-7}$$

depending on the two functions $y_1(x)$ and $y_2(x)$ and the boundary conditions

$$y_1(x_0) = y_{10}, \qquad y_2(x_0) = y_{20}, \qquad y_1(x_1) = y_{11}, \qquad y_2(x_1) = y_{21}$$

are

$$F_{y_1} - \frac{d}{dx} F_{y_1'} = 0 \qquad F_{y_2} - \frac{d}{dx} F_{y_2'} = 0 \tag{9-8}$$

When n functions are involved, that is, when

$$\Omega(y_1, y_2, y_3, \ldots, y_n) = \int_{x_0}^{x_1} F(x, y_1, y_2, y_3, \ldots, y_n, y_1', y_2', y_3', \ldots, y_n')\, dx \tag{9-9}$$

[27] See, for example, Reference 75.

TABLE 9-1

Functionals — Fixed Boundaries

1 $\Omega(y(x)) = \int_{x_0}^{x_1} F(x, y, y')\,dx \qquad y(x_0) = y_0 \quad y(x_1) = y_1$

2 $\Omega(y_1(x), y_2(x)) = \int_{x_0}^{x_1} F(x, y_1, y_2, y_1', y_2')\,dx$

$y_1(x_0) = y_{10}, \quad y_2(x_0) = y_{20}, \quad y_1(x_1) = y_{11}, \quad y_2(x_1) = y_{21}$

3 $\Omega(y_1, y_2, \ldots, y_n) = \int_{x_0}^{x_1} F(x, y_1, y_2, \ldots, y_n, y_1', y_2', \ldots, y_n')\,dx$

$y_1(x_0) = y_{10}, \quad y_2(x_0) = y_{20}, \ldots, y_n(x_0) = y_{n0}$

$y_1(x_1) = y_{11}, \quad y_2(x_1) = y_{21}, \ldots, y_n(x_1) = y_{n1}$

4 $\Omega(y(x)) = \int_{x_0}^{x_1} F(x, y, y', y'', \ldots, y^{(n)})\,dx$

$y(x_0) = y_0, \quad y'(x_0) = y_0', \ldots, y^{(n-1)}(x_0) = y_0^{(n-1)}$

$y(x_1) = y_1, \quad y'(x_1) = y_1', \ldots, y^{(n-1)}(x_1) = y_1^{(n-1)}$

5 $\Omega(y(x), z(x)) = \int_{x_0}^{x_1} F(x, y, y', \ldots, y^{(n)}, z, z', \ldots, z^{(m)})\,dz$

6 $\Omega(y_1, y_2, \ldots, y_m) = \int_{x_0}^{x_1} F(x, y_1\, y_1', \ldots, y_1^{(n_1)}, y_2, y_2', \ldots, y_2^{(n_2)}, \ldots, y_m, y_m', \ldots, y_m^{(n_m)})$

7 $\Omega(z(x, y)) = \iint_D F\left(x, y, z, \frac{\partial z}{\partial x}, \frac{\partial z}{\partial y}\right) dx\,dy$

8 $\Omega(z(x_1, x_2, \ldots, x_n)) = \iint_D \cdots \int F(x_1, x_2, \ldots, x_n, z, p_1, p_2, \ldots, p_n)\,dx_1\,dx_2 \cdots dx_n$

where $\quad p_i = \dfrac{\partial z}{\partial x_i}$

9 $\Omega = \iint_D \left(\left(\frac{\partial^2 z}{\partial x^2}\right)^2 + \left(\frac{\partial^2 z}{\partial y^2}\right)^2 + 2\left(\frac{\partial^2 z}{\partial x \partial y}\right)^2\right) dx\,dy$

TABLE 9-1 *(continued)*

Necessary Conditions

1 $F_y - \frac{d}{dx} F_{y'} = 0$ (Euler equation)

2 $F_{y_1} - \frac{d}{dx} F_{y_1'} = 0, \qquad F_{y_2} - \frac{d}{dx} F_{y_2'} = 0$

3 $F_{y_i} - \frac{d}{dx} F_{y_i'} = 0 \qquad i = 1, 2, 3, \ldots, n$

4 $F_y - \frac{d}{dx} F_{y'} + \frac{d^2}{dx^2} F_{y''} + \cdots + (-1)^{(n)} \frac{d^n}{dx^n} F_{y^{(n)}} = 0$

(Euler-Poisson equation)

5 $F_y - \frac{d}{dx} F_{y'} + \cdots + (-1)^n \frac{d^n}{dx^n} F_{y^{(n)}} = 0 \qquad F_z - \frac{d}{dx} F_{z'} + \cdots + (-1)^m \frac{d^m}{dx^m} F_{z^{(m)}} = 0$

6 $F_{y_i} - \frac{d}{dx} F_{y_i'} + \cdots + (-1)^{n_i} \frac{d^{n_i}}{dx^{n_i}} F_{y_i^{(n_i)}} = 0 \qquad i = 1, 2, \ldots, m$

7 $F_z - \frac{\partial}{\partial x} \{F_p\} - \frac{\partial}{\partial y} \{F_q\} = 0 \qquad p = \frac{\partial z}{\partial x}, \qquad q = \frac{\partial z}{\partial y}$

(The second and third terms are total partial derivatives with respect to x and y, respectively.)

8 $F_z - \sum_{i=1}^{n} \frac{\partial}{\partial x_i} \{F_{p_i}\} = 0$

9 $\frac{\partial^4 z}{\partial x^4} + 2 \frac{\partial^4 z}{\partial x^2 \partial y^2} + \frac{\partial^4 z}{\partial y^4} = 0$ (Biharmonic equation)

with boundary conditions

$$\begin{aligned} y_1(x_0) &= y_{10}, \quad y_2(x_0) = y_{20}, \quad y_3(x_0) = y_{30}, \ldots, y_n(x_0) = y_{n0} \\ y_1(x_1) &= y_{11}, \quad y_2(x_1) = y_{21}, \quad y_3(x_1) = y_{31}, \ldots, y_n(x_1) = y_{n1} \end{aligned} \tag{9-10}$$

then the Euler equations are written as

$$F_{y_i} - \frac{d}{dx} F_{y_i}{}' = 0 \qquad i = 1, 2, 3, \ldots, n \tag{9-11}$$

The derivation of the Euler equations can be carried out by keeping all functions constant except one, whose variation is then set equal to zero.

The necessary conditions for an extremum of various types of functionals with fixed boundaries are shown in Table 9-1. The solution of variational problems could be also given in parametric representation. In such cases the functions and derivatives will involve time. An important and well-known variational principle involving time is Hamilton's principle. This principle can be written as

$$\Omega = \int_{t_0}^{t_1} (T - U)\, dt \tag{9-12}$$

where T and U are, respectively, the kinetic and potential energies.

9-4 FUNCTIONALS WITH MOVABLE BOUNDARIES

In the preceding discussion, the end points A and B of a functional were assumed to be fixed. In this case, one or both of the boundary points are assumed movable. Thus the family of curves forming such functionals have common end points, but are permitted to expand because the end points are variable.

Consider again the functional of the simplest form

$$\Omega = \int_{x_0}^{x_1} F(x, y, y')\, dx \tag{9-13}$$

When the boundary points are movable, the Euler equation

$$F_y - \frac{d}{dx} F_{y'} = 0 \tag{9-14}$$

should again be satisfied in order that the functional in Eq. 9-13 can take on an extremum. Equation 9-14 may be satisfied by a family of curves of the form

$$y = y(x, C_1, C_2) \qquad (9\text{-}15)$$

where C_1 and C_2 are constants of integration.

In the case of fixed boundaries, the constants C_1 and C_2 in Eq. 9-15 can be determined from the boundary conditions

$$y(x_0) = y_0 \qquad y(x_1) = y_1$$

When one or both boundary points are movable, then one or both of the boundary conditions do not hold true. Thus to solve this problem additional conditions must be developed that take into consideration the boundary movements. This can be accomplished by incorporating the movement of the boundaries in the variation $\delta\Omega$ of the functional and making $\delta\Omega = 0$. In this manner, the necessary conditions for selecting curves that cause the functional to have an extremum can be derived.

By considering the functional given by Eq. 9-13 and assuming that end A is fixed while end B is movable, the extremals that pass through point $A(x_0, y_0)$ form a pencil of extremals $y = y(x, C_1)$ as shown in Fig. 9-2. If the boundary end B has moved from position (x_1, y_1) to position $(x_1 + \delta x_1, y_1 + \delta y_1)$, as shown in Fig. 9-2, the variation $\delta\Omega$ of the functional $\Omega(y(x, C_1))$ taken along extremals of the pencil $y = y(x, C_1)$ can be computed. It should be noted, however, that the functional, taken only along the curves of the pencil, turns into a function of x_1 and y_1, hence its variation turns into the differential of this function. Therefore, by making

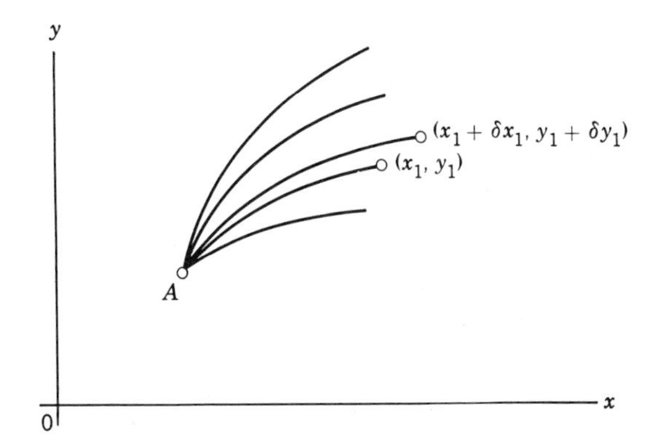

Fig. 9-2

$\delta\Omega = 0$, the fundamental necessary condition for an extremum becomes

$$(F - y'F_{y'})|_{x=x_1}\delta x_1 + F_{y'}|_{x=x_1}\delta y_1 = 0 \tag{9-16}$$

If the variations δx_1 and δy_1 are independent, then

$$(F - y'F_{y'})|_{x=x_1} = 0 \quad \text{and} \quad F_{y'}|_{x=x_1} = 0 \tag{9-17}$$

However, it is often required that the variations δx_1 and δy_1 be independent. For example, if the end $B(x_1, y_1)$ is permitted to move along the curve $y_1 = \Phi(x_1)$, then $\delta y_1 = \Phi'(x_1)\,\delta x_1$, and Eq. 9-16 becomes

$$[F + (\Phi' - y')F_{y'}]|_{x=x_1} = 0 \tag{9-18}$$

which is known as the *transversality condition*. This condition, along with $y_1 = \Phi(x_1)$, provides, in general, a way to distinguish one or more extremals from the pencil $y(x, C)$ that can give an extremum. If the end point $A(x_0, y_0)$ can move along the curve $y = \psi(x)$, then the transversality condition for this point can be written as

$$[F + (\psi' - y')F_{y'}]|_{x=x_0} = 0 \tag{9-19}$$

In a similar manner the transversality conditions for other forms of functionals can be obtained. By assuming the point $B(x_1, y_1, z_1)$ to be movable and point $A(x_0, y_0, z_0)$ to be fixed, the transversality condition for determining extrema of the functional

$$\Omega = \int_{x_0}^{x_1} F(x, y, z, y', z')\, dx \tag{9-20}$$

can be written as

$$[F + (\Phi' - y')F_{y'} + (\psi' - z')F_{z'}]_{x=x_1} = 0 \tag{9-21}$$

where

$$y_1 = \Phi(x_1) \qquad z_1 = \psi(x_1) \tag{9-22}$$

Equations 9-21 and 9-22 are not sufficient to evaluate all constants from the solution of the Euler equations

$$F_y - \frac{d}{dx}F_{y'} = 0 \qquad F_z - \frac{d}{dx}F_{z'} = 0 \tag{9-23}$$

If point $B(x_1, y_1, z_1)$ is made to move on a surface $z_1 = \Phi(x_1, y_1)$, then for independent δx_1 and δy_1,

$$[F - y'F_{y'} + (\Phi'_x - z')F_{z'}]_{x=x_1} = 0$$

$$[F_{y'} + F_{z'}\Phi'_y]_{x=x_1} = 0 \tag{9-24}$$

The two expressions, together with $z_1 = \Phi(x_1, y_1)$, are usually enough to determine two arbitrary constants from the general solution of the expressions in Eq. 9-23. Similar equations can be written when end $A(x_0, y_0, z_0)$ is movable while end $B(x_1, y_1, z_1)$ is fixed.

For functionals of the form

$$\Omega = \int_{x_0}^{x_1} F(x, y, y', y'')\, dx \tag{9-25}$$

the necessary condition for an extremum, when the boundaries are fixed, is given by the Euler-Poisson equation, Table 9-1. That is,

$$F_y - \frac{d}{dx}F_{y'} + \frac{d^2}{dx^2}F_{y''} = 0 \tag{9-26}$$

Four arbitrary constants are to be determined from the solution of Eq. 9-26. When the boundaries are movable, additional expressions are needed, which can be obtained in a manner similar to that followed previously.

If, for example, the end $B(x_1, y_1)$ is movable and end $A(x_0, y_0)$ is fixed, then the condition $\delta\Omega = 0$ yields

$$\left[F - y'F_{y'} - y''F_{y''} + y'\frac{d}{dx}(F_{y''})\right]_{x=x_1} \delta x_1$$

$$+\left[F_{y'} - \frac{d}{dx}(F_{y''})\right]_{x=x_1} \delta y_1 + F_{y''}\,|_{x=x_1} \delta y_1' = 0 \tag{9-27}$$

When δx_1, δy_1, and $\delta y_1'$ are independent, then

$$\left[F - y'F_{y'} - y''F_{y''} + y'\frac{d}{dx}(F_{y''})\right]_{x=x_1} = 0$$

$$\left[F_{y'} - \frac{d}{dx}(F_{y''})\right]_{x=x_1} = 0 \tag{9-28}$$

$$F_{y''}\,|_{x=x_1} = 0$$

If

$$y_1=\Phi(x_1) \qquad y_1'=\psi(x_1) \tag{9-29}$$

then Eq. 9-27 becomes

$$\left[F - y'F_{y'} - y''F_{y''} + y'\frac{d}{dx}(F_{y''}) + \left(F_{y'} - \frac{d}{dx}F_{y''}\right)\Phi' + F_{y''}\psi'\right]_{x=x_1} = 0 \tag{9-30}$$

This equation together with the two expressions in Eq. 9-29, are usually enough to determine x_1, y_1, and y_1'. Similar conditions can be written when end $A(x_0, y_0)$ is variable.

9-5 VIBRATION OF STRINGS AND RODS

Let it be assumed that it is required to determine the differential equation describing the free vibration of an elastic string. The dynamic deflection curve $y(x, t)$ during motion is as shown in Fig. 9-3. In the deformed position, an element dx of the string will have the length

$$ds = \sqrt{1+y_x^2}\, dx \tag{9-31}$$

By using Taylor's formula,

$$\sqrt{1+y_x^2} \cong 1 + \tfrac{1}{2}y_x^2 \tag{9-32}$$

If the string is assumed to be perfectly elastic, the potential energy dU of an element of the string is proportional to the longitudinal displacement. Thus

$$dU = \tfrac{1}{2}Ay_x^2\, dx \tag{9-33}$$

where A is a certain coefficient representing the tension force in the string. By considering all the elements of the string, its total potential energy U is

$$U = \tfrac{1}{2}\int_0^L Ay_x^2\, dx \tag{9-34}$$

Hamilton's principle given by Eq. 9-12 takes the form

$$\Omega=\int_{t_0}^{t_1}\int_0^L\left(\tfrac{1}{2}my_t-\tfrac{1}{2}Ay_x^2\right)dx\,dt \tag{9-35}$$

The first term within the integral sign represents the kinetic energy T, where m is the mass per unit of length of the string. This functional is of the form

$$\Omega(y(x,t))=\int_{t_0}^{t_1}\int_0^L F\left(\tfrac{1}{2}m\dot{y}-\tfrac{1}{2}Ay'^2\right)dx\,dt \tag{9-36}$$

which is similar to item 7 in Table 9-1. Thus the differential equation of motion of the string is the Euler-Lagrange equation:

$$\frac{\partial}{\partial t}(m\dot{y})-\frac{\partial}{\partial x}(Ay')=0 \tag{9-37}$$

The deflection curve $y(x, t)$ that satisfies Eq. 9-37 is the one that makes the functional in Eq. 9-36 have an extremum.

For a homogeneous string, m and A are constants, and Eq. 9-37 can be written as

$$\frac{\partial^2 y}{\partial t^2}-\frac{A}{m}\frac{\partial^2 y}{\partial x^2}=0 \tag{9-38}$$

If it is assumed that the string is acted upon by a force $q(x,t)$ that is proportional to the mass of an element of the string, its work on the element is

$$mq(x,t)y\,dx \tag{9-39}$$

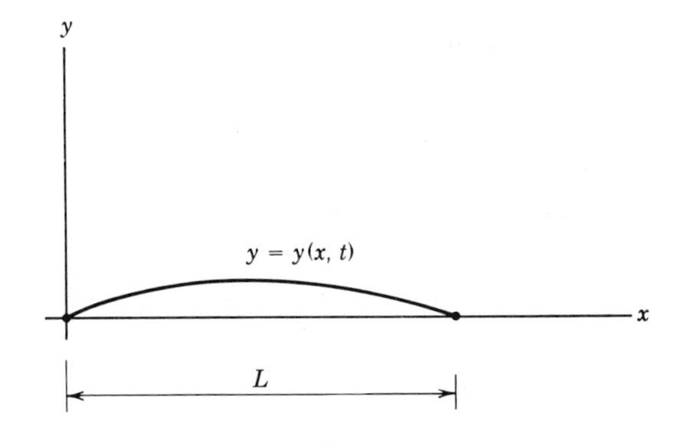

Fig. 9-3

On this basis, Eq. 9-12 yields

$$\Omega=\int_{t_0}^{t_1}\int_0^L\left(\tfrac{1}{2}my_t^2-\tfrac{1}{2}Ay_x^2+mq(x,t)\,y\right)dx\,dt \tag{9-40}$$

and the equation of forced vibration is

$$\frac{\partial}{\partial t}(my_t)-\frac{\partial}{\partial x}(Ay_x)-mq(x,t)=0$$

or

$$\frac{\partial^2 y}{\partial t^2}-\frac{A}{m}\frac{\partial^2 y}{\partial x^2}=q(x,t) \tag{9-41}$$

For a rod with deflection $y(x,t)$ and coordinates x, y as in Fig. 9-3, its kinetic energy T during free vibration is

$$T=\tfrac{1}{2}\int_0^L my_t^2\,dx$$

where L is its length. If it is assumed that the rod will not stretch, its potential energy is proportional to the square of its curvature. Hence

$$U=\frac{1}{2}\int_0^L A\left\{\frac{\partial^2 y/\partial x^2}{[1+(\partial y/\partial x)^2]^{3/2}}\right\}^2 dx \tag{9-42}$$

where A in this case is the stiffness EI of the rod. For small deflections,

$$U=\frac{1}{2}\int_0^L A\left(\frac{\partial^2 y}{\partial x^2}\right)dx \tag{9-43}$$

Hamilton's integral yields

$$\Omega=\int_{t_0}^{t_1}\int_0^L\left[\tfrac{1}{2}my_t^2-\tfrac{1}{2}Ay_{xx}^2\right]dx\,dt \tag{9-44}$$

The equation of motion of the rod is

$$\frac{\partial}{\partial t}(my_t)+\frac{\partial^2}{\partial x^2}(Ay_{xx})=0$$

or, for a homogeneous rod,

$$\frac{\partial^2 y}{\partial t^2}+\frac{A}{m}\frac{\partial^4 y}{\partial x^4}=0 \tag{9-45}$$

9-6 FREE TRANSVERSE VIBRATION OF BEAMS

Consider, for example, the uniform simply supported beam in Fig. 9-4 and let it be assumed that it is required to determine its natural frequencies of vibration. In this case, the kinetic energy dT of an element of the beam is

$$dT = \tfrac{1}{2} m\dot{y}^2\, dx \tag{9-46}$$

The total kinetic energy T is

$$T = \int_0^L \frac{m\dot{y}^2}{2}\, dx \tag{9-47}$$

where m is the mass per unit of length, $y(x, t)$ is the dynamic deflection, and L is its length.

The potential energy dU of an element of the beam is

$$dU = \frac{EI}{2}\left(\frac{d^2y}{dx^2}\right)^2 dx \tag{9-48}$$

and its total potential energy U is

$$U = \int_0^L \frac{EI}{2}\left(\frac{d^2y}{dx^2}\right)^2 dx \tag{9-49}$$

Hamilton's integral, Eq. 9-12, yields

$$\Omega = \int_0^t \int_0^L \left[\frac{m\dot{y}^2}{2} - \frac{EI}{2}\left(\frac{d^2y}{dx^2}\right)^2\right] dx\, dt \tag{9-50}$$

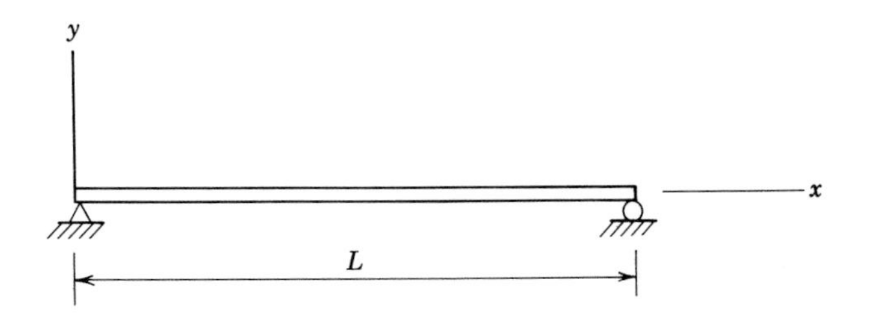

Fig. 9-4

The equation above is represented by the functional of the general form

$$\Omega(y(x,t))=\int_0^t\int_0^L F\left(t,x,y,\frac{\partial y}{\partial x},\frac{\partial y}{\partial t},\frac{\partial^2 y}{\partial x^2},\frac{\partial^2 y}{\partial t^2},\frac{\partial^2 y}{\partial x\partial y}\right)dx\,dt \tag{9-51}$$

Equation 9-51 will yield an extremum, that is, $\delta\Omega=0$, if the following equation is satisfied by $y(x,t)$.

$$F_y-\frac{\partial F_{y'}}{\partial x}-\frac{\partial F_{\dot{y}}}{\partial t}+\frac{\partial^2 F_{y''}}{\partial x^2}+\frac{\partial^2 F_s}{\partial x\partial t}+\frac{\partial^2 F_{\ddot{y}}}{\partial t^2}=0 \tag{9-52}$$

where

$$s=\frac{\partial^2 y}{\partial x\partial t}$$

Equation 9-52, by substitution, yields

$$-m\ddot{y}-EIy^{\mathrm{IV}}=0$$

or

$$\frac{\partial^4 y}{\partial x^4}+\frac{m}{EI}\frac{\partial^2 y}{\partial t^2}=0 \tag{9-53}$$

This is the familiar general differential equation for the free transverse vibration of beams. Its solution will yield $y(x,t)$, as well as the natural frequencies of the simply supported beam.

A solution regarding the natural frequencies of the uniform simply supported beam can be easily obtained by using a direct approach to the variational problem. For example, let it be assumed that a solution $y=f(x)g(t)$ exists which satisfies the boundary conditions $y=y''=0$ at $x=0$ and $x=L$. In this solution, $f(x)$ is a function of x only and $g(t)$ is a function of time. Also let

$$f(x)=C\sin\frac{n\pi x}{L}=C\sin\lambda x \tag{9-54}$$

where

$$\lambda=\frac{n\pi}{L}\qquad n=1,2,3,\ldots \tag{9-55}$$

On this basis,

$$y = (C \sin \lambda x)\, g$$

$$\dot{y} = (C \sin \lambda x)\, \dot{g}$$

$$\frac{\partial^2 y}{\partial x^2} = (-C \lambda^2 \sin \lambda x)\, g$$

Substituting into Eq. 9-50,

$$\Omega = \int_0^t \int_0^L \left[\frac{m}{2} C^2 \sin^2 \lambda x (\dot{g})^2 - \frac{EI}{2} C^2 \lambda^4 \sin^2 \lambda x (g)^2 \right] dx\, dt$$

$$= C^2 \int_0^t \left\{ \left[\frac{m}{2} (\dot{g})^2 - \frac{EI}{2} \lambda^4 (g)^2 \right] \int_0^L \sin^2 \lambda x\, dx \right\} dt$$

By integration

$$\Omega = C^2 \int_0^t \left[\frac{mL}{4} (\dot{g})^2 - \frac{EIL}{4} \lambda^4 (g)^2 \right] dt \qquad (9\text{-}56)$$

The necessary condition that makes Eq. 9-56 have an extremum, $\delta\Omega = 0$, is Euler's equation

$$F_g - \frac{dF\dot{g}}{dt} = 0 \qquad (9\text{-}57)$$

Thus

$$-\frac{EIL\lambda^4}{2} g - \frac{d}{dt}\left(\frac{mL}{2} \dot{g} \right) = 0$$

or

$$\ddot{g} + \frac{EI\lambda^4}{m} g = 0 \qquad (9\text{-}58)$$

The solution of Eq. 9-58 yields

$$g = A \cos \sqrt{EI/m}\, \lambda^2 t + B \sin \sqrt{EI/m}\, \lambda^2 t \qquad (9\text{-}59)$$

where λ is given by Eq. 9-55. The natural frequencies ω_n of the uniform simply supported beam are

$$\omega_n = \sqrt{EI/m}\,\lambda^2 = \sqrt{EI/m}\,\frac{n^2\pi^2}{L^2}$$

or

$$\omega_n = \frac{n^2\pi}{L^2}\sqrt{EI/m} \qquad n = 1, 2, 3, \ldots \tag{9-60}$$

The beam has infinite degrees of freedom, and, therefore, infinite frequencies ω_n are given by Eq. 9-60. Other boundary conditions can be treated in a similar manner

9-7 LONGITUDINAL VIBRATIONS OF UNIFORM ELASTIC BEAMS

Beams can oscillate longitudinally, and for this type of vibration sections will again be assumed to remain planar, provided that elastic conditions are not violated. If the cantilever beam in Fig. 9-5 is assumed to vibrate longitudinally, the kinetic energy dT of an element dx is

$$dT = \tfrac{1}{2}\rho\dot{u}\,dx \tag{9-61}$$

where ρ is the mass density per unit of length and u is the longitudinal displacement of a cross section of the beam. By considering all elements, its total kinetic energy T is

$$T = \int_0^L \frac{\rho\dot{u}^2}{2}\,dx \tag{9-62}$$

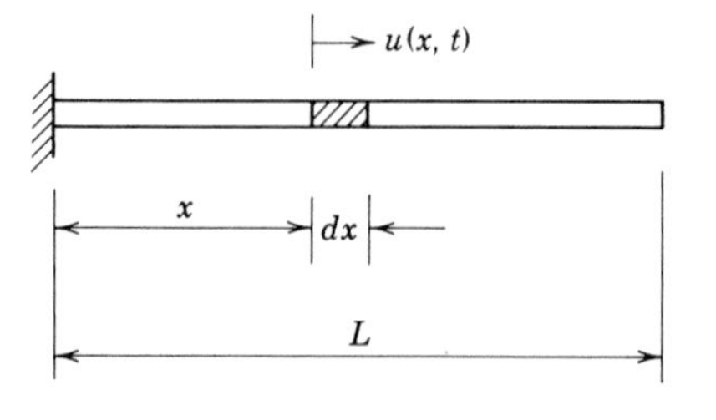

Fig. 9-5

The potential energy dU of this element is

$$dU = \frac{Eu'^2 A}{2}\, dx \tag{9-63}$$

and the total potential energy U is

$$U = \int_0^L \frac{EAu'^2}{2}\, dx \tag{9-64}$$

where A is the cross-sectional area of the beam. Hamilton's integral, Eq. 9-12, yields

$$\Omega = \int_0^t \int_0^L \left(\frac{\rho \dot{u}^2}{2} - \frac{EAu'^2}{2} \right) dx\, dt \tag{9-65}$$

The functional is of the form

$$\Omega = \int_0^t \int_0^L F\left(t, x, u, \frac{\partial u}{\partial t}, \frac{\partial u}{\partial x} \right) dx\, dt \tag{9-66}$$

The necessary condition for an extremum, $\delta\Omega = 0$, yields the equation

$$F_u - \frac{\partial F_p}{\partial x} - \frac{\partial F_g}{\partial t} = 0 \tag{9-67}$$

where

$$p = \frac{\partial u}{\partial x} \quad \text{and} \quad g = \frac{\partial u}{\partial t}$$

By substitution, Eq. 9-67 yields

$$EAu'' - \ddot{u} = 0$$

or

$$\frac{\partial^2 u}{\partial x^2} - \frac{\rho}{EA} \frac{\partial^2 u}{\partial t^2} = 0 \tag{9-68}$$

Equation 9-68 is the familiar general differential equation for the longitudinal vibration of a beam. For given boundary conditions, its solution will yield the longitudinal displacements $u(x, t)$, the natural frequencies of vibration, and the corresponding mode shapes. For example, by letting $u=f(x)g(t)$, Eq. 9-68 yields

$$f\,\ddot{g} - \frac{EA}{\rho} f''\, g = 0$$

or, by separating variables,

$$\frac{\ddot{g}}{g} = \frac{EA}{\rho}\frac{f''}{f} = \pm\Psi \tag{9-69}$$

where Ψ is an arbitrary constant.

Equation 9-69 yields the following two differential equations:

$$\ddot{g} + \Psi g = 0 \tag{9-70}$$

$$a^2 f'' + \Psi f = 0 \tag{9-71}$$

where

$$a^2 = \frac{EA}{\rho} \tag{9-72}$$

The solution of the two equations is

$$g = C_1 \cos\sqrt{\Psi}\, t + C_2 \sin\sqrt{\Psi}\, t$$

$$f = C_3 \cos\frac{\sqrt{\Psi}}{a}x + C_4 \sin\frac{\sqrt{\Psi}}{a}x$$

Thus

$$u(x,t) = \left(C_1 \cos\sqrt{\Psi}\, t + C_2 \sin\sqrt{\Psi}\, t\right)\left(C_3 \cos\frac{\sqrt{\Psi}}{a}x + C_4 \sin\frac{\sqrt{\Psi}}{a}x\right) \tag{9-73}$$

By applying the boundary condition $u=0$ at $x=0$, we have $C_3=0$. The boundary condition $u'=0$ at $x=L$, for a nontrivial solution, yields

$$\cos\frac{\sqrt{\Psi}}{a}L = 0 \tag{9-74}$$

Equation 9-74 will be satisfied if

$$\frac{\sqrt{\Psi}}{a}L = \frac{(2n-1)\pi}{2}$$

or

$$\sqrt{\Psi} = \omega_n = a\frac{(2n-1)\pi}{2L}$$

Thus the natural frequencies ω_n of the beam in Fig. 9-5 can be evaluated from the expression

$$\omega_n = \frac{(2n-1)\pi}{2L}\sqrt{EA/\rho} \qquad n = 1, 2, 3, \ldots \tag{9-75}$$

Other boundary conditions can be treated in a similar manner.

9-8 VIBRATION OF PLATES

Lateral vibrations of plates are considered in this section. The general differential equation of motion is derived by using variational principles. The plate is assumed to be perfectly elastic, and the deflections are considered to be small compared to the thickness h of the plate. It is further assumed that the plate is uniform, the middle surface does not stretch, and cross sections are always perpendicular to the middle surface.

For an element of a plate, Fig. 9-6, the symbols u, v, and w denote the

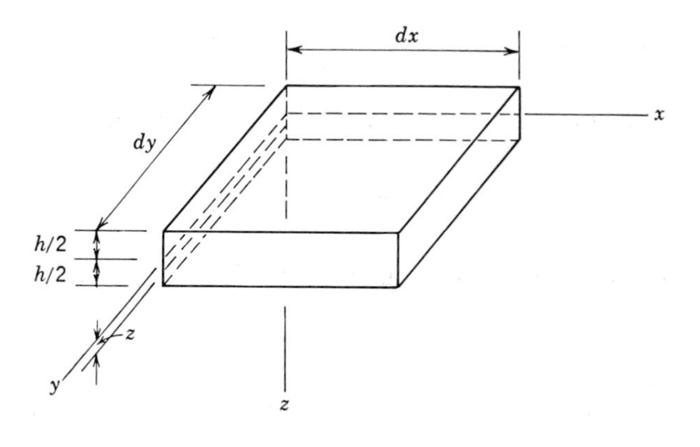

Fig. 9-6

displacements in the x, y, and z directions, respectively. The plate is assumed to vibrate in the z direction only. During vibration, the potential energy U of the plate, due to bending, is given by the known expression

$$U=\frac{1}{2}\iiint_V [\sigma_x\epsilon_x+\sigma_y\epsilon_y+\sigma_z\epsilon_z+\gamma_{xy}\tau_{xy}+\gamma_{xz}\tau_{xz}+\gamma_{yz}\tau_{yz}]\,dx\,dy\,dz \tag{9-76}$$

where σ_x, σ_y, σ_z are the normal stresses in the x, y, and z directions, respectively; ϵ_x, ϵ_y, ϵ_z are the strains in the same directions; τ_{xy}, τ_{xz}, τ_{yz} are the shear stresses; and γ_{xy}, γ_{xz}, γ_{yz} are the corresponding shear strains.

The symbol V associated with the integral sign in Eq. 9-76 denotes volume. From strength of materials, based on the assumptions above,

$$\epsilon_x=\frac{\partial u}{\partial x}=-z\frac{\partial^2 w}{\partial x^2}$$

$$\epsilon_y=\frac{\partial v}{\partial y}=-z\frac{\partial^2 w}{\partial y^2}$$

$$\gamma_{xy}=\frac{\partial u}{\partial y}+\frac{\partial v}{\partial x}=-2z\frac{\partial^2 w}{\partial x\,\partial y}$$

$$\epsilon_z=\gamma_{xz}=\gamma_{yz}=0$$

$$\sigma_x=\frac{E}{1-\mu^2}(\epsilon_x+\mu\epsilon_y)=-\frac{Ez}{1-\mu^2}\left(\frac{\partial^2 w}{\partial x^2}+\mu\frac{\partial^2 w}{\partial y^2}\right)$$

$$\sigma_y=\frac{E}{1-\mu^2}(\epsilon_y+\mu\epsilon_x)=-\frac{Ez}{1-\mu^2}\left(\frac{\partial^2 w}{\partial y^2}+\mu\frac{\partial^2 w}{\partial x^2}\right)$$

$$\tau_{xy}=G\gamma_{xy}=-\frac{Ez}{1+\mu}\frac{\partial^2 w}{\partial x\,\partial y}\qquad G=\frac{E}{2(1+\mu)}$$

where μ is the Poisson ratio, G is the shear modulus, and

$$u=-z\frac{\partial w}{\partial x}\qquad v=-z\frac{\partial w}{\partial y}$$

Substitution and integration of z between the limits $-h/2\leqslant z\leqslant h/2$, yields from Eq. 9-76

$$U=\frac{D}{2}\iint[{w_{xx}}^2+{w_{yy}}^2+2\mu w_{xx}w_{yy}+2(1-\mu){w_{xy}}^2]\,dx\,dy \tag{9-77}$$

where

$$D=\frac{Eh^3}{12(1-\mu^2)}$$

is the plate stiffness.

The kinetic energy T of the vibrating plate is

$$T=\frac{m}{2}\int\int \dot{w}^2\,dx\,dy \tag{9-78}$$

where m is the mass per unit area of the plate. Thus

$$U-T=\int\int\left\{\frac{D}{2}\left[\left(w_{xx}{}^2+w_{yy}{}^2\right)+2\mu w_{xx}w_{yy}\right.\right.$$

$$\left.\left.+2(1-\mu)w_{xy}{}^2\right]-\frac{m}{2}w_t^2\right\}dx\,dy \tag{9-79}$$

This is a functional of the form

$$\Omega=\int\int\int F(x,y,w,t,w_{xx},w_{yy},w_{xy},w_t)\,dx\,dy\,dt \tag{9-80}$$

The necessary condition that makes Eq. 9-80 have an extremum, $\delta\Omega=0$, yields

$$D\left(\frac{\partial^4 w}{\partial x^4}+2\frac{\partial^4 w}{\partial x^2\partial y^2}+\frac{\partial^4 w}{\partial y^4}\right)=-m\ddot{w} \tag{9-81}$$

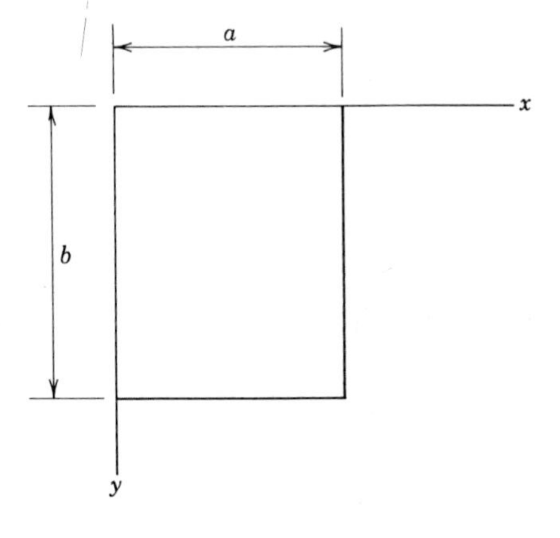

Fig. 9-7

Equation 9-81 is the differential equation of motion that can be used to determine the vibration of plates. Note that $m\ddot{w}$ is the inertia force during free vibration.

In the case of a rectangular plate with simply supported edges, Fig. 9-7, a solution of Eq. 9-81 can be obtained by assuming that

$$w = \sin\frac{p\pi x}{a}\sin\frac{n\pi y}{b}\cos\omega t \qquad p = 1, 2, 3, \ldots \qquad n = 1, 2, 3, \ldots \tag{9-82}$$

which satisfies the boundary conditions

$$w = 0 \qquad \frac{d^2 w}{dx^2} = 0 \qquad \frac{d^2 w}{dy^2} = 0$$

at the edges of the plate. The values of the natural frequencies ω of the plate, by substituting into Eq. 9-81 and performing the required computations, are found to be given by the expression.

$$\omega = \pi^2\left(\frac{p^2}{a^2} + \frac{n^2}{b^2}\right)\sqrt{D/m} \qquad p = 1, 2, 3, \ldots \qquad n = 1, 2, 3, \ldots \tag{9-83}$$

The natural frequency of the lowest mode, as well as the frequencies of higher modes of vibration, can be obtained by assigning values to p and n. Other boundary conditions can be treated in a similar manner. It should be pointed out, however, that elaborate computations are usually involved in the solution of plate problems.

PROBLEMS

9-1 By using the appropriate Euler equation, examine the given boundary conditions and determine the value of a for which the functional

$$\Omega(y(x)) = \int_0^1 (y^2 + x^2 y')\,dx$$

$$y(0) = 0 \qquad y(1) = a$$

can have an extremum.

9-2 Find a curve joining two given points A and B so that a particle moving along this curve, starting at point A, reaches B in the shortest time. Neglect friction and resistance of the medium.

9-3 The boundary conditions of the functional

$$\Omega(y(x)) = \int_0^{\pi/2} [(y'')^2 - y^2 + x^2]\, dx$$

are $y(0)=1$, $y'(0)=0$, $y(\pi/2)=0$, and $y'(\pi/2)=-1$. Determine the extremal of this functional.

9-4 Find the extremal of the functional

$$\Omega(y(x), z(x)) = \int_{x_0}^{x_1} [2yz - 2y^2 + (y')^2 - (z')^2]\, dx$$

9-5 The homogeneous elastic string in Fig. 9-3 is subjected to a harmonic force $q(x,t) = q\cos\omega_f t$ that is uniformly distributed along the length of the string. Determine its differential equation of motion.

9-6 Repeat Problem 9-5 when $q(x,t) = q\sin\omega_f t$.

9-7 Determine the natural frequencies and mode shapes of a uniform steel rod with free-free ends.

9-8 Determine the natural frequencies and mode shapes of the steel beams in Problem 4-1 by following the procedure discussed in Section 9-6. The stiffness EI is 30×10^6 kip-in.^{2} and the weight w per unit of length is 0.40 kips/ft.

9-9 By applying the procedure discussed in Section 9-7, determine the expression for the longitudinal displacement $u(x,t)$, the natural frequencies of vibration, and the corresponding mode shapes of a uniform simply supported beam of length L and mass m per unit of length.

9-10 Repeat Problem 9-9 for a uniform beam that is hinged at the left end and elastically supported at the other end by a linear spring of constant k.

9-11 Derive Eq. 9-83 by using Eqs. 9-81 and 9-82 and applying the boundary conditions of a rectangular plate with simply supported edges.

10

APPROXIMATE METHODS FOR DYNAMIC RESPONSE

10-1 INTRODUCTION

In the preceding chapters, rigorous and approximate methods of analysis were used to determine the dynamic response of structures subjected to dynamic forces. The application of these methods was limited to all-elastic response, except in Section 2-8 where elastoplastic systems were considered. There are practical situations, however, where the stresses and displacements of a structure could extend beyond the elastic range, depending on the degree of damage a structure is permitted to undertake. Structures that may be subjected to blast and earthquake are samples of such types of problems.

The purpose in this chapter is to provide simplified methods of analysis that can be used to determine with reasonable accuracy the dynamic response of structures in the elastic, elastoplastic, and plastic ranges. These methods are general and can be easily applied to component elements of a structure as well as to multielement structures.

10-2 FUNDAMENTAL CONCEPTS

In earlier chapters, satisfactory solutions in the elastic range were obtained by using the method of modal analysis and considering the response of only a few normal modes of the system. For many cases, the fundamental mode was usually sufficient for an accurate solution to a given problem. The simplified methods in this chapter are not based on the fundamental mode, but on the derivation of approximate mathematical models, often called equivalent systems, which provide a reasonable approximation of the dynamic behavior of multidegree structural elements. In fact, the methodology is general and permits one to determine the dynamic response of a structural system by treating independently each stage, that is, elastic, elastoplastic, and plastic, as is explained later in this chapter.

By virtue of this method, the dynamic behavior of a structural element, or a component element of a multielement structure, can be approximated by an equivalent single degree of freedom system. For example, the structures in Fig. 10-1 can be approximated by the equivalent spring-mass systems as shown, provided that there is a way to determine the equivalent parameters k_e, m_e, and P_e. Plate of various shapes and boundary conditions could be also represented by similar mathematical models.

For a structural element, say the simply supported beam in Fig. 10-1*a*, the derivation of an equivalent single-degree spring-mass system involving the parameters k_e, m_e, and P_e requires the assumption of an appropriate shape for the deflection of the element. In addition, the time variation of the dynamic load $q(t)$ must be known. An assumed deflection shape for the element establishes an equation that relates the relative deflection of all points, thus permitting the motion of the element to be specified by any point on the element. An equivalent one-degree system is usually obtained by making the displacement of its mass m_e the same as that of a significant point on the structural element. For example, in Figs. 10-1*a* and 10-1*b*, the

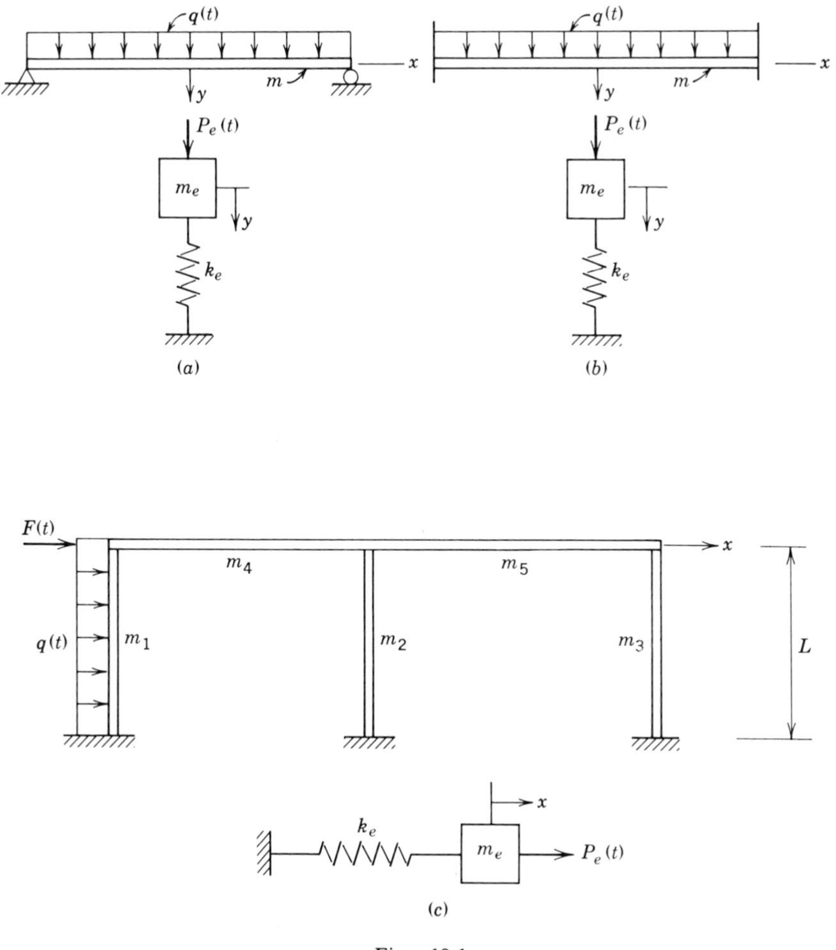

Fig. 10-1

deflection of m_e can be made equal to the one at the midspan of the beam. In Fig. 10-1c it can be made equal to the horizontal deflection of the top of the frame.

In the analysis of this chapter, the assumed deflected shape is taken to be that produced by the static application of the dynamic load, and the parameters of the equivalent system are computed on this basis. Once the equivalent one-degree system is determined, the analysis for dynamic response can be carried out by applying the methods discussed in Chapter 2. It should be pointed out, however, that the actual structure is usually a multidegree system. Earlier, such structures, when subjected to dynamic loads, were found to have the response of one of the modes, usually the first mode, predominating over that of the other modes.

The above-mentioned assumption regarding the shape of the deflection curve of the actual structure is not meant to approximate the shape of the fundamental mode of vibration. In fact, the shape of this mode is usually different. This procedure is intended to be used primarily for cases where the shapes of the normal modes are not easily determined. On the other hand, investigations[28] have shown that the two procedures differ only slightly, except that the one suggested here is somewhat more accurate, because it takes into account some of the contributions of the higher modes.

10-3 DERIVATION OF TRANSFORMATION FACTORS

The derivation of an equivalent one degree of freedom system to replace a given structural element of continuous mass and elasticity can be made easier if certain *transformation factors* are used. These factors are denoted by the symbol K and are defined as follows.

Load Factor

The load factor is designated as K_L and is defined as the factor by which the total load acting on a structural element must be multiplied in order to obtain the equivalent concentrated load P_e for the equivalent one-degree system. Thus

$$K_L = \frac{P_e}{P_t} \tag{10-1}$$

where P_t is the total actual load acting on the structural element.

To compute $P_e(t)$, the results in Sections 5-2 and 5-4 can be used. In these sections, it was found that for a normal mode the equivalent force $P_e(t)$ is

$$P_e = \sum_{i=1}^{r} P_i \beta_i \tag{10-2}$$

when the structure is loaded with concentrated loads, and

$$P_e = \int q(x)\beta(x)\,dx \tag{10-3}$$

[28] See Reference 41, Manual Number EM1110-345-416.

when the actual load is distributed. Equations 10-2 and 10-3 can be used to determine $P_e(t)$ in Eq. 10-1, provided that β_i and $\beta(x)$ in these equations represent the deflection shape produced in the structure by the static application of the dynamic loads.

Consider for example the simply supported beam in Fig. 10-1*a*. In the *elastic range*, the static application of the dynamic load yields the static elastic line

$$y(x) = \frac{qx}{24EI}(L^3 - 2Lx^2 + x^3) \qquad (10\text{-}4)$$

where L is the length of the beam and EI is assumed to be constant. At midspan,

$$y_{x=L/2} = y_c = \frac{5qL^4}{384EI} \qquad (10\text{-}5)$$

From this equation,

$$q = \frac{384EI}{5L^4} y_c$$

and Eq. 10-4 becomes

$$y(x) = \frac{16}{5L^4}(L^3x - 2Lx^3 + x^4)y_c \qquad (10\text{-}6)$$

If y_c above is taken to be equal to unity, the deflection shape $\beta(x)$ in Eq. 10-3 is given by the expression

$$\beta(x) = \frac{16}{5L^4}(L^3x - 2Lx^3 + x^4) \qquad (10\text{-}7)$$

Thus, from Eq. 10-3,

$$P_e = \frac{16q}{5L^4}\int_0^L (L^3x - 2Lx^3 + x^4)\,dx$$

$$= \frac{16qL}{25} \qquad (10\text{-}8)$$

The total load P_t in Eq. 10-1 is equal to qL. Thus Eq. 10-1 yields

$$K_L = \frac{P_e}{P_t} = \frac{16qL}{25qL} = \frac{16}{25} \text{ (elastic)} \qquad (10\text{-}9)$$

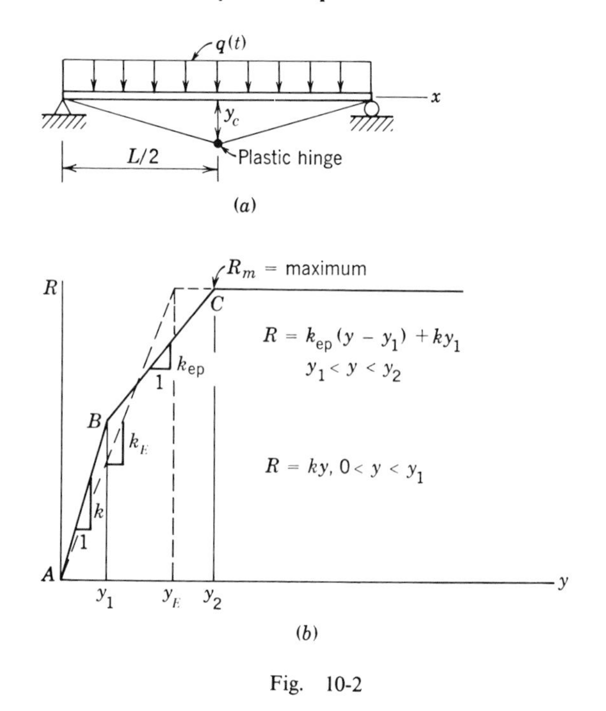

Fig. 10-2

In the *plastic range*, when the plastic *hinge* at the center of the beam is formed, the shape of the elastic line consists of the two straight lines shown in Fig. 10-2*a*. Thus, with $y_c = 1$, the shape $\beta(x)$ is given by the expression

$$\beta(x) = \frac{2x}{L} \qquad x < \frac{L}{2} \tag{10-10}$$

Equation 10-3 yields

$$P_e = 2\int_0^{L/2} q\left(\frac{2x}{L}\right)dx = \frac{qL}{2}$$

and Eq. 10-1 gives

$$K_L = \frac{qL/2}{qL} = \frac{1}{2} \text{ (plastic)} \tag{10-11}$$

The time variation of the dynamic load for the actual and the equivalent systems is the same, because K_L is not affected by the nature of the time variation of the dynamic load.

Mass Factor

The mass factor is denoted by the symbol K_M and is defined as the factor by which the total mass of the structural element should be multiplied in order to yield the equivalent mass m_e of the equivalent one-degree system. Thus

$$K_M = \frac{m_e}{m_t} \tag{10-12}$$

where m_t is the total mass of the actual structural element.

By following the results in Sections 5-2 and 5-4, the mass factor m_e is

$$m_e = \sum_{i=1}^{r} m_i \beta_i^2 \tag{10-13}$$

for lumped parameter systems, and

$$m_e = \int m\beta^2(x)\, dx \tag{10-14}$$

for structures with distributed mass. In the equations above, the β's are determined from the assumed shape of the deflection curve.

For the simply supported beam in Fig. 10-1*a*, $\beta(x)$ in the elastic range is given by Eq. 10-7. Thus, for the elastic range,

$$K_M = \frac{m_e}{m_t} = \frac{\int_0^L m\beta^2(x)\, dx}{mL} = \frac{1}{2} \quad (\text{elastic}) \tag{10-15}$$

For the plastic range, $\beta(x)$ is given by Eq. 10-10. Thus

$$K_M = \frac{m_e}{m_t} = \frac{2\int_0^{L/2} m(2x/L)^2 dx}{mL} = \frac{1}{3} \quad (\text{plastic}) \tag{10-16}$$

Resistance Factor

The resistance factor is designated by K_R and is the factor by which the resistance of the actual structural element should be multiplied in order to obtain the resistance R_e of the equivalent one-degree system. Thus

$$K_R = \frac{R_e}{R} \tag{10-17}$$

where R is the computed resistance of the structural element.

Considerable discussion regarding the resistance R is given in Section 2-8. In general, the resistance R of a structure is defined as the internal force that tends to restore the structure to its equilibrium position. Possible shapes of this function were shown in Fig. 2-10*d* and they were idealized by the bilinear form in Fig. 2-10*b* and 2-10*c*. The same idealization is used here, except that in certain cases this function is not bilinear. For example, the resistance function for the simply supported beam in Fig. 10-1*a* can be represented by the bilinear form in Fig. 2-10*c*, where R_m is the maximum value of the load qL that the beam can support and k is the value of the load qL required to produce a deflection equal to unity at midspan. Thus for the actual structure $R = ky$ and for the equivalent one-degree system $R_e = k_e y$.

The resistance function R of the beam in Fig. 10-1*b* is idealized by the form shown by solid lines in Fig. 10-2*b*. In this case, a plastic hinge is first developed simultaneously at each end of the member; the resistance function in this elastic range is represented by the straight line AB in Fig. 10-2*b*. As the load q increases, another plastic hinge will eventually be formed at midspan. This range is called elastoplastic and the resistance function for this range is the straight line BC in Fig. 10-2*b*. The third line is the plastic range, and the resistance function attains its maximum value as soon as this range is reached. This resistance function in Fig. 10-2*b* could be also approximated by the bilinear form shown by a dashed line, where the slope k_E is selected so that the areas under the two shapes are the same. This will not yield appreciable error in the computation of the dynamic deflections, because the energy absorbed in both cases is the same.

The discussion and definitions above make it clear that the resistance factor K_R should be equal to the load factor K_L. Thus

$$K_R = K_L = \frac{R_e}{R} \tag{10-18}$$

and

$$K_R = K_L = \frac{k_e}{k} \tag{10-19}$$

where k is the value of the load distribution on the structural element which is required to produce a deflection equal to unity at the point where the deflection is made equal to that of the equivalent system.

Load-Mass Factor

This factor is designated as K_{LM} and is merely a combination of the load and mass factors. This factor is usually needed for cases where it is convenient to write the equations of motion in terms of this factor. For example, the equation of motion for the one-degree system can be written as

$$P_e(t) - R_e = m_e \ddot{y} \tag{10-20}$$

By utilizing Eqs. 10-1, 10-17, and 10-18, Eq. 10-20 yields

$$K_L P_t(t) - K_L R = K_M m_t \ddot{y} \tag{10-21}$$

or

$$P_t(t) - R = \left(\frac{K_M}{K_L}\right) m_t \ddot{y} \tag{10-22}$$

The load-mass factor K_{LM} is defined by the expression

$$K_{LM} = \frac{K_M}{K_L} \tag{10-23}$$

In addition, the natural frequency ω of both the actual and idealized system is given by the expression

$$\omega^2 = \frac{k_e}{m_e} = \frac{k}{m_t}\frac{K_L}{K_M} = \frac{k}{m_t K_{LM}} \tag{10-24}$$

The transformation factors for various types of structural elements with various boundary conditions can be determined by using the definitions and concepts given here; they are tabulated as shown in the next section.

10-4 TABULATION OF TRANSFORMATION FACTORS

The transformation factors for several types of structural elements that are needed in the derivation of equivalent systems are conveniently tabulated as shown in Tables 10-1 through 10-8. Here the elastic, elastoplastic, and plastic ranges are considered by including also various loadings and boundary conditions. In Tables 10-1, 10-2, and 10-3 the transformation factors for beams and one-way slabs are shown. The computations are based on the static deflection shape and are carried out as discussed in the preceding section for the simply supported beam. Note that the values given by Eqs. 10-9, 10-11, 10-15, and 10-16 are shown in Table 10-1. The term M_P in Table 10-1 is the plastic or ultimate bending strength of the structural element. In Tables 10-2 and 10-3 the term M_{Ps} is the plastic resisting moment at support, and M_{Pm} is the one at the center.

TABLE 10-1
TRANSFORMATION FACTORS: BEAMS AND ONE-WAY SLABS

Simply-supported

Loading Diagram	Strain Range	Load Factor K_L	Mass Factor K_M: Concentrated Mass*	Mass Factor K_M: Uniform Mass	Load-Mass Factor K_{LM}: Concentrated Mass*	Load-Mass Factor K_{LM}: Uniform Mass	Maximum Resistance R_m	Spring Constant k	Dynamic Reaction V
$P = pL$ (uniform load, span L)	Elastic	0.64		0.50		0.78	$8M_P/L$	$384EI/5L^3$	$0.39R+0.11P$
	Plastic	0.50		0.33		0.66	$8M_P/L$	0	$0.38R_m+0.12P$
P (at midspan, $\frac{L}{2}$, $\frac{L}{2}$)	Elastic	1	1.0	0.49	1.0	0.49	$4M_P/L$	$48EI/L^3$	$0.78R-0.28P$
	Plastic	1	1.0	0.33	1.0	0.33	$4M_P/L$	0	$0.75R_m-0.25P$
$\frac{P}{2}$, $\frac{P}{2}$ (at third points, $\frac{L}{3}$, $\frac{L}{3}$, $\frac{L}{3}$)	Elastic	0.87	0.76	0.52	0.87	0.60	$6M_P/L$	$56.4EI/L^3$	$0.62R-0.12P$
	Plastic	1	1.0	0.56	1.0	0.56	$6M_P/L$	0	$0.75R_m-0.25P$

From Reference 41 *Equal parts of the concentrated mass are lumped at each concentrated load.

TABLE 10-2
TRANSFORMATION FACTORS: BEAMS AND ONE-WAY SLABS

Loading Diagram	Strain Range	Load Factor K_L	Mass Factor K_M		Load-Mass Factor K_{LM}		Maximum Resistance R_m	Spring Constant k	Effective Spring Constant k_E		Dynamic Reaction V
			Concentrated Mass*	Uniform Mass	Concentrated Mass*	Uniform Mass			Elastic	Plastic	
$P = pL$, L	Elastic	0.53		0.41		0.77	$12M_{Ps}/L$	$384EI/L^3$	$264EI/L^3$	$307EI/L^3$	$0.36R+0.14P$
	Elastoplastic	0.64		0.50		0.78	$(8/L)(M_{Ps}+M_{Pm})$	$384EI/5L^3$	$R_{mf}=22M_P/L$		$0.39R+0.11P$
	Plastic	0.50		0.33		0.66	$(8/L)(M_{Ps}+M_{Pm})$	0			$0.38R_m+0.12P$
P, $\frac{L}{2}$, $\frac{L}{2}$	Elastic	1	1.0	0.37	1.0	0.37	$(4/L)(M_{Ps}+M_{Pm})$	$192EI/L^3$			$0.71R-0.21P$
	Plastic	1	1.0	0.33	1.0	0.33	$(4/L)(M_{Ps}+M_{Pm})$	0			$0.75R_m-0.25P$

From Reference 41. *Concentrated mass is lumped at the concentrated load.

TABLE 10-3
TRANSFORMATION FACTORS: BEAMS AND ONE-WAY SLABS

Simply-supported and fixed

Loading Diagram	Strain Range	Load Factor K_L	Mass Factor K_M Concentrated Mass*	Mass Factor K_M Uniform Mass	Load-Mass Factor K_{LM} Concentrated Mass*	Load-Mass Factor K_{LM} Uniform Mass	Maximum Resistance R_m	Spring Constant k	Effective Spring Constant k_E Elastic	Effective Spring Constant k_E Plastic	Dynamic Reaction V
$P = pL$ (uniform load)	Elastic	0.58		0.45		0.78	$8M_{Ps}/L$	$185EI/L^3$	$153EI/L^3$		$V_1=0.26R+0.12P$ $V_2=0.43R+0.19P$
	Elastoplastic	0.64		0.50		0.78	$(4/L)(M_{Ps}+2M_{Pm})$	$(384/5)EI/L^3$		$160EI/L^3$	$V_1=V_2=0.39R+0.11P$
	Plastic	0.50		0.33		0.66	$(4/L)(M_{Ps}+2M_{Pm})$	0	$R_{mf}=14.6M_P/L$		$V_1=V_2=0.38R_m+0.12P$
P at $L/2$	Elastic	1.0	1.0	0.43	1.0	0.43	$16M_{Ps}/3L$	$107EI/L^3$	$104EI/L^3$		$V_1=0.54R+0.14P$ $V_2=0.25R+0.07P$
	Elastoplastic	1	1.0	0.49	1.0	0.49	$(2/L)(M_{Ps}+2M_{Pm})$	$48EI/L^3$		$106EI/L^3$	$V_1=V_2=0.78R-0.28P$
	Plastic	1	1.0	0.33	1.0	0.33	$(2/L)(M_{Ps}+2M_{Pm})$	0	$R_{mf}=6.63M_P/L$		$V_1=V_2=0.75R_m-0.25P$
$\frac{P}{2}$, $\frac{P}{2}$ at third points	Elastic	0.81	0.67	0.45	0.83	0.55	$6M_{Ps}/L$	$132EI/L^3$	$117.5EI/L^3$		$V_1=0.17R+0.17P$ $V_2=0.33R+0.33P$
	Elastoplastic	0.87	0.76	0.52	0.87	0.60	$(2/L)(M_{Ps}+3M_{Pm})$	$56EI/L^3$		$122EI/L^3$	$V_1=V_2=0.62R-0.12P$
	Plastic	1	1.0	0.56	1.0	0.56	$(2/L)(M_{Ps}+3M_{Pm})$		$R_{mf}=9.52M_P/L$		$V_1=0.56R_m-0.25P$ $V_2=0.56R_m+0.13P$

From Reference 41. *Equal parts of the concentrated mass are lumped at each concentrated load.

TABLE 10-4
TRANSFORMATION FACTORS:
TWO-WAY SLABS—SIMPLE SUPPORTS, FOUR SIDES, UNIFORM LOAD

Simple support

a

b

Strain Range	a/b	Load Factor K_L	Mass Factor K_M	Load-Mass Factor K_{LM}	Maximum Resistance	Spring Constant k	Dynamic Reactions	
							V_A	V_B
Elastic	1.0	0.45	0.31	0.68	$(12/a)(M_{Pfa}+M_{Pfb})$	$271EI_a/a^2$	$0.07P+0.18R$	$0.07P+0.18R$
	0.9	0.47	0.33	0.70	$(1/a)(12M_{Pfa}+11M_{Pfb})$	$248EI_a/a^2$	$0.06P+0.16R$	$0.08P+0.20R$
	0.8	0.49	0.35	0.71	$(1/a)(12M_{Pfa}+10.3M_{Pfb})$	$228EI_a/a^2$	$0.06P+0.14R$	$0.08P+0.22R$
	0.7	0.51	0.37	0.73	$(1/a)(12M_{Pfa}+9.8M_{Pfb})$	$216EI_a/a^2$	$0.05P+0.13R$	$0.08P+0.24R$
	0.6	0.53	0.39	0.74	$(1/a)(12M_{Pfa}+9.3M_{Pfb})$	$212EI_a/a^2$	$0.04P+0.11R$	$0.09P+0.26R$
	0.5	0.55	0.41	0.75	$(1/a)(12M_{Pfa}+9.0M_{Pfb})$	$216EI_a/a^2$	$0.04P+0.09R$	$0.09P+0.28R$
Plastic	1.0	0.33	0.17	0.51	$(12/a)(M_{Pfa}+M_{Pfb})$	0	$0.09P+0.16R_m$	$0.09P+0.16R_m$
	0.9	0.35	0.18	0.51	$(1/a)(12M_{Pfa}+11M_{Pfb})$	0	$0.08P+0.15R_m$	$0.09P+0.18R_m$
	0.8	0.37	0.20	0.54	$(1/a)(12M_{Pfa}+10.3M_{Pfb})$	0	$0.07P+0.13R_m$	$0.01P+0.20R_m$
	0.7	0.38	0.22	0.58	$(1/a)(12M_{Pfa}+9.8M_{Pfb})$	0	$0.06P+0.12R_m$	$0.10P+0.22R_m$
	0.6	0.40	0.23	0.58	$(1/a)(12M_{Pfa}+9.3M_{Pfb})$	0	$0.05P+0.10R_m$	$0.10P+0.25R_m$
	0.5	0.42	0.25	0.59	$(1/a)(12M_{Pfa}+9.0M_{Pfb})$	0	$0.04P+0.08R_m$	$0.11P+0.27R_m$

From Reference 41.

TABLE 10-5
TRANSFORMATION FACTORS:
TWO-WAY SLABS—FIXED FOUR SIDES, UNIFORM LOAD

Fixed
a
b

Strain Range	a/b	Load Factor K_L	Mass Factor K_M	Load-Mass Factor K_{LM}	Maximum Resistance	Spring Constant k	Dynamic Reactions V_A	Dynamic Reactions V_B
1								
Elastic	1.0	0.33	0.21	0.63	$30.2M^0_{Psb}$	$870EI_a/a^2$	$0.10P+0.15R$	$0.10P+0.15R$
	0.9	0.34	0.23	0.68	$27.8M^0_{Psb}$	$798EI_a/a^2$	$0.09P+0.14R$	$0.10P+0.17R$
	0.8	0.36	0.25	0.69	$26.0M^0_{Psb}$	$757EI_a/a^2$	$0.08P+0.12R$	$0.11P+0.19R$
	0.7	0.38	0.27	0.71	$26.0M^0_{Psb}$	$744EI_a/a^2$	$0.07P+0.11R$	$0.11P+0.21R$
	0.6	0.41	0.29	0.71	$26.4M^0_{Psb}$	$778EI_a/a^2$	$0.06P+0.09R$	$0.12P+0.23R$
	0.5	0.43	0.31	0.72	$25.0M^0_{Psb}$	$866EI_a/a^2$	$0.05P+0.08R$	$0.12P+0.25R$
Elasto-plastic	1.0	0.46	0.31	0.67	$(1/a)[12(M_{Pfa}+M_{Psa})+12(M_{Pfb}+M_{Psb})]$	$271EI_a/a^2$	$0.07P+0.18R$	$0.07P+0.18R$
	0.9	0.47	0.33	0.70	$(1/a)[12(M_{Pfa}+M_{Psa})+11(M_{Pfb}+M_{Psb})]$	$248EI_a/a^2$	$0.06P+0.16R$	$0.08P+0.20R$
	0.8	0.49	0.35	0.71	$(1/a)[12(M_{Pfa}+M_{Psa})+10.3(M_{Pfb}+M_{Psb})]$	$228EI_a/a^2$	$0.06P+0.14R$	$0.08P+0.22R$
	0.7	0.51	0.37	0.73	$(1/a)[12(M_{Pfa}+M_{Psa})+9.8(M_{Pfb}+M_{Psb})]$	$216EI_a/a^2$	$0.05P+0.13R$	$0.08P+0.24R$
	0.6	0.53	0.39	0.74	$(1/a)[12(M_{Pfa}+M_{Psa})+9.3(M_{Pfb}+M_{Psb})]$	$212EI_a/a^2$	$0.04P+0.11R$	$0.09P+0.26R$
	0.5	0.55	0.41	0.75	$(1/a)[12(M_{Pfa}+M_{Psa})+9.0(M_{Pfb}+M_{Psb})]$	$216EI_a/a^2$	$0.04P+0.09R$	$0.09P+0.28R$
Plastic	1.0	0.33	0.17	0.51	$(1/a)[12(M_{Pfa}+M_{Psa})+12(M_{Pfb}+M_{Psb})]$	0	$0.09P+0.16R_m$	$0.09P+0.16R_m$
	0.9	0.35	0.18	0.51	$(1/a)[12(M_{Pfa}+M_{Psa})+11(M_{Pfb}+M_{Psb})]$	0	$0.08P+0.15R_m$	$0.09P+0.18R_m$
	0.8	0.37	0.20	0.54	$(1/a)[M_{Pfa}+M_{Psa})+10.3(M_{Pfb}+M_{Psb})]$	0	$0.07P+0.13R_m$	$0.10P+0.20R_m$
	0.7	0.38	0.22	0.58	$(1/a)[12(M_{Pfa}+M_{Psa})+9.8(M_{Pfb}+M_{Psb})]$	0	$0.06P+0.12R_m$	$0.10P+0.22R_m$
	0.6	0.40	0.23	0.58	$(1/a)[12(M_{Pfa}+M_{Psa})+9.3(M_{Pfb}+M_{Psb})]$	0	$0.05P+0.10R_m$	$0.10P+0.25R_m$
	0.5	0.42	0.25	0.59	$(1/a)[12(M_{Pfa}+M_{Psa})+9.0(M_{Pfb}+M_{Psb})]$	0	$0.04P+0.08R_m$	$0.11P+0.27R_m$

From Reference 41.

TABLE 10-6
TRANSFORMATION FACTORS:
TWO-WAY SLABS–SIMPLE SUPPORTS, TWO LONG SIDES; FIXED SUPPORTS,
TWO SHORT SIDES, UNIFORM LOAD

Simple support
Fixed
a
b

Strain Range	a/b	Load Factor K_L	Mass Factor K_M	Load-Mass Factor K_{LM}	Maximum Resistance	Spring Constant k	Dynamic Reactions V_A	Dynamic Reactions V_B
Elastic	1.0	0.39	0.26	0.67	$20.4M^0_{Psa}$	$575EI_a/a^2$	$0.09P+0.16R$	$0.07P+0.18R$
	0.9	0.41	0.28	0.68	$10.2M^0_{Psa}+(11/a)M_{Pfb}$	$476EI_a/a^2$	$0.08P+0.14R$	$0.08P+0.20R$
	0.8	0.44	0.30	0.68	$10.2M^0_{Psa}+(10.3/a)M_{Pfb}$	$396EI_a/a^2$	$0.08P+0.12R$	$0.08P+0.22R$
	0.7	0.46	0.33	0.72	$9.3M^0_{Psb}+(9.7/a)M_{Pfb}$	$328EI_a/a^2$	$0.07P+0.11R$	$0.08P+0.24R$
	0.6	0.48	0.35	0.73	$8.5M^0_{Psb}+(9.3/a)M_{Pfb}$	$283EI_a/a^2$	$0.06P+0.09R$	$0.09P+0.26R$
	0.5	0.51	0.37	0.73	$7.4M^0_{Psb}+(9.0/a)M_{Pfb}$	$243EI_a/a^2$	$0.05P+0.08R$	$0.09P+0.28R$
Elastoplastic	1.0	0.46	0.31	0.67	$(1/a)[12(M_{Pfa}+M_{Psa})+12(M_{Pfb})]$	$271EI_a/a^2$	$0.07P+0.18R$	$0.07P+0.18R$
	0.9	0.47	0.33	0.70	$(1/a)[12(M_{Pfa}+M_{Psa})+11(M_{Pfb})]$	$248EI_a/a^2$	$0.06P+0.16R$	$0.08P+0.20R$
	0.8	0.49	0.35	0.71	$(1/a)[12(M_{Pfa}+M_{Psa})+10.3(M_{Pfb})]$	$228EI_a/a^2$	$0.06P+0.14R$	$0.08P+0.22R$
	0.7	0.51	0.37	0.72	$(1/a)[12(M_{Pfa}+M_{Psa})+9.7(M_{Pfb})]$	$216EI_a/a^2$	$0.05P+0.13R$	$0.08P+0.24R$
	0.6	0.53	0.37	0.70	$(1/a)[12(M_{Pfa}+M_{Psa})+9.3(M_{Pfb})]$	$212EI_a/a^2$	$0.04P+0.11R$	$0.09P+0.26R$
	0.5	0.55	0.41	0.74	$(1/a)[12(M_{Pfa}+M_{Psa})+9.0(M_{Pfb})]$	$216EI_a/a^2$	$0.04P+0.09R$	$0.09R+0.28R$
Plastic	1.0	0.33	0.17	0.51	$(1/a)[12(M_{Pfa}+M_{Psa})+12M_{Pfb}]$	0	$0.09P+0.16R_m$	$0.09P+0.16R_m$
	0.9	0.35	0.18	0.51	$(1/a)[12(M_{Pfa}+M_{Psa})+11M_{Pfb}]$	0	$0.08P+0.15R_m$	$0.09P+0.18R_m$
	0.8	0.37	0.20	0.54	$(1/a)[12(M_{Pfa}+M_{Psa})+10.3M_{Pfb}]$	0	$0.07P+0.13R_m$	$0.10P+0.20R_m$
	0.7	0.38	0.22	0.58	$(1/a)[12(M_{Pfa}+M_{Psa})+9.7M_{Pfb}]$	0	$0.06P+0.12R_m$	$0.10P+0.22R_m$
	0.6	0.40	0.23	0.58	$(1/a)[12(M_{Pfa}+M_{Psa})+9.3M_{Pfb}]$	0	$0.05P+0.10R_m$	$0.10P+0.25R_m$
	0.5	0.42	0.25	0.59	$(1/a)[12(M_{Pfa}+M_{Psa})+9.0M_{Pfb}]$	0	$0.04P+0.08R_m$	$0.11P+0.27R_m$

From Reference 41.

TABLE 10-7
TRANSFORMATION FACTORS: TWO-WAY SLABS—SIMPLE SUPPORTS, TWO SHORT SIDES; FIXED SUPPORTS, TWO LONG SIDES, UNIFORM LOAD

Strain Range	a/b	Load Factor K_L	Mass Factor K_M	Load-Mass Factor K_{LM}	Maximum resistance	Spring Constant k	Dynamic Reactions V_A	Dynamic Reactions V_B
Elastic	1.0	0.39	0.26	0.67	$20.4M^0_{Psb}$	$575EI_a/a^2$	$0.07P+0.18R$	$0.09P+0.16R$
	0.9	0.40	0.28	0.70	$19.5M^0_{Psb}$	$600EI_a/a^2$	$0.06P+0.16R$	$0.10P+0.18R$
	0.8	0.42	0.29	0.69	$19.5M^0_{Psb}$	$610EI_a/a^2$	$0.06P+0.14R$	$0.11P+0.19R$
	0.7	0.43	0.31	0.71	$20.2M^0_{Psb}$	$662EI_a/a^2$	$0.05P+0.13R$	$0.11P+0.21R$
	0.6	0.45	0.33	0.73	$21.2M^0_{Psb}$	$731EI_a/a^2$	$0.04P+0.11R$	$0.12P+0.23R$
	0.5	0.47	0.34	0.72	$22.2M^0_{Psb}$	$850EI_a/a^2$	$0.04P+0.09R$	$0.12P+0.25R$
Elasto-plastic	1.0	0.46	0.31	0.67	$(1/a)[12M_{Pfa}+12(M_{Psb}+M_{Pfb})]$	$271EI_a/a^2$	$0.07P+0.18R$	$0.07P+0.18R$
	0.9	0.47	0.33	0.70	$(1/a)[12M_{Pfa}+11(M_{Psb}+M_{Pfb})]$	$248EI_a/a^2$	$0.06P+0.16R$	$0.08P+0.20R$
	0.8	0.49	0.35	0.71	$(1/a)[12M_{Pfa}+10.3(M_{Psb}+M_{Pfb})]$	$228EI_a/a^2$	$0.06P+0.14R$	$0.08P+0.22R$
	0.7	0.51	0.37	0.73	$(1/a)[12M_{Pfa}+9.8(M_{Psb}+M_{Pfb})]$	$216EI_a/a^2$	$0.05P+0.13R$	$0.08P+0.24R$
	0.6	0.53	0.39	0.74	$(1/a)[12M_{Pfa}+9.3(M_{Psb}+M_{Pfb})]$	$212EI_a/a^2$	$0.04P+0.11R$	$0.09P+0.26R$
	0.5	0.55	0.41	0.74	$(1/a)[12M_{Pfa}+9.0(M_{Psb}+M_{Pfb})]$	$216EI_a/a^2$	$0.04P+0.09R$	$0.09P+0.26R$
Plastic	1.0	0.33	0.17	0.51	$(1/a)[12M_{Pfa}+12(M_{Psb}+M_{Pfb})]$	0	$0.09P+0.16R_m$	$0.09P+0.16R_m$
	0.9	0.35	0.18	0.51	$(1/a)[12M_{Pfa}+11(M_{Psb}+M_{Pfb})]$	0	$0.08P+0.15R_m$	$0.09P+0.18R_m$
	0.8	0.37	0.20	0.54	$(1/a)[12M_{Pfa}+10.3(M_{Psb}+M_{Pfb})]$	0	$0.07P+0.13R_m$	$0.10P+0.20R_m$
	0.7	0.38	0.22	0.58	$(1/a)[12M_{Pfa}+9.8(M_{Psb}+M_{Pfb})]$	0	$0.06P+0.12R_m$	$0.10P+0.22R_m$
	0.6	0.40	0.23	0.58	$(1/a)[12M_{Pfa}+9.3(M_{Psb}+M_{Pfb})]$	0	$0.05P+0.10R_m$	$0.10P+0.25R_m$
	0.5	0.42	0.25	0.59	$(1/a)[12M_{Pfa}+9.0(M_{Psb}+M_{Pfb})]$	0	$0.04P+0.08R_m$	$0.11P+0.27R_m$

From Reference 41.

INTERIOR, UNIFORM LOAD

Strain Phase	d/a	Load Factor K_L	Mass Factor K_M	Load-Mass Factor K_{LM}	Spring Constant k (kips/ft)	Maximum Resistance R_m (kips)	Dynamic Column Load V_c (kips)
Elastic	0.05	8/15	0.34	0.64	$1.45EI_a/a^2$	$4.2\sum M_P$	$0.16P + 0.84R + pd^2$
	0.10	8/15	0.34	0.64	$1.60EI_a/a^2$	$4.4\sum M_P$	
	0,15	8/15	0.34	0.64	$1.75EI_a/a^2$	$4.6\sum M_P$	
	0.20	8/15	0.34	0.64	$1.92EI_a/a^2$	$4.8\sum M_P$	
	0.25	8/15	0.34	0.64	$2.10EI_a/a^2$	$5.0\sum M_P$	
Plastic	0.05	1/2	7/24	7/12	0	$4.2\sum M_P$	$0.14P + 0.86R_m + pd^2$
	0.10	1/2	7/24	7/12	0	$4.4\sum M_P$	
	0.15	1/2	7/24	7/12	0	$4.6\sum M_P$	
	0.20	1/2	7/24	7/12	0	$4.8\sum M_P$	
	0.25	1/2	7/24	7/12	0	$5.0\sum M_P$	

From Reference 41.

Notes. a = column spacing, ft.

E = compressive modulus of elasticity of concrete, ksi.

I_a = average of gross and transformed moments of inertia per unit width, equal in both directions, in.^{4}/ft.

P = total load on one slab panel, excluding capitals.

R = total resistance of one slab panel, excluding capitals.

$\sum M_P = M_{Pmp} + M_{Pmn} + M_{Pcp} + M_{Pcn}$.

pd^2 = load on capital

These resistances are those that occur at the upper limit of each range. The effective spring constant k_E in the plastic range is the one defined in Fig. 10-2*b*. If the design is limited to the elastoplastic range, an elastic k_E is computed, which is based on a fictitious maximum resistance R_{mf} whose values are given in these tables. They are computed so that the areas under the resistance-deflection diagrams up to the elastoplastic value y_{ep} of the deflection are equal. Thus

$$k_E \text{ (elastic)} = \frac{R_{mf}}{y_c} \qquad (10\text{-}25)$$

where y_c is the same as the deflection y_2 in Fig. 10-2*b*. It should be emphasized, however, that Eq. 10-25 can be used only for cases where the static stresses of the system are negligible.[29]

The transformation factors for two-way slabs with uniform loading and indicated boundary conditions are given in Tables 10-4 through 10-7. Various ratios of the lengths and sides are included, and they are derived in a manner similar to the one used for beams. In these tables, M_{Pfa} is the total positive plastic bending moment capacity along the midspan section which is parallel to the short edge; M_{Pfb} is the total positive plastic bending moment capacity along the midspan section which is parallel to the long edge; M_{Psa} is the total negative plastic bending moment capacity along the short edge; M_{Psb} is the total negative plastic moment capacity along the long edge; M^0_{Psa} and M^0_{Psb} are the plastic negative bending moment capacities per unit width at center of short and long edges, respectively; and I_a in the expression for the spring constants is the moment of inertia per unit width.

It should be noted, however, that the maximum resistance and spring constant are for the total load on the slab. In addition, the derivation of these factors in the elastic range is based on the classic theory of plates. The yield-line theory is used in the plastic range. In the elastoplastic range, for slabs fixed on any number of sides, the factors for slabs simply supported on four sides are assumed to apply.

In these tables the values of the dynamic reactions are also included, because they are needed to determine the design shear forces and stresses as well as the dynamic loads acting on the supporting elements. For slabs, the symbol V_A is used to denote the total dynamic reaction along the short edge and V_B to denote the total dynamic reaction along the long edge.

In Table 10-8 the transformation factors for square interior flat slabs with uniform loads are shown. These factors are given only for the elastic and plastic ranges. The elastoplastic range has been neglected. In this table, M_{Pmp} and M_{Pmn} are the positive and negative plastic bending moment

[29] Additional information regarding the use of an elastic k_E can be found in Reference 41, Manual Number EM1110-345-416, page 8.

capacities, respectively, per unit width in middle strip of width $a/2$, and M_{Pcp} and M_{Pcn} are the positive and negative plastic bending moment capacities, respectively, per unit width in column strip of width $a/2$. Transformation factors for tee-beams, deep beams, and trusses can be found in Reference 41, Manual Number M1110-345-416. For other cases, a similar procedure can be used to derive the required transformation factors.

For *frames*, the transformation factors can be easily obtained. The equivalent one-degree system in Fig. 10-1*c*, for example can be obtained by using a load factor of unity for the concentrated load $F(t)$ at the roof, and the load factor of one-half for the distributed load along the side wall. Thus

$$P_e(t) = F(t) + q(t)\frac{L}{2} \tag{10-26}$$

The equivalent mass m_e can be taken to be equal to the total mass at the roof level plus one-third[30] of the mass of the walls.

The spring constant can be determined as discussed in Chapter 3, and the procedures given there can be used to compute the dynamic response. Similar procedure can be followed for multistory frames. If the girders of the frame are also subjected to dynamic loads, they could be analyzed as separate beams, provided that reasonable assumptions are made regarding their end boundary conditions. The design of the columns, however, should include the dynamic reactions of the girders. This problem is discussed more extensively later in this chapter.

10-5 CHARTS FOR ELASTOPLASTIC RESPONSE

In practice, the criteria used in the design of structures for dynamic response depend usually on the character of the dynamic load and the amount of damage a structure is permitted to undertake. The criterion often used is the ductility ratio η, which is defined as the ratio of the maximum deflection y_m to the elastic deflection y_{el}. That is,

$$\eta = \frac{y_m}{y_{el}} \tag{10-27}$$

If a structure is designed to remain elastic when the loads are applied, then η is equal to unity. If the dynamic loads are expected to occur only a few times

[30] In Chapter 3, the value of one-half of the mass of the wall was arbitrarily used. It should be pointed out, however, that none of these values is exact. The value of $\frac{1}{3}$ was obtained by assuming that the columns of the frame remain straight during deformation which is a rather crude approximation. It is believed, however, that the value of $\frac{1}{3}$ is somewhat more accurate than $\frac{1}{2}$.

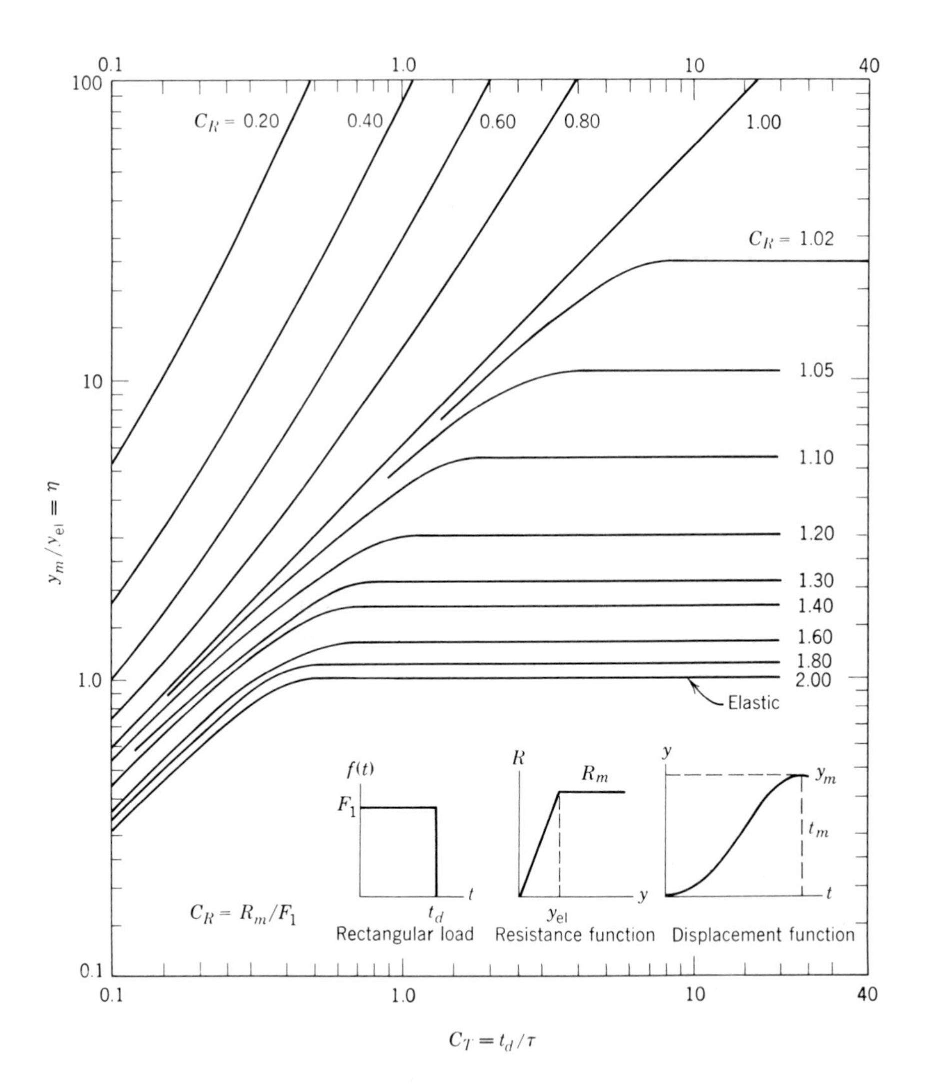

Fig. 10-3. The y_m/y_{el} curves for elastoplastic system, triangular load. (Reference 41.)

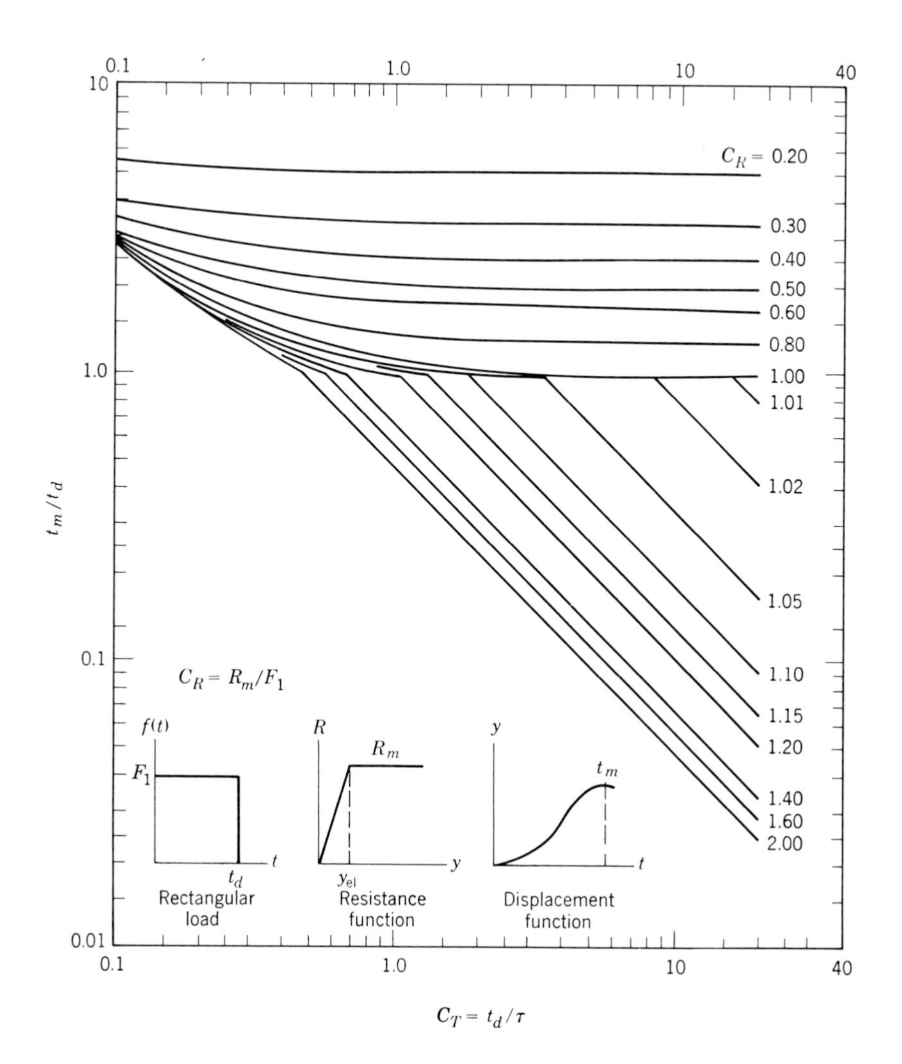

Fig. 10-4. The t_m/t_d curves for elastoplastic system, rectangular load. (Reference 41.)

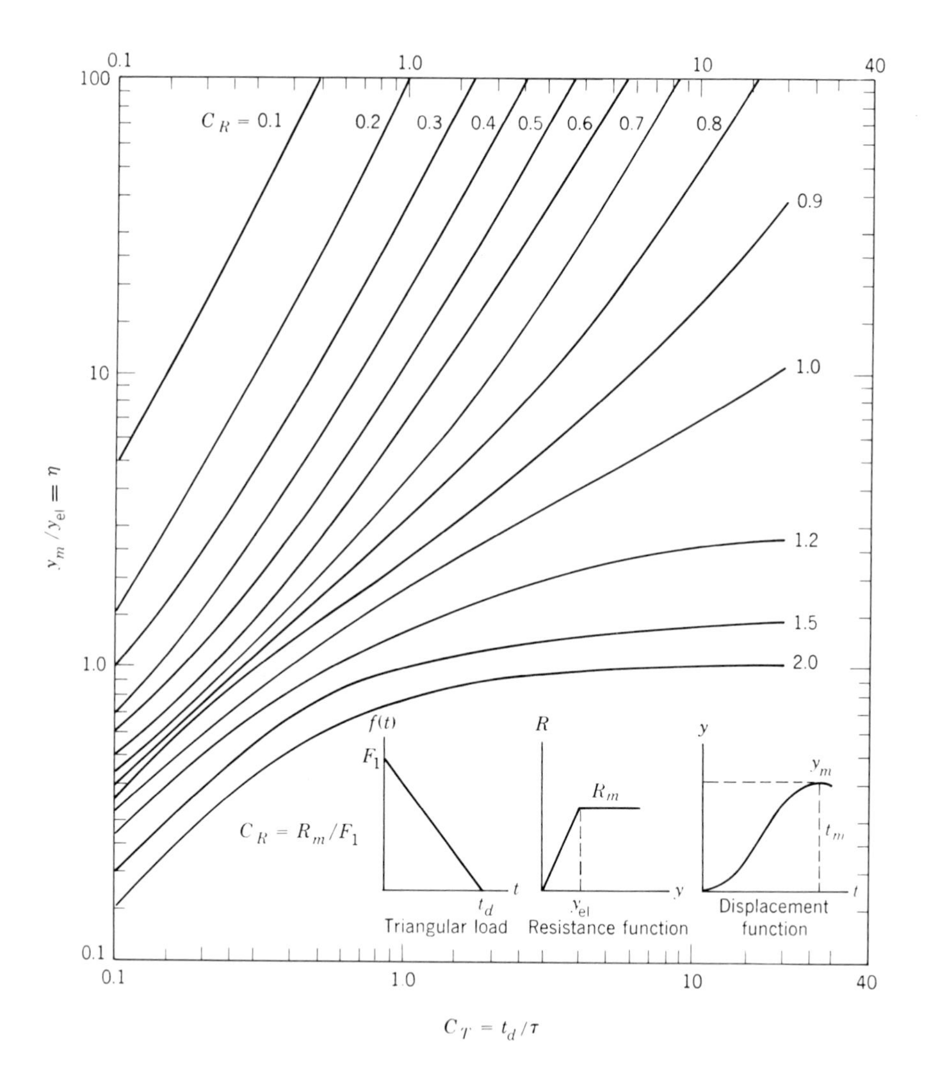

Fig. 10-5. The y_m/y_{el} curves for elastoplastic system, triangular load. (Reference 41.)

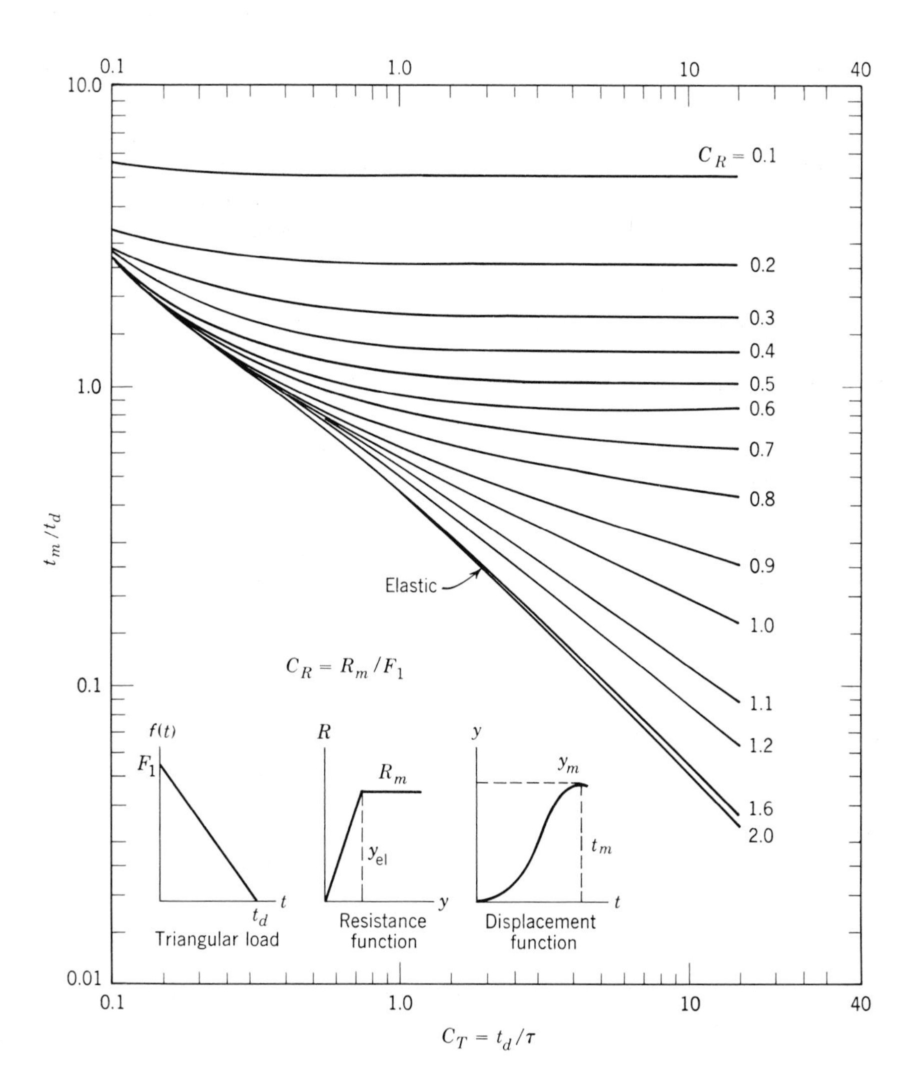

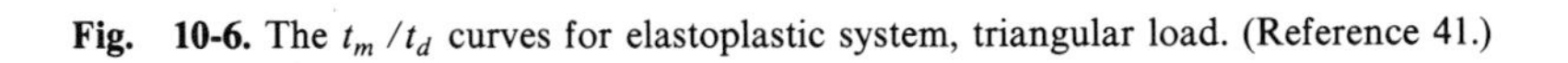
Fig. 10-6. The t_m/t_d curves for elastoplastic system, triangular load. (Reference 41.)

during the life span of a structure and a certain amount of damage can be tolerated, values of η equal to 2, 3, 4, or 5 can be used. Higher values of η are sometimes used when more severe damage is permitted to the structure.

It was pointed out earlier that when the equivalent one-degree system is obtained, the methods in Chapter 2 can be utilized for dynamic response computations. The analysis of the elastoplastic systems in Section 2-8 becomes more convenient if the graphs in Figs. 10-3 through 10-6 are applied. The results in these graphs can be used for one-degree undamped systems and they are based on a bilinear resistance function. In addition, the initial motion of the system is assumed to be zero. These charts provide maximum response only, and they apply in cases where the design requirements are based on maximum displacement. Although they are limited to two types of load-time variations, many practical situations can be approximated by a rectangular or a triangular time variation.

Figures 10-3 and 10-4 are for rectangular loads and Figs. 10-5 and 10-6 for triangular variations. The explanation of the symbols is given in the figures, except τ, which denotes the undamped natural period of vibration of the one-degree system. It should be noted, however, that the response for $R_m/F_1 = 2$ is completely elastic. Thus for values of R_m/F_1 larger than 2 the response curves in Figs. 2-5 through 2-7 should be used.

10-6 DYNAMIC RESPONSE OF BEAMS

The transformation factors and the approximate methods for dynamic analysis discussed in the preceding sections of this chapter are now used to determine the dynamic response of beams. As a first application, let it be assumed that it is required to determine the maximum elastic deflection and stress of the simply supported beam in Fig. 10-7*a*. This beam supports a total dead weight of 1 kip/ft and is acted upon by the dynamic load $q(t)$ whose time variation is shown in Fig. 10-7*b*. The modulus of elasticity E of the beam is 30×10^6 psi and the moment of inertia I is 1153.9 in.^{4}

The total mass m_t, which includes the mass of the steel beam, is

$$m_t = \frac{(1{,}000)(20)}{386} \doteq 51.8 \text{ lb-sec}^2/\text{in.}$$

The total peak dynamic load P_t is (5000)(20) = 100,000 lb. From Table 10-1, for elastic response, we have,

$$k = \frac{384EI}{5L^3} = \frac{(384)(30)(10^6)(1153.9)}{(5)(20^3)(12^3)} = 76.75 \times 10^6 \text{ lb/in.}$$

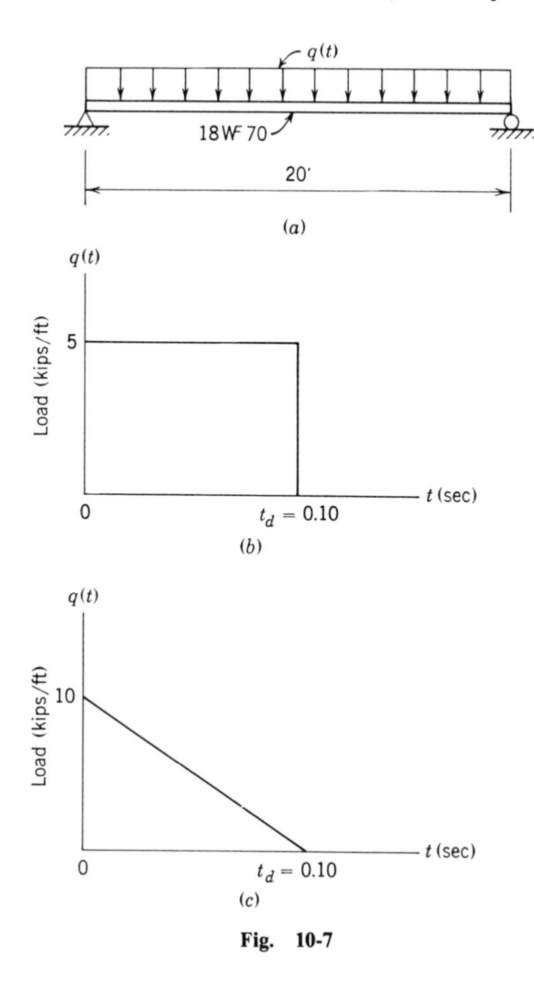

Fig. 10-7

From the same table

$$K_L = 0.64 \qquad K_M = 0.50$$

Thus, for the equivalent system,

$$m_e = m_t K_M = (51.8)(0.50) = 25.9 \text{ lb-sec}^2/\text{in.}$$

$$P_e = P_t K_L = (100{,}000)(0.64) = 64{,}000 \text{ lb}$$

$$k_e = kK_R = kK_L = (19.25)(10^4)(0.64) = 12.35 \times 10^4 \text{ lb/in.}$$

The natural period of vibration τ is

$$\tau = 2\pi\sqrt{m_e/k_e} = 2\pi\sqrt{25.9/(12.35)(10^4)} = 0.091 \text{ sec}$$

From Fig. 2-6 with

$$\frac{t_d}{\tau} = \frac{0.10}{0.091} = 1.10$$

the values of Γ_{max} and t_m are

$$\Gamma_{max} = 2.0 \qquad t_m = 0.0455 \text{ sec}$$

Therefore, the resistance R_e of the equivalent system is

$$R_e = (\Gamma_{max}) P_e = (2.0)(64{,}000) = 128{,}000 \text{ lb}$$

and the maximum displacement y_m is

$$y_m = \frac{R_e}{k_e} = \frac{128{,}000}{(12.35)(10^4)} = 1.030 \text{ in.}$$

For the real beam, the resistance R_m is

$$R_m = \frac{R_e}{K_R} = \frac{128{,}000}{0.64} = 200{,}000 \text{ lb}$$

The maximum bending moment M of the beam is

$$M = \frac{RL}{8} = \frac{(200{,}000)(20)}{8} = 500{,}000 \text{ lb-ft}$$

and the maximum bending stress σ_{max} is

$$\sigma_{max} = \frac{M}{S} = \frac{(500{,}000)(12)}{128.2} = 46{,}700 \text{ psi}$$

To this stress, the bending stress due to the dead weight should be added. The resulting total stress is well above the dynamic yield[31] stress of 41,600 psi. Thus, in order to stay within the elastic region, the procedure should be repeated by using a larger beam section.

The shear in the beam is equal to the dynamic reaction V and can be found from Table 10-1. From this table, for the elastic range,

$$V = 0.39R + 0.11P = (0.39)(200{,}000) + (0.11)(100{,}000) = 89{,}000 \text{ lb}$$

It should be pointed out, however, that the shear varies with time. The maximum value here is assumed to occur at t_m where $R_m = 200{,}000$ lb and

[31] Dynamic properties of materials can be found in Reference 41.

$P = 100{,}000$ lb. To this shear, the shear due to the dead weight which is 10,000 lb should be added. A reasonable value of the shear stress can be found by dividing the total shear force by the cross-sectional area of the web. Thus

$$\text{maximum shear stress} = \frac{89{,}000 + 10{,}000}{(18)(0.438)} = 12{,}550 \text{ psi}$$

As an example of plastic analysis, let it be assumed that the time variation of the dynamic load is as shown in Fig. 10-7*c*. The dynamic yield stress σ_{dy} is 41,600 psi and the maximum bending resistance M_P of the cross section can be determined from the expression[32]

$$M_P = 1.05\sigma_{\text{dy}} S \tag{10-28}$$

where S is the section modulus of the steel beam. Thus

$$M_P = (1.05)(41{,}600)(128.2) = 5.62 \times 10^6 \text{ lb-in.}$$

From the plastic range in Table 10-1, the maximum resistance of the beam is

$$R_m = \frac{8M_P}{L} = \frac{(8)(5.62)(10^6)}{(20)(12)} = 187{,}000 \text{ lb}$$

The limiting elastic deflection y_{el} is

$$y_{\text{el}} = \frac{R_m}{k} = \frac{187{,}000}{(19.25)(10^4)} = 0.973 \text{in.}$$

By using an average between the elastic and plastic values, we have transformation factors

$$K_L = 0.57 \qquad K_M = 0.415$$

Thus, for the equivalent system, the peak value P_e of the dynamic load is

$$P_e = (0.57)(10{,}000)(20) = 114{,}000 \text{ lb}$$

In addition,

$$m_e = (0.415)(51.8) = 21.5 \text{lb-sec}^2/\text{in.}$$

$$\tau = 2\pi\sqrt{21.5/(11.0)(10^4)} = 0.088 \text{ sec}$$

$$R_e = (0.57)(187{,}000) = 106{,}500 \text{ lb}$$

[32] This expression can be found in References 41 and 45.

From Fig. 10-5,

$$\frac{t_d}{\tau}=\frac{0.10}{0.088}=1.14$$

$$C_R=\frac{R_e}{P_e}=\frac{106{,}500}{114{,}000}=0.935$$

Thus

$$\eta=\frac{y_m}{y_{\text{el}}}=2.4$$

or

$$y_m=\eta y_{\text{el}}=(2.4)(0.973)=2.33\text{ in.}$$

From Fig. 10-6, the time t_m of maximum response is

$$t_m=0.57t_d=0.057\text{ sec}$$

At time t_m, the maximum total load P_t on the beam is

$$P_t=\frac{(0.10-0.057)}{0.10}(10{,}000)(20)=86{,}000\text{ lb}$$

The maximum dynamic reaction, from the plastic range in Table 10-1, can be determined from the expression

$$V_{\max}=0.38R_m+0.12P_t=(0.38)(187{,}000)+(0.12)(86{,}000)$$

or

$$V_{\max}=81{,}300\text{ lb}$$

Therefore, the maximum shear stress is

$$\text{maximum shear stress}=\frac{81{,}300}{(18)(0.438)}=10{,}330\text{ psi}$$

The shear stress due to the dead weight is not included in this value.
Other cases of beam problems can be solved in a similar manner.

10-7 DYNAMIC RESPONSE OF CONCRETE SLABS

Let it be assumed that it is required to design a simply supported one-way reinforced concrete slab which is acted upon by a uniform dynamic load

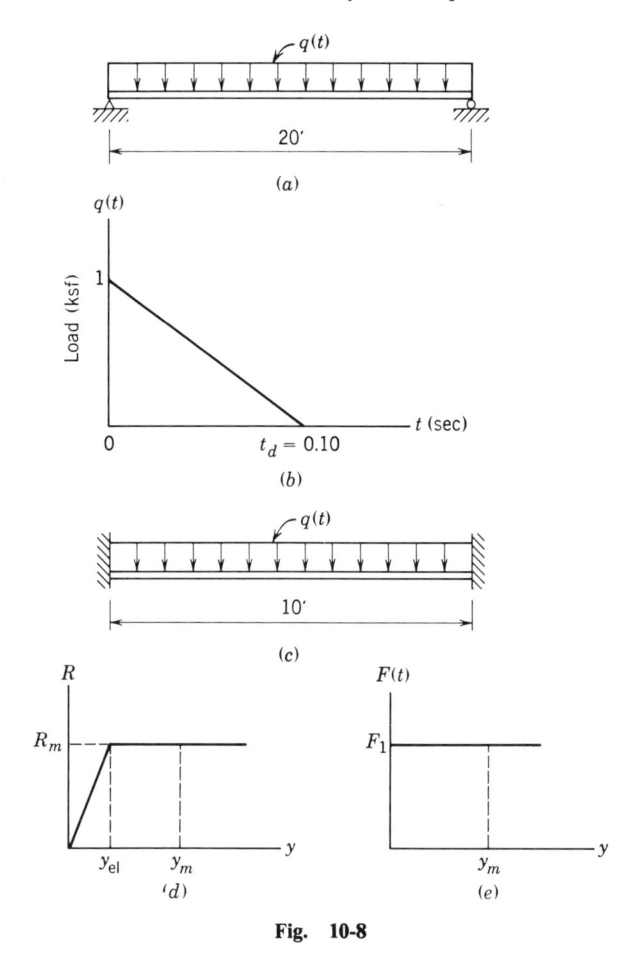

Fig. 10-8

$q(t)$ whose time variation is shown in Fig. 10-8*b*. The slab is designed for a ductility ratio $\eta=3$. The stresses due to the dead weight of the slab are neglected. The dynamic yield strength σ_{dy} of the steel is 50,000 psi, the dynamic compressive strength of concrete σ_{dc}' is 4000 psi, and the steel ratio p is taken to be equal to 0.015.

For preliminary design, let it be assumed that

$$C_R = \frac{R_m}{F_1} = 0.6$$

Thus the required R_m is

$$R_m = (0.6)(1.0)(20) = 12.0 \text{ kips}$$

and, from Table 10-1, the required M_P is

$$M_P = R_m \frac{L}{8} = (12) \frac{(20)(12)}{8} = 360.0 \text{ kip-in.}$$

By using the given values for the dynamic stresses and steel ratio p, the ultimate bending strength M_P can be determined from the known expression

$$M_P = p\sigma_{\text{dy}} bd^2 \left(1 - \frac{p\sigma_{\text{dy}}}{1.7\sigma'_{\text{dc}}}\right) \tag{10-29}$$

or

$$M_P = (0.015)(50)(12)d^2 \left[1 - \frac{(0.015)(50)}{(1.7)(4)}\right] = 8.08d^2 \text{ kip-in.}$$

By equating the required and provided M_P, we have

$$8.08d^2 = 360$$

or

$$d = 6.67 \text{ in.}$$

A trial slab of depth $h = 8.5$ in. and $d = 7$ in. is here analyzed for maximum deflection.

For the actual slab,

$$m_t = (0.15)\left(\frac{8.5}{12}\right)(20)\left(\frac{1}{386}\right) = 0.00552 \text{ kip-sec}^2/\text{in.}$$

The moment of inertia is considered to be the average of the cracked and uncracked transform sections, and is given by the approximate expression

$$I_a = \frac{bd^3}{2}(0.083 + 5.5p) \tag{10-30}$$

Thus

$$I_a = \frac{(12)(7)^3}{2}[0.083 + (5.5)(0.015)] = 338.0 \text{ in.}^4$$

Also

$$k = \frac{384EI}{5L^3} = \frac{(384)(3)(10^3)(338)}{(5)(20^3)(1728)} = 5.63 \text{ kips/in.}$$

$$M_P = 8.08d^2 = (8.08)(7)^2 = 396 \text{ kip-in.}$$

$$R_m = \frac{8M_P}{L} = \frac{(8)(396)}{(20)(12)} = 13.20 \text{ kips}$$

For the equivalent system,

$$m_e = (0.33)(0.00552) = 0.00182 \text{ kip-sec}^2/\text{in.}$$

$$k_e = (0.50)(5.63) = 2.82 \text{ kips/in.}$$

$$P_e = (0.50)(20) = 10.00 \text{ kips}$$

$$\tau = 2\pi\sqrt{0.00182/2.82} = 0.158 \text{ sec}$$

$$R_e = (0.50)(13.20) = 6.60 \text{ kips}$$

Thus

$$\frac{t_d}{\tau} = \frac{0.10}{0.158} = 0.633$$

$$C_R = \frac{R_e}{P_e} = \frac{6.60}{10} = 0.66$$

and Fig. 10-5 yields,

$$\eta = \frac{y_m}{y_{\text{el}}} = 3.3$$

This is somewhat higher than $\eta = 3$. Therefore, a slightly thicker slab should be used. If desired, the procedure can be repeated.

The time t_m of maximum response can be determined from Fig. 10-6:

$$\frac{t_m}{t_d} = 0.95$$

and

$$t_m = (0.95)(0.10) = 0.095 \text{ sec}$$

From Table 10-1,

$$V = 0.38R_m + 0.12P_t = (0.38)(13.20) + (0.12)(20) = 7.42 \text{ kips}$$

In reality, the time at which the maximum reaction occurs is not known because the time of maximum resistance R_m is not known. This can only be determined by a numerical analysis or a rigorous solution. In computing the above value of V, it was assumed that the maximum value of P_t was not appreciably reduced when R_m was reached. This is a conservative estimate of V. The maximum shear stress is determined from the formula

$$\text{maximum shear stress} = \frac{V}{bjd} = \frac{7420}{(12)\left(\frac{7}{8}\right)(7)} = 101 \text{ psi}$$

As another example, let it be assumed that the one-way concrete slab in Fig. 10-8c is fixed at the ends, and that it is required to design it so that the behavior is within the elastoplastic range. In other words, the plastic moment at midspan can be reached, but it is not to be exceeded when the plastic hinges at the supports are formed. The resistance function in this case is trilinear, Fig. 10-2b, but an elastic effective stiffness k_E will be used. The time variation of the load is as shown in Fig. 10-8b.

The transformation factors K_L and K_M are taken as the average value of the elastic and elastoplastic ranges. Thus

$$K_L = 0.585 \qquad K_M = 0.455$$

Inspection of the response chart in Fig. 2-5 shows that the maximum magnification factor $\Gamma_{\max}$ would not be larger than 1.2, because t_d/τ would be probably about 1/2. So let it be assumed that $\Gamma_{\max} = 1.2$. On this basis, the required R_M is

$$R_m = (1.2)(1.0)(10) = 12.0 \text{ kips}$$

From Table 10-2, by assuming that $M_{Ps} = M_{Pm} = M_P$, the required plastic moment M_P is

$$M_P = R_m \frac{L}{16} = (12)\frac{(10)(12)}{16} = 90.20 \text{ kip-in.}$$

From Eq. 10-29, with p, σ_{dy}, and σ'_{dc} as given in this problem,

$$M_P = 8.08d^2$$

By equating the two values of M_P, we have

$$8.08d^2 = 90.20$$

$$d = 3.39 \text{ in.}$$

Use a trial slab thickness $h=4.75$ in. with $d=3.50$ in. Thus, for the actual slab,

$$M_P = (8.08)(3.50)^2 = 99.0 \text{ kip-in.}$$

$$R_m = 16\frac{M_P}{L} = \frac{(16)(99)}{(10)(12)} = 13.15 \text{ kips}$$

$$I_a = \frac{(12)(3.5)^3}{2}[0.083 + (5.5)(0.015)] = 42.5 \text{ in.}^4$$

Also, from Table 10-2, the effective stiffness k_E is

$$k_E = \frac{264EI}{L^3} = \frac{(264)(3)(10^3)(42.5)}{(10)^3(1728)} = 19.50 \text{ kips/in.}$$

The total mass m_t is

$$m_t = \frac{(4.75)(0.15)(10)}{(12)(386)} = 0.00154 \text{ kip-sec}^2\text{/in.}$$

For the equivalent system,

$$m_e = (0.455)(1.54)(10^{-3}) = 0.00070 \text{ kip-sec}^2\text{/in.}$$

$$k_e = (0.585)(19.50) = 11.40 \text{ kips/in.}$$

$$P_e = (1.0)(10)(0.585) = 5.85 \text{ kips}$$

$$\tau = 2\pi\sqrt{0.00070/11.40} = 0.049$$

Thus

$$\frac{t_d}{\tau} = \frac{0.10}{0.049} = 2.04$$

Figure 2-5*a* shows that Γ_{max} should be about 1.72, indicating that the design is not quite adequate. The procedure can be repeated until the assumed Γ_{max} and the one found have approximately the same value.

In the design of structures to resist the effects of dynamic loads it is important to understand first the various parameters involved. For example, if the duration of the load compared to the natural period of the system is short, the applied dynamic load can be considered to be an impulse, representing an amount of energy to be absorbed by the structure. This is easily observed by inspecting Figs. 10-3 and 10-5. In such cases, when the natural period is five or more times the duration of the dynamic load, the shape of the load impulse is of no importance.

Another important conclusion can be drawn by considering the strain energy U for plastic design. From Fig. 10-8d, this energy is given by the expression

$$U = R_m\left(y_m - \frac{y_{el}}{2}\right) \tag{10-31}$$

If it is assumed that the load-deflection variation is as shown in Fig. 10-8e, then the work W_e is

$$W_e = F_1 y_m \tag{10-32}$$

Thus, by equating the expressions in Eqs. 10-31 and 10-32 and keeping in mind that $\eta = y_m / y_{el}$,

$$R_m = F_1\left[\frac{1}{1 - \dfrac{1}{2\eta}}\right] \tag{10-33}$$

and

$$\eta = 1/2\left(1 - \frac{F_1}{R_m}\right) \tag{10-34}$$

The equations above can be used for the analysis and design of structural elements only when the analysis extends beyond the elastic range. From Eq. 10-33, one may easily note that if η is large, then R_m tends to be equal to F_1. When η is unity, then R_m is twice the value of F_1.

PROBLEMS

10-1 Derive the transformation factors K_L, K_M, and K_{LM} for a uniform beam that is fixed at both ends and acted upon by a uniformly distributed load. Show all computations and compare the results with the ones given in Table 10-2.

10-2 By using the transformation factors derived in Problem 10-1, determine the natural frequency of vibration of the equivalent one-degree system, and compare it with the exact value of the fundamental mode of the uniform fixed-ended beam.

10-3 Derive the transformation factors K_L, K_M, and K_{LM} for a uniform cantilever beam acted upon by a uniformly distributed load. Use the elastic static deflection curve as the assumed shape.

10-4 Determine the mass, load, and stiffness of the equivalent one-degree system for a uniform cantilever beam acted upon by a uniformly distributed dynamic load $q(t)$.

10-5 Determine k_e, m_e, and $P_e(t)$ for the equivalent one-degree systems in Figs. 10-1*b* and 10-1*c*.

10-6 The simply supported beam in Fig. 10-7*a* supports a total dead weight of 2 kips/ft and is acted upon by a uniformly distributed dynamic load $q(t)$ of 3.0 kips/ft which is applied suddenly at $t=0$ and of time duration t_d equal to 0.02 sec. Determine the maximum elastic deflection and the maximum bending stress. The modulus of elasticity E is 30×10^6 psi and the moment of inertia I is 1,153.9 in.^{4} Assume that the dynamic yield stress is equal to 41,600 psi.

10-7 Repeat Problem 10-6 when $q(t)$ is applied suddenly at $t=0$ and decreases linearly to zero at $t=0.02$ sec.

10-8 The simply supported beam in Fig. 10-7*a* supports a total dead weight of 1200 lb/ft and is acted upon by a uniformly distributed dynamic load $q(t)$ of 12 kips/ft which is applied suddenly at $t=0$ and decreases linearly to zero at $t=0.015$ sec. The dynamic yield stress σ_{dy} is 50,000 psi, the modulus of elasticity E is 30×10^6 psi, and the moment of inertia I is 1153.9 in.^{4} Determine the maximum deflection y_m at the center of the beam and the value of the ductility ratio η.

10-9 Design the simply supported beam in Problem 10-8 for a ductility ratio η not to exceed the value 2. The length of the beam and the loading are the same as in Problem 10-8.

10-10 The dynamic load $q(t)$ acting on the simply supported one-way reinforced concrete slab in Fig. 10-8*a* is a force of 2.0 ksf applied suddenly at $t=0$ and decreasing linearly to zero at $t=0.08$ sec. The dynamic yield strength σ_{dy} of the steel is 50,000 psi, the dynamic compressive strength σ'_{dc} of concrete is 4000 psi, and the steel ratio p is taken to be equal to 0.015. Design the slab for a ductility ratio $\eta=3$.

10-11 Repeat Problem 10-10 when $q(t)$ is a constant force of 2.0 ksf applied suddenly at $t=0$ and of duration $t_d=0.08$ sec.

10-12 Repeat Problem 10-10 for a slab length of 30 ft.

10-13 Repeat Problem 10-8 when the dynamic load is a constant force of 80.0 kips applied suddenly at midspan at $t=0$ and of duration $t_d=0.015$ sec.

10-14 Repeat Problem 10-10 when the variation of the dynamic force $q(t)$ is represented by the solid line in Fig. 11-24. Note that the maximum value of $q(t)$ is 4.90 psi of plate.

10-15 The dynamic load $q(t)$ acting on the one-way fixed-end reinforced concrete slab in Fig. 10-8*c* is a force of 1.5 ksf applied suddenly at $t=0$ and decreasing linearly to zero at $t=0.15$ sec. The dynamic yield strength σ_{dy} of the steel is 50,000 psi, the compressive strength σ_{dc}' of concrete is 4000 psi, and the steel ratio p is 0.015. Design the slab so that the behavior is within the elastoplastic range.

10-16 Repeat Problem 10-15 with $q(t)$ as a constant pressure of 1.0 ksf applied suddenly at $t=0$ and of duration $t_d=0.15$ sec.

10-17 Repeat Problem 10-15 for a concrete slab length of 20.0 ft.

10-18 Repeat Problem 10-15 for the case where the dynamic pressure $q(t)$ is represented by the solid line in Fig. 11-24.

10-19 A 20×15-ft two-way concrete slab is fixed on all edges and is acted upon by a uniformly distributed dynamic pressure of 1.5 ksf which is applied suddenly at $t=0$ and decreasing linearly to zero at $t=0.15$ sec. The weight of the slab is 90 psf, I_a is 35 in.^{4}/ft, the modulus of elasticity E of concrete is 3×10^6 psi, and $M_{Pfa}=M_{Pfb}=M_{Psa}=M_{Psb}$ $=3000$ kip-in. Determine the maximum deflection at the center of the slab by using appropriate transformation factors.

10-20 Repeat Problem 10-19 for a 16×16-ft two-way concrete slab fixed on all edges.

11

BLAST AND EARTHQUAKE

11-1 INTRODUCTION

In the past few decades considerable emphasis has been given to the problems of blast and earthquake. The earthquake problem is rather old, but most knowledge on this subject has been accumulated during the past 30 years. The blast problem is rather new; information about the developments in this field is made available mostly through publications of the Army

Corps of Engineers, Department of Defense, U.S. Air Force, and other governmental offices and public institutions. Much of the work is done by the Massachusetts Institute of Technology, The University of Illinois, and other leading educational institutions and engineering firms.

Despite much recent progress, the complexities in these fields are great and require experience and sound judgment on the part of the engineer. The purpose in this chapter is to provide the fundamental knowledge that can serve as a basis for further study and research of these important subjects.

11-2 DYNAMIC EFFECTS OF NUCLEAR EXPLOSIONS[33]

In general, an explosion is the result of a very rapid release of large amounts of energy within a limited space. In the case of conventional explosions this energy comes from chemical reactions that involve a rearrangement among the atoms. In a nuclear explosion, however, the energy arises from the formation of different atomic nuclei. The sudden release of energy initiates a pressure wave in the surrounding medium, known as a shock wave, with a sudden increase of pressure at the front followed by a gradual decrease as shown in Fig. 11-1*a*. The destructive action of nuclear weapons is much more severe than that of a conventional weapon and is due mainly to blast or shock. In a typical air burst at an altitude below 100,000 ft, an approximate distribution of energy would consist of 50% blast and shock, 35% thermal radiation, 10% residual nuclear radiation, and 5% initial nuclear radiation.

The phenomena immediately associated with nuclear explosions vary with the location of the point of burst in relation to the surface of the earth. The types of burst are usually classified as *(a)* air burst, *(b)* high-altitude burst, *(c)* underwater burst, *(d)* underground burst, and *(e)* surface burst. The discussion in this section is limited to air-burst or surface-burst weapons; factual information on the characteristics of the blast pressures produced by such nuclear explosions is provided. This information is then used to determine the dynamic loads on surface[34] structures that are subjected to such blast pressures and to design them accordingly. It should be pointed out that a surface structure cannot be protected from a direct hit by a nuclear bomb; it can, however, be designed to resist the blast pressures when it is located at some distance from the point of burst.

When an explosion takes place, the expansion of the hot gasses produces a pressure wave in the surrounding air. As this wave moves away from the

[33] The analysis in this section is based on U.S. Government studies presented in Reference 40 and Manual EM1110-345-413 in Reference 41. For more information, the reader should consult these references.

[34] Buried or semiburied structures are discussed in Manual EM1110-345-421 in Reference 41.

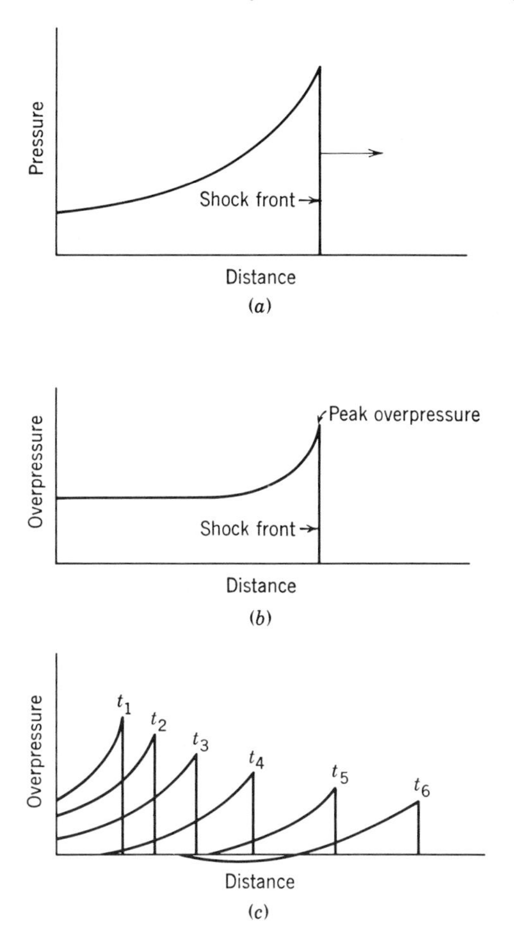

Fig. 11-1. (*a*) Variation of pressure with distance in a shock wave. (*b*) Formation of shock front. (*c*) Variation of overpressure with distance from center of explosion at various times.

center of the explosion, the inner part moves through the region that was previously compressed and is now heated by the leading part of the wave. As the pressure wave moves with the velocity of sound, the temperature and pressure of the air cause this velocity to increase. The inner part of the wave starts to move faster and gradually overtakes the leading part of the wave. After a short period of time the pressure wavefront becomes abrupt, thus forming a shock front somewhat similar to the one illustrated in Fig. 11-1*b*. The maximum overpressure occurs at the shock front and is called the peak overpressure. Behind the shock front, the overpressure drops very rapidly to about one-half the peak overpressure and remains almost uniform in the central region of the explosion.

As the expansion proceeds, the overpressure in the shock front decreases steadily; the pressure behind the front does not remain constant but, instead, falls off in a regular manner. After a short time, at a certain distance from the center of the explosion, the pressure behind the shock front becomes smaller than that of the surrounding atmosphere and the so-called negative phase, or suction, develops. The front of the blast wave weakens as it progresses outward, and its velocity drops toward the velocity of sound in the undisturbed atmosphere. This sequence of events is shown in Fig. 11-1*c*. The overpressures at times $t_1, t_2, \ldots, t_6$ are indicated. In the curves marked t_1 through t_5, the pressure in the blast wave has not fallen below that of the atmosphere. In the curve marked t_6, at some distance behind the shock front, the overpressure becomes negative. This is better illustrated in Fig. 11-2*a*.

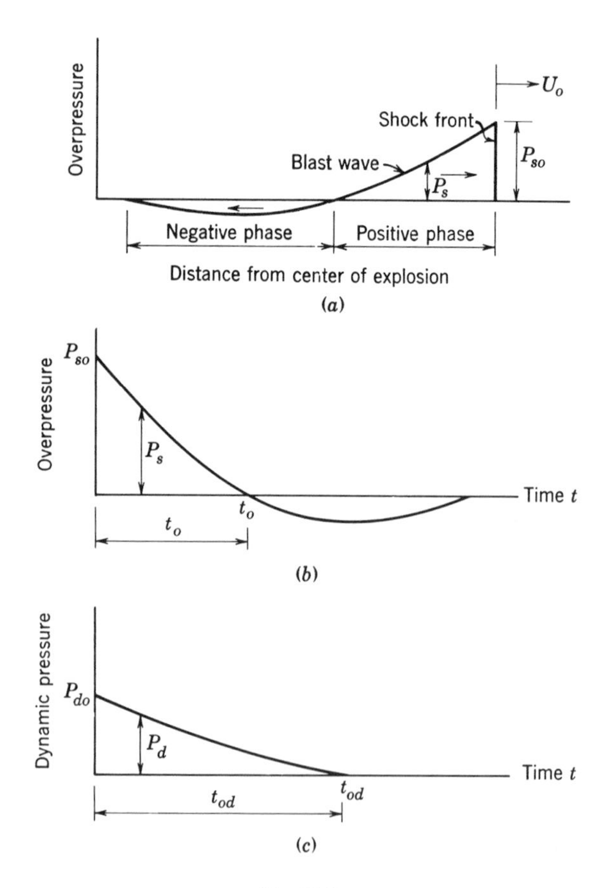

Fig. 11-2

Figure 11-2*a* shows the variation of overpressure with distance at a given time. Here U_o is the velocity of the shock front, P_{so} is the peak overpressure, P_s is the overpressure behind the shock front, and the arrows under the curve indicate the direction of the air mass movement. The time variation of the same blast wave at a given distance from the explosion is also shown in Fig. 11-2*b*; t_o indicates the time duration of the positive phase and also the time at the end of the positive phase.

Another quantity of equivalent importance is the force that is developed from the strong winds accompanying the blast wave. Known as the dynamic pressure, this is proportional to the square of the wind velocity and the density of the air behind the shock front. Its variation at a given distance from the explosion is shown in Fig. 11-2*c*. Here P_{do} is the dynamic peak pressure, P_d is the dynamic pressure at any time t, and t_{od} is as indicated in the figure. Mathematically, the dynamic pressure P_d is expressed as

$$P_d = \tfrac{1}{2}\rho u^2 \tag{11-1}$$

where u is the velocity of the air particles and ρ is the air density.

The peak dynamic pressure decreases with increasing distance from the center of the explosion, but the rate of decrease is different from that of the peak overpressure. At a given distance from the explosion, the time variation of the dynamic pressure P_d behind the shock front is somewhat similar to that of the overpressure P_s, but the rate of decrease is usually different. For design purposes, the negative phase of the overpressure in Fig. 11-2*a* or Fig. 11-2*b* is not important and can be ignored.

In Fig. 11-2*b*, the variation of the overpressure P_s in the positive phase only, for overpressures of about 10 psi or less, is given by the approximate expression

$$P_s = P_{so}\left(1 - \frac{t}{t_o}\right)e^{-t/t_o} \tag{11-2}$$

where e is the base of natural system of logarithms and is equal to 2.7182. The equation shows that the rate of decay of the overpressure behind the shock front is a function of the peak overpressure. This can be expressed mathematically by a series of equations similar to Eq. 11-2. A set of such equations is represented by the normalized curves in Fig. 11-3, where t is the time after arrival of the shock front.

The corresponding expression for the variation of the dynamic pressure P_d in Fig. 11-2*c* is given by

$$P_d = P_{do}\left(1 - \frac{t}{t_{od}}\right)e^{-2t/t_{od}} \tag{11-3}$$

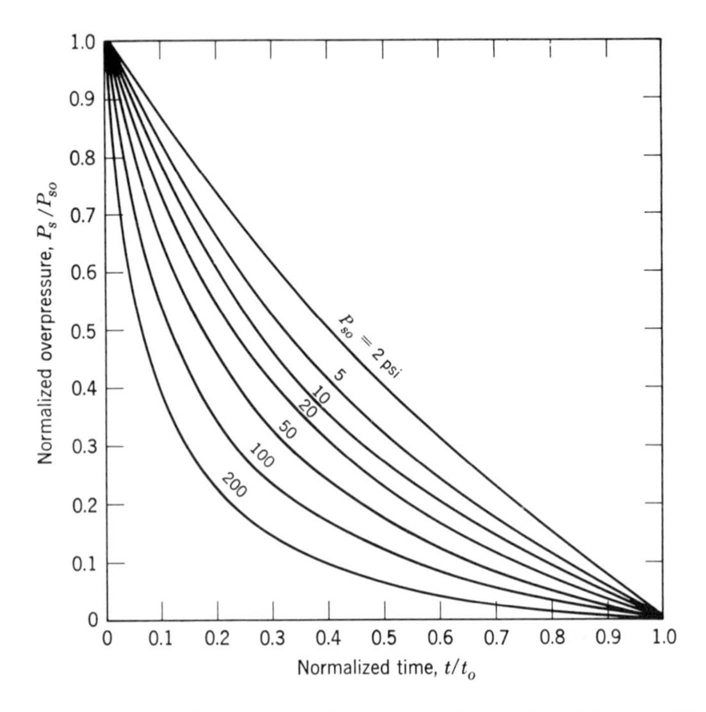

Fig. 11-3. Rate of decay of overpressure for various values of P_{so}. (Reference 40.)

where t_{od} is the duration of the dynamic pressure. The decay of the normalized dynamic pressure, however, is dependent on the overpressure. A set of normalized curves for several values of the peak overpressure P_{so} are shown in Fig. 11-4. It should be noted that the dynamic pressure decays faster than the overpressure.

In Fig. 11-2*a*, the velocity U_o of the shock front depends on the peak overpressure and is given by the expression

$$U_o = U_s\left(1+\frac{6P_{so}}{7P_o}\right)^{\frac{1}{2}} \tag{11-4}$$

where P_o is the atmospheric pressure and U_s is the velocity of sound. At sea level and for normal atmospheric conditions, the equation yields

$$U_o = 1117\left[1+\frac{6P_{so}}{(7)(14.7)}\right]^{\frac{1}{2}} \tag{11-5}$$

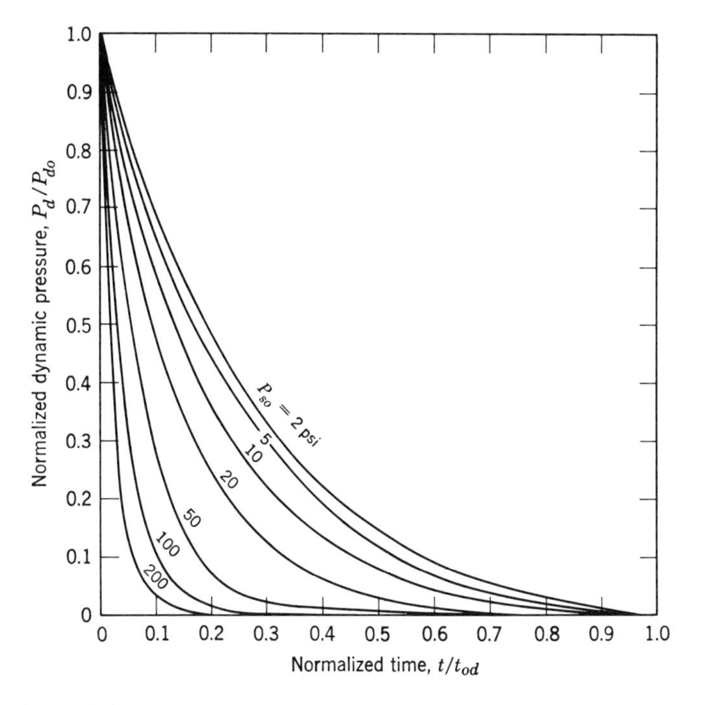

Fig. 11-4. Rate of decay of dynamic pressure for various values of P_{so}. (Reference 40.)

In Fig. 11-2, the variation of the peak overpressure P_{so} and peak dynamic pressure P_{do} with distance from the point of explosion for 1-kiloton surface burst in a standard sea-level atmosphere is given by the curves in Fig. 11-5. If the yield is not 1 kiloton, the range to which a given peak overpressure or peak dynamic pressure extends can be determined from the expression

$$d = d_1 (Z)^{\frac{1}{3}} \tag{11-6}$$

Here d_1 is the distance from the explosion point for 1-kiloton surface burst and d is the distance from the same point for Z kilotons. For example, the range d for 1000 kilotons (one megaton) would be ten times that of a 1-kiloton yield.

The duration t_o of the positive phase of the blast wave in Fig. 11-2*b* is a function of the peak overpressure P_{so} and the total energy yield of the weapon. Values of t_o for 1-kiloton, 100-kiloton, 1-megaton, and 10-megaton surface bursts are plotted in Fig. 11-6. For other weapon yields, the positive phase duration of the overpressure can be determined from the expression

$$t_{o2} = t_{o1} (Z)^{\frac{1}{3}} \tag{11-7}$$

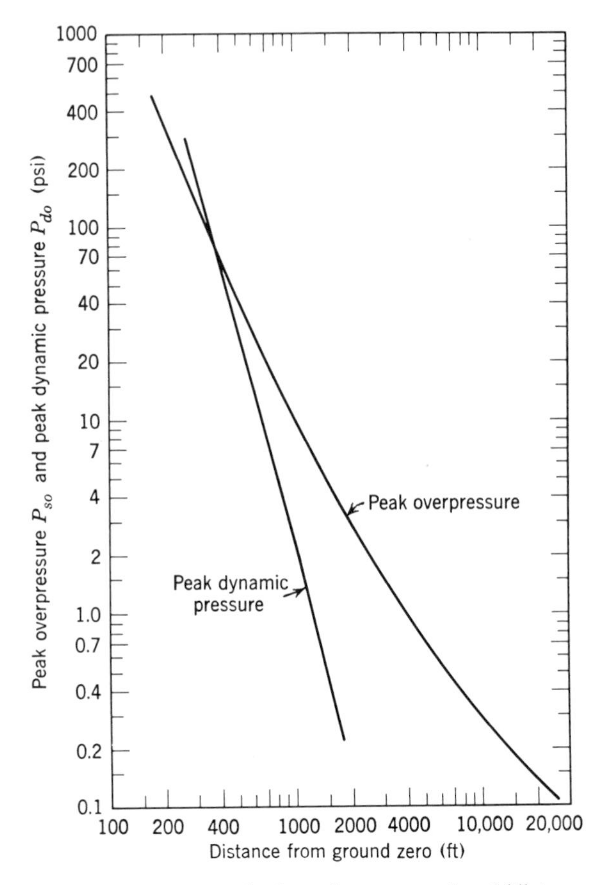

Fig. 11-5. Peak overpressure and peak dynamic pressure for 1-kiloton surface burst. (Reference 40.)

In this equation, t_{o1} is the positive phase duration for one of the weapon yields in Fig. 11-6 and t_{o2} is the positive phase duration for Z weapon yields. For example, if t_{o1} is the positive phase duration for 1-kiloton weapon yield, then t_{o2} is the positive phase duration for Z kilotons.

The curves in Fig. 11-7 show the duration t_o on the ground of the positive phase of the overpressure, Fig. 11-2*b*, and the dynamic pressure (in parentheses), Fig. 11-2*c*, for 1-kiloton burst in a standard sea-level atmosphere, as a function of the distance from ground zero and height of burst. For other weapon yields, the positive phase durations can be determined from the expression

$$\frac{d}{d_1} = \frac{h}{h_1} = \frac{t_{o2}}{t_{o1}} = (Z)^{\frac{1}{3}} \tag{11-8}$$

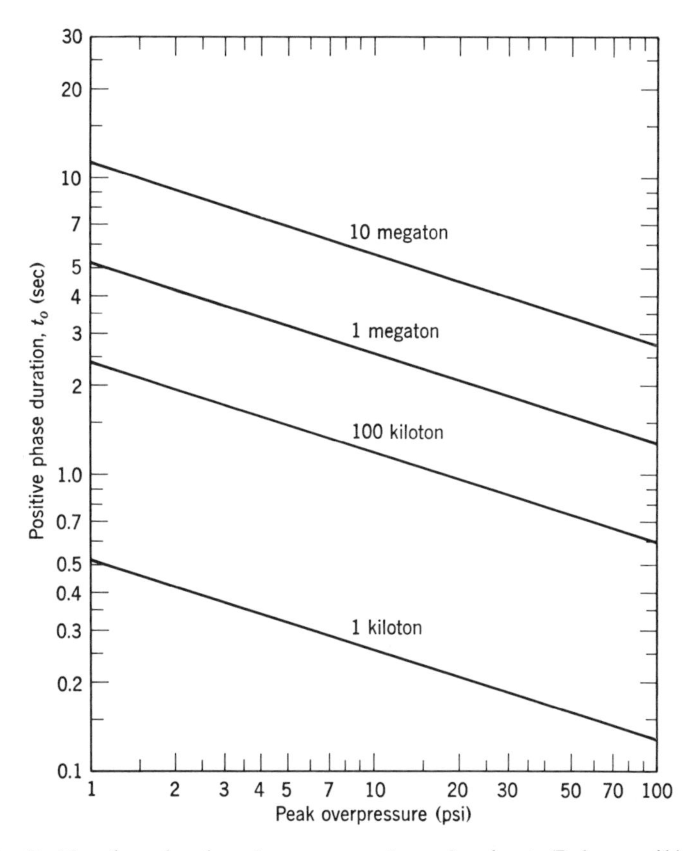

Fig. 11-6. Positive phase duration of overpressure for surface burst. (Reference 41.)

Here d_1, h_1, and t_{o1} are the distance from ground zero, the height of burst, and duration, respectively, for 1-kiloton weapon yield, while d, h, and t_{o2} are the corresponding quantities for Z kilotons. It should be pointed out, however, that the analysis in this chapter is limited to surface bursts only and the curves in Fig. 11-7 will be used only for $h = h_1 = 0$. Furthermore, it is suggested to use this figure for computing the positive phase duration t_{od} of the dynamic pressure only, because the positive phase duration t_o of the overpressure can be conveniently determined by using Fig. 11-6 and Eq. 11-7.

When the shock front of a blast wave strikes a more dense medium, such as a solid wall surface, the pressure of the shock front increases instantaneously because of the formation of a reflected wave. The exact value of the peak reflected overpressure P_r will depend on the incident blast overpressure P_{so} and the angle α at which it strikes the surface. The angle α

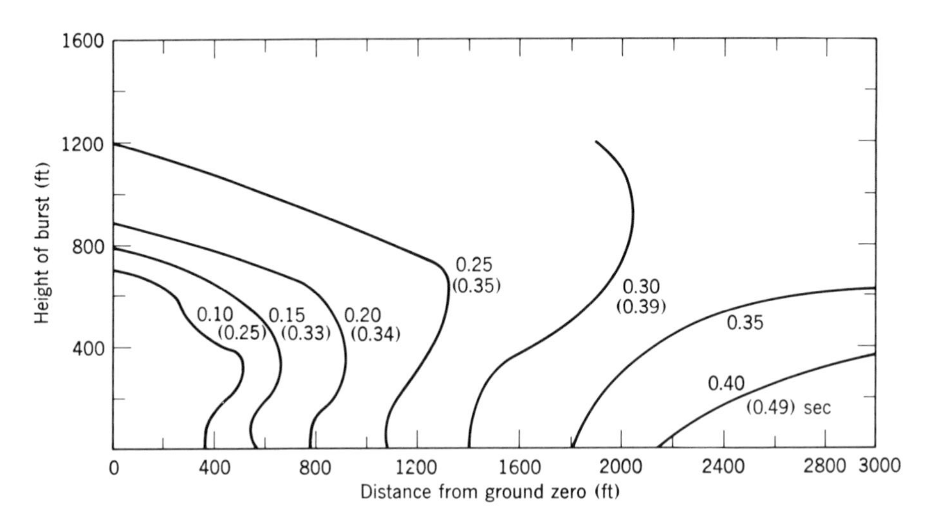

Fig. 11-7. Positive phase duration of overpressure and dynamic pressure (in parentheses) for 1-kiloton burst. (Reference 40.)

is known as the angle of incidence and is the angle between the shock front of a blast wave and the reflecting surface. If the reflecting surface is normal to the direction of travel of the shock front, that is, $\alpha = 0$, the peak reflected overpressure P_r is given by the expression

$$P_r = 2P_{so}\left(\frac{7P_o + 4P_{so}}{7P_o + P_{so}}\right) \tag{11-9}$$

where P_o is the atmospheric pressure and P_{so} the peak overpressure.

The value of P_r approaches $8P_{so}$ for strong shocks and tends toward $2P_{so}$ for weak shocks. The increase of P_r beyond the expected value of $2P_{so}$ is due to the dynamic or wind pressure. At sea level and under normal atmospheric conditions, Eq. 11-9 yields

$$P_r = 2P_{so}\left(\frac{103 + 4P_{so}}{103 + P_{so}}\right) \tag{11-10}$$

If the angle of incidence α is other than zero, that is, if the surface is inclined, the peak reflected overpressure can be found from Fig. 11-8. In this figure, the overpressure ratio P_r/P_{so} is plotted as a function of α for various values of peak overpressure. Since the reflection effects will diminish after a short period of time, it would be required to know the clearing time t_c and the way these effects are diminishing. A reasonable approximation would be to assume that the reflection effects diminish linearly. The clearing time t_c

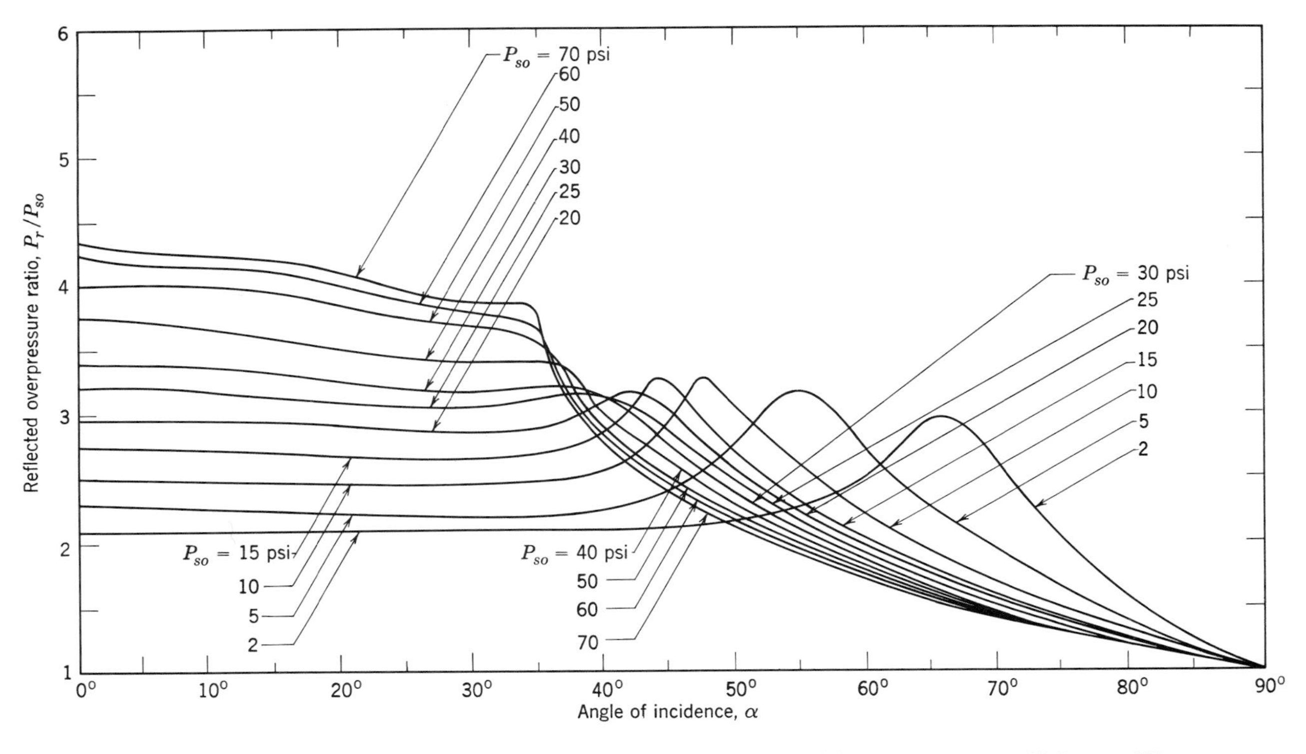

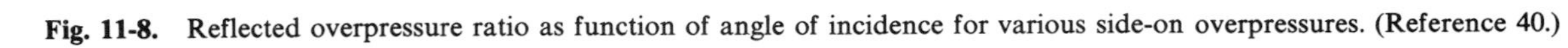

Fig. 11-8. Reflected overpressure ratio as function of angle of incidence for various side-on overpressures. (Reference 40.)

for a solid flat surface, such as the wall of a building, can be determined from the expression

$$t_c = \frac{3S}{U_r} \tag{11-11}$$

Here S is either the height of the reflecting surface or one-half its width, whichever is smaller, and U_r is the velocity of sound in the region of the reflected overpressure. For a surface burst, U_r can be taken equal to U_o and can be computed from Eq. 11-4 or Eq. 11-5. The velocity U_r can be also computed from the expression[35]

$$U_r = 422\left(\frac{1.0088P_{so} \ +70P_{so} \ +720}{102.9+6P_{so}}\right)^{\frac{1}{2}} \tag{11-12}$$

11-3 DYNAMIC LOADING ON CLOSED RECTANGULAR STRUCTURES

In the preceding section, the general characteristics and effects of an explosion were discussed. To be able to determine the dynamic response of a structure subjected to the shock effects of an explosion, the blast loading acting on the structure must first be computed. By following the discussion of the preceding section, the total loading on a structure can be assumed to consist of three main parts: *(a)* the effects of the initial reflected overpressures, *(b)* the effects of the general overpressure, and *(c)* the drag loading. The drag loading includes the effects of the drag pressure which is related to the dynamic pressure P_d by the expression

$$\text{drag pressure} = C_d P_d \tag{11-13}$$

where C_d is the drag coefficient. The drag coefficient can be either positive or negative and is dependent on the size, shape, and orientation of the structure.

In the present section, this theory is used to determine the blast wave loads on closed aboveground rectangular structures. Such a structure is ideally represented by the closed cube in Fig. 11-9*a*, where the front face is oriented in a direction normal to the direction of propagation of the shock front. This orientation produces the most severe loading on the component elements of the structure.

The average loading P_{front} on the front face of the structure will consist of the reflected pressure P_r up to the clearing time t_c, Fig. 11-9*b*, and the

[35] Equation 11-12 is given in Reference 41, Manual Number EM1110-345-413

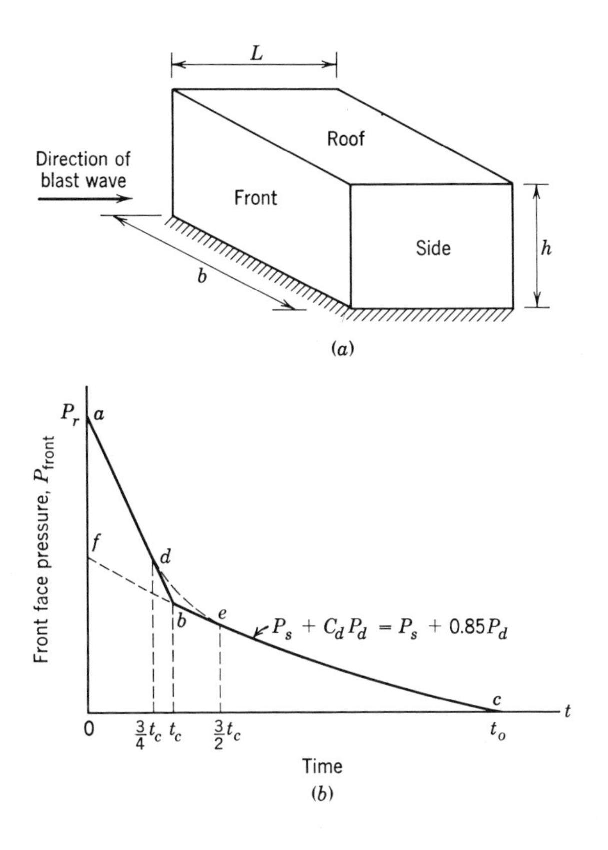

Fig. 11-9. (*a*) Closed rectangular structure. (*b*) Average front face pressure versus time. (Reference 41.)

summation

$$P_c = P_s + C_d P_d \tag{11-14}$$

from time t_c to time t_o. The reflected overpressure P_r is given by Eq. 11-10, and the decay of the overpressure P_s and dynamic pressure P_d can be determined from Figs. 11-3 and 11-4, respectively. The drag coefficient C_d can be taken to be equal to 0.85 and the clearing time t_c can be determined from Eq. 11-11. Thus the average pressure on the front face of a rectangular structure is given by the curve *abc* in Fig. 11-9*b*. The incompatible discontinuity at point *b*, if preferred, can be smoothed out by the fairing curve *de* as shown in the same figure.

Average Loading P_{back} on Back Face

At the back face, the shock front arrives at time $t_d = L/U_o$, where U_o is given by Eq. 11-5 and L is the length of the rectangular structure. When the shock front arrives at the back wall, it requires an additional time $t_b = 4S/U_o$ for the average pressure to build up to its maximum value $(P_{back})_{max}$, where S is either the full height h of the back wall or one-half its width $(b/2)$, whichever is smaller.

The peak value $(P_{back})_{max}$ of the average pressure on the back face after the buildup is completed can be determined from the expression

$$(P_{back})_{max} = \frac{P_{sb}}{2}\left[1 + (1-\beta)e^{-\beta}\right] \tag{11-15}$$

In the expression above, P_{sb} is the blast wave overpressure at time $t - t_d = t_b$, $\beta = 0.5P_{so}/14.7$, e is the base of natural logarithms, and $(P_{back})_{max}$ occurs at $t - t_d = t_b$. The average back face pressure versus time is shown in Fig. 11-10. The variation of P_{back} for times in excess of $t - t_d = t_b$ is given by the expression

$$\frac{P_{back}}{P_s} = \frac{(P_{back})_{max}}{P_{sb}} + \left[1 - \frac{(P_{back})_{max}}{P_{sb}}\right]\left[\frac{t - (t_d + t_b)}{t_o - t_b}\right]^2 \tag{11-16}$$

where t_o is the duration of the positive phase of the overpressure P_s. In Fig. 11-10, the average back face pressure is assumed to be linear in the time interval between t_d and $t_b + t_d$. It is further assumed that the peak overpressure P_{so} does not reduce in strength in the time interval t_d where the blast wave passes over the structure.

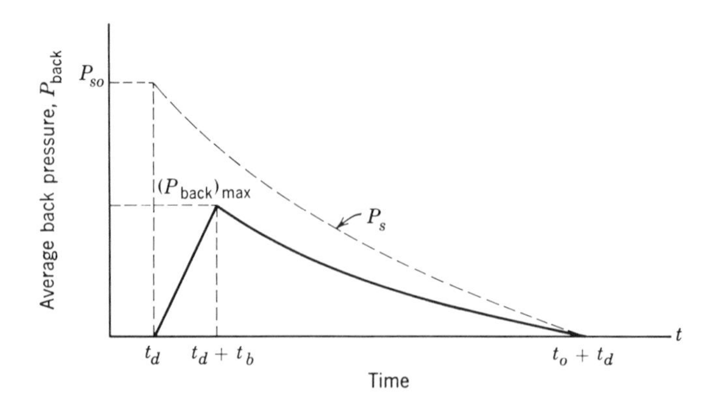

Fig. 11-10. Average back face pressure versus time. (Reference 41.)

In applying Eq. 11-15, the overpressure P_{sb} at time $t - t_d = t_b$, can be determined from the curves in Fig. 11-3. If P_{so} is 10 psi or less, Eq. 11-2 with $t = t_b$ can be also used to determine P_{sb}. In Eq. 11-16, the values of P_{back}/P_s can be computed for a sequence of times in excess of $t = t_d + t_b$. Then, for the equivalent times, P_s is computed from either Eq. 11-2, provided that P_{so} is 10 psi or less, or from the curves in Fig. 11-3. The procedure will become convenient if the computations are tabulated by using the following tabular headings:

(1)	(2)	(3)	(4)	(5)	(6)	(7)
time, t	$t - t_d$	$(t - t_d)/t_o$	P_s/P_{so}	P_s	P_{back}/P_s	P_{back}

In column 1, a sequence of times ranging from $t = t_b + t_d$ to $t = t_o + t_d$ is selected. Columns 2 and 3 are self explanatory. For the selected times, the values in the remaining columns are determined as explained above. A typical plot of P_{back} versus time is shown in Fig. 11-10.

The average *net horizontal loading* P_{net} on a rectangular structure is equal to the front face loading P_{front} minus the back face loading P_{back}. That is,

$$P_{net} = P_{front} - P_{back} \tag{11-17}$$

The algebraic sum is illustrated in Fig. 11-11.

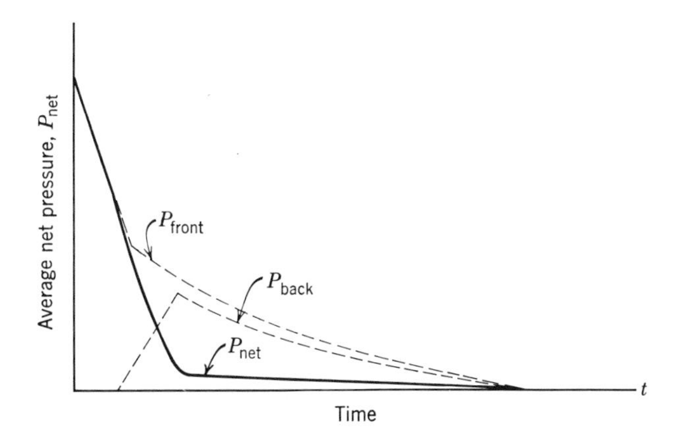

Fig. 11-11. Average net horizontal pressure versus time. (Reference 41.)

Average Pressure on Roof and Sides

The variation of the average pressure on the roof and sides of a closed aboveground rectangular structure can be approximated by the curve in Fig. 11-12. The sides and roof are not fully loaded until the blast wave has traveled the length L of the building. At this time, the average pressure gains its maximum value P_m and is equal to the algebraic sum of the overpressure P_s and drag pressure $C_d P_d$ at the distance $L/2$ from the front face of the structure. Thus, at $t = L/U_o$,

$$P_m = P_s(L/2U_o) + C_d P_d(L/2U_o) \tag{11-18}$$

where U_o can be computed from either Eq. 11-4 or Eq. 11-5.

The average pressure P_a at times $L/U_o \leqslant t \leqslant t_o + L/2U_o$ can be computed from the expression

$$P_a = P_s(t - t_s) + C_d P_d(t - t_s) \tag{11-19}$$

where t_s is equal to $L/2U_o$ and P_s and P_d are the overpressure and dynamic pressure, respectively, at time $t - t_s$. The value of the drag coefficient C_d can be obtained from the following table:

Dynamic Pressure (psi)	Drag Coefficient C_d
0- 25	−0.4
25- 50	−0.3
50-130	−0.2

11-4 COMPUTATION OF DYNAMIC LOADING ON CLOSED RECTANGULAR STRUCTURES

As an illustration, let it be assumed that it is required to determine the average net horizontal pressure P_{net} that acts on the rectangular aboveground building in Fig. 11-13. The dimensions of the building are shown in the same figure. The structure is assumed to be subjected to a peak overpressure $P_{so} = 5.45$ psi, which is produced by a weapon yield of 1.2 kilotons.

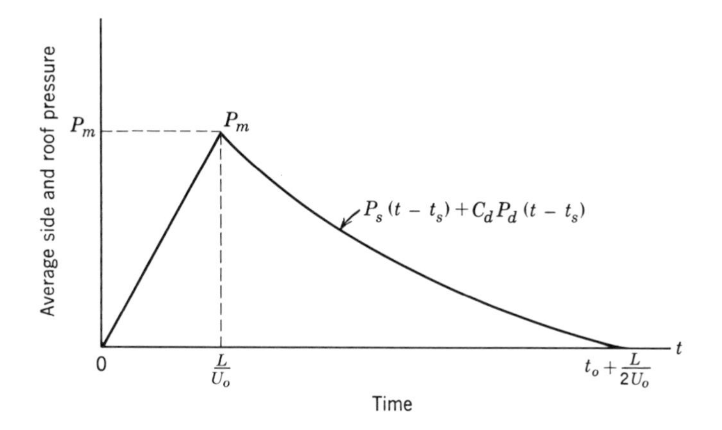

Fig. 11-12. Average side and roof pressure versus time. (Reference 40.)

From Eq. 11-5, the shock front velocity U_o is

$$U_o = 1117\left[1+\frac{(6)(5.45)}{(7)(14.7)}\right]^{\frac{1}{2}} = 1{,}283 \text{ fps}$$

From Fig. 11-6, the positive phase duration t_{o1} of the overpressure P_s for 1-kiloton weapon yield is 0.314 sec. For 1.2 kilotons, the positive phase duration t_{o2} can be computed from Eq. 11-7. That is,

$$t_{o2} = (0.314)(1.2)^{\frac{1}{3}} = 0.33 \text{ sec}$$

The peak dynamic pressure P_{do} can be determined from Fig. 11-5. With $P_{so} = 5.45$ psi, the range d_1 for a 1-kiloton weapon is 1400.00 ft. In the same figure, for the same range, the peak dynamic pressure P_{do} is 0.65 psi. For 1.2 kilotons, the range d can be found from Eq. 11-6. That is,

$$d = (1400)(1.2)^{\frac{1}{3}} = 1485 \text{ ft}$$

The positive phase duration t_{od_1} of the dynamic pressure P_d for 1-kiloton weapon yield can be found from Fig. 11-7. Thus, with $d_1 = 1400$ ft, Fig. 11-7 yields $t_{od_1} = 0.39$ sec. From Eq. 11-7, the positive phase duration t_{od} for 1.2 kilotons is

$$t_{od} = (0.39)(1.2)^{\frac{1}{3}} = 0.41 \text{ sec}$$

The peak reflected overpressure P_r is given by Eq. 11-9. Thus

$$P_r = (2)(5.45)\left[\frac{(7)(14.7) + (4)(5.45)}{(7)(14.7) + 5.45}\right] = 12.56 \text{ psi}$$

From Eq. 11-11, with $U_r = U_o = 1{,}283$ fps, the clearing time t_c is

$$t_c = \frac{(3)(8.33)}{1283} = 0.0195 \text{ sec}$$

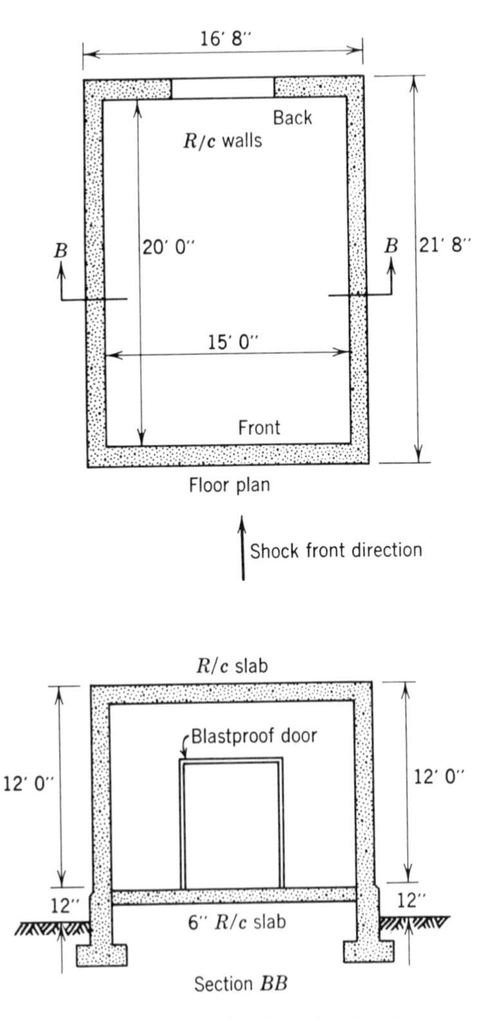

Fig. 11-13. Closed rectangular structure.

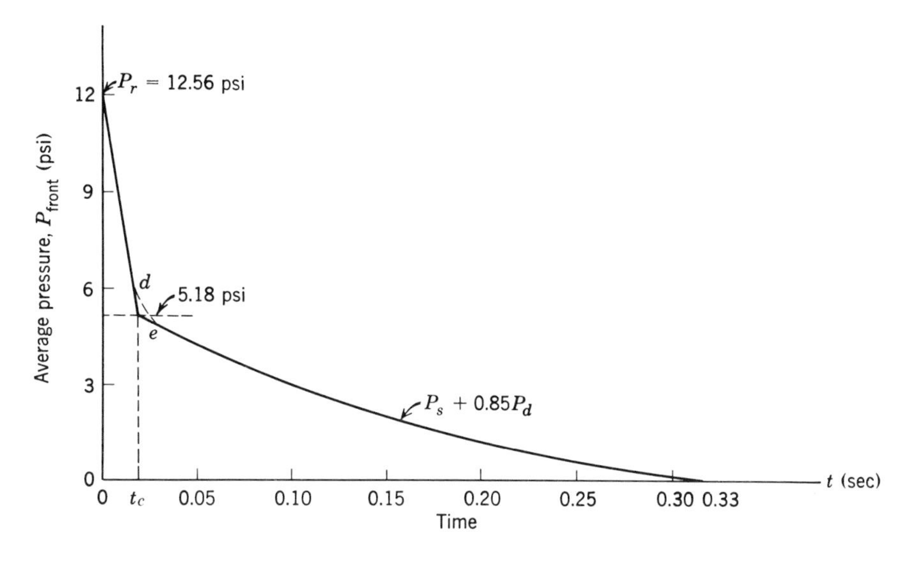

Fig. 11-14. Average front face pressure versus time.

Average Pressure on Front Face

With these results in mind, the variation of the average pressure P_{front} of the front face can be easily determined. The results are shown in Fig. 11-14. For simplicity, it is assumed that $t_{o2} = t_{od} = 0.33$ sec. This is a fair approximation, because the difference between t_{o2} and t_{od} is not large and the dynamic pressure P_d is rather small compared to the overpressure P_s. At $t = 0$, the pressure P_{front} in Fig. 11-14 is equal to P_r. At $t = t_c = 0.0195$ sec, the values of P_s and P_d can be determined from Figs. 11-3 and 11-4, respectively. If preferred, Eqs. 11-2 and 11-3 can be used, because P_{so} and P_{do} are less than 10 psi. Thus at $t = t_c = 0.0195$ sec, P_s and P_d are 4.75 and 0.51 psi, respectively. Therefore, at $t = t_c$,

$$P_{front} = P_s + 0.85P_d = 4.75 + (0.85)(0.51) = 5.18 \text{ psi}$$

The values of P_{front} for the time interval between $t = t_c$ and $t = 0.33$ sec are equal to $P_s + 0.85P_d$. In Fig. 11-14, the variation of P_{front} between $t = 0$ and $t = t_c$ is assumed to be linear. The incompatible discontinuity at $t = t_c$, if preferred, can be smoothed out by the fairing curve *de* shown by a dashed line.

Average Pressure on Back Face

The time t_d at which the shock front arrives at the back face of the building is

$$t_d = \frac{L}{U_o} = \frac{21.67}{1283} = 0.017 \text{ sec}$$

The time t_b that is required for the average pressure P_{back} to build up to its maximum value $(P_{\text{back}})_{\max}$ is

$$t_b = \frac{4S}{U_o} = \frac{(4)(8.33)}{1283} = 0.026 \text{ sec}$$

With $t = t_b = 0.026$ sec and $t_o = t_{o2} = 0.33$ sec, the value P_{sb} of the overpressure P_s can be determined from either Eq. 11-2 or Fig. 11-3. The result is

$$P_{sb} = 0.82 P_{so} = (0.82)(5.45) = 4.47 \text{ psi}$$

In Eq. 11-15,

$$\beta = \frac{(0.5) P_{so}}{14.7} = \frac{(0.5)(5.45)}{14.7} = 0.186$$

Thus Eq. 11-15 yields

$$(P_{\text{back}})_{\max} = \frac{4.47}{2}\left[1 + (1 - 0.186)\, e^{-0.186}\right] = 3.75 \text{ psi}$$

The variation of P_{back} for the time interval $t_d \leqslant t \leqslant t_d + t_b$ is assumed to be linear. The variation in the time interval $t_d + t_b < t \leqslant t_{o2} + t_d$ can be computed from Eq. 11-16. For example, at $t = 0.070$ sec, Eq. 11-16 yields

$$\frac{P_{\text{back}}}{P_s} = \frac{3.75}{4.47} + \left[1 - \frac{3.75}{4.47}\right]\left[\frac{0.070 - (0.017 + 0.026)}{0.330 - 0.026}\right]^2 = 0.839$$

or

$$P_{\text{back}} = 0.839 P_s$$

From Fig. 11-3, with $t/t_o = t - t_d/t_{o2} = 0.070 - 0.017/0.330 = 0.161$, the value of the overpressure P_s is

$$P_s = 0.716 P_{so} = (0.716)(5.45) = 3.90 \text{ psi}$$

Thus, at $t = 0.070$ sec,

$$P_{\text{back}} = 0.839P_s = (0.839)(3.90) = 3.27 \text{ psi}$$

In a similar manner, the values of P_{back} for other times t can be obtained. The complete results are shown plotted in Fig. 11-15.

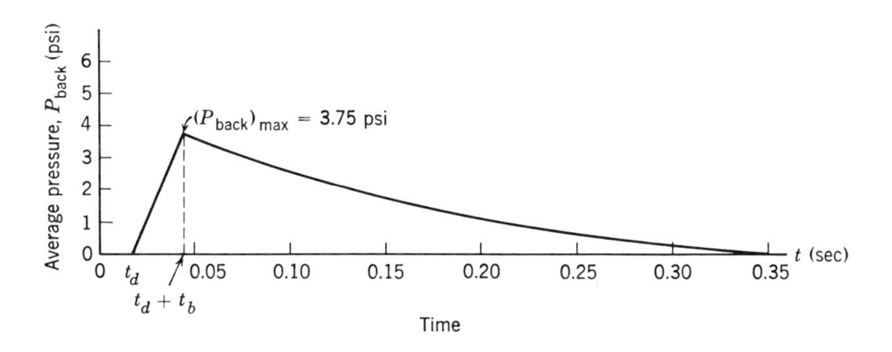

Fig. 11-15. Average back face pressure versus time.

Average Net Horizontal Pressure

The average net horizontal pressure P_{net} is given by Eq. 11-17. Thus, if the values of the pressure in Fig. 11-15 are subtracted from the corresponding values in Fig. 11-14, the variation of the average net horizontal pressure in Fig. 11-16 is obtained. Note that the values of P_{net} at times larger than about 0.06 sec are small compared to those between $t = 0$ and $t = 0.06$ sec.

Average Pressure on Roof and Sides

For the rectangular building in Fig. 11-13,

$$\frac{L}{U_o} = \frac{21.67}{1283} = 0.017 \text{ sec}$$

and

$$\frac{L}{2U_o} = \frac{21.67}{(2)(1283)} = 0.0085 \text{ sec}$$

From Eq. 11-18,

$$P_m = P_s(t = 0.0085) + C_d P_d(t = 0.0085)$$

$$= 5.12 + (-0.4)(0.56) = 4.90 \text{ psi}$$

where P_s and P_d are computed by using Figs. 11-3 and 11-4, respectively.

The average pressure P_a for times exceeding $L/U_o = 0.017$ sec are obtained from Eq. 11-19. The complete variation of the average pressure on the sides and roof of the rectangular building in Fig. 11-13 versus time is shown in Fig. 11-17.

The dynamic loading on other closed rectangular structures can be determined in a similar manner. With known dynamic loading, the dynamic response of a structure can be found by applying either the simplified methods of analysis in Chapter 10 or other suitable methods discussed in earlier chapters of this book.

11-5 DYNAMIC LOADING ON RECTANGULAR STRUCTURES WITH OPENINGS

The structures included in this category are the ones whose front and back faces have at least 30% of openings or window area and have no interior

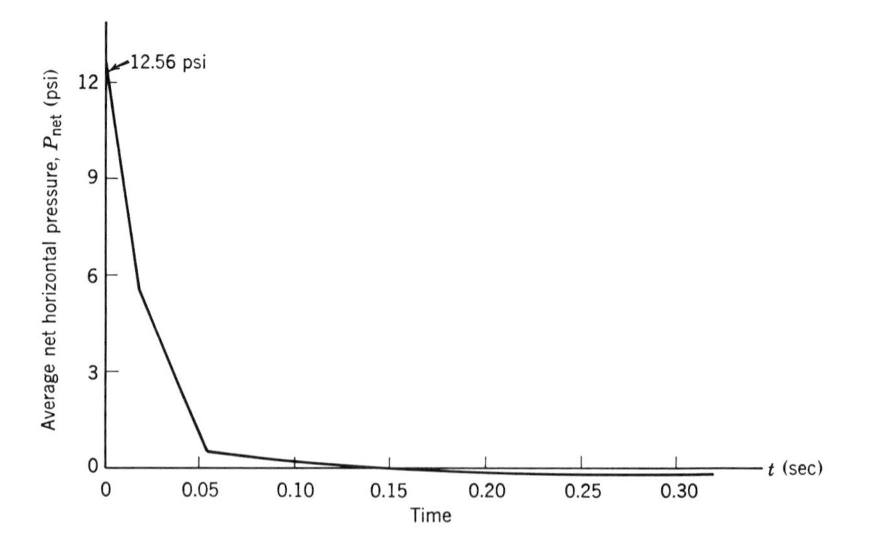

Fig. 11-16. Average net horizontal pressure versus time.

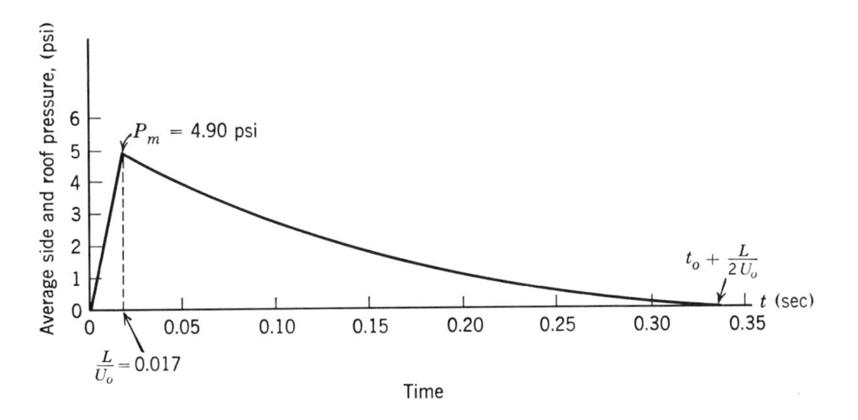

Fig. 11-17. Average side and roof pressure versus time.

partitions to block or influence the passage of the blast wave. The average pressure on the outside of the front face can be computed in the same manner as for closed rectangular structures, except that S in Eq. 11-11 is replaced by S'. The quantity S' is the weighted average distance that the rarefaction wave must travel in order to cover the wall once, provided that the blast wave has immediate access to the interior of the structure. This is not an unreasonable assumption, because the windows and doors usually break before the clearing of the reflected overpressure is completed.

The distance S' can be computed from the expression

$$S' = \sum \frac{\delta_n h_n A_n}{A_f} \leqslant S \tag{11-20}$$

To understand the meaning of the symbols in Eq. 11-20, consider the front face in Fig. 11-18*a* which has two openings. The face is divided into rectangular areas as shown. The areas cleared from two opposite sides are marked *a*, while *b* and *c* are the areas cleared from two adjacent sides and one side, respectively. The remaining areas are marked *d*. On this basis, A_f in Eq. 11-20 is the area of the front face less the area of the openings; A_n is the area of each of the portions of the subdivided front face, except openings; h_n, for areas *a*, is the average distance between the sides from which clearing occurs; for areas *b* and *d*, h_n is the average height or width, whichever is smaller, and for areas *c*, h_n is the average distance between the side from which clearing occurs and the side opposite. In the same equation, δ_n is the clearing factor that has the value of 1/2 for areas *a*, and it is equal to unity for areas *b*, *c*, and *d*.

The average pressure on the inside of the front face is zero at $t = 0$ and it

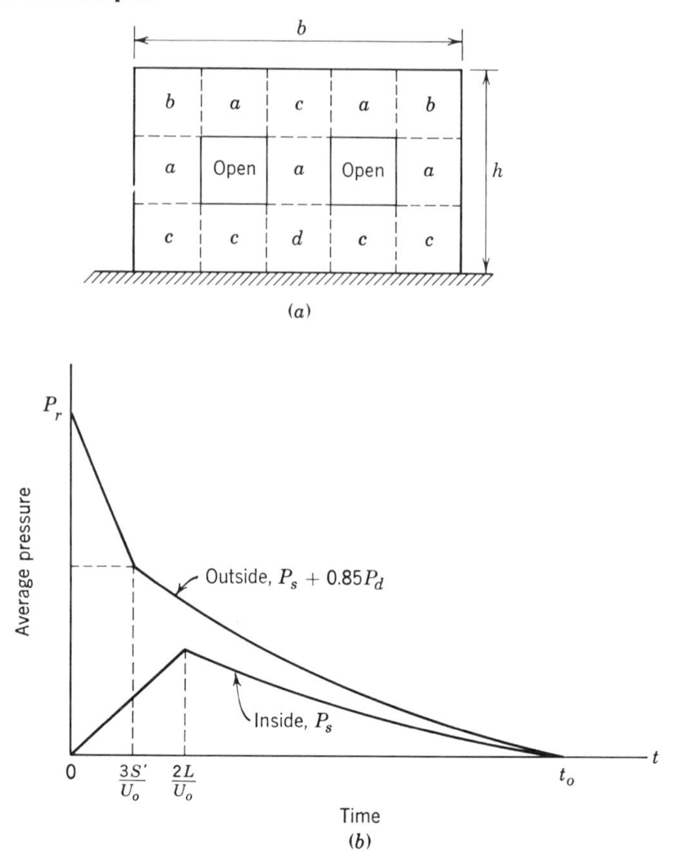

Fig. 11-18. (*a*) Subdivision of a typical wall with openings. (*b*) Time variation of front face average pressure. (Reference 40.)

takes a time $2L/U_o$ to reach the value P_s of the blast wave overpressure. The dynamic pressures P_d are assumed to be negligible on the interior of the structure. The variation with time of the average pressures of the inside and the outside of the front face are shown in Fig. 11-18*b*.

For the sides and top, the outside average pressures are obtained as for a closed structure. The inside pressures, as for the front face, require a time $2L/U_o$ to attain the overpressure of the blast wave. Here again the dynamic pressures in the interior are neglected. The variations of the pressures with time are shown in Fig. 11-19*a*.

For the back face, the outside pressure is the same as for the closed structure, but with S replaced by S' where S' is given by Eq. 11-20. The inside pressure is reflected from the inside of the back face and it takes a time equal to L/U_o to reach the same value as the blast overpressure. For

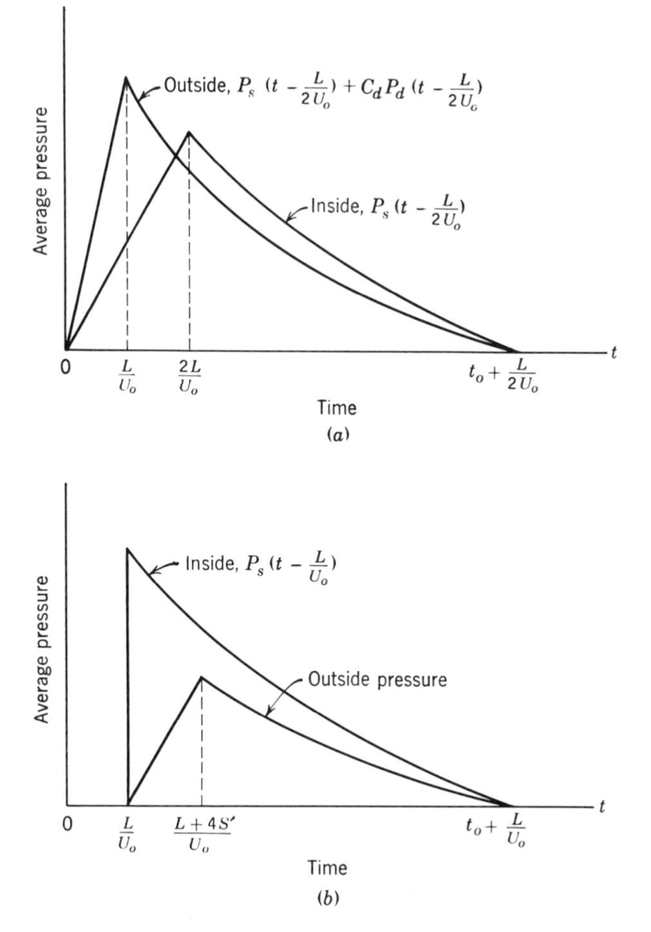

Fig. 11-19. (*a*) Time variation of side and top average pressures. (*b*) Time variation of back face average pressure. (Reference 40.)

times in excess of L/U_o, the inside pressure decays as $P_s\,(t - L/U_o)$. The dynamic pressure is assumed to be negligible. The variations of these pressures with time are shown in Fig. 11-19*b*.

The net horizontal loading is equal to the net front face loading, that is, outside minus inside, and minus the net back face loading.

11-6 DYNAMIC LOADING ON OPEN-FRAME STRUCTURES

An open-frame structure is one whose structural elements are exposed to a blast wave. For example, steel-frame office buildings whose wall areas are mostly glass, truss bridges, and so on, are classified as open-frame structures. Before the frangible material breaks, it will transmit some loading to the frame. This loading is assumed to be negligible if the frangible material is glass, provided that the blast loading is sufficiently large to fracture the glass. If the frangible material is asbestos, corrugated steel, or aluminum paneling, an approximate value of the load transmitted to the frame is an impulse of 0.04 lb-sec/in.^{2} When the frangible material breaks, the frames of the structure are directly subjected to the effects of the blast wave.

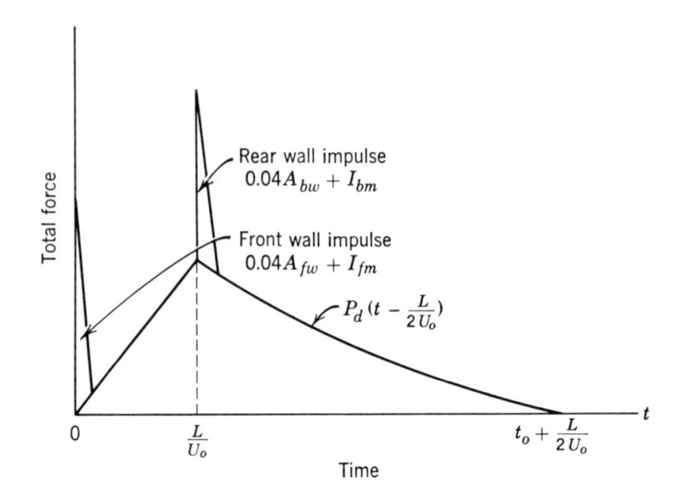

Fig. 11-20. Net horizontal loading of an open frame structure. (Reference 40.)

A simplification of the problem would be to treat the overpressure loading as an impulse. The value of this impulse is first computed for an average member in the same way as for a closed structure, and the result is multiplied by the number of members. This impulse is assumed to be delivered as soon as the shock front strikes the structure. If preferred, it can be separated into two impulses—one for the front face and one for the back face as shown in Fig. 11-20. The symbols A_{fw} and A_{bw} represent the areas of the front and back walls, respectively, which transmit loads before failure, and I_{fm} and I_{bm} are the overpressure loading impulses on front and back members.

The major portion of the loading on an open-frame structure is the drag loading. The drag coefficient C_d for an individual member in the open whose section is an I-beam, channel, angle, or rectangle, is about 1.5. When the whole frame is considered, the average drag coefficient is reduced to 1.0, because the various members shield one another to a certain extent from the effects of the full blast loading. Thus, on an individual member, the force F, that is, pressure multiplied by area, is given by the expression

$$F_{\text{member}} = C_d P_d A_i \tag{11-21}$$

where A_i is the area of the member that is projected perpendicular to the direction of propagation of the blast wave, and C_d is 1.5.

For the loading on a frame, the force F is

$$F_{\text{frame}} = C_d P_d \sum A_i = C_d P_d A \tag{11-22}$$

where C_d is 1.0 and $\sum A_i = A$ is the sum of the projected areas of all the members. This loading, versus time, for a frame of length L and with major areas in the planes of both the front and back walls, is shown in Fig. 11-20. The drag force attains its full value at the time L/U_o, that is, when the blast wave reaches the end of the structure.

11-7 DYNAMIC LOADING ON STRUCTURES WITH CYLINDRICAL SURFACES

In this category are structures of circular cross section such as telephone poles and smokestacks, arched structures with semicircular cross section, and a rough approximation to dome-shaped or spherical structures. The treatment is limited to peak overpressures that are less than 25 psi.

Consider the case of a cylindrical or semicylindrical structure where the direction of propagation of the blast wave is perpendicular to the axis of the cylinder. In particular, the pressure-time curves to be developed are those for an arched structure with a semicircular cross section. The results can be applied also to a cylindrical structure, because they consist of two semicylinders with identical loading on each half. Such an arched structure is depicted in Fig. 11-21*a*, where H is the height of the arch, or the radius of the cylinder, and P is a point on the surface. The horizontal projection X of the curved segment AP is equal to $H(1 - \cos\alpha)$ and is the horizontal distance in the direction of propagation of the blast wave. The angle between the horizontal and the line joining P to the center of curvature is designated by the symbol α.

On a flat surface, the pressure rises instantaneously to the reflected value and then soon drops to the stagnation pressure which is the sum of the incident overpressure and drag pressure. On curved surfaces, however, because of the vortex formation that occurs just after the reflection, there may be a temporary sharp drop before the stagnation pressure is reached.

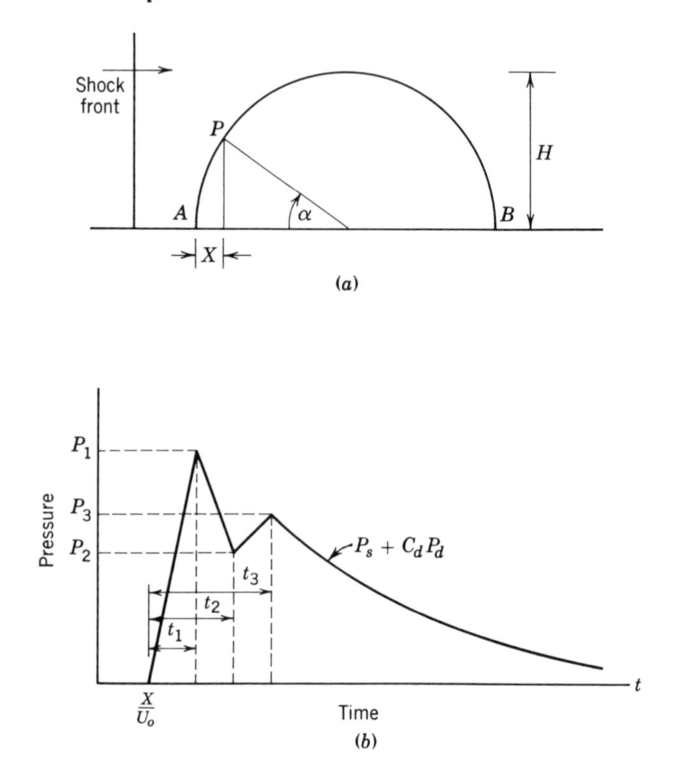

Fig. 11-21. (*a*) Typical semicircular arched structure. (*b*) Typical pressure variation at a point on an arched structure. (Reference 40.)

The variation with time of the pressure at any point P on the curved surface in Fig. 11-21*a* is shown in Fig. 11-21*b*. The shock front of the blast wave strikes first the base A of the arch at time $t = 0$ and arrives at point P at time X/U_o, regardless of whether point P is in the front or back half of the arch. In the time interval t_1, the overpressure rises sharply to the reflected value P_1. Because of vortex formation, the pressure drops to the value P_2, and again increases to the value P_3 which is the stagnation pressure. The stagnation pressure is equal to $P_s + C_d P_d$ and decays in the normal manner.

The values of P_1, P_2, and the drag coefficient C_d are dependent on the angle α, and they can be obtained from Fig. 11-22. For example, at the base A of the arch the angle α is zero and the pressure P_1 is equal to the pressure P_r, where P_r is the ideal reflected overpressure for a flat surface. The values of P_1 and P_2 in Fig. 11-22 are expressed as ratios of P_r. In the same figure, the rise time t_1 and the time intervals t_2 and t_3 in terms of the time unit H/U_o are also included. It should be noted, however, that t_2 and t_3 are independent of the angle α.

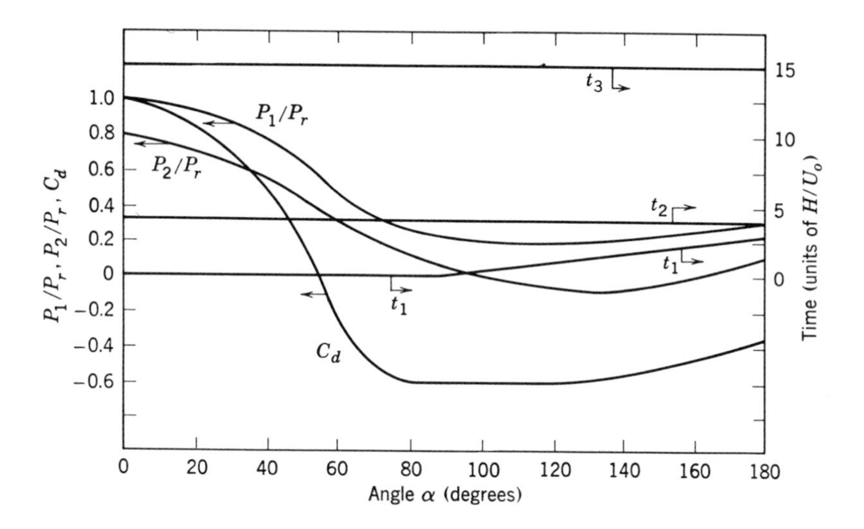

Fig. 11-22. Variation of pressure ratios, drag coefficient, and time intervals for an arched structure. (Reference 40.)

This procedure gives the normal loads on an arbitrary point P on the surface. The net horizontal loading can be determined by totaling first the horizontal components of the loads in the front face of the arch and then those in the back face of the arch. The algebraic sum of these two resultant loads yields the net horizontal loading. Usually an approximate procedure is used in practice to perform these summations. It should be pointed out,

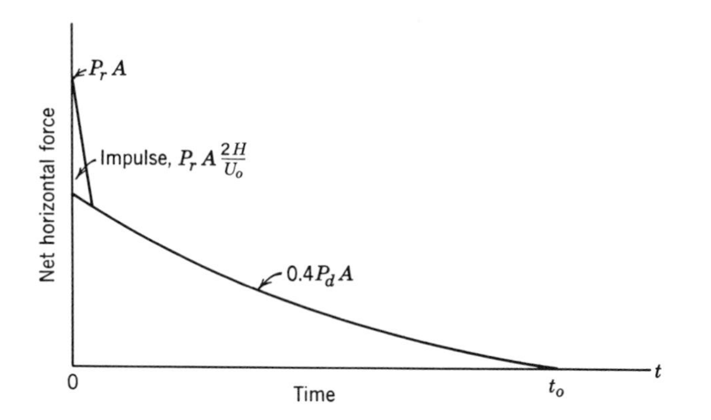

Fig. 11-23. Approximate equivalent net horizontal force loading on semicylindrical structure. (Reference 40.)

however, that in certain instances, particularly for large structures, it is the local loading rather than the net loading that is used as a criterion of damage.

As an approximate equivalent net horizontal loading, the one shown in Fig. 11-23 can be used. The loading consists of an initial impulse $P_r A\ (2H/U_o)$, where A is the projected area normal to the direction of the blast wave propagation and $2H/U_o$ is the time required for the shock front to transverse the structure. The remainder of the net horizontal loading is represented by the force $0.4P_d A$, where 0.4 is the value of the drag coefficient C_d, P_d is the dynamic pressure, and A is the area as defined above. When a frame is made up of a number of circular elements, the procedure to determine the loading is similar to that for a open-frame structure, but with C_d equal to 0.2.

11-8 DYNAMIC ANALYSIS OF STRUCTURES SUBJECTED TO BLAST LOADINGS

As an example, let it be assumed that it is required to design the roof of the one-story building in Fig. 11-13, which is subjected to an average blast wave pressure given in Fig. 11-17. This loading is the result of a 1.2-kiloton nuclear explosion with 5.45 psi peak overpressure, and it has been computed as discussed in Section 11-4.

In Chapter 10 it was pointed out that a reasonable solution for the dynamic response of a multielement structure would be to treat each component element independently. The building in Fig. 11-13 is a reinforced-concrete monolithic construction and it would be reasonable to assume that the roof is a rectangular two-way reinforced concrete slab, fixed on all edges. The unsupported lengths of the long and short sides of the slab

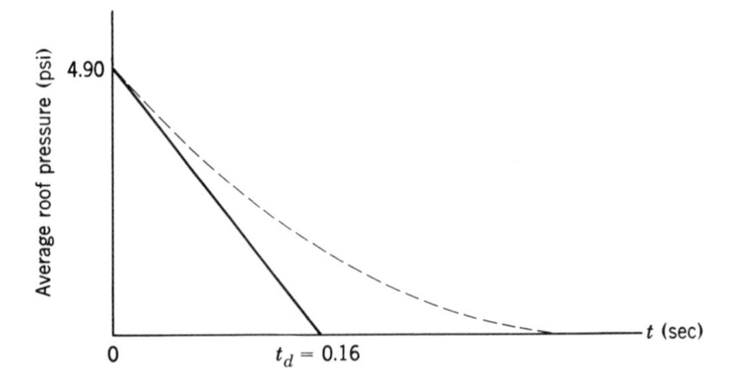

Fig. 11-24

are 20 and 15 ft, respectively, and, for convenience, let it be assumed that these are the dimensions of the roof slab. The load-time variation in Fig. 11-17 will be approximated by the triangular load in Fig. 11-24, yielding a time duration t_d equal to 0.16 sec. The rise time $L/U_o = 0.017$ sec is rather small and can be neglected. The pressure across the roof, although not quite correct, is assumed to be uniformly distributed. The slab is designed for $\eta = y_m/y_{el} \cong 4$.

For this problem, as a first approximation of the required slab strength, Eq. 10-33 is used. The use of Eq. 10-33 implies that the duration t_d of the dynamic load is assumed to be infinite, which is reasonable as a first approximation of the required resistance. Thus, from Eq. 10-33, the required resistance R_m is,

$$R_m = 4.9\left[\frac{1}{1-\dfrac{1}{(2)(4)}}\right] = 5.63 \text{ psi}$$

The required total slab resistance is

$$\text{required } R_m = (20)(15)(144)(5.63)$$

$$= 244{,}000 \text{ lb}$$

The ratio a/b of the sides of the slab is 0.75. For convenience in the computations, let it be assumed that $a/b = 0.8$. From Table 10-5, the maximum resistance R_m for $a/b = 0.8$ is

$$R_m = \frac{1}{a}[12(M_{Pfa} + M_{Psa}) + 10.3(M_{Pfb} + M_{Psb})]$$

$$= \frac{1}{15}[(12)(2)(15)M_P + (10.3)(2)(20)M_P]$$

$$= 51.40 M_P$$

Here it is assumed that $M_{Pfa} = M_{Psa} = M_{Pfb} = M_{Psb} = M_P$, where M_P is the bending resistance for a unit width. Equating the two values of R_m found above, the required M_P is

$$M_P = \frac{244{,}000}{51.40} = 4750 \text{ lb-ft/ft}$$

For bending strength, Eq. 10-29 can be used. This expression, for $\sigma_{dy} = 50$ ksi, $\sigma'_{dc} = 4$ ksi, $P = 0.015$, and $b = 12$ in., yields

$$M_P = 8.08 d^2 \text{ kip-in./in.}$$

or

$$M_P = 672d^2 \text{ lb-ft/ft}$$

By equating the two values of M_P, we have

$$672d^2 = 4750$$

or

$$d = \sqrt{4750/672} = 2.66 \text{ in.}$$

Thus consider a slab of total thickness h equal to 4.5 in. and $d = 3$ in. From Eq. 10-30, the moment of inertia I_a per inch of width b is

$$I_a = \frac{(1)(3)^3}{2}[0.083 + (5.5)(0.015)] = 2.23 \text{ in.}^4/\text{in.}$$

The trilinear resistance function can be replaced by a bilinear resistance as shown in Fig. 10-2*b*. The effective stiffness k_E can be determined by equating the area under the trilinear form up to the deflection y_2, with the area under the bilinear form up to the same point. This, for $a/b = 0.8$, yields

$$y_E = 0.130a^2 \frac{M^0_{Psb}}{EI_a} \tag{11-23}$$

$$k_E = \frac{R_m}{y_E} = \frac{430EI_a}{a^2}$$

where a is the length of the short side of the slab and I_a is the moment of inertia per inch of width.

For the problem in question,

$$k_E = \frac{(430)(3)(10^6)(2.23)}{(15^2)(144)} = 89{,}000 \text{ lb/in.}$$

The total mass m_t of the slab is

$$m_t = \frac{(4.5/12)(150)(15)(20)}{386} = 43.8 \text{ lb-sec}^2/\text{in.}$$

From Table 10-5, the load-mass factor K_{LM} for the elastic range is 0.69 and for the plastic range 0.54. A proper value for $\eta = 4$ would be about 0.58. Thus, from Eq. 10-24,

$$\tau = \frac{2\pi}{\omega} = 2\pi\sqrt{K_{LM}m_t/k_E} = 2\pi\sqrt{(0.58)(43.8)/89{,}000} = 0.106 \text{ sec}$$

Thus from Fig. 10-5

$$\frac{t_d}{\tau} = \frac{0.16}{0.106} = 1.51$$

$$\frac{R_m}{F_1} = \frac{5.63}{4.9} = 1.15$$

and

$$\eta = \frac{y_m}{y_{el}} = 1.8$$

The ratio η is lower than $\eta = 4$ and a smaller slab thickness is required. If desired, the procedure can be repeated with a lower value of the required resistance R_m. The trial-and-error procedure can be repeated as many times as is necessary to obtain the specified design value of the ductility ratio η. For this problem, one or two additional trials should be sufficient.

The design of the other component elements of the building could be carried out independently by following similar procedure and reasoning. Other problems can be solved in a similar manner.

11-9 THE EARTHQUAKE PROBLEM

The earthquake problem is probably as old as the earth itself, and very likely there is no part of the United States and the world that has not experienced, to some degree, the effects of an earthquake shock. In 1931, the earthquake activity of the United States for the year 1929 was published by N. H. Heck and R. R. Bodle of the U.S. Coast and Geodetic Survey. Typical of any year, it shows that 24 states had experienced the effects of minor, moderate, and severe earthquake shocks. Major and severe earthquakes in that year were observed in California, Ohio, Oklahoma, South Dakota, Alaska, and the Hawaiian Islands. Thus no area in the world seems to be immune from the possibility of earthquake damage.

The theories[36] of the earth's structure hold that the central core of the earth is molten nickel iron 4350 mi in diameter. The core is enveloped first by three concentric layers of dense material totaling 1050 mi, then by a mantle of highly elastic substance 700 mi thick, and finally by the outer part of the earth that is a shell of rock known as the *crust*.

The planes of cleavage of the crust of the earth, known as *faults*, are innumerable and divide this crust into very large irregular blocks called the *fault blocks*.

[36] See, for example, References 46 and 48.

The surface trace of a fault plane is known as *rift*. The fault blocks are normally in equilibrium, but there is always a possibility that a relative motion known as *slip* may take place. Such a motion can be in any direction and at any point well below the earth's surface.

If the slip is gradual, there is a similar adjustment in the continuous strata and the consequences are not usually observable at the time they occur. If, however, the slip is abrupt, a series of local movements begin to propagate through the earth; the surface vibration caused by this disturbance is known as an *earthquake*. These types of earthquakes are called tectonic. When an abrupt slip occurs, it is expected that after a certain period of time the equilibrium of the adjacent fault blocks will be restored and no further movement will occur. If, however, the forces that produced the previous slip continue to act, such movements may again take place in the future.

The design of a structure to resist the effects of an earthquake is a rather difficult task, primarily because it is almost impossible to predict the character and intensity of the earthquake to which the structure will be subjected during its lifetime. Some indications can be drawn from earthquakes that have been recorded as they occurred during the past decades. The intensities of such earthquakes were obtained by measuring their respective accelerations as a function of time. Records regarding the accelerations of such motions can be found in terms of both vertical and horizontal components.With known accelerations, the ground velocities and displacements as a function of time can be determined by integration. A sample of what is considered to be a strong earthquake is shown in Fig. 11-25. Note the maximum values of the ground acceleration, velocity, and displacement.

Such records provide a useful history of past earthquake activity that can be used to establish some reasonable criteria for the type and intensity of an earthquake that the structure should be designed to resist. However, a structure should never be located over a known fault or rift, even in the absence of all signs of seismic activity. In the event of a slip, such a structure will be destroyed. Thus, when a structural system is designed to resist the effects of an earthquake, it is assumed that provisions have been made to locate it at some distance from any existing fault or rift.

Another consideration in earthquake analysis is the additional costs. Earthquakes are not an everyday activity and a structure could be subjected to such a dynamic activity only a few times in its life span. Thus it would be highly uneconomical to design it so that it remains completely elastic. Therefore, as in blast analysis, the structure could be permitted to undergo some plastic deformation if an earthquake occurs.

The remaining sections of this chapter provide basic knowledge on methods of analysis that can be used to analyze structures for earthquake response. The stochastic approach is more extensively treated in Chapter 12.

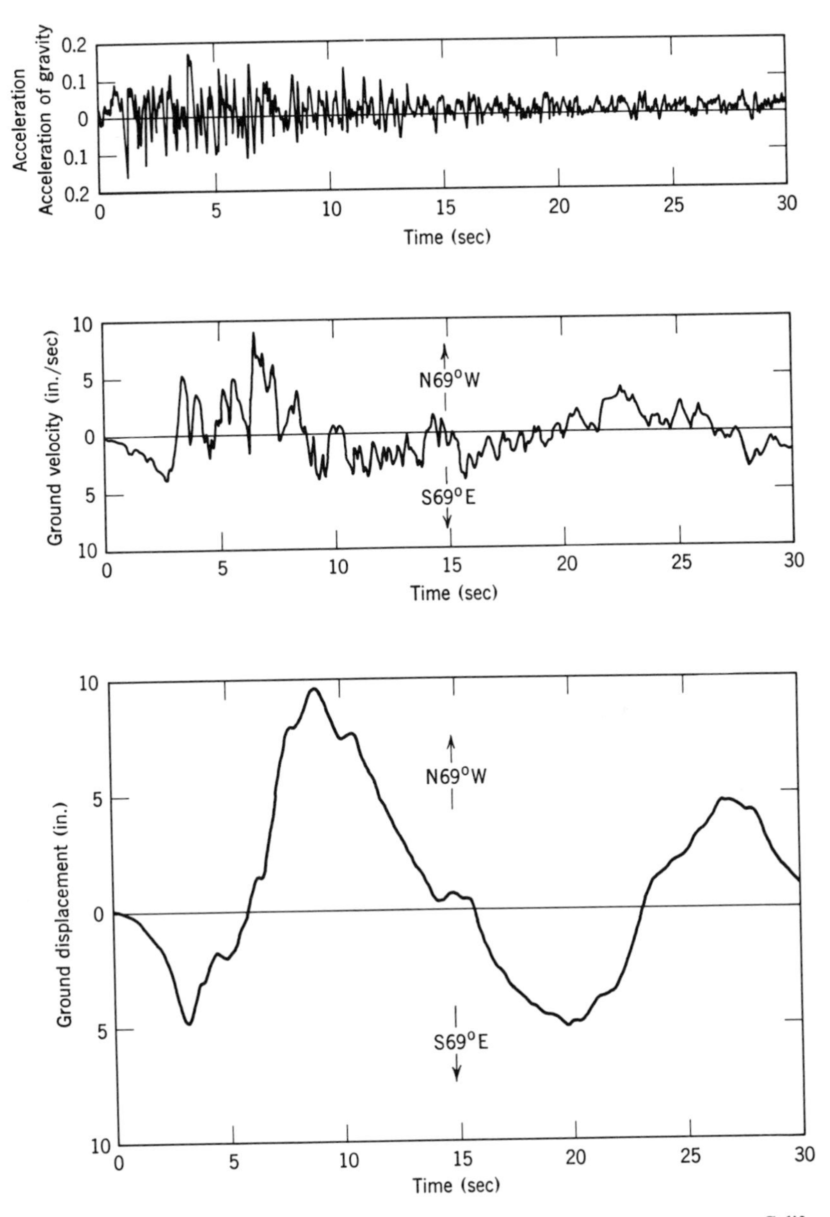

Fig. 11-25. Ground accelerations at a distance of 35 miles from the Kern County, California, earthquake, and the corresponding velocities and displacements. (Adapted from Berg and Housner, Reference 48. Accelerations record courtesy of the United States Coast and Geodetic Survey.)

11-10 EARTHQUAKE INTENSITY SCALES

In practice, the intensity of an earthquake is estimated on the basis of noninstrumental observation of damage to man-made structures and the sensations of persons who have experienced its effects. On this basis, several scales have been developed during the past decades that classify earthquakes by degrees of intensity. Thus in 1904 A. Cancani introduced a dynamic intensity scale having 12 degrees of intensity. This scale is shown in Table 11-1, where each degree is represented by the maximum ground accelerations involved. Later it was found that the correlation between earthquake damage and maximum ground acceleration was rather poor,[37] and scales based on more reasonable criteria were developed. An example of such a scale is given in Table 11-2. This is a modification and condensation of the Mercalli-Cancani-Sieberg scale which was worked out by H. Wood and F. Neumann of the U.S. Coast and Geodetic Survey.[38]

Moderate earthquakes are much more frequent than the higher-intensity ones and from the point of view of economics, structures subjected to such earthquakes can be designed for completely elastic response. Stronger earthquakes are less frequent and, since a structure will experience such earthquake intensities only a few times, if at all, during its life span, some

TABLE 11-1 CANCANI DYNAMIC INTENSITY SCALE

	Degree	Maximum Acceleration (mm/sec^2)
I.	Instrumental	0 – 2.5
II.	Very slight	2.5 – 5
III.	Slight	5 – 10
IV.	Moderate	10 – 25
V.	Rather strong	25 – 50
VI.	Strong	50 – 100
VII.	Very strong	100 – 250
VIII.	Ruinous	250 – 500
IX.	Disastrous	500 – 1,000
X.	Very disastrous	1,000 – 2,500
XI.	Catastrophe	2,500 – 5,500
XII.	Enormous catastrophe	5,000 – 10,000

($g = 9806\ mm/sec^2$)

[37] See, for example, Reference 51.

[38] Additional information regarding this scale is given in Reference 52.

structural damage could be tolerated. For such cases, it would be more appropriate to consider elastic-plastic response, provided a ductility ratio that is not hazardous to life is used.

TABLE 11-2 MODIFIED MERCALLI INTENSITY SCALE

I. Not felt except under particularly favorable circumstances.

II. Felt by persons at rest favorably placed, such as upper floors.

III. Felt indoors; hanging objects swing. Vibration is like that from a passing light truck, and people may not recognize the seismic nature of the disturbance. Duration estimated.

IV. Vibration like that from the passing of heavy trucks; hanging objects swing. Dishes, windows, doors rattle; sensation like heavy truck striking building. Walls crack.

V. Felt outdoors; many awakened; direction estimated. Some dishes, windows, and so on, broken; pendulum clocks may stop, start, change rate. A few instances of cracked plaster; unstable objects overturned.

VI. Felt by all; many frightened and run outdoors. Persons walk unsteadily; some heavy furniture moved. Damage slight.

VII. Difficult to stand; everybody runs outdoors. Fall of plaster; slight to moderate damage in well-built ordinary structures; considerable damage in poorly built or badly designed structures; negligible damage to buildings of good design and construction. Large church bells ring. Noticed by persons driving cars.

VIII. Steering of motor cars affected. Partial collapse of ordinary buildings; great damage in poorly built structures; slight damage in specially designed brick structures. Fall of factory stacks, columns, chimneys, monuments, towers, and walls. Heavy furniture overturned. Cracks in wet ground and on steep slopes.

IX. General panic. Considerable damage in ordinary substantial buildings with partial collapse; great damage in poorly built structures; slight damage in specially designed brick structures. Serious damage to reservoirs. Underground pipes broken. Conspicuous cracks in ground, and buildings shifted off foundations.

X. Most masonry and frame structures destroyed with their foundations. Some well-built wooden structures and bridges destroyed. Rails bent and ground badly cracked. Sand and mud shifted horizontally on beaches and flat land. Large landslides. Water splashed over banks.

XI. Underground pipelines completely out of order. Rails bent greatly. Bridges destroyed and few if any masonry structures remain standing.

XII. Nearly total damage. Objects thrown into the air. Large rock masses displaced. Lines of sight and level destroyed.

11-11 EARTHQUAKE RESPONSE OF SINGLE-STORY STRUCTURES

Consider the single-story frame structure in Fig. 11-26*a* that is subjected to a ground motion represented by the support displacement u_s shown in the figure. The displacement of the top of the frame relative to the ground is u. On this basis, the total horizontal displacement x of the top of the frame is

$$x = u_s + u \tag{11-24}$$

The idealized one-degree system is determined by assuming that the girders are infinitely stiff as compared to the columns; it is shown in Fig. 11-26*b*. Under the influence of viscous damping, the free-body diagram of the mass m is shown in Fig. 11-26*c*. Applying dynamic equilibrium, the equation of motion is

$$m\ddot{x} + ku + c\dot{u} = 0$$

or, by Eq. 11-24,

$$m\ddot{u} + ku + c\dot{u} = -m\ddot{u}_s \tag{11-25}$$

The equation can be written as

$$\ddot{u} + 2\omega\zeta\dot{u} + \omega^2 u = -\ddot{u}_s \tag{11-26}$$

where $\omega = \sqrt{k/m}$ is the undamped natural frequency of the system and $\zeta = c/c_{cr} = c/2m\omega$ is the damping ratio.

The solution of Eq. 11-26 can be obtained directly from Eq. 2-30 in Section 2-5 by using as $F(T)$ the inertia force $m\ddot{u}_s$. Thus, with initial conditions $u_o = \dot{u}_o = 0$ at $t = 0$, Eq. 2-30 yields

$$u(t) = \frac{1}{\omega_d}\int_0^t [-\ddot{u}_s(T)]\, e^{-\mu(t-T)} \sin\omega_d(t-T)\,dT \tag{11-27}$$

where $\omega_d = \omega(1-\zeta^2)^{1/2}$ is the damped frequency of the system and $\mu = c/2m = \zeta\omega$. For the majority of structures ζ is usually small, and ω_d could be taken as equal to ω without introducing any appreciable error in the solution for dynamic response.

Equation 11-27 gives the horizontal displacement u of the top of the frame relative to the ground. The ground acceleration $\ddot{u}_s$ under the integral sign can be selected to represent the earthquake that the frame structure should be designed to resist. The type and intensity of the acceleration function $\ddot{u}_s$ can be selected from measurements of actual earthquakes that occurred during the past decades. A sample of high-intensity earthquake is shown in Fig. 11-25.

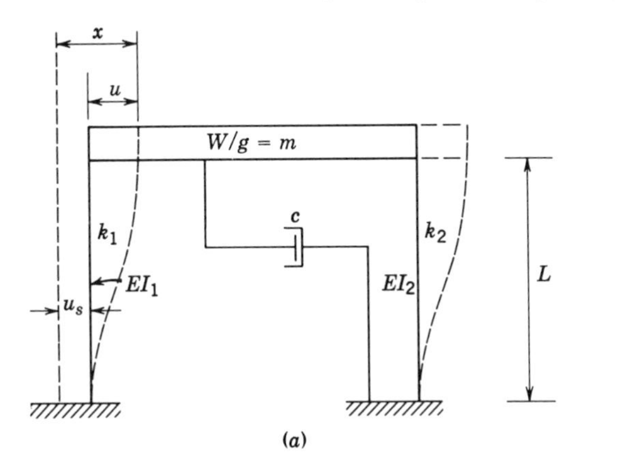

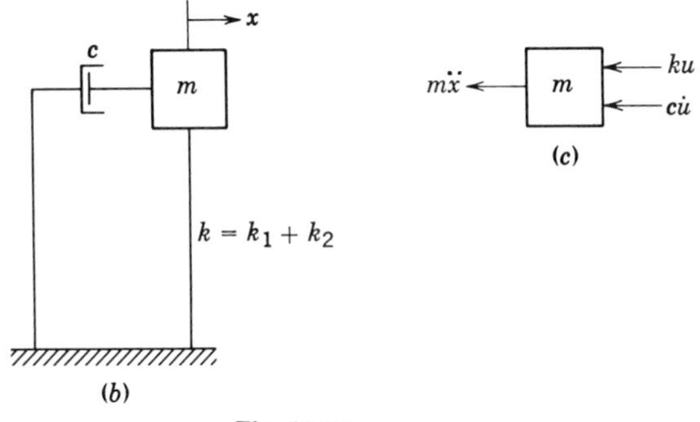

Fig. 11-26

The total shear force V at the column tops of the frame is

$$V = ku \tag{11-28}$$

or

$$V = \frac{k}{\omega_d} \int_0^t (-\ddot{u}_s)\, e^{-\mu(t-T)} \sin \omega_d (t-T)\, dT \tag{11-29}$$

The shear force is a function of time, and its value at any time t is distributed among the columns of the frame in proportion to their k's. With known shear forces, the bending moments and bending stresses can be determined in the usual way.

For many structures damping is usually small and can be neglected without appreciable loss of accuracy. Thus, if damping is neglected, Eq. 11-27 yields

$$u(t) = \frac{1}{\omega} \int_0^t [-\ddot{u}_s(T)] \sin \omega (t-T)\, dT \tag{11-30}$$

and the total shear force V at the column tops is

$$V = ku$$

$$= \frac{k}{\omega} \int_0^t [-\ddot{u}_s(T)] \sin \omega (t - T) \, dT \tag{11-31}$$

As an illustration, let it be assumed that it is required to analyze the frame in Fig. 11-26 for an earthquake with acceleration $\ddot{u}_s$ given by the expression

$$\ddot{u}_s = 0.2 g \sin \omega_s T \tag{11-32}$$

where g is the acceleration of gravity and T is time.

The period τ_s of this acceleration is

$$\tau_s = \frac{2\pi}{\omega_s} = 0.5 \text{ sec} \tag{11-33}$$

and the duration of $\ddot{u}_s$ is 1 sec. It is further assumed that k_1 and k_2 in Fig. 11-26*a* are equal and that damping is zero.

With $k = m\omega^2$, Eq. 11-31 yields

$$V = -0.2 gm\omega \int_0^t \sin \omega_s T \sin \omega (t - T) \, dT \tag{11-34}$$

In this expression, the quantity

$$G = -0.2 g\omega \int_0^t \sin \omega_s T \sin \omega (t - T) \, dT \tag{11-35}$$

is known as the effective acceleration.

The maximum numerical value of V corresponds to the maximum absolute value of G. For a practical problem, only the maximum numerical value of the shear force, denoted as $V_{\max}$, is usually of importance. Thus

$$V_{\max} = mG_m \tag{11-36}$$

where G_m denotes the maximum absolute value of G. The representation of G_m as a function of the natural period of vibration τ is known as the *acceleration spectrum.*

In Eq. 11-35, the natural frequency ω is equal to $2\pi/\tau$. Keeping this in mind and using Eq. 11-33, we have the expression for G, Eq. 11-35, yielding

$$G = -0.2 g \frac{2\pi}{\tau} \int_0^t \sin \frac{2\pi}{\tau_s} T \sin \frac{2\pi}{\tau} (t - T) \, dT \tag{11-37}$$

The acceleration spectrum for this problem can be obtained by assuming various values of τ in Eq. 11-37 and determining the corresponding G_m. The plot of G_m versus τ is the acceleration spectrum. If for a given earthquake such a spectrum is known, the maximum shear force is readily determined from Eq. 11-36. For the earthquake given by Eq. 11-32, the acceleration spectrum can be obtained from Eq. 11-37 as discussed above. With known V_{max}, the maximum bending moments and bending stresses can be determined in the usual way.

The use of Eq. 11-30 for earthquake analysis and design was first introduced by M. A. Biot.[39] The application of this method to multistory building is discussed in the next section.

11-12 EARTHQUAKE RESPONSE OF MULTISTORY BUILDINGS

The procedure discussed in the preceding section, after certain modifications and simplifying assumptions, can be used to determine the dynamic response of multistory buildings. It should be pointed out, however, that the analysis is based on elastic response and is appropriate only in the cases of moderate earthquake ground motions.

As an illustration, consider the 12-story building in Fig. 11-27*a* that is subjected to an earthquake ground motion of moderate intensity. The floors are assumed to be identical and the girders are infinitely stiff as compared to the columns. These are not severe limitations, because for many buildings the plans are symmetrical and the mass and rigidity are often nearly uniform throughout the building, in order to prevent rotation about a vertical axis.

Based on these assumptions, the building can be replaced by a hypothetical prismatic beam[40] of homogeneous material that has elasticity only in shear. This beam is shown in Fig. 11-27*b*, where u_s is the support displacement due to the ground motion and u is the shear displacement at any cross section *mn*. Thus the total horizontal displacement x of the beam at any location y is

$$x = u_s + u \tag{11-38}$$

The shear strain at the cross section *mn*, Fig. 11-27*b*, is $\partial u/\partial y$ and the shear force V at the same cross section is

$$V = kh\frac{\partial u}{\partial y} \tag{11-39}$$

where k is the sum of the k 's of the columns in each floor and h is their

[39] See M. A. Biot, *Proceedings of the National Academy of Sciences*, Volume 19, 1933. See also Reference 47, Chapter 12.

[40] Considerable additional information on this subject can be found in Reference 53.

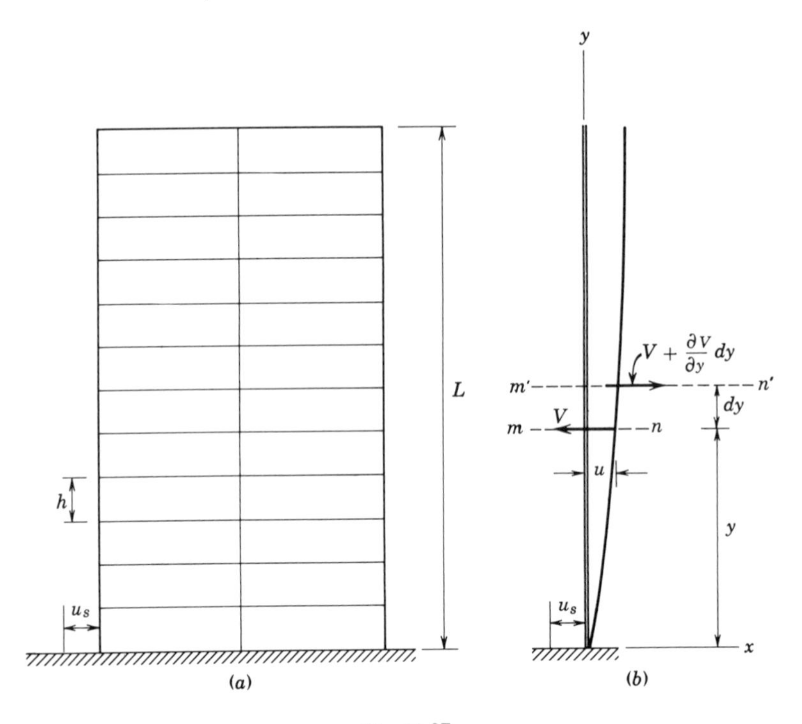

Fig. 11-27

length. Equations 3-2 and 3-3 in Section 3-3 can be used to determine k.

The free-body diagram of the element dy in Fig. 11-27b reveals that the difference in shear force between sections mn and $m'n'$ must be equilibrated by the inertia force of the element dy. Thus, if q is the weight per unit length of the hypothetical beam, the differential equation of motion is

$$kh\frac{\partial^2 u}{\partial y^2} = \frac{q}{g}(\ddot{u}_s + \ddot{u})$$

or

$$kh\frac{\partial^2 u}{\partial y^2} - \frac{q}{g}(\ddot{u}_s + \ddot{u}) = 0 \tag{11-40}$$

Here $\ddot{u}_s$ is the ground acceleration produced by the earthquake. With

$$\Psi^2 = \frac{khg}{q} \tag{11-41}$$

Eq. 11-40 yields

$$\Psi^2\frac{\partial^2 u}{\partial y^2} - \ddot{u} = \ddot{u}_s \tag{11-42}$$

The natural frequencies and mode shapes of the hypothetical beam can be determined from the differential equation

$$\Psi^2 \frac{\partial^2 u}{\partial y^2} - \ddot{u} = 0 \tag{11-43}$$

By assuming

$$u = X \sin \omega t \tag{11-44}$$

Eq. 11-43 yields

$$\Psi^2 \frac{\partial^2 X}{\partial y^2} + \omega^2 X = 0 \tag{11-45}$$

The general solution of Eq. 11-45 is

$$X = A \sin \frac{\omega y}{\Psi} + B \cos \frac{\omega y}{\Psi} \tag{11-46}$$

where A and B are constants. The boundary condition $u=0$ at $y=0$ yields $B=0$. The boundary condition $\partial u/\partial y=0$ at $y=L$, for a solution other than the trivial one, requires

$$\cos \frac{\omega L}{\Psi} = 0$$

or

$$\frac{\omega L}{\Psi} = \frac{\pi}{2}, \frac{3\pi}{2}, \frac{5\pi}{2}, \ldots$$

For any mode n, where $n = 1, 3, 5, \ldots$, these results yield

$$\omega_n = \frac{n\pi\psi}{2L} \qquad \tau_n = \frac{2\pi}{\omega_n} = \frac{4L}{n\psi} \tag{11-47}$$

$$X_n = A_n \sin \frac{n\pi y}{2L} \tag{11-48}$$

and, from Eq. 11-44,

$$u_n = A_n \sin \frac{n\pi y}{2L} \sin \frac{n\pi\psi t}{2L} \tag{11-49}$$

The solution of Eq. 11-42 can be taken as

$$u_n = f_n(t) \sin \frac{n\pi y}{2L} \qquad n = 1, 3, 5, \ldots \tag{11-50}$$

where $f_n(t)$ is a function of time that needs to be determined. By substituting Eq. 11-50 into Eq. 11-42,

$$\ddot{f}_n \sin\frac{n\pi y}{2L} + \frac{n^2\pi^2\Psi^2}{4L^2} f_n \sin\frac{n\pi y}{2L} = -\ddot{u}_s$$

Multiplying by $\sin(n\pi y/2L)dy$ and then integrating from $y=0$ to $y=L$, leads to

$$\ddot{f}_n + \omega_n^2 f_n = -\frac{4}{n\pi}\ddot{u}_s \tag{11-51}$$

where ω_n is given by Eq. 11-47.

Equation 11-51 has the same form as Eq. 11-26 with $\zeta=0$, and its solution is similar to that given by Eq. 11-30. Thus, with $F(T)$ in Eq. 11-30 replaced by $(-4/n\pi)\ddot{u}_s$, the solution of Eq. 11-51 can be written as

$$f_n = \frac{8L}{n^2\pi^2\Psi}\int_0^t [-\ddot{u}_s(T)] \sin\frac{n\pi\Psi}{2L}(t-T)\,dT \tag{11-52}$$

Therefore, Eq. 11-50 yields

$$u_n = \left\{\frac{8L}{n^2\Psi\pi^2}\int_0^t [-\ddot{u}_s(T)] \sin\frac{n\pi\Psi}{2L}(t-T)\,dT\right\} \sin\frac{n\pi y}{2L} \tag{11-53}$$

Equation 11-53 gives the dynamic response of the hypothetical beam in the nth mode. By superimposing all modes, the total response $u(t)$ is

$$u(t) = \frac{8L}{\pi^2\Psi}\sum_{n=1,3,5,\ldots}\left\{\frac{1}{n^2}\int_0^t [-u_s(T)]\sin\frac{n\pi\Psi}{2L}(t-T)\,dT\right\}\sin\frac{n\pi y}{2L} \tag{11-54}$$

The corresponding shear force V at any cross section of the hypothetical beam can be obtained from Eq. 11-39. That is,

$$V = \frac{4kh}{\pi\Psi}\sum_{n=1,3,5}\left\{\frac{1}{n}\int_0^t [-\ddot{u}_s(T)] \sin\frac{n\pi\Psi}{2L}(t-T)\,dT\right\}\cos\frac{n\pi y}{2L} \tag{11-55}$$

In terms of the coordinate y, Eq. 11-55 shows that the maximum shear force occurs at $y=0$. It should be pointed out, however, that $u_n(t)$ in Eq. 11-53 or $u(t)$ in Eq. 11-54 vary with time, hence the shear force in Eq. 11-55 will also vary with time. The time at which V becomes maximum thus should be also determined.

In practice, one way to determine a maximum value for the shear force V is to follow the concept of the effective acceleration as discussed in the preceding section. By utilizing Eq. 11-47, the modal response in Eq. 11-53 can be written as

$$u_n = \left\{ \frac{16L^2}{n^3\pi^3\Psi^2} \frac{2}{\tau_n} \int_0^t [-\ddot{u}_s(T)] \sin \frac{n\pi\Psi}{2L}(t-T)\,dT \right\} \sin \frac{n\pi y}{2L} \tag{11-56}$$

From this equation, the effective acceleration G_n is

$$G_n = \frac{2\pi}{\tau_n} \int_0^t [-\ddot{u}_s(T)] \sin \frac{n\pi\Psi}{2L}(t-T)\,dT \tag{11-57}$$

If the maximum absolute value of G_n is denoted by $(G_n)_{\max}$, the sum of all mode responses yields the total absolute maximum response

$$u = \frac{16L^2}{\pi^3\Psi^2} \sum_{n=1,3,5,\ldots} \frac{(G_n)_{\max}}{n^3} \sin \frac{n\pi y}{2L} \tag{11-58}$$

Thus

$$V = \frac{8kh}{\pi^2\Psi^2} \sum_{n=1,3,5,\ldots} \frac{(G_n)_{\max}}{n^2} \cos \frac{n\pi y}{2L} \tag{11-59}$$

At $y=0$ the shear force V gains its maximum value $V_{\max}$, and Eq. 11-59 becomes

$$V_{\max} = \frac{8kh}{\pi^2\Psi^2} \sum_{n=1,3,5,\ldots} \frac{(G_n)_{\max}}{n^2} \tag{11-60}$$

Thus, for a given acceleration spectrum, the absolute maximum value of the shear force can be determined from Eq. 11-60. With known shear forces, the dynamic stresses can be determined in the usual way. It should be noted, however, that damping is neglected and that $V_{\max}$ in Eq. 11-60 is only an upper limit of the maximum shear force.

11-13 MODAL ANALYSIS OF EARTHQUAKE RESPONSE

The method of modal analysis as developed in Chapter 5 can be used to determine the dynamic response of multiple degree of freedom systems subjected to earthquake ground motions. For a lumped-mass system under the action of external dynamic forces, the modal equation of motion is

$$\ddot{Y}_p + \omega_p{}^2 Y_p = g(t) \frac{\sum_{i=1}^{r} F_i \beta_{ip}}{\sum_{i=1}^{r} m_i \beta_{ip}^2} \tag{11-61}$$

where $g(t)$ is the time function of the applied dynamic forces and $F_i g(t)$ is the dynamic force acting on the mass m_i.

The effects of viscous damping can be easily included in Eq. 11-61 by adding a third term in its right-hand side as follows:

$$\ddot{Y}_p + \omega_p{}^2 Y_p + 2\mu \dot{Y}_p = g(t) \frac{\sum_{i=1}^{r} F_i \beta_{ip}}{\sum_{i=1}^{r} m_i \beta_{ip}^2} \tag{11-62}$$

where the damping coefficient μ is equal to $\zeta\omega$.

For a single degree of freedom system such as that in Fig. 11-26*b*, the differential equation of motion is given by Eq. 11-26 and its solution by Eq. 11-27. This solution, by assuming[41] that $\omega_d = \omega$, is written again as

$$u(t) = \frac{1}{\omega} \int_0^t (-\ddot{u}_s) e^{-\mu(t-T)} \sin\omega(t-T)\, dT \tag{11-63}$$

where $\ddot{u}_s$ is the applied ground or support acceleration and $u(t)$ is the displacement of the mass m relative to the ground. In this equation, the support acceleration $\ddot{u}_s$ can be written as

$$\ddot{u}_s = \ddot{u}_{so} f_s(t) \tag{11-64}$$

where $\ddot{u}_{so}$ is the maximum support acceleration and $f_s(t)$ is the time variation. Thus Eq. 11-63 becomes

$$u(t) = -\frac{\ddot{u}_{so}}{\omega} \int_0^t f_s(T)\, e^{-\mu(t-T)} \sin\omega(t-T)\, dT \tag{11-65}$$

[41] In practical situations damping is usually light, and it is reasonable to assume that the damped frequency ω_d is equal to the natural undamped frequency ω of the system. If preferred, it can be retained as ω_d.

If there is no damping, μ in the expression above is zero, and Eq. 11-65 becomes

$$u(t) = -\frac{\ddot{u}_{so}}{\omega}\int_0^t f_s(T)\sin\omega(t-T)\,dT \qquad (11\text{-}66)$$

If the right-hand side of Eq. 11-65 is multiplied and divided by the product $m\omega$ and k is substituted for $m\omega^2$, we have

$$u(t) = -\frac{m\ddot{u}_{so}\omega}{k}\int_0^t f_s(T)\,e^{-\mu(t-T)}\sin\omega(t-T)\,dT \qquad (11\text{-}67)$$

This equation can also be written as

$$u(t) = -u_{st}\Gamma = -\frac{\ddot{u}_{so}}{\omega^2}\Gamma \qquad (11\text{-}68)$$

where

$$u_{st} = \frac{m\ddot{u}_{so}}{k} = \frac{\ddot{u}_{so}}{\omega^2} \qquad (11\text{-}69)$$

and

$$\Gamma = \omega\int_0^t f_s(T)\,e^{-\mu(t-T)}\sin\omega(t-T)\,dT \qquad (11\text{-}70)$$

The static deflection u_{st} and the magnification factor Γ in Eq. 11-68 serve the same purpose as y_{st} and Γ in Chapter 5. When there is no damping, Eq. 11-70 yields

$$\Gamma = \omega\int_0^t f_s(T)\sin\omega(t-T)\,dT \qquad (11\text{-}71)$$

From these solutions, it is easily observed that support motions produce responses that are equivalent to those produced by the applied forces $-m\ddot{u}_s$, where $\ddot{u}_s$ can be expressed as shown in Eq. 11-64. For multidegree systems, the applied force on the mass m_i is $-m_i\ddot{u}_s$. Thus, by replacing $F_i g(t)$ in Eqs. 11-61 and 11-62 by $-m_i\ddot{u}_s$ and also using Eq. 11-64, we have

$$\ddot{Y}_p + \omega_p^2 Y_p = -f_s(t)\,\ddot{u}_{so}\frac{\sum_{i=1}^{r} m_i\beta_{ip}}{\sum_{i=1}^{r} m_i\beta_{ip}^2} \qquad (11\text{-}72)$$

and

$$\ddot{Y}_p + \omega_p{}^2 Y_p + 2\mu \dot{Y}_p = -f_s(t)\ddot{u}_{so} \frac{\sum_{i=1}^{r} m_i \beta_{ip}}{\sum_{i=1}^{r} m_i \beta_{ip}^2} \tag{11-73}$$

Equations 11-72 and 11-73 are the modal equations for a mode p in the case of support motion. Note that Eq. 11-73 takes into consideration the effects of viscous damping.

In the last two equations, the summations on the right-hand side are constants. In practice, the factor Λ_p defined as

$$\Lambda_p = \frac{\sum_{i=1}^{r} m_i \beta_{ip}}{\sum_{i=1}^{r} m_i \beta_{ip}^2} \tag{11-74}$$

is known as the *modal participation factor.* Thus the modal displacement $Y_p(t)$ relative to the support can be determined from the expression

$$Y_p(t) = \Lambda_p u_p(t) = -\Lambda_p \frac{\ddot{u}_{so}}{\omega_p{}^2} \Gamma_p \tag{11-75}$$

where $u_p(t)$ is given by Eq. 11-68. For any natural frequency ω_p, the magnification factor can be determined from Eq. 11-70 when viscous damping is considered and from Eq. 11-71 for the undamped case.

For the mass m_i, the modal displacement $u_{ip}(t)$ is given by the expression

$$u_{ip}(t) = -\Lambda_p \Gamma_p \beta_{ip} \frac{\ddot{u}_{so}}{\omega_p{}^2} \tag{11-76}$$

By superimposing the responses of all modes, the total relative displacement $u_i(t)$

$$u_i(t) = -\sum_{p=1}^{N} \Lambda_p \Gamma_p \beta_{ip} \frac{\ddot{u}_{so}}{\omega_p{}^2} \tag{11-77}$$

where N on the summation sign is the total number of modes.

As an illustration of the application of the modal analysis method, consider again the two-story rigid frame building in Fig. 3-5a and let it be

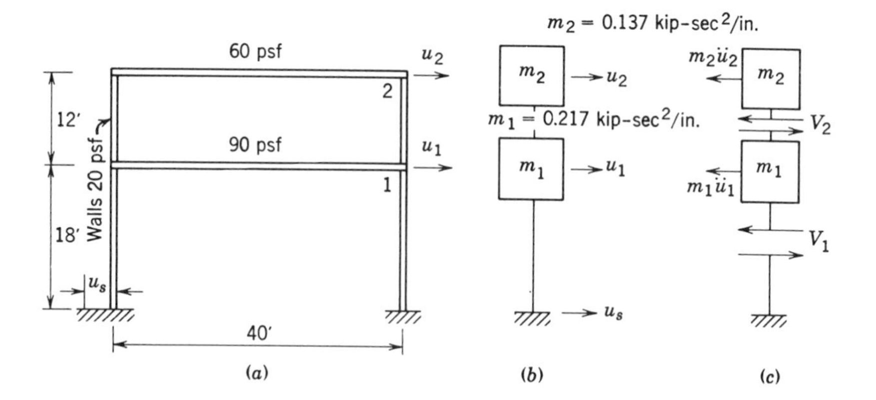

Fig. 11-28

assumed that it is subjected to an artificial earthquake of acceleration $\ddot{u}_s$ given by the expression

$$\ddot{u}_s = \ddot{u}_{so} f_s(t) = 0.1\, g \sin \omega_s t \tag{11-78}$$

Here $\ddot{u}_{so}$ is equal to 0.1 g, where g is the acceleration of gravity, and $\sin \omega_s t$ is the time function $f_s(t)$. The duration of $\ddot{u}_s$ is 1 sec and ω_s is equal to 4π.

The elastic properties of the building in Fig. 3-5*a* are uniform throughout its length, and its dynamic response due to the ground motion $\ddot{u}_s$ will be computed by considering only the interior frame in Fig. 11-28*a* together with the associated wall and floor areas. The steel sections of the girders and columns are 21 W 62 and 10 W 25, respectively.

By taking into consideration the flexibility of the girders, the undamped natural frequencies and the corresponding mode shapes of the idealized two-degree system, Fig. 11-28*b*, are determined in Section 5-14:

$$\omega_1 = 4.65 \text{ rps}$$

$$\omega_2 = 17.90 \text{ rps}$$

First mode	*Second mode*
$\beta_{11} = 1.00$	$\beta_{12} = 1.00$
$\beta_{21} = 1.21$	$\beta_{22} = -1.30$

By considering the case of completely elastic response, the modal participation factors, Eq. 11-74, are

$$\Lambda_1 = \frac{(0.217)(1.00) + (0.137)(1.21)}{(0.217)(1.00)^2 + (0.137)(1.21)^2} = 0.915$$

$$\Lambda_2 = \frac{(0.217)(1.00) + (0.137)(-1.30)}{(0.217)(1.00)^2 + (0.137)(-1.30)^2} = 0.087$$

The magnification factors can be determined from Eq. 11-71. For a sinusoidal time function $f_s(t)$, the expression for these factors is given in Table 2-1. That is,

$$\Gamma_p = \frac{1}{(1-\omega_s^2/\omega^2)}\left(\sin\omega_s t - \frac{\omega_s}{\omega}\sin\omega t\right) \qquad (11\text{-}79)$$

Therefore,

$$\Gamma_1 = -0.159\sin 4\pi t + 0.429\sin 4.65t$$

$$\Gamma_2 = 1.965\sin 4\pi t - 1.389\sin 17.9t$$

From Eq. 11-77, the relative horizontal displacements $u_1(t)$ and $u_2(t)$ of the masses m_1 and m_2, respectively, are

$$u_1(t) = 0.239\sin 4\pi t - 0.703\sin 4.65t + 0.0145\sin 17.9t$$

$$u_2(t) = 0.342\sin 4\pi t - 0.850\sin 4.65t - 0.0188\sin 17.9t$$

The displacement $u_1(t)$ above gives the horizontal movement of the tops of the first-story columns relative to their lower end, which in this case is fixed. The horizontal displacement of the tops of the second-story columns relative to the column tops of the floor below is $u_2(t) - u_1(t)$. If this displacement is designated as $u_{r2}(t)$, the result is

$$\begin{aligned} u_{r2}(t) &= u_2(t) - u_1(t) \\ &= 0.103\sin 4\pi t - 0.147\sin 4.65t - 0.0333\sin 17.9t \end{aligned}$$

The k_{ij} stiffness coefficients for this frame are computed in Section 1-15:

$$k_{11} = 36.4 \text{ kips/in.} \qquad k_{12} = -25.8 \text{ kips/in.}$$

$$k_{21} = -25.8 \text{ kips/in.} \qquad k_{22} = 24.2 \text{ kips/in.}$$

They are shown in Fig. 5-10. The shear forces V_1 and V_2, Fig. 11-28*c*, are

$$V_1 = (36.4 - 25.8)u_1(t) - (25.8 - 24.2)u_2(t)$$

$$= 10.6u_1(t) - 1.6u_2(t)$$

$$V_2 = 24.2u_2(t) - 25.8u_1(t)$$

The shear forces above vary with time. Their maximum values could be easily obtained graphically by plotting V_1 and V_2 versus time.

With known displacements u_1 and u_2, the bending moments at the top and bottom of each column can be computed by multiplying the moments due to unit deflections in Fig. 5-10 by the actual deflections. This is shown in Section 5-14. When the bending moments are known, the bending stresses can be determined in the usual way. On the other hand, since the shear forces are related to the bending moments, known principles of structural analysis can be also used to determine the moments in the columns as well as the stresses involved.

It should be pointed out that in practice the superposition of the mode responses is usually done in one of three ways. The most conservative approach that yields an upper limit is to add numerically the responses of the modes. This approach yields reasonable results for cases where the contribution of the fundamental mode predominates. For many problems this is usually true.

A less conservative approach is to take the sum of the fundamental mode response plus the square root of the sum of the squares of the higher modes. This will yield more reasonable results for cases where the contributions of the higher modes are appreciable. A third approach is to obtain a total maximum response by taking the root mean square, that is, the square root of the sum of the squares of the modal maximum responses. This is based on the assumption that the modal components are random variables. This assumption is reasonable, since the input is random. The accuracy of this approach is dependent on the number of mode responses considered.

11-14 INELASTIC RESPONSE OF MULTISTORY STRUCTURES

The analysis in the preceding sections was based on elastic response. For strong earthquakes, however, an elastoplastic analysis would be more appropriate because such earthquakes are experienced by a structure only a few times in its life span, if at all, and some damage to the structure can be tolerated.

When a multistory structure is permitted to become inelastic, it would be reasonable to neglect the flexibility of the girders by assuming that their stiffness is infinite as compared to the columns. In such cases, yielding usually occurs in the story that is comparatively speaking the weakest in transmitting the magnitudes of the shear forces. Investigations[42] have shown that in many cases this yielding occurs near the base of the structure. Then the magnitudes of the shear forces in the upper part of the structure are reduced as compared with the values obtained by an elastic analysis for the same base earthquake motion. Consequently, if the design of a structure is based on some fraction of the maximum value of the critical shear obtained by an elastic analysis, yielding will occur in the weakest story and the shear forces in the remaining parts of the structure will have appropriately revised values.

Consider for example the two-story frame in Fig. 11-28*a* and assume that the girders are infinitely stiff as compared to the columns. The idealized two-degree system is shown in Fig. 11-29*a*, where x_1 and x_2 are the first- and second-story horizontal displacements, respectively, and u_s is the support earthquake motion. The spring constants k_1 and k_2 are determined in Section 3-4 and they are as shown. The resistance functions R_1 and R_2 are characterized by the total shear forces V_1 and V_2, respectively, and they are shown in Fig. 11-29*b*. That is,

$$R_1 = k_1(x_1 - u_s) \qquad R_2 = k_2(x_2 - x_1) \tag{11-80}$$

The differential equations of motions can be derived by using Fig. 11-29*b* and applying dynamic equilibrium in the usual way. This yields

$$m_1\ddot{x}_1 + R_1 - R_2 = 0$$

$$m_2\ddot{x}_2 + R_2 = 0 \tag{11-81}$$

where R_1 and R_2 are as given by Eq. 11-80.

The elastoplastic analysis to determine the dynamic response of the system may be initiated by assuming that the resistance functions R_1 and R_2 are bilinear, and assigning maximum plastic story resistances $(R_1)_m$ and $(R_2)_m$ for R_1 and R_2, respectively. If an elastic analysis is previously executed, (see, for example, Section 11-13), the maximum plastic story resistances $(R_1)_m$ and $(R_2)_m$ could be taken as one-half of the maximum shears computed by the elastic analysis. This is equivalent to selecting elastic limits for the story shears.

With the resistances defined as above, the dynamic response of the nonlinear system can be determined easily by using the acceleration impulse extrapolation method to solve the differential equations of motion given in Eq. 11-81. For single-degree systems, this method is discussed in Section 2-8. Its extension to systems with two or more degrees of freedom can be done in

[42] Additional information on this subject can be found in References 42, 53, and 76.

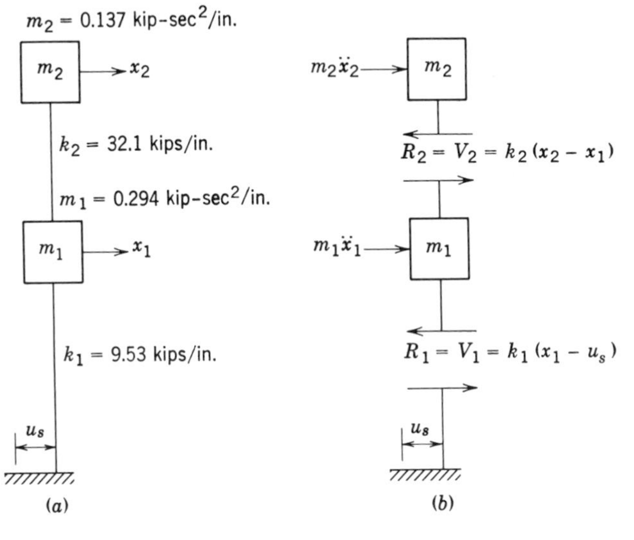

Fig. 11-29

a way similar to the one used in Section 3-4. In the solution of these equations, R_1 and R_2 are assumed to remain constant after the maximum values $(R_1)_m$ and $(R_2)_m$ are attained.

This analysis presupposes that some plastic distortions are permitted to be undertaken by the structure. The extent of this distortion can be controlled by the selection of the maximum plastic story resistances and the ductility ratio η defined as the ratio x_m/x_{el}, where x_m is the maximum displacement and x_{el} is the elastic limit displacement. When a building consists of many stories, the use of a digital computer will facilitate the computations involved.

It was pointed out above that an elastic analysis can be used as a guidance regarding the selection of the maximum plastic story resistances. If, however, such an analysis is not carried out, a trial-and-error procedure can be used until a satisfactory design is obtained.

11-15 RANDOM ANALYSIS FOR EARTHQUAKE RESPONSE

The ground accelerations that are produced by an earthquake are essentially random. (See, for example, Fig. 11-25, where the positive and negative peaks occur at irregular times and have various amplitudes.) In practice, three methods are commonly used in the design of structures for earthquake response. All have advantages and disadvantages, but none has been developed to a desirable extent.

The first approach utilizes a standard earthquake of certain amplitude of acceleration and time variation, and the analysis for earthquake response is carried out on this basis. The disadvantages of this method are that actual earthquakes vary in intensity and time variation, while the response of a structure is dependent on these properties.

The second approach is considered by many to be the most practical and involves the use of earthquake response spectra[43] as discussed in preceding sections of this chapter. This is a reasonable simplification of the problem and is considered by many to be one of the most significant contributions in earthquake engineering. This idea was first introduced by M. A. Biot, E. C. Robinson, G. Housner, and others, and response spectra from real earthquake records have been determined that can be used directly for earthquake analysis.

The third approach is to treat earthquake ground motions as random variables. This is a more logical way of thinking, since the input is nondeterministic. This analysis involves the element of prediction, and the stochastic methods discussed in Chapter 12 should be used. For example, if a structure is to be built in a region of known earthquake history, the object would be to design the structure for the least favorable among all foreseeable excitations. On this basis, a structure could be designed so that the probabilities that the maximum stresses and maximum relative displacements of the structure would not exceed specified values.

It is the opinion of the author that the reader should make himself familiar with all three approaches, and use the one most appropriate in a given physical situation.

11-16 PRACTICAL CONSIDERATIONS OF EARTHQUAKE DESIGN

The dynamic analysis in the preceding sections involved methods that can be used to design a structure to resist the effects of earthquakes. It should be pointed out, however, that there are disadvantages and limitations to the application of these methods. Therefore, considerable emphasis should be placed also on engineering judgment in making a decision regarding the type of analysis and design that should be followed in a given situation. Although many uncertainties are involved in carrying out this task, there is reasonable assurance that a structure can be designed for both elastic and inelastic behavior without endangering the lives of people.

In practice, the design of many structures to resist the effects of earthquakes is based on engineering design codes. The development of such codes is based on experience gained from previous observations of

[43] See, for example, the excellent works of Blume, Newmark, and Corning in Reference 42.

earthquake damage, as well as on current developments in the field of earthquake design. These codes are prepared to provide a simple but adequate design for many types of structures. For the more specialized cases, however, a more elaborate analysis is required.

The general objectives in a design code are to provide construction and design requirements that should be followed by the practicing engineer in order to protect the structures from undesirable damage and prevent loss of life and injury from earthquake activity. These requirements vary to some extent among the various codes, because geological formations, earthquake intensities, and engineering technology vary from one locality to another. No attempt is made, however, to describe these codes here, because this is beyond the scope of this section. On the other hand, for the reasons stated above, the engineers of a certain locality usually have their own ideas about which particular code they like to follow for earthquake design. Some of the widely used codes are the ones prepared by the Structural Engineers Association of California (SEAOC), the Uniform Building Code (UBC), the National Japanese Building Code (JBC), and the Joint Committee Code (JCC). In these codes, the maximum conditions of dynamic response are usually converted into a set of equivalent static forces, hence the actual design is carried out on the basis of a static analysis. In this manner, a simplified analysis is obtained.

PROBLEMS

11-1 The aboveground rectangular building in Fig. 11-13 is subjected to a peak overpressure P_{so} of 10 psi, produced by a weapon yield of 10 kilotons. Determine and plot the time variation of the following pressures:

(*a*) Average front face pressure
(*b*) Average back face pressure
(*c*) Net horizontal pressure

11-2 For the building in Problem 11-1, determine and plot the average side and roof pressures. The peak overpressure and weapon yield are the same as in Problem 11-1.

11-3 Solve Problem 11-1 for the case where the blastproof door in the back face of the building in Fig. 11-13 is assumed to be removed before the shock front reaches the building. The door opening is symmetrically located with respect to the sides of the building and has a height of 9.0 ft and a width of 6.0 ft.

11-4 Determine and plot the time variations of the average pressures acting on the sides and top of the rectangular structure of Problem 11-3.

11-5 Repeat Problem 11-3 when both front and back faces of the structure have the indicated opening.

11-6 Repeat Problem 11-1 for a peak overpressure P_{so} of 20 psi that is produced by a weapon yield of 500 kilotons.

11-7 Repeat Problem 11-3 for a peak overpressure P_{so} of 20 psi produced by a weapon yield of 500 kilotons.

11-8 The semicircular arched structure in Fig. 11-21*a* is assumed to have a width equal to unity, $H = 20$ ft, and is subjected to a peak overpressure P_{so} of 10 psi, produced by a weapon yield of 10 kilotons. Determine and plot the pressure variation at point P when $\alpha = 30°$.

11-9 Repeat Problem 11-8 when $\alpha = 120°$.

11-10 Determine the net horizontal loading acting on the semicircular arched structure in Problem 11-8.

11-11 Determine an approximate equivalent net horizontal loading acting on the semicircular arched structure in Problem 11-8.

11-12 Repeat Problem 11-8 for $H = 10$ ft, $P_{so} = 20$ psi, and $\alpha = 30°$.

11-13 Repeat Problem 11-8 for $\alpha = 30$, 60, 90, 120, and 150°.

11-14 Repeat Problem 11-12 for $\alpha = 90$ and 120°.

11-15 The aboveground rectangular building in Fig. 11-13 is subjected to a peak overpressure P_{so} of 8 psi produced by a weapon yield of 10 kilotons. Design the roof of this building so that the ductility ratio $\eta = 3$ is not exceeded. Assume that the roof is a rectangular two-way reinforced concrete slab fixed on all edges. Assume that the steel ratio p is equal to 0.015.

11-16 Repeat Problem 11-15 for the case where the roof is assumed to be a two-way reinforced concrete slab simply supported on all edges.

11-17 Repeat Problem 11-15 for the case where the roof is assumed to be a two-way reinforced concrete slab simply supported on the two long sides and fixed on the two short sides.

11-18 The steel frames shown in Fig. P11-18 are subjected to an earthquake ground acceleration $\ddot{u}_s$ equal to $0.10 g \sin 4\pi T$, where g is the acceleration of gravity and T is time. The duration of the acceleration $\ddot{u}_s$ is 2 sec. By following the procedure explained in Section 11-11, determine the maximum horizontal displacement of the top of each frame relative to the ground. Neglect damping. The modulus of elasticity E is 30×10^6 psi.

11-19 Repeat Problem 11-18 by assuming that the frames are also subjected to 10% viscous damping.

11-20 The one-story steel frames in Problem 11-18 are subjected to an earthquake ground acceleration $\ddot{u}_s$ equal to $0.15\,g\sin 8\pi T$, where g is the acceleration of gravity and T is time. The duration of the acceleration $\ddot{u}_s$ is 1 sec. By following the procedure explained in Section 11-11, determine the maximum bending stress in the columns of each frame. Neglect damping. The modulus of elasticity E is 30×10^6 psi.

11-21 Repeat Problem 11-20 by assuming that the steel frames are also subjected to 10% viscous damping.

11-22 The two-story rigid frame building in Fig. 3-5*a* is subjected to an artificial earthquake of ground acceleration $\ddot{u}_s$ equal to $0.05\,g\sin$

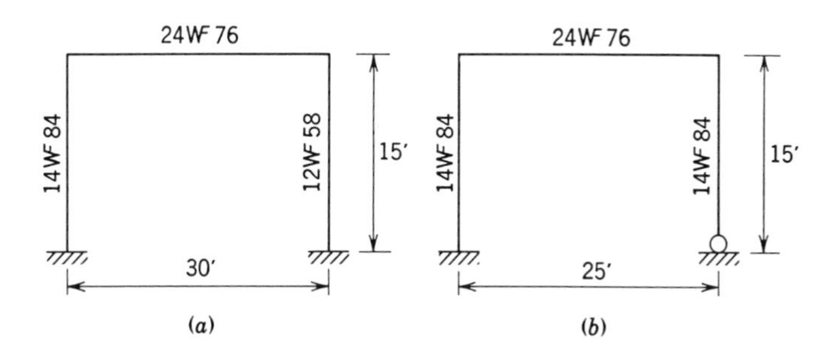

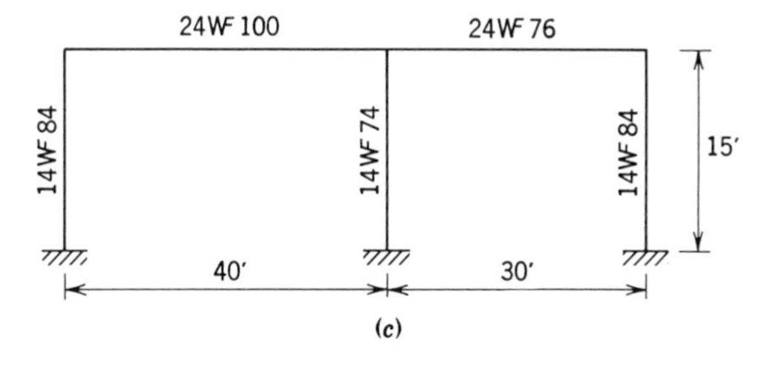

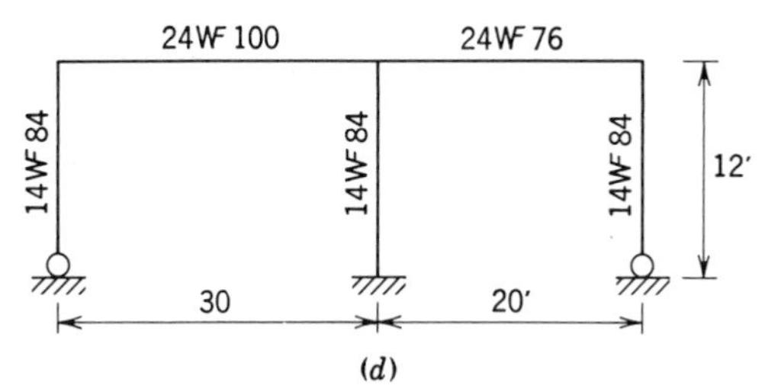

Fig. P11-18

$4\pi t$, where g is the acceleration of gravity and t is time. The duration of the acceleration $\ddot{u}_s$ is 1 sec. The steel sections of the girders and columns are 24 W̲F 76 and 12 W̲F 58, respectively. By considering the frame shown in Fig. 11-28*a* and applying the method of modal analysis as discussed in Section 11-13, determine the maximum bending stress in the columns of the frame. Assume that the girders of the frame are infinitely rigid compared to the columns. The modulus of elasticity E is 30×10^6 psi. Neglect damping.

11-23 Repeat Problem 11-22 by considering the flexibility of the girders.

11-24 The steel frames in Problem 1-22 are subjected to an artificial earthquake of ground acceleration $\ddot{u}_s$ equal to $0.10\, g \sin 4\pi t$, where g is the acceleration of gravity and t is time. The duration of $\ddot{u}_s$ is 1 sec. By applying the method of modal analysis, determine the maximum bending stresses in the columns of each frame. Assume that the girders of each frame are infinitely rigid compared to the columns. The modulus of elasticity E is 30×10^6 psi. Neglect damping.

11-25 Repeat Problem 11-24 by considering the flexibility of the girders.

12

STOCHASTIC APPROACH TO STRUCTURAL DYNAMICS

12-1 INTRODUCTION

The dynamic response studies discussed in the preceding chapters involved structural systems that are subjected to well defined time-varying dynamic forces. These forces may be periodic or nonperiodic, and they are usually called *deterministic* because they are known forcing functions. In physical situations, however, the time-varying forces are not always known, and the analysis must then be based on predictions. The prediction of such excitations is usually based on experience, or on experimental results, and the design of a system for such types of excitations depends on how closely an actual situation is predicted. Excitations that involve the element of prediction are called *nondeterministic*, and there is no guarantee that they will actually occur. Random excitations, such as those produced by earthquake, blast, or gust, are nondeterministic.

The purpose in this chapter is to introduce the subject of stochastic structural dynamics with applications to simple structural problems. The first sections discuss the basic concepts of such topics as probability and random variables to help the reader to better comprehend the material that follows.

12-2 PROBABILITY, RANDOM VARIABLES, AND DISTRIBUTION FUNCTIONS

In situations where the input disturbances are not well defined, probability and statistics can aid in predicting the dynamic response of structures subjected to this type of excitation. Earthquake ground motions are examples of such nondeterministic excitations. In this section, basic aspects of probability, random variables, and distribution functions are discussed to the extent that they are needed for comprehension of the work in the sections that follow.

To illustrate probability, we consider an experiment where an event denoted by Q is certain to occur. When two events a and b are given, the certain event $Q=a+b$ will occur when a, b, or both a and b occur. The events a and b are mutually exclusive when the occurrence of one excludes the occurrence of the other. For example, in the tossing of a coin the certain event Q to occur is either head or tail. If the event a is head and b is tail, the events a and b are mutually exclusive.

The probability of an event a is the number $P(a)$ assigned to this event which obeys the following laws:

$$P(a) \geqslant 0 \qquad P(Q)=1 \tag{12-1}$$

If two events a and b are mutually exclusive, then

$$P(a+b)=P(a)+P(b) \tag{12-2}$$

For a given experiment, the set of all possible outcomes is called the sample space.

If there are n mutually exclusive, exhaustive, and equally likely cases, and if m of these events are favorable to an event A, then the probability of A is defined as

$$P(A)=\frac{m}{n} \tag{12-3}$$

In the tossing of a coin, the probability of the event head or tail to occur is 1/2.

A process is random if it is not possible to predict its final state from its initial state as in the case of tossing a coin or rolling a die. This process is defined by its sample space which is the totality of sample points associated with a given experiment. A random variable is a function X that is defined within the sample space. If a number x_k $(k=1, 2, 3, \ldots, k)$ is associated with a particular outcome of an experiment, then the random variable X can assume all possible values of x_k. Thus, by assigning a random variable to an experiment, the notation in the following discussion is simplified. For example, if a random variable X is assigned to the rolling of a die, the possible values of the outcome are $x_1=1$ for one dot, $x_2=2$ for two dots, and so on. In this manner, the probability of rolling a six, for example, can be written as $P(X=x_6)$.

The frequency function[44] $p(X=x_k)$, or simply $p(x_k)$, denotes the probability of the random variable X for a value x_k. The frequency function in the rolling of a die, for example, is

$$p(x_1)=p(x_2)=p(x_3)=p(x_4)=p(x_5)=p(x_6)=\frac{1}{6}$$

The probability for a random variable to be less than or equal to some specified value x_j is called the distribution function and is denoted by $P(X \leqslant x_j)$. Mathematically, the distribution is written as

$$P(X \leqslant x_j)=\sum_{k=-\infty}^{j} p(x_k) \tag{12-4}$$

The frequency and distribution functions for the rolling of a die are shown in Figs. 12-1*a* and 12-1*b*, respectively.

In the discussion above, only one random variable was involved. This case of probability is known as one-dimensional. When two random variables X and Y are considered, two-dimensional probability is involved. For this case, the distribution function $P(X \leqslant x_j, Y \leqslant y_i)$ is defined as

$$P(X \leqslant x_f, Y \leqslant y_i)=\sum_{k=-\infty}^{j} \sum_{l=-\infty}^{i} p(x_k, y_l) \tag{12-5}$$

[44] The frequency function is also known as density function.

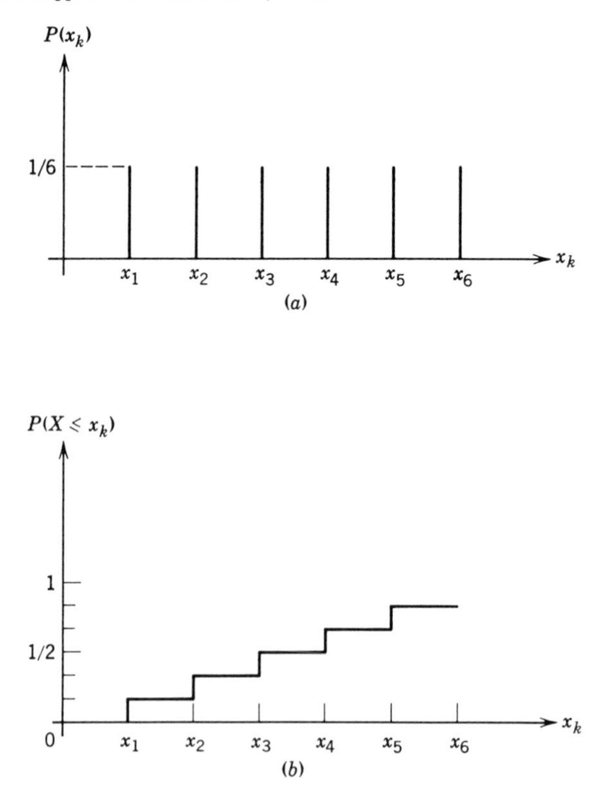

Fig. 12-1. (*a*) Frequency function. (*b*) Distribution function.

and the frequency function $p(x_k)$ is

$$p(x_k) = \sum_{l=-\infty}^{\infty} p(x_k, y_l) \tag{12-6}$$

For continuous one-dimensional probability, the distribution function is defined as the probability of a random variable X that assumes a value in the interval between any two values x_1 and x_2. That is,

$$P(x_1 \leqslant X \leqslant x_2) = \int_{x_1}^{x_2} p(x)\, dx \tag{12-7}$$

where $p(x)$ is the probability frequency function for the continuous case that is defined as

$$p(x) = \frac{dP(x_1 \leqslant X \leqslant x_2)}{dx} \tag{12-8}$$

12-3 EXPECTATIONS

In applying probability and statistics, an important parameter to be determined is the expected value or mean of a random variable X which is denoted as $E\{X\}$. If, for example, an experiment is repeated n times and the observed outcomes are x_1, x_2, $x_3, \ldots, x_m$, then for each outcome the random variable X assumes a numerical value. Thus, for the indicated outcomes, the random variable X assumes the values $X(x_1)$, $X(x_2)$. $X(x_3), \ldots$ $, X(x_m)$. If for the n repetitions of the experiment the symbol d_1 denotes the number of times that $X(x_1)$ occurs, n_2 the number of times that $X(x_2)$ occurs, and so on, the arithmetic average X_{av} of the $X(x_m)$ values, $m=1, 2, 3, \ldots$, is

$$X_{av} = \frac{n_1 X(x_1) + n_2 X(x_2) + \cdots + n_m X(x_m)}{n} \tag{12-9}$$

or

$$X_{av} = \frac{1}{n} \sum_{m=1}^{m} n_m X(x_m) \tag{12-10}$$

It should be noted that $n = n_1 + n_2 + \cdots + n_m$.

If the number of repetitions n approaches infinity, then n_1/n approximates the probability $p(x_1)$, n_2/n approximates the probability $p(x_2)$, and so on. The statistical average or expected value $E\{X\}$ of a random variable X is given by

$$E\{X\} = \sum_{m=1}^{m} p(x_m) X(x_m) \tag{12-11}$$

Equation 12-11 is also known as the first moment of the random variable X. The expression for the nth moment $E\{X^n\}$ of the random variable X is

$$E\{X^n\} = \sum_{m=1}^{m} p(x_m) [X(x_m)]^n \tag{12-12}$$

For the continuous case, the analogous expression for the nth moment is

$$E\{X^n\} = \int_{-\infty}^{\infty} x^n p(x)\, dx \tag{12-13}$$

The second moment $E\{X^2\}$ yields the mean square value of X. That is,

$$E\{X^2\} = \overline{X}^2 = \int_{-\infty}^{\infty} x^2 p(x)\, dx \tag{12-14}$$

The broadness or spread of the density function is called the variance of the random variable and is denoted by σ^2. The variance σ^2 is defined as

$$\sigma^2 = E([X - E\{X\}]^2) \tag{12-15}$$

or

$$\sigma^2 = E\{X^2\} - [E\{X\}]^2 \tag{12-16}$$

12-4 CORRELATION FUNCTIONS

The probability frequency function discussed earlier provides amplitude variations of a random process and, therefore, is not a unique function of the process. In other words, two sample functions describing two random processes can have the same probability frequency function $p(x)$. Hence it becomes necessary to define a correlation function that will be used to distinguish the two processes.

Consider, for example, two random processes described by the random variables X and Y at times t_1 and t_2, respectively. The cross-correlation function $R(t_1, t_2)$ of these two processes is defined as

$$R(t_1, t_2) = E\{XY\} = \int_{-\infty}^{\infty}\int_{-\infty}^{\infty} xyp(x, y)\, dx\, dy \tag{12-17}$$

where x and y within the integral sign are the assigned variables of the random variables X and Y, respectively, and $p(x,y)$ is the two-dimensional frequency function.

Equation 12-17 is a measure of the correlation of the two random processes. It can also be written as

$$R_{xy}(t) = \lim_{\tau\to\infty} \frac{1}{2\tau}\int_{-\tau}^{\tau} x(\rho)\, y(t+\rho)\, d\rho \tag{12-18}$$

where ρ is a dummy variable of integration and τ is the period. Equation 12-18 gives the cross-correlation of the random functions $x(t)$ and $y(t)$. The cross-correlation $R_{yx}(t)$ of $y(t)$ on $x(t)$ is

$$R_{yx}(t) = R_{xy}(-t) \tag{12-19}$$

If the random variables X and Y describe the same random processes at their respective times t_1 and t_2, where $t_2 = t_1 + \rho$ and $|\rho|$ is large enough so that X and Y can be considered independent, then the expected value $E\{XY\}$ is given by

$$E\{XY\} = E\{X\}E\{Y\} \tag{12-20}$$

If $E\{X\}$ or $E\{Y\}$ is zero, then $E\{XY\}$ is zero and consequently $R(t_1, t_2)$ in Eq. 12-17 is zero.

If the dummy variable ρ above is approaching zero, then the random variables X and Y describe the same process at time $t_1 = t_2$. On this basis, Eq. 12-20 can be written as

$$R(t_1, t_1) = E\{XY\} = E\{X^2\} \tag{12-21}$$

It is easily noted that Eq. 12-21 represents the second moment of the random process. In addition, this equation is also the autocorrelation function of the random process.

A random process is ergotic if all types of ensemble averages, or expected values, are interchangeable with the corresponding time average. For ergotic processes, an ensemble average can be replaced by a time average involving a single sample function $x_i(t)$ of a random process $x(t)$. On this basis, Eq. 12-20 yields

$$E\{XY\} = \langle x_i(t_1) x_i(t_2) \rangle \tag{12-22}$$

where $E\{X\}$ and $E\{Y\}$ in Eq. 12-20 are replaced by $x_i(t_1)$ and $x_i(t_2)$, respectively, in Eq. 12-22.

If in Eq. 12-22 $t_2 = t_1 + \rho$, where ρ is a dummy variable, the autocorrelation function described by Eq. 12-22 or Eq. 12-21 is

$$R_{xx}(t_1) = \lim_{\tau \to \infty} \frac{1}{2\tau} \int_{-\tau}^{\tau} x(\rho) x(t_1 + \rho) d\rho$$

or simply

$$R_{xx}(t) = \lim_{\tau \to \infty} \frac{1}{2\tau} \int_{-\tau}^{\tau} x(\rho) x(t + \rho) d\rho \tag{12-23}$$

This equation is a measure of the regularity of a random process. Thus the cross-correlation and the autocorrelation functions can be used to distinguish any two random processes that possess similar characteristics. In addition, these functions are used in power spectra analysis, as shown later in this chapter.

12-5 POWER SPECTRA ANALYSIS

A random process $x(t)$ is called stationary in the strict sense if its statistics are not affected by a shift ϵ of the time origin. In other words, $x(t)$ and $x(t + \epsilon)$ have the same statistics. The power spectrum $\Phi(\omega)$ of a stationary random process $x(t)$ is the Fourier transform of its autocorrelation function.

To prove this concept, consider a random function $x(t)$. The total energy content of this function is given by the integral

$$\int_{-\tau}^{\tau} x^2(t)\, dt \tag{12-24}$$

Since power is defined as energy per unit time, the average power P_{av} over the interval $-\tau \leqslant t \leqslant \tau$ is

$$P_{\text{av}} = \frac{1}{2\tau} \int_{-\tau}^{\tau} x^2(t)\, dt \tag{12-25}$$

At $t = 0$, the autocorrelation function given by Eq. 12-23 yields

$$R_{xx}(0) = \lim_{\tau \to \infty} \frac{1}{2\tau} \int_{-\tau}^{\tau} x^2(\rho)\, d\rho \tag{12-26}$$

Equation 12-26 indicates that the autocorrelation function can be related to the power of $x(t)$, because the limit of Eq. 12-25 as $\tau \to \infty$ is identical to Eq. 12-26.

From Eq. 8-23, the Fourier transform $X(\omega)$ of the function $x(t)$ is

$$X(\omega) = \int_{-\infty}^{\infty} x(t)\, e^{-j\omega t}\, dt \tag{12-27}$$

At a frequency $\omega = \omega_0$, Eq. 12-27 yields

$$X(\omega_0) = A \cos \omega_0 - B \sin \omega_0$$

$$= \sqrt{A^2 + B^2}\; e^{-j(\arctan B/A \tan \omega_0)} \tag{12-28}$$

where A and B are the parts of $x(t)$ that travel with frequency ω_0 in the cosine and sine modes, respectively.

The amount of energy that travels at ω_0, by Eq. 12-24, is simply $A^2 + B^2$. This result can be verified by multiplying Eq. 12-28 by its conjugate. The average power at ω_0 as $\tau \to \infty$ is

$$P_{\text{av}} = \frac{A^2 + B^2}{2\tau} \tag{12-29}$$

From Eq. 12-29, the average power $\Phi(\omega)$ as a function of the frequency ω can be written as

$$\Phi(\omega) = \lim_{\tau \to \infty} \frac{1}{2\tau} X(\omega) X^*(\omega) \tag{12-30}$$

The asterisk (*) denotes complex conjugate, and $\Phi(\omega)$ is known as the power spectrum of $x(t)$.

The Fourier transform of the autocorrelation function of $x(t)$, Eq. 12-23, is

$$\int_{-\infty}^{\infty} R_{xx}(t)\, e^{-j\omega t}\, dt = \lim_{T\to\infty} \frac{1}{2T} \int_{-T}^{T} x(\rho) \int_{-\infty}^{\infty} e^{-j\omega t} x(\rho+t)\, d\rho\, dt \qquad (12\text{-}31)$$

If the variables on the right-hand side of Eq. 12-31 are changed so that $u=\rho+t$ and $du=dt$, Eq. 12-31 yields

$$\int_{-\infty}^{\infty} R_{xx}(t)\, e^{-j\omega t}\, dt = \lim_{T\to\infty} \frac{1}{2T} \int_{-T}^{T} x(\rho)\, e^{j\omega\rho}\, d\rho X(\omega) \qquad (12\text{-}32)$$

where

$$X(\omega) = \int_{-\infty}^{\infty} x(u)\, e^{-j\omega u}\, du$$

Equation 12-32 can be rewritten as

$$R_{xx}(\omega) = \lim_{T\to\infty} \frac{1}{2T} X(\omega) X^*(\omega) \qquad (12\text{-}33)$$

where

$$R_{xx}(\omega) = \int_{-\infty}^{\infty} R_{xx}(t)\, e^{-j\omega t}\, dt$$

and

$$X^*(\omega) = \int_{-\infty}^{\infty} x(\rho)\, e^{j\omega\rho}\, d\rho$$

Equations 12-33 and 12-30 are identical. Therefore, it may be stated that the Fourier transform of the autocorrelation function $R_{xx}(t)$ is the power spectrum density $\Phi_{xx}(\omega)$ of the stationary random function $x(t)$.

The autocorrelation function is symmetrical only with respect to the vertical axis. On this basis, the Fourier transform $\Phi_{xx}(\omega)$ of the autocorrelation function $R_{xx}(t)$ can be written as

$$\Phi_{xx}(\omega) = \int_{-\infty}^{\infty} R_{xx}(t)\, e^{-j\omega t}\, dt$$

$$= \int_{-\infty}^{\infty} R_{xx}(t) \cos\omega t\, dt - j\int_{-\infty}^{\infty} R_{xx}(t) \sin\omega t\, dt$$

$$= 2\int_{0}^{\infty} R_{xx}(t) \cos\omega t\, dt \qquad (12\text{-}34)$$

TABLE 12-1

$x(t)$	$R_{xx}(t)$	$\Phi_{xx}(\omega)$
$Ax(t)$	$\|A\|^2 R_{xx}(t)$	$\|A\|^2 \Phi_{xx}(\omega)$
$\frac{dx(t)}{dt}$	$-\frac{d^2 R_{xx}(t)}{dt^2}$	$\omega^2 \Phi_{xx}(\omega)$
$\frac{d^n x(t)}{dt^n}$	$(-1)^n \frac{d^{2n} R_{xx}(t)}{dt^{2n}}$	$\omega^{2n} \Phi_{xx}(\omega)$
$x(t) e^{\pm j\omega_0 t}$	$R_{xx}(t) e^{\pm j\omega_0 t}$	$\Phi_{xx}(\omega \mp \omega_0)$

It should be noted that because of symmetry,

$$\int_{-\infty}^{\infty} R_{xx}(t) \sin \omega t \, dt = 0$$

Equation 12-34 shows that the power spectrum of a stationary random function $x(t)$ can be computed by using only a cosine transformation.

In a similar manner it can be proved that the cross-power spectrum $\Phi_{xy}(\omega)$ of two jointly stationary random processes $x(t)$ and $y(t)$ is the Fourier transform of their cross-correlation function $R_{xy}(t)$. The cross-correlation function $R_{xy}(t)$ is given by Eq. 12-18. Thus

$$\Phi_{xy}(\omega) = \int_{-\infty}^{\infty} R_{xy}(t) \, e^{-j\omega t} \, dt \tag{12-35}$$

Table 12-1 shows the correlation of $x(t)$, $R_{xx}(t)$, and $\Phi_{xx}(\omega)$ for various functions $x(t)$. Figure 12-2 illustrates a number of autocorrelations and their Fourier transforms.

As an illustration, let it be assumed that it is required to determine the spectrum density of a random process $x(t)$ if the autocorrelation function $R_{xx}(t)$ is $e^{-2\lambda|\rho|}$. The power spectrum can be obtained directly by taking the Fourier transform $\Phi_{xx}(\omega)$ of the autocorrelation function. That is,

$$\Phi_{xx}(\omega) = \int_{-\infty}^{\infty} e^{-2\lambda|\rho|} e^{-j\omega\rho} \, d\rho$$

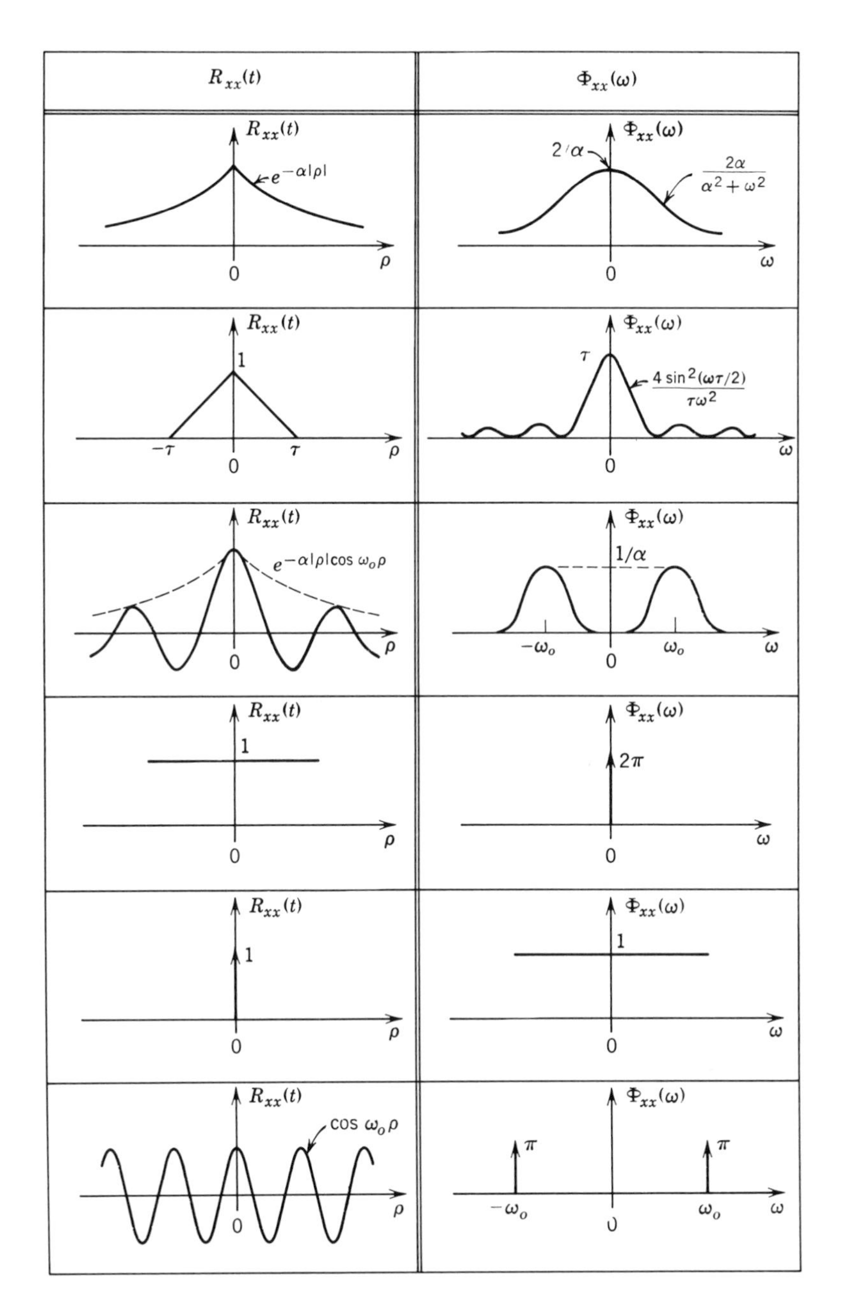

$R_{xx}(t)$
$\Phi_{xx}(\omega)$
$R_{xx}(t)$
$e^{-\alpha|\rho|}$
0
ρ
$\Phi_{xx}(\omega)$
$2/\alpha$
$\frac{2\alpha}{\alpha^2+\omega^2}$
0
ω
$R_{xx}(t)$
1
$-\tau$
0
τ
ρ
$\Phi_{xx}(\omega)$
τ
$\frac{4\sin^2(\omega\tau/2)}{\tau\omega^2}$
0
ω
$R_{xx}(t)$
$e^{-\alpha|\rho|}\cos\omega_o\rho$
0
ρ
$\Phi_{xx}(\omega)$
$1/\alpha$
$-\omega_o$
0
ω_o
ω
$R_{xx}(t)$
1
0
ρ
$\Phi_{xx}(\omega)$
2π
0
ω
$R_{xx}(t)$
1
0
ρ
$\Phi_{xx}(\omega)$
1
0
ω
$R_{xx}(t)$
$\cos\omega_o\rho$
0
ρ
$\Phi_{xx}(\omega)$
π
π
$-\omega_o$
0
ω_o
ω

Fig. 12-2

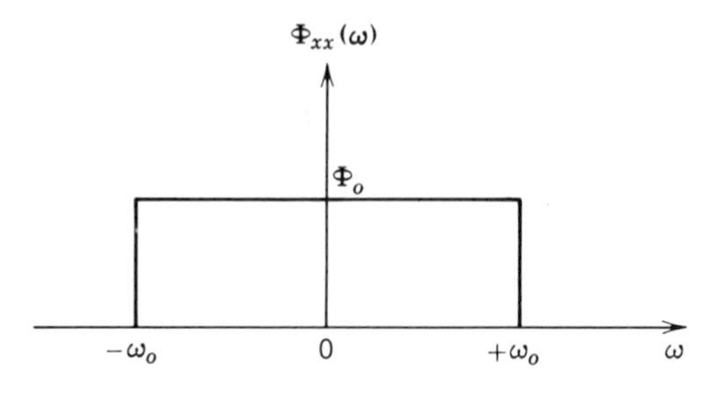

Fig. 12-3

or

$$\Phi_{xx}(\omega)=\int_{-\infty}^{0} e^{2\lambda\rho}e^{-j\omega\rho}\,d\rho+\int_{0}^{\infty} e^{-2\lambda\rho}e^{-j\omega\rho}\,d\rho$$

Evaluation of this expression yields

$$\Phi_{xx}(\omega)=\frac{4\lambda}{4\lambda^2+\omega^2}$$

As a second example, consider a process that has the power spectrum density function shown in Fig. 12-3 and let it be assumed that it is required to determine its autocorrelation function. This can be done by taking the inverse transform $R_{xx}(t)$ of the power spectrum density. That is,

$$R_{xx}(t)=\frac{1}{2\pi}\int_{-\omega_0}^{\omega_0} \Phi_0 e^{j\omega\rho}\,d\omega$$

The expression yields

$$R_{xx}(t)=\frac{\omega_0\Phi_0}{\pi}\,\frac{\sin\omega_0\rho}{\omega_0\rho}$$

12-6 DYNAMIC RESPONSE OF STRUCTURAL SYSTEMS DUE TO RANDOM EXCITATIONS

The previous discussion is here extended to include the computation of the dynamic response of structural systems subjected to random excitations. For this purpose, the required equations are first derived, and then used to determine the dynamic response of systems with one degree of freedom.

The development of the required equations is initiated by considering the block diagram of a system, Fig. 12-4. Here $H(\omega)$ represents the impulse

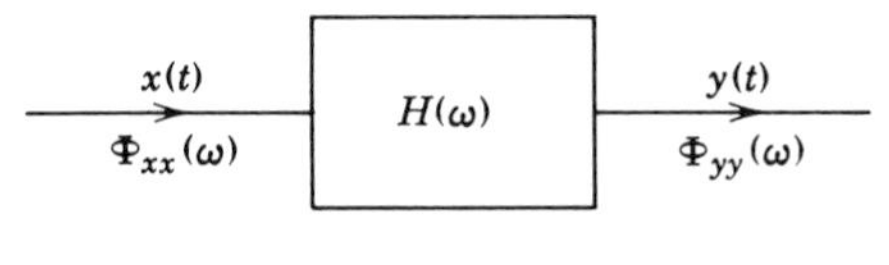

Fig. 12-4

response, $x(t)$ is the random forcing function applied to the system, and $y(t)$ is the random output function. In the same figure, $\Phi_{xx}(\omega)$ is the power spectra of the input and $\Phi_{yy}(\omega)$ is the power spectra of the output.

The output $y(t)$, by Eq. 8-94, can be written as

$$y(t)=\int_{-\infty}^{\infty} h(\rho)x(t-\rho)\,d\rho \tag{12-36}$$

where $h(\rho)$ is the time domain of the system's response $H(\omega)$. The output $y(t)$ is a stationary random process if the input $x(t)$ is a random stationary.

By multiplying the conjugate of Eq. 12-36 by $x(t_1)$, we have

$$x(t_1)y^*(t)=\int_{-\infty}^{\infty} x(t_1)x^*(t-\rho)h^*(\rho)\,d\rho \tag{12-37}$$

The expected value of Eq. 12-37 is

$$E\{x(t_1)y^*(t)\}=\int_{-\infty}^{\infty} E\{x(t_1)x^*(t-\rho)h^*(\rho)\}\,d\rho \tag{12-38}$$

Using the result of Eq. 12-17, Eq. 12-38 at $t=t_2$ yields the expression

$$\begin{aligned} R_{xy}(t_1,t_2) &= \int_{-\infty}^{\infty} R_{xx}(t_1,t_2-\rho)h^*(\rho)\,d\rho \\ &= R_{xx}(t_1,t_2)*h^*(t_2) \end{aligned} \tag{12-39}$$

Equation 12-39 indicates that the autocorrelation $R_{xx}(t_1,t_2)$ considered as a function of time t_2, is convolved with the impulse response $h^*(t_2)$ while the variable t_1 is treated as a parameter. In addition, Eq. 12-39 yields the cross-correlation function R_{xy} of the input $x(t)$ and output $y(t)$.

The autocorrelation function of the output $y(t_1, t_2)$ can be obtained by multiplying Eq. 12-36 by $y^*(t_2)$ and proceeding in a similar manner. The result is

$$\begin{aligned} R_{yy}(t_1,t_2) &= \int_{-\infty}^{\infty} R_{xy}(t_1-\rho)h(\rho)\,d\rho \\ &= R_{yx}(t_1,t_2)*h(t_1) \end{aligned} \tag{12-40}$$

Since this process is stationary, it follows that $R_{xx}(t_1, t_2 - \rho) = R_{xx}(\beta + \rho)$, where $t_1 - t_2 = \beta$. On this basis, Eq. 12-39 yields

$$R_{xy}(t_1, t_2) = \int_{-\infty}^{\infty} R_{xx}(\beta + \rho) h^*(\rho)\, d\rho \tag{12-41}$$

Equation 12-41 depends only on β and represents a convolution integral. That is,

$$R_{xy}(\beta) = R_{xx}(\beta) * h^*(\beta) \tag{12-42}$$

In a similar manner, Eq. 12-40 yields

$$R_{yy}(\beta) = R_{yx}(\beta) * h(\beta) \tag{12-43}$$

By using Eqs. 12-19, 12-42, and 12-43, the following expression can be obtained:

$$R_{yy}(\beta) = R_{xx}(\beta) * h^*(-\beta) * h(\beta) \tag{12-44}$$

It was shown earlier that the power spectrum density of a function is the Fourier transform of the autocorrelation function. On this basis,

$$\Phi_{yy}(\omega) = \int_{-\infty}^{\infty} R_{yy}(\beta) e^{-j\omega\beta}\, d\beta \tag{12-45}$$

Thus Eqs. 12-42 and 12-43 can be written as

$$\Phi_{xy}(\omega) = \Phi_{xx}(\omega) H^*(\omega) \tag{12-46}$$

and

$$\Phi_{yy}(\omega) = \Phi_{yx}(\omega) H(\omega) \tag{12-47}$$

The frequency domain of Eq. 12-44 is

$$\Phi_{yy}(\omega) = \Phi_{xx}(\omega) H^*(-\omega) H(\omega)$$

or

$$\Phi_{yy}(\omega) = \Phi_{xx}(\omega) |H(\omega)|^2 \tag{12-48}$$

By using the definition given by Eq. 12-21, as well as Eq. 12-48, the expected value of the input can be written as

$$E\{|x(t)|^2\} = R_{xx}(0) = \frac{1}{2\pi} \int_{-\infty}^{\infty} \Phi_{xx}(\omega) |H(\omega)|^2 d\omega \tag{12-49}$$

In a similar manner, the expected value of the output can be written as

$$E\{|y(t)|^2\} = R_{yy}(0) = \frac{1}{2\pi}\int_{-\infty}^{\infty} \Phi_{yy}(\omega)|H(\omega)|^2 d\omega$$

$$= \frac{1}{2\pi}\int_{-\infty}^{\infty} \Phi(\omega)|H(\omega)|^2 d\omega \qquad (12\text{-}50)$$

This equation will be used to determine the expected value of the output of a linear system under a random excitation $x(t)$.

As an example, let it be assumed that it is required to determine the expected value $E\{|y(t)|^2\}$ of the output for the single degree of freedom system in Fig. 8-3 which is excited by the random dynamic force $F(t) = F_0 f(t) = F_0 \sin \omega_f t$.

The autocorrelation function of this random process can be determined by using Eq. 12-23. That is,

$$R_{ff}(t) = \lim_{T\to\infty} \frac{1}{2T}\int_{-T}^{T} F_0{}^2 \sin \omega_f \rho \sin \omega_f (t+\rho)\, d\rho \qquad (12\text{-}51)$$

Evaluation of the integral yields

$$R_{ff}(t) = \frac{F_0{}^2}{2} \cos \omega_f \rho \qquad (12\text{-}52)$$

From Fig. 12-2, the power spectrum density $\Phi_{ff}(\omega)$ is

$$\Phi_{ff}(\omega) = \frac{\pi F_0{}^2}{2} \delta(\omega - \omega_f) \qquad (12\text{-}53)$$

where $\delta(\omega)$ is the Dirac delta function in the frequency domain.

The transfer function[45] $H(s)$ of the single-degree system, thus, can be taken as

$$H(s) = \frac{1}{mZ(s)}$$

where $Z(s) = s^2 + 2\omega\zeta s + \omega^2$. With $s = j\omega_f$, the expression yields

$$H(j\omega_f) = \frac{1}{m(\omega^2 - \omega_f^2 + 2j\omega\omega_f\zeta)} \qquad (12\text{-}54)$$

This equation gives the impulse response of the system.

[45] For more information see Chapter 8.

If the numerator and the denominator on the right-hand side of Eq. 12-54 is multiplied by the conjugate of the denominator, the absolute value $|H(\omega_f)|$ of the resulting expression is

$$|H(\omega_f)| = \frac{1}{m[(\omega^2-\omega_f^2)^2+(2\omega\omega_f\zeta)^2]^{\frac{1}{2}}} \tag{12-55}$$

By using Eqs. 12-50, 12-53, and 12-55, the expected value $E\{|y(t)|^2\}$ of the output $y(t)$ is

$$E\{|y(t)|^2\} = \frac{F_0{}^2}{4m^2}\int_{-\infty}^{\infty}\frac{\delta(\omega-\omega_f)\,d\omega_f}{(\omega^2-\omega_f^2)^2+(2\omega\omega_f\zeta)^2} \tag{12-56}$$

The integral can be evaluated by using the following property of the delta[46] function:

$$\int_{-\infty}^{\infty}\delta(\omega-\omega_f)\,g(\omega_f)\,d\omega_f = g(\omega) \tag{12-57}$$

Equation 12-56 then yields

$$E\{|y(t)|^2\} = \frac{F_0{}^2}{4}\,\frac{1}{4m^2\omega^2\zeta^2}$$

$$= \frac{F_0{}^2}{16m^2\omega^2\zeta^2} \tag{12-58}$$

As a second example, let it be assumed that the single degree of freedom system in Fig. 8-3 is excited by white noise. A white noise here is defined as any random process whose power spectra is constant. In such a case the spectra are independent of the frequency. That is, $\Phi_{ff}(\omega)=\Phi_{ff}=\text{constant}$.

It was pointed out earlier that the autocorrelation function $R_{ff}(t)$ is the inverse Fourier transform of the power spectra function. That is,

$$R_{ff}(t) = \int_{-\infty}^{\infty}\Phi_{ff}e^{j\omega\rho}\,d\omega$$

This yields

$$R_{ff}(t) = \Phi_{ff}\delta(\rho) \tag{12-59}$$

Equation 12-59 shows that the autocorrelation function for white noise is an impulse, indicating that correlation exists only when $\rho=0$. This observation suggests that Eq. 12-50 can be used to determine the expected

[46]Information regarding delta functions can be found in Reference 68.

value $E\{|y(t)|^2\}$ of the output $y(t)$ when the single-degree system is excited by a white noise. That is,

$$E\{|y(t)|^2\} = \frac{1}{2\pi}\int_{-\infty}^{\infty} \Phi_{ff}|H(\omega_f)|^2 d\omega_f \tag{12-60}$$

By using Eqs. 12-55 and 12-60, we obtain

$$E\{|y(t)|^2\} = \frac{\Phi_{ff}}{2\pi m^2}\int_{-\infty}^{\infty} \frac{d\omega_f}{(\omega^2-\omega_f^2)^2+(2\omega\omega_f\zeta)^2} \tag{12-61}$$

The integral can be evaluated by using the method of residues.[47] If the integral in Eq. 12-61 is treated as a function of the complex variable z, that is, the real variable ω_f is replaced by the complex variable z, the resulting expression is

$$E\{|y(t)|^2\} = \frac{\Phi_{ff}}{2\pi m^2}\int_{-\infty}^{\infty} \frac{dz}{(\omega^2-z^2)^2+(2\omega z\zeta)^2} \tag{12-62}$$

The evaluation of the integral is obtained by using the expression

$$E\{|y(t)|^2\} = 2\pi j \sum[\text{of residues}] \tag{12-63}$$

The integrand in Eq. 12-62 has two simple poles in the upper half plane:

$$\begin{aligned} z_1 &= \sqrt{1-\zeta^2}\,\omega - j\,\zeta\omega \\ z_2 &= -\sqrt{1-\zeta^2}\,\omega + j\,\zeta\omega \end{aligned} \tag{12-64}$$

By using Eqs. 12-63 and 12-64 and working in a manner similar to that of partial fractions, Eq. 12-62 is written as

$$E\{|y(t)|^2\} = \frac{j\,\Phi_{ff}}{m^2}\left[(z-z_1)f(z)|_{z=z_1} + (z-z_2)f(z)|_{z=z_2}\right] \tag{12-65}$$

where

$$f(z) = \frac{1}{(\omega^2-z^2)^2+(2\omega z\zeta)^2} \tag{12-66}$$

[47] For information regarding this method see Reference 68.

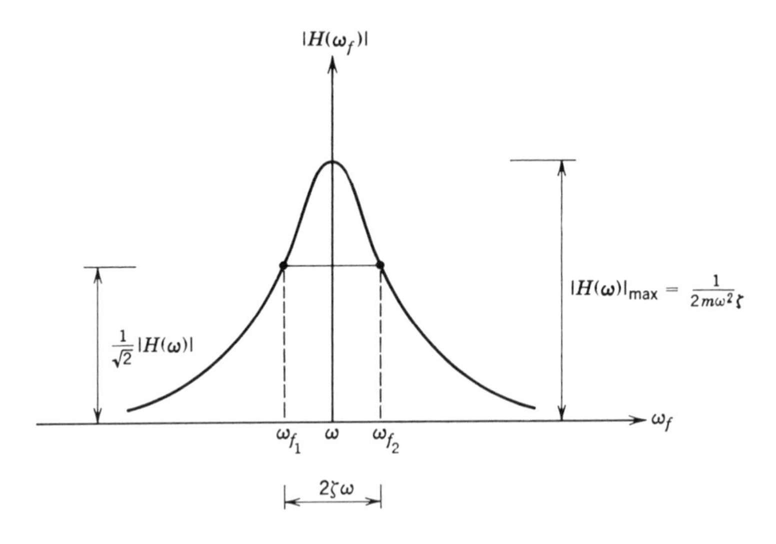

Fig. 12-5

Evaluation of this equation yields

$$E\{|y(t)|^2\} = \frac{\Phi_{ff}}{2\zeta\omega^3 m^2} \qquad (12\text{-}67)$$

or

$$E\{|y(t)|^2\} = \frac{\omega\Phi_{ff}}{2\zeta k^2} \qquad (12\text{-}68)$$

Equation 12-68 is the expected value of the output $y(t)$ and depends on the natural frequency ω of the system and the damping factor ζ. To design a system for a given input with power spectra density function $\Phi_{ff}(\omega_f)$, the quantities ω and ζ of the system would have to be determined for an optimum performance.

As an illustration, consider again the one-degree system in Fig. 8-3 whose absolute value of the impulse response $|H(\omega_f)|$ is given by Eq. 12-55. If Eq. 12-55 is plotted for various values of ω_f, it will have the shape shown in Fig. 12-5. In this figure, the peak value of $|H(\omega_f)|$ occurs at the resonant frequency $\omega = \omega_f$. That is,

$$|H(\omega_f)|_{max} = \frac{1}{2m\omega^2\zeta} = \frac{1}{2k\,\zeta} \qquad (12\text{-}69)$$

The frequencies $\omega = \omega_{f_1}$ and $\omega = \omega_{f_2}$ that satisfy the relationship

$$|H(\omega_{f_1})|^2 = |H(\omega_{f_2})|^2 = \frac{1}{2}|H(\omega)|^2 \qquad (12\text{-}70)$$

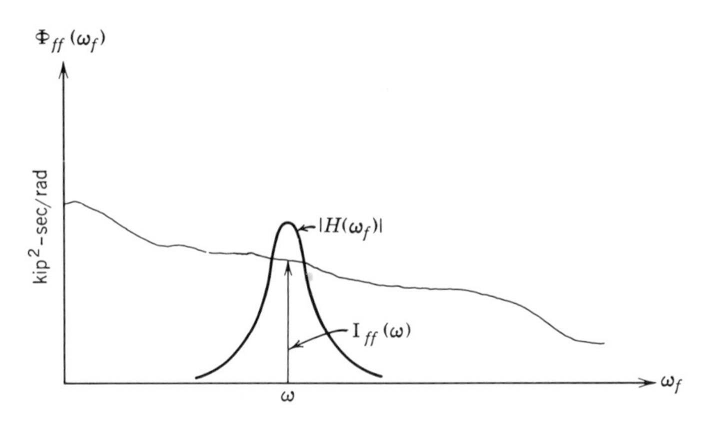

Fig. 12-6

yield the location of the so-called half-power points in Fig. 12-5. The difference $\omega_{f_2}-\omega_{f_1}$ is the bandwidth of the complex frequency response function $|H(\omega_f)|$ of the system, which for the case of viscous damping is equal to $2\zeta\omega$.

Let it now be assumed that the system is lightly damped and that the power spectra density function $\Phi_{ff}(\omega)$ of the exciting force is as shown in Fig. 12-6. The frequency response $|H(\omega)|$ of the one-degree system is also shown. This figure indicates that the contribution of the power spectra to the response of the system is large in the neighborhood of the natural frequency of the system and very small in the region outside this frequency. With this in mind, Eq. 12-50 can be modified as shown below, without introducing any appreciable error in the final results. That is,

$$E\{|y(t)|^2\}=\frac{\Phi_{ff}(\omega)}{2\pi m^2}\int_{-\infty}^{\infty}|H(\omega_f)|^2 d\omega_f \tag{12-71}$$

The integral is the same as that of Eq. 12-61. Thus Eq. 12-71 yields

$$E\{|y(t)|^2\}=\frac{\omega\Phi_{ff}(\omega)}{2\zeta k^2} \tag{12-72}$$

The equation shows that the power spectra can be expressed as a function of the natural frequency of the system. Thus, for a given random excitation, it is possible to design a system for a desired response by varying its natural frequency.

Based on the theoretical observations above, let it now be assumed that the natural frequency ω of the system is 6.45 rps, $k=4.57$ kips/in., and $\zeta=0.1$. This case is shown in Fig. 12-7. If $\Phi_{ff}(\omega)=0.60$ kips^2-sec/rad, the

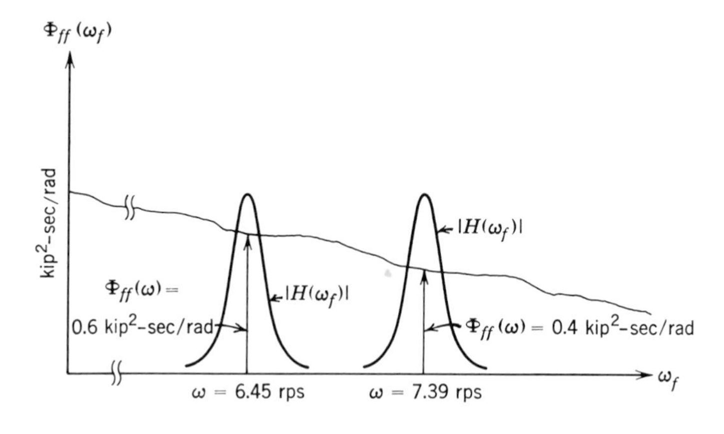

Fig. 12-7

expected value $E\{|y(t)|^2\}$ of the displacement can be determined from Eq. 12-72. That is,

$$E\{|y(t)|^2\} = \frac{(6.45)(0.60)}{(2)(0.1)(4.57)^2} = 0.93 \text{ in.}^2 \tag{12-73}$$

This value of the displacements represents the mean square value of the expected displacement of the mass m of the system.

The significance of the result given by Eq. 12-73 becomes more meaningful for practical applications if it is expressed in terms of a probability density function. Such a function could then be used to determine the probability that an event, such as that given by Eq. 12-73, can happen. This is discussed in the next section.

12-7 PROBABILITY AS A DESIGN PROCESS

Random excitations such as those caused by an earthquake or blast, are of approximately normal[48] distribution. For linear systems subjected to such excitations, the response $y(t)$ is also of normal distribution. The normal probability function will be explained by using a special case of the central limit theorem[49]. That is, if $X_1, X_2, X_3, \ldots, X_n$ are n independent random variables, and if each random variable is distributed in accordance to a given probability density function, then the probability density $p(y)$ of the

[48] Normal distribution is also called Gaussian distribution.

[49] Information regarding the central limit theorem can be found in Reference 71.

composite random variable Y, that is,

$$Y = X_1 + X_2 + X_3 + \cdots + X_n$$

will approach the Gaussian or normal density function if n is large. That is,

$$p(y) = \frac{1}{\sqrt{2\pi\sigma^2}} e^{-(y-a)^2/2\sigma^2} \tag{12-74}$$

Here $a = \overline{Y}$, where $\overline{Y}$ is the mean value of the process.

Earthquake, for example, is a physical process that consists of a summation of a large number of individual similar events. The sum of such events will approach a Gaussian density function. The response $y(t)$ due to this random excitation is also Gaussian with zero mean and variance $\sigma^2 = E\{|y(t)|^2\}$. For the case given by Eq. 12-73,

$$\sigma^2 = E\{|y(t)|^2\} = 0.93 \text{ in.}$$

$$a = \overline{Y} = 0$$

and the normal distribution function $p(y)$ is

$$p(y) = \frac{1}{\sqrt{2\pi(0.93)}} e^{-y^2/2(0.93)} \tag{12-75}$$

To be practical, let it now be assumed that the displacement $y(t)$ of the mass of the single-degree system discussed in the preceding section is not supposed to exceed a certain value, say 2.0 in. The probability that the absolute value of $y(t)$ will exceed 2.0 in. can be determined from Eq. 12-7. That is,

$$P(|y(t)|>2) = \int_{-\infty}^{-2} p(y)\,dy + \int_{2}^{\infty} p(y)\,dy \tag{12-76}$$

where $p(y)$ is given by Eq. 12-75.

By using Eqs. 12-75 and 12-76, and making note of the symmetry involved in Eq. 12-76, we obtain

$$P(|y(t)|>2) = \frac{2}{\sqrt{2\pi(0.93)}} \int_{2}^{\infty} e^{-y^2/2(0.93)}\,dy \tag{12-77}$$

Making the transformations $t^2/2 = y^2/2(0.93)$, or $t = y/\sqrt{0.93}$ and

$dt = dy/\sqrt{0.93}$, Eq. 12-77 can be written as

$$P(|y(t)|>2) = \frac{2}{\sqrt{2\pi}} \int_{t=2/\sqrt{0.93}}^{\infty} e^{-t^2/2} dt \tag{12-78}$$

The evaluation of the integral can be found in mathematical tables,[50] or $P\,(|\,y(t)\,|>2)$ can be determined from the expression

$$P(|y(t)|>2) = \frac{2}{\sqrt{2\pi}} e^{-t^2} \left[\frac{\sqrt{2}}{2t} - \frac{2\sqrt{2}}{2^2 t^3} + \frac{(1)(3)(4\sqrt{2})}{2^3 t^7} - \cdots \right] \tag{12-79}$$

Equation 12-79 is obtained by using repetitive integration by parts. Either of the two approaches leads to

$$P(|y(t)|>2) = 0.0384$$

which is the probability that the displacement of 2.0 in. will be exceeded.

If this probability is considered to be large, a smaller one can be obtained by redesigning the system. For example, if $\omega = 7.39$ rps, $k = 6.0$ kips/in., and $\Phi_{ff}(\omega) = 0.4$, Fig. 12-7, Eq. 12-72 yields,

$$E\{|y(t)|^2\} = \frac{(7.39)(0.4)}{(2)(0.1)(6.0)^2} = 0.41 \text{ in.}^2$$

Thus, by evaluating $p(y)$ from Eq. 12-74 and using Eq. 12-76, we have

$$P(|y(t)|>2) = 0.00171$$

which is a much smaller probability than that obtained in the previous design.

This procedure can be applied also to systems with two or more degrees of freedom. This can be accomplished by obtaining the power spectra density functions of the excitations and the system's frequency response. On this basis, the expected values of the output displacements can be derived in a manner similar to that used to derive Eq. 12-50. The development of these equations is beyond the scope of the material in this chapter.

[50] See, for example, Reference 74.

PROBLEMS

12-1 For the autocorrelation functions $R_{xx}(t)$ given in Fig. 12-2, determine in each case the power spectra $\Phi_{xx}(\omega)$.

12-2 For the power spectra $\Phi_{xx}(\omega)$ given in Fig. 12-2, determine in each case the autocorrelation function $R_{xx}(t)$.

12-3 The expected value of the input for a general system is given by Eq. 12-49. By following the procedure used in Section 12-6 to derive Eq. 12-49, verify Eq. 12-50 which represents the expected value of the output.

12-4 By using the power spectra $\Phi_{xx}(\omega)$ derived in Problem 12-1, determine in each case the expected value $E\{|y(t)|^2\}$ of the output for a single-degree spring-mass system.

12-5 For the spring-mass system in Fig. 8-3, $k = 4.57$ kips/in., the natural frequency ω is 6.45 rps, $\zeta = 0.1$ and the power spectra $\Phi_{xx}(\omega)$ of the exciting force $F(t)$ is $\Phi_{xx}(\omega) = 0.8$ kips^2-sec/rad. Determine the probability that the displacement of the mass of the system will not exceed 2.0 in.

12-6 Design the system in Problem 12-5 so that the probability to exceed the displacement of 2.0 in. is between 0.0002 and 0.0001.

12-7 Repeat Problem 12-5 for $\Phi_{xx}(\omega) = 0.4$ kips^2-sec/rad.

12-8 Repeat Problem 12-5 for the probability that the displacement of the mass of the system will not exceed 1.0 in.

12-9 For the spring-mass system in Fig. 8-3, $k = 6.0$ kips/in., the natural frequency ω is 7.39 rps, $\zeta = 0.1$, and the power spectra $\Phi_{xx}(\omega)$ of the exciting force $F(t)$ is $\Phi_{xx}(\omega) = 0.5$ kips^2-sec/rad. Determine the probability that the displacement of the mass of the system will not exceed 2.0 in.

APPENDIX BASIC PRINCIPLES OF MATRIX ALGEBRA

Consider the system of linear equations

$$
\begin{aligned}
y_1 &= a_{11}x_1 + a_{12}x_2 + a_{13}x_3 + \cdots + a_{1n}x_n \\
y_2 &= a_{21}x_1 + a_{22}x_2 + a_{23}x_3 + \cdots + a_{2n}x_n \\
y_3 &= a_{31}x_1 + a_{32}x_2 + a_{33}x_3 + \cdots + a_{3n}x_n \\
&\cdots\cdots\cdots\cdots\cdots\cdots\cdots \\
y_m &= a_{m1}x_1 + a_{m2}x_2 + a_{m3}x_3 + \cdots + a_{mn}x_n
\end{aligned}
\tag{A-1}
$$

These equations can be expressed in matrix form as

$$\begin{bmatrix} y_1 \\ y_2 \\ y_3 \\ \vdots \\ y_m \end{bmatrix} = \begin{bmatrix} a_{11} & a_{12} & a_{13} & \cdots & a_{1n} \\ a_{21} & a_{22} & a_{23} & \cdots & a_{2n} \\ a_{31} & a_{32} & a_{33} & \cdots & a_{3n} \\ \cdots & \cdots & \cdots & \cdots & \cdots \\ a_{m1} & a_{m2} & a_{m3} & \cdots & a_{mn} \end{bmatrix} \begin{bmatrix} x_1 \\ x_2 \\ x_3 \\ \cdot \\ \cdot \\ \cdot \\ x_n \end{bmatrix} \tag{A-2}$$

With

$$\{y\} = \begin{bmatrix} y_1 \\ y_2 \\ y_3 \\ \vdots \\ y_m \end{bmatrix} \qquad \{x\} = \begin{bmatrix} x_1 \\ x_2 \\ x_3 \\ \vdots \\ x_n \end{bmatrix}$$

and

$$\left[{}^{m}\overset{n}{M} \right] = \begin{bmatrix} a_{11} & a_{12} & a_{13} & \cdots & a_{1n} \\ a_{21} & a_{22} & a_{23} & \cdots & a_{2n} \\ a_{31} & a_{32} & a_{33} & \cdots & a_{3n} \\ \cdots & \cdots & \cdots & \cdots & \cdots \\ a_{m1} & a_{m2} & a_{m3} & \cdots & a_{mn} \end{bmatrix}$$

Equation A-2 can be written in a more compact form as

$$\{y\} = \left[{}^{m}\overset{n}{M} \right] \{x\} \tag{A-3}$$

The matrix $[{}^{m}\overset{n}{M}]$ is of the order (m,n) and consists of m rows and n columns. The elements of this matrix are the a_{ij} coefficients of Eq. A-1. The column vectors $\{y\}$ and $\{x\}$ are also matrices of order $(m,1)$ and $(n,1)$, respectively. A row vector consists of one row and n columns. The order of such a vector is $(1,n)$. When the number of rows of a matrix equals the number of its columns, the result is a square matrix of order (m,m).

A matrix whose elements are all zero is called a null matrix. The components of a null vector, row or column, are also zero. When the elements of a square matrix are all zero except those of the main diagonal, the resulting matrix is known as a diagonal matrix. For example, the square matrix

$$\left[\overset{m\ m}{M}\right] = \begin{bmatrix} a_{11} & 0 & 0 & \cdots & 0 \\ 0 & a_{22} & 0 & \cdots & 0 \\ 0 & 0 & a_{33} & \cdots & 0 \\ \cdots & \cdots & \cdots & \cdots & \cdots \\ 0 & 0 & 0 & \cdots & a_{mm} \end{bmatrix}$$

is a diagonal matrix. A unit matrix is a diagonal matrix whose diagonal elements are all equal to unity. For example,

$$\left[\overset{mm}{I}\right] = \begin{bmatrix} 1 & 0 & 0 & \cdots & 0 \\ 0 & 1 & 0 & \cdots & 0 \\ 0 & 0 & 1 & \cdots & 0 \\ \cdots & \cdots & \cdots & \cdots & \cdots \\ 0 & 0 & 0 & \cdots & 1 \end{bmatrix}$$

is a unit matrix.

A square matrix is symmetric about its main diagonal when the elements satisfy the relation $a_{ij}=a_{ji}$. If it is symmetric about its cross diagonal, it is called a cross-symmetric matrix. That is,

$$\left[\overset{3\ 3}{M}\right] = \begin{bmatrix} 1 & 2 & 4 \\ 2 & 3 & 6 \\ 4 & 6 & 7 \end{bmatrix}$$

is a symmetric matrix and

$$\begin{bmatrix} {}^{3}\overset{3}{M} \end{bmatrix} = \begin{bmatrix} 4 & 6 & 7 \\ 2 & 3 & 6 \\ 1 & 2 & 4 \end{bmatrix}$$

is a cross-symmetric matrix.

The transpose of a matrix $[M]$ is the matrix $[M']$ that is formed by writing the ith row of $[M]$ as the ith column of $[M']$. For example, the transpose of the matrix

$$\begin{bmatrix} {}^{2}\overset{3}{M} \end{bmatrix} = \begin{bmatrix} a_{11} & a_{12} & a_{13} \\ a_{21} & a_{22} & a_{23} \end{bmatrix}$$

is the matrix

$$\begin{bmatrix} {}^{3}\overset{2}{M'} \end{bmatrix} = \begin{bmatrix} a_{11} & a_{21} \\ a_{12} & a_{22} \\ a_{13} & a_{23} \end{bmatrix}$$

The transpose of $[M']$ is again $[M]$. The transpose of a square matrix that is symmetrical about its main diagonal is the same as the square-symmetric matrix.

Matrices may be added or subtracted by adding algebraically their corresponding elements. For example,

$$\begin{bmatrix} {}^{4}\overset{3}{M} \end{bmatrix} = \begin{bmatrix} a_{11} & a_{12} & a_{13} \\ a_{21} & a_{22} & a_{23} \\ a_{31} & a_{32} & a_{33} \\ a_{41} & a_{42} & a_{43} \end{bmatrix} \quad \text{and} \quad \begin{bmatrix} {}^{4}\overset{3}{N} \end{bmatrix} = \begin{bmatrix} b_{11} & b_{12} & b_{13} \\ b_{21} & b_{22} & b_{23} \\ b_{31} & b_{32} & b_{33} \\ b_{41} & b_{42} & b_{43} \end{bmatrix}$$

may be added or subtracted to form the matrix

$$\left[\overset{4\ 3}{C}\right] = \left[\overset{4\ 3}{M}\right] \pm \left[\overset{4\ 3}{N}\right] = \begin{bmatrix} (a_{11} \pm b_{11}) & (a_{12} \pm b_{12}) & (a_{13} \pm b_{13}) \\ (a_{21} \pm b_{21}) & (a_{22} \pm b_{22}) & (a_{23} \pm b_{23}) \\ (a_{31} \pm b_{31}) & (a_{32} \pm b_{32}) & (a_{33} \pm b_{33}) \\ (a_{41} \pm b_{41}) & (a_{42} \pm b_{42}) & (a_{43} \pm b_{43}) \end{bmatrix}$$

$$= \begin{bmatrix} c_{11} & c_{12} & c_{13} \\ c_{21} & c_{22} & c_{23} \\ c_{31} & c_{32} & c_{33} \\ c_{41} & c_{42} & c_{43} \end{bmatrix}$$

One may easily verify that

$$[M] + [N] = [N] + [M]$$

and

$$([M] + [N])' = [M'] + [N']$$

Also,

$$[A] - ([B] + [C]) = ([A] - [B]) - [C]$$

MULTIPLICATION OF MATRICES

Multiplication of a matrix $[M]$ by a scalar quantity β yields the matrix $[N]$ consisting of the elements of $[M]$ multiplied individually by the scalar quantity β.

Two matrices

$$\left[\overset{m\ n}{M}\right] = \begin{bmatrix} a_{11} & a_{12} & a_{13} & \cdots & a_{1n} \\ a_{21} & a_{22} & a_{23} & \cdots & a_{2n} \\ \cdots & \cdots & \cdots & \cdots & \cdots \\ a_{i1} & a_{i2} & a_{i3} & \cdots & a_{in} \\ \cdots & \cdots & \cdots & \cdots & \cdots \\ a_{m1} & a_{m2} & a_{m3} & \cdots & a_{mn} \end{bmatrix}$$

and

$$\left[{}^{n}\overset{P}{N}\right]=\begin{bmatrix} b_{11} & b_{12} & b_{13} & \dots & b_{1p} \\ b_{21} & b_{22} & b_{23} & \dots & b_{2p} \\ \dots & \dots & \dots & \dots & \dots \\ b_{n1} & b_{n2} & b_{n3} & \dots & b_{np} \end{bmatrix}$$

where the number of columns of $[{}^{m}\overset{n}{M}]$ is equal to the number of rows of $[{}^{n}\overset{P}{N}]$, may be multiplied to form a new matrix $[{}^{m}\overset{P}{C}]$ of order (m, p). The element c_{i1} of the new matrix is obtained by multiplying each element in row i of $[M]$ by the corresponding element in column 1 of $[N]$ and summing the products. That is,

$$c_{i1}=a_{i1}b_{11}+a_{i2}b_{21}+a_{i3}b_{31}+\cdots+a_{in}b_{n1}$$

In a similar manner,

$$c_{i2}=a_{i1}b_{12}+a_{i2}b_{22}+a_{i3}b_{32}+\cdots+a_{in}b_{n2}$$

$$c_{i3}=a_{i1}b_{13}+a_{i2}b_{23}+a_{i3}b_{33}+\cdots+a_{in}b_{n3}$$

$$\dots\dots\dots\dots\dots\dots$$

$$c_{in}=a_{i1}b_{1p}+a_{i2}b_{2p}+a_{i3}b_{3p}+\cdots+a_{in}b_{np}$$

The elements of any other row of $[{}^{m}\overset{p}{C}]$ may be found in a similar manner.

Consider, for example, the matrices

$$\left[{}^{2}\overset{3}{M}\right]=\begin{bmatrix} 4 & -2 & 0 \\ 3 & 1 & 5 \end{bmatrix} \quad \text{and} \quad \left[{}^{3}\overset{3}{N}\right]=\begin{bmatrix} 3 & 2 & -1 \\ 0 & 4 & 0 \\ 1 & -2 & 1 \end{bmatrix}$$

Their product $[{}^{2}\overset{3}{M}][{}^{3}\overset{3}{N}]$ is the matrix

$$\left[{}^{2}\overset{3}{C}\right]=\left[{}^{2}\overset{3}{M}\right]\left[{}^{3}\overset{3}{N}\right]=\begin{bmatrix} 12 & 0 & -4 \\ 14 & 0 & 2 \end{bmatrix}$$

When many matrices are to be multiplied, it is convenient to follow the multiplication procedure shown in the following example.

Suppose that it is required to multiply the matrices

$$\left[\overset{3\,2}{K}\right]=\begin{bmatrix} 1 & -3 \\ 4 & 0 \\ -3 & 5 \end{bmatrix} \qquad \left[\overset{2\,3}{M}\right]=\begin{bmatrix} 4 & -2 & 0 \\ 3 & 1 & 5 \end{bmatrix}$$

and

$$\left[\overset{3\,3}{N}\right]=\begin{bmatrix} 3 & 2 & -1 \\ 0 & 4 & 0 \\ 1 & -2 & 1 \end{bmatrix}$$

in order to determine the product

$$\left[\overset{3\,3}{D}\right]=\left[\overset{3\,2}{K}\right]\left[\overset{2\,3}{M}\right]\left[\overset{3\,3}{N}\right]$$

The multiplication arrangement is as follows:

$$\begin{bmatrix} 3 & 2 & -1 \\ 0 & 4 & 0 \\ 1 & -2 & 1 \end{bmatrix}=\left[\overset{3\,3}{N}\right]$$

$$\left[\overset{2\,3}{M}\right]=\begin{bmatrix} 4 & -2 & 0 \\ 3 & 1 & 5 \end{bmatrix}\begin{bmatrix} 12 & 0 & -4 \\ 14 & 0 & 2 \end{bmatrix}=\left[\overset{2\,3}{M}\right]\left[\overset{3\,3}{N}\right]=\left[\overset{2\,3}{C}\right]$$

$$\text{(A-4)}$$

$$\left[\overset{3\,2}{K}\right]=\begin{bmatrix} 1 & -3 \\ 4 & 0 \\ -3 & 5 \end{bmatrix}\begin{bmatrix} -30 & 0 & -10 \\ 48 & 0 & -16 \\ 34 & 0 & 22 \end{bmatrix}=\left[\overset{3\,2}{K}\right]\left[\overset{2\,3}{C}\right]=\left[\overset{3\,3}{D}\right]$$

Here the matrices $[M]$ and $[N]$ are first multiplied to find the product

$$\left[\overset{2\ 3}{C}\right]=\left[\overset{2\ 3}{M}\right]\left[\overset{3\ 3}{N}\right]$$

Then, matrices $[K]$ and $[C]$ are multiplied, yielding

$$\left[\overset{3\ 3}{D}\right]=\left[\overset{3\ 2}{K}\right]\left[\overset{2\ 3}{C}\right]=\left[\overset{3\ 2}{K}\right]\left[\overset{2\ 3}{M}\right]\left[\overset{3\ 3}{N}\right]=\begin{bmatrix} -30 & 0 & -10 \\ 48 & 0 & -16 \\ 34 & 0 & 22 \end{bmatrix}$$

The commutative law of multiplication is not valid for matrices. In general

$$[M][N] \neq [N][M]$$

The distributive law

$$[M]([N]+[K])=[M][N]+[M][K]$$

or

$$([N]+[K])[M]=[N][M]+[K][M]$$

for matrices $[N]$ and $[K]$ of the same order, as well as the associative law

$$[M]([N][K])=([M][N])[K]$$

are valid for matrices.

In calculating the transpose of the product of matrices, the following relation may be used:

$$([A][B]\cdots[M][N])'=[N'][M']\cdots[B'][A'] \qquad \text{(A-5)}$$

The matrix product of two symmetric matrices does not, in general, yield a symmetric matrix.

INVERSE OF SQUARE MATRICES

The inverse of a square matrix $[M]$ is designated by the symbol $[M^{-1}]$ and is defined as

$$[M][M^{-1}]=[I] \qquad \text{(A-6)}$$

Equation A-3, with $m = n$, yields

$$\{y\} = \left[\overset{n\ n}{M}\right]\{x\} \tag{A-7}$$

One application of Eq. A-6 is to express $\{x\}$ in terms of $\{y\}$ in Eq. A-7.
For example, premultiplying both sides of Eq. A-7 by $[M^{-1}]$,

$$[M^{-1}]\{y\} = [M^{-1}][M]\{x\} \tag{A-8}$$

Substitution of Eq. A-6 into Eq. A-8 yields

$$[M^{-1}]\{y\} = \{x\} \tag{A-9}$$

which expresses $\{x\}$ in terms of $\{y\}$. In this discussion, the term premultiplication of a matrix $[M]$ by a matrix $[N]$ means the product $[N][M]$. The term postmultiplication of $[M]$ by $[N]$ means the product $[M][N]$.

Many approaches have been suggested in the literature for the calculation of the inverse $[M^{-1}]$ of a square matrix $[M]$. The one used here is the expression

$$[M^{-1}] = \frac{[N]}{|M|} \tag{A-10}$$

The elements N_{ki} of the matrix $[N]$ are the subdeterminants, or minors, of the matrix $[M]$ formed by deleting row k and column i. The plus or minus sign by which the N_{ki} elements should be multiplied, can be obtained from the identity

$$(-1)^{i+k} \quad i = 1, 2, 3, \ldots \qquad k = 1, 2, 3, \ldots \tag{A-11}$$

where k and i are the deleted row and column, respectively. The elements of the determinant $|M|$ are the elements of the matrix $[M]$.

Consider, for example, the matrix

$$\left[\overset{3\ 3}{M}\right] = \begin{bmatrix} a_{11} & a_{12} & a_{13} \\ a_{21} & a_{22} & a_{23} \\ a_{31} & a_{32} & a_{33} \end{bmatrix} = \begin{bmatrix} 1 & 2 & 1 \\ 1 & 1 & 2 \\ 3 & 1 & 2 \end{bmatrix}$$

By applying Eq. A-10,

$$[M^{-1}] = \frac{\begin{bmatrix} \begin{vmatrix} a_{22} & a_{23} \\ a_{32} & a_{33} \end{vmatrix} & -\begin{vmatrix} a_{12} & a_{13} \\ a_{32} & a_{33} \end{vmatrix} & \begin{vmatrix} a_{12} & a_{13} \\ a_{22} & a_{23} \end{vmatrix} \\ -\begin{vmatrix} a_{21} & a_{23} \\ a_{31} & a_{33} \end{vmatrix} & \begin{vmatrix} a_{11} & a_{13} \\ a_{31} & a_{33} \end{vmatrix} & -\begin{vmatrix} a_{11} & a_{13} \\ a_{21} & a_{23} \end{vmatrix} \\ \begin{vmatrix} a_{21} & a_{22} \\ a_{31} & a_{32} \end{vmatrix} & -\begin{vmatrix} a_{11} & a_{12} \\ a_{31} & a_{32} \end{vmatrix} & \begin{vmatrix} a_{11} & a_{12} \\ a_{21} & a_{22} \end{vmatrix} \end{bmatrix}}{\begin{vmatrix} a_{11} & a_{12} & a_{13} \\ a_{21} & a_{22} & a_{23} \\ a_{31} & a_{32} & a_{33} \end{vmatrix}}$$

or

$$[M^{-1}] = \frac{\begin{bmatrix} 0 & -3 & 3 \\ 4 & -1 & -1 \\ -2 & 5 & -1 \end{bmatrix}}{\begin{vmatrix} 1 & 2 & 1 \\ 1 & 1 & 2 \\ 3 & 1 & 2 \end{vmatrix}} = \frac{1}{6}\begin{bmatrix} 0 & -3 & 3 \\ 4 & -1 & -1 \\ -2 & 5 & -1 \end{bmatrix}$$

$$= \begin{bmatrix} 0 & -\frac{1}{2} & \frac{1}{2} \\ \frac{2}{3} & -\frac{1}{6} & -\frac{1}{6} \\ -\frac{1}{3} & \frac{5}{6} & -\frac{1}{6} \end{bmatrix}$$

It may be noted that

$$[M][M^{-1}] = \begin{bmatrix} 1 & 2 & 1 \\ 1 & 1 & 2 \\ 3 & 1 & 2 \end{bmatrix}\begin{bmatrix} 0 & -\frac{1}{2} & \frac{1}{2} \\ \frac{2}{3} & -\frac{1}{6} & -\frac{1}{6} \\ -\frac{1}{3} & \frac{5}{6} & -\frac{1}{6} \end{bmatrix} = \begin{bmatrix} 1 & 0 & 0 \\ 0 & 1 & 0 \\ 0 & 0 & 1 \end{bmatrix}$$

$$= [I]$$

which verifies the definition for the inverse of a square matrix given by Eq. A-6.

The inverse of a unit matrix is also a unit matrix. If $a_{11}, a_{22}, a_{33}, \ldots, a_{nn}$ are the elements of the main diagonal of a diagonal matrix, its inverse is also a diagonal matrix with corresponding elements equal to $1/a_{11}, 1/a_{22}, 1/a_{33}, \ldots, 1/a_{nn}$. If $[M]$ is a symmetric matrix, then its inverse is also symmetric. The inverse of the transpose of a matrix $[M]$ is equal to the transpose of its inverse. That is,

$$[(M')^{-1}] = [(M^{-1})'] \tag{A-12}$$

The inverse of the product of two square matrices may be found by determining the corresponding inverse of each matrix, if it exists, and multiplying them in reverse order. Suppose that $[C]$ is the product

$$[C] = [M][N] \tag{A-13}$$

Then

$$[C^{-1}] = [N^{-1}][M^{-1}] \tag{A-14}$$

provided the $[N^{-1}]$ and $[M^{-1}]$ exist. If the inverse of a square matrix does not exist, the square matrix is called singular. This means that a row or column of such matrix is a linear combination of other rows or columns of the matrix.

PARTITION OF MATRICES

In many cases it becomes advantageous for computational purposes to partition a matrix $[M]$ into a number of smaller submatrices. Consider, for example a matrix $[{}^{m}\overset{n}{M}]$ of order (m,n). This matrix can be partitioned so that

$$\left[{}^{m}\overset{n}{M}\right] = \left[\begin{array}{c|c|c|c} \left[{}^{m_1}\overset{n_1}{M_1}\right] & \left[{}^{m_1}\overset{n_2}{M_2}\right] & \left[{}^{m_1}\overset{n_3}{M_3}\right] & \left[{}^{m_1}\overset{n_4}{M_4}\right] \\ \hline \left[{}^{m_2}\overset{n_1}{M_5}\right] & \left[{}^{m_2}\overset{n_2}{M_6}\right] & \left[{}^{m_2}\overset{n_3}{M_7}\right] & \left[{}^{m_2}\overset{n_4}{M_8}\right] \\ \hline \left[{}^{m_3}\overset{n_1}{M_9}\right] & \left[{}^{m_3}\overset{n_2}{M_{10}}\right] & \left[{}^{m_3}\overset{n_3}{M_{11}}\right] & \left[{}^{m_3}\overset{n_4}{M_{12}}\right] \end{array}\right] \tag{A-15}$$

where

$$m = m_1 + m_2 + m_3 \qquad \text{and} \qquad n = n_1 + n_2 + n_3 + n_4$$

The multiplication of partitioned matrices is carried out in the same way as that of regular matrices, but the submatrices are treated as if they were elements. If $[M]$ and $[N]$ are partitioned matrices to be multiplied, the partitioning of $[M]$ with respect to columns should be made in the same way as the partitioning of $[N]$ with respect to rows in order to be able to perform the multiplication.

Suppose that it is required to determine the product

$$[C] = [M][N]$$

where

$$[M] = \begin{bmatrix} 2 & 0 & 2 & 0 \\ 0 & 1 & 0 & 1 \\ 1 & 1 & 1 & 2 \end{bmatrix} \qquad [N] = \begin{bmatrix} 1 & 0 & 0 \\ 0 & 2 & 1 \\ 1 & 0 & 1 \\ 0 & 1 & 2 \end{bmatrix}$$

Matrices $[M]$ and $[N]$ can be partitioned as follows:

$$[M] = \left[\begin{array}{ccc|c} 2 & 0 & 2 & 0 \\ 0 & 1 & 0 & 1 \\ 1 & 1 & 1 & 2 \end{array}\right] \qquad [N] = \left[\begin{array}{ccc} 1 & 0 & 0 \\ 0 & 2 & 1 \\ 1 & 0 & 1 \\ \hline 0 & 1 & 2 \end{array}\right]$$

Multiplication of $[M]$ by $[N]$ yields

$$[C] = \begin{bmatrix} 2 & 0 & 2 \\ 0 & 1 & 0 \\ 1 & 1 & 1 \end{bmatrix} \begin{bmatrix} 1 & 0 & 0 \\ 0 & 2 & 1 \\ 1 & 0 & 1 \end{bmatrix} + \begin{bmatrix} 0 \\ 1 \\ 2 \end{bmatrix} \begin{bmatrix} 0 & 1 & 2 \end{bmatrix}$$

$$= \begin{bmatrix} 4 & 0 & 2 \\ 0 & 2 & 1 \\ 2 & 2 & 2 \end{bmatrix} + \begin{bmatrix} 0 & 0 & 0 \\ 0 & 1 & 2 \\ 0 & 2 & 4 \end{bmatrix} = \begin{bmatrix} 4 & 0 & 2 \\ 0 & 3 & 3 \\ 2 & 4 & 6 \end{bmatrix}$$

Matrices

$$[M] = \left[\begin{array}{cc|cc|cc} 1 & 0 & 0 & 0 & 0 & 0 \\ 0 & 3 & 0 & 0 & 0 & 0 \\ \hline 0 & 0 & 2 & 0 & 0 & 0 \\ 0 & 0 & 0 & 1 & 0 & 0 \\ \hline 0 & 0 & 0 & 0 & 4 & 0 \\ 0 & 0 & 0 & 0 & 0 & 2 \end{array}\right]$$

$$[N] = \left[\begin{array}{cc|cc} 1 & 0 & 0 & 0 \\ 0 & 2 & 0 & 0 \\ \hline 0 & 0 & 5 & 0 \\ 0 & 0 & 0 & 4 \\ \hline 0 & 0 & 1 & 0 \\ 0 & 0 & 0 & 1 \end{array}\right]$$

are partitioned as shown by the dashed lines.

SOME RULES REGARDING DETERMINANTS

Since determinants have been introduced in the preceding discussion, it is appropriate to give some of the rules associated with their evaluation. These rules permit one to interchange the rows and columns of a determinant without altering its value. That is,

$$|M| = |M'| \tag{A-16}$$

The sign of a determinant is altered if two rows or two columns are interchanged. A determinant remains unaltered if a column or a row is changed by subtracting from or adding to its elements the corresponding elements or any other column or row. If there exists a common factor β between the elements of a column or row of a determinant, then β may be placed outside the determinant as a multiplier. A determinant is zero when at least one of its rows or columns is a linear combination of the other rows or columns. Products of square matrices satisfy relations of the form

$$|[M][N][L]| = |[N][M][L]| = |[L]||[M]||[N]| \qquad \text{(A-17)}$$

DIFFERENTIATION AND INTEGRATION OF MATRICES

The elements of a matrix can be a function of some parameter x. In this case, differentiation and integration of a matrix [M] are defined as follows:

$$\frac{d}{dx}[M] = \begin{bmatrix} \frac{da_{11}}{dx} & \frac{da_{12}}{dx} & \frac{da_{13}}{dx} & \cdots & \frac{da_{1n}}{dx} \\ \frac{da_{21}}{dx} & \frac{da_{22}}{dx} & \frac{da_{23}}{dx} & \cdots & \frac{da_{2n}}{dx} \\ \cdots & \cdots & \cdots & \cdots & \cdots \\ \frac{da_{n1}}{dx} & \frac{da_{n2}}{dx} & \frac{da_{n3}}{dx} & \cdots & \frac{da_{nn}}{dx} \end{bmatrix} \qquad \text{(A-18)}$$

$$\int [M]\,dx = \begin{bmatrix} \int a_{11}dx & \int a_{12}dx & \int a_{13}dx & \cdots & \int a_{1n}dx \\ \int a_{21}dx & \int a_{22}dx & \int a_{23}dx & \cdots & \int a_{2n}dx \\ \cdots & \cdots & \cdots & \cdots & \cdots \\ \int a_{n1}dx & \int a_{n2}dx & \int a_{n3}dx & \cdots & \int a_{nn}dx \end{bmatrix}$$

$$\text{(A-19)}$$

$$\frac{d}{dx}\{y\} = \left\{ \frac{dy_1}{dx}\ \frac{dy_2}{dx}\ \frac{dy_3}{dx} \cdots \frac{dy_n}{dx} \right\} \qquad \text{(A-20)}$$

$$\int \{y\}\,dx = \left\{ \int y_1 dx \int y_2 dx \int y_3 dx \ldots \int y_n dx \right\} \qquad \text{(A-21)}$$

REFERENCES AND BIBLIOGRAPHY

1. J. C. Maxwell, "On the Calculation of the Equilibrium and Stiffness of Frames," *The London, Edinburgh, and Dublin Philosophical Magazine*, Vol. 27, p. 294, 1864.
2. L. C. Maugh, *Statically Indeterminate Structures*, John Wiley and Sons, New York, 1951.
3. R. A. Frazer and W. J. Duncan, *Elementary Matrices*, Cambridge University Press, London, 1938.
4. F. R. Gantmacher, *The Theory of Matrices*, Vol. I and II, Chelsey Publishing Company, New York, 1959.
5. J. T. Oden, *Mechanics of Elastic Structures*, McGraw-Hill Book Company, New York, 1967.
6. R. K. Vierck, *Vibration Analysis*, International Textbook Company, Scranton, Pa., 1967.

7. S. P. Timoshenko and J. N. Goodier, *Theory of Elasticity*, McGraw-Hill Book Company, New York, 1970.
8. U.S. Army Corps of Engineers, "Design of Structures to Resist the Effects of Atomic Weapons," Manual EM-1110-345-415, 1957.
9. C. E. Massonnet and M. A. Save, *Plastic Analysis and Design*, Vol. 1, Blaisdell Publishing Company, New York, Toronto, London, 1965.
10. J. P. Den Hartog, *Mechanical Vibrations*, 4th Edition, McGraw-Hill Book Company, New York, 1956.
11. A. H. Church, *Mechanical Vibrations*, John Wiley and Sons, New York, 1957.
12. F. B. Hindebrand, *Methods of Applied Mathematics*, Prentice-Hall, Englewood Cliffs, N. J., 1952.
13. S. Timoshenko and D. H. Young, *Advanced Dynamics*, McGraw-Hill Book Company, 1948.
14. J. M. Bigs, *Introduction to Structural Dynamics*, McGraw-Hill Book Company, 1964.
15. E. C. Pestel and F. A. Leckie, *Matrix Methods in Elastomechanics*, McGraw-Hill Book Company, 1963.
16. L. S. Jacobsen and R. S. Ayre, *Engineering Vibrations*, McGraw-Hill Book Company, New York, 1958.
17. S. Timoshenko, *Vibration Problems in Engineering*, D. Van Nostrand Company, Princeton, N. J., 1955.
18. R. S. Burington, *Handbook of Mathematical Tables and Formulas*, 3rd Edition, Handbook Publishers, Sandusky, Ohio, 1949.
19. E. E. Inglis, *A Mathematical Treatise on Vibration in Railway Bridges*, Cambridge University Press, London, 1934.
20. E. Saibel and E. D'Appolonia, "Forced Vibrations of Continuous Beams," *Transactions, American Society of Civil Engineers*, Vol. 117, pp. 1075–1090, 1952.
21. A. S. Veletsos and N. M. Newmark, "Natural Frequencies of Continuous Flexural Members," *Proceedings, American Society of Civil Engineers*, Paper No. 735, July 1955.
22. Lord Rayleigh, *Theory of Sound*, Dover Publications, New York, 1945 (reprint).
23. E. P. Popov, *Introduction to Mechanics of Solids*, Prentice-Hall, Englewood Cliffs, N. J., 1968.
24. D. G. Fertis and E. C. Zobel, *Transverse Vibration Theory*, The Ronald Press Company, New York, 1961.
25. D. G. Fertis, "Dynamic Hinge Concept for Beam Vibrations," *Journal of the Structural Division, Proceedings of the American Society of Civil Engineers*, Paper 4689, February 1966.
26. S. H. Crandall, *Engineering Analysis*, McGraw-Hill Book Company, New York, 1956.
27. D. G. Fertis and E. C. Zobel, "Equivalent Systems for the Deflection of Variable Stiffness Members," *Journal of the Structural Division, Proceedings of the American Society of Civil Engineers*, Paper 1820, October 1958.
28. D. G. Fertis, "Contribution to the Deflection and Free Vibration of Uniform and Variable Stiffness Members," Doctoral Dissertation, The National Technical University of Athens, Athens, Greece, 1964.
29. D. G. Fertis and H. Cunningham, "Equivalent Systems for Shear Deflections of Variable Thickness Members," *Industrial Mathematics*, Vol. 12, Part I, The Industrial Mathematics Society, Detroit, Mich., 1962.

30. A. Kozma and D. G. Fertis, "Solution of the Deflection of Variable Thickness Plates by the Method of the Equivalent Systems," *Industrial Mathematics*, Vol. 12, Part I, The Industrial Mathematics Society, Detroit, Mich., 1962.

31. M. Hetenyi, "Deflection of Beams of Varying Cross Section," *Journal of Applied Mechanics*, The American Society of Mechanical Engineers, Vol. 4, June 1937.

32. H. Cross, "Analysis of Continuous Frames by Distributing Fixed End Moments," *Transactions of the American Society of Civil Engineers*, Vol. 96, pp. 1–10, 1932.

33. N. O. Mykelstad, "A New Method of Calculating Natural Modes of Uncoupled Bending Vibration of Airplane Wings and Other Types of Beams," *Journal of Aeronautical Sciences*, pp. 153–162, April 1944.

34. M. A. Prohl, "A General Method for Calculating Critical Speeds of Flexible Rotors," *Transactions of the American Society of Mechanical Engineers,* A-142, September 1945.

35. W. T. Thomson, "Matrix Solution for the Vibration of Nonuniform Beams," *Journal of Applied Mechanics*, Vol. 17, No. 3, pp. 337–339, 1950.

36. J. H. Argyris, "Energy Theorems and Structural Analysis,"*Aircraft Engineering*, Vol. 27, p. 345, 1955, and subsequent issues.

37. J. H. Argyris, "Die Matrizentheorie der Statik," *Ingenieur-Archiv*, Vol. 25, p. 174, 1957.

38. J. H. Argyris and S. Kelsey, "Structural Analysis by the Matrix Force Method with Applications to Aircraft Wings," *Jahrbuch der Wissenschaftliche Gesellschaft für Luftfahrt*, pp. 78–98, Vieweg-Verlag, Brunswick, Germany, 1956.

39. J. H. Argyris and S. Kelsey, "The Analysis of Fuselages of Arbitrary Cross-section and Taper," *Aircraft Engineering*, Vol. 31, p. 62, 1959; Vol. 33, p. 34, 1961; and subsequent issues.

40. *The Effects of Nuclear Weapons*, prepared by the U.S. Department of Defense and published by the U.S. Atomic Energy Commission, April 1962.

41. U.S. Army Corps of Engineers, *Design of Structures to Resist the Effects of Atomic Weapons*, Manual numbers.

    EM-1110-345-413, Weapon Effects Data
    EM-1110-345-414, Strength of Materials and Structural Elements
    EM-1110-345-415, Principles of Dynamic Analysis and Design
    EM-1110-345-416, Structural Elements Subjected to Dynamic Loads
    EM-1110-345-417, Single-Story Frame Buildings
    EM-1110-345-418, Multistory Frame Buildings
    EM-1110-345-419, Shear Wall Structures
    EM-1110-345-420, Arches and Domes
    EM-1110-345-421, Buried and Semiburied Structures

    Manual 413 was published on July 1, 1959; Manuals 414, 415 and 416 were published on March 15, 1957; Manuals 417 and 419 were published on January 15, 1958; and Manuals 418, 420 and 421 were published on January 15, 1960.

42. J. A. Blume, N. M. Newmark, and L. H. Corning, *Design of Multistory Reinforced Concrete Buildings for Earthquake Motions*, Portland Cement Association, Chicago, Ill., 1961.

43. U.S. Department of Defense, *Protective Construction Review Guide*, Vol. 1, June 1961. The work was performed by Newmark, Hansen and Associates under Contract SD 52.

44. J. W. Melin and S. Sutcliffe, "Development of Procedures for Rapid Computations of Dynamic Structural Response," Final Report, Contract AF 33(600)-24994, University of Illinois for U.S. Air Force, Urbana, Ill., January 1959.

45. C. H. Noris *et al., Structural Design for Dynamic Loads*, McGraw-Hill Book Company, New York, 1959.

46. J. B. Macelwane, "The Interior of the Earth," *Bulletin of the Seismological Society of America*, June 1924.
47. S. P. Timoshenko and D. H. Young, *Theory of Structures*, 2nd Edition, McGraw-Hill Book Company, New York, 1965.
48. J. H. Hodgson, *Earthquakes and Earth Structure*, Prentice-Hall, Englewood Cliffs, N. J., 1964.
49. G. V. Berg, *The Skopje Yugoslavia Earthquake*, American Iron and Steel Institute, New York, 1964.
50. U.S. Coast and Geodetic Survey, *United States Earthquake*, U.S. Government Printing Office, Washington, D. C., various years.
51. J. Hershberger, "A Comparison of Earthquake Acceleration with Intensity Rating," *Bulletin of the Seismological Society of America*, Vol. 46, pp. 317–320, October 1956.
52. H. O. Wood and F. Neumann, "Modified Mercalli Intensity Scale, 1931," *Bulletin of the Seismological Society of America*, Vol. 21, pp. 277–283, December 1931.
53. D. G. Fertis, "Proposed Method for the Design of Earthquake Resistant Structures," M. S. Thesis, Michigan State University, 1955.
54. G. V. Berg and S. S. Thomaides, "Discussion of Electrical Analog for Earthquake Yield Spectra," *Journal of the Engineering Mechanics Division of the American Society of Civil Engineers*, Vol. 86, No. EM2, April 1960.
55. M. A. Biot, "Analytical and Experimental Methods in Engineering Seismology," *Transactions of the American Society of Civil Engineers*, Vol. 108, pp. 365–408, 1943.
56. D. E. Hudson, "Response Spectrum Techniques in Engineering Seismology," *Proceedings, World Conference on Earthquake Engineering*, Earthquake Engineering Resea·ch Institute, Berkeley, Calif., pp. 4-1 to 4-12, 1956.
57. G. W. Housner, "Limit Design of Structures to Resist Earthquakes," *Proceedings, World Conference on Earthquake Engineering*, Earthquake Engineering Research Institute, Berkeley, Calif. pp. 5-1 to 5-13, 1956.
58. D. E. Hudson and G. W. Housner, "Structural Vibrations Produced by Ground Motion," *Transactions of the American Society of Civil Engineers*, pp. 705–721, 1957.
59. J. A. Blume, "Structural Dynamics in Earthquake-Resistant Design," *Transactions of the American Society of Civil Engineers*, Vol. 125, pp. 1088–1139, 1960.
60. R. L. Jennings and N. M. Newmark, "Elastic Response of Multi-Story Shear Beam Type Structures Subjected to Strong Ground Motion," *Proceedings, Second World Conference on Earthquake Engineering*, Tokyo, Vol. II, pp. 699–718, 1960.
61. F. A. Leckie, "The Application of Transfer Matrices to Plate Vibration," *Ingenieur-Archiv*, Vol. 30, 1962.
62. E. C. Pestel, "Application of the Transfer Matrix Method to Cylindrical Shells," *International Journal of Mechanical Science*, Vol. 5, 1963.
63. S. H. Randall *et al.*, *Random Vibration*, The M.I.T. Press, Cambridge, Mass., and John Wiley and Sons, New York, 1958.
64. S. H. Crandall et al., *Random Vibration*, Vol. 2, The M.I.T. Press, Cambridge, Mass., 1963.
65. A. K. Chopra, "Earthquake Response of Concrete Gravity Dams," *Proceedings of the ASCE Journal of the Engineering Mechanics Division*, Paper No. 7485, August 1970.
66. R. F. Drenick, "Model-free Design of Aseismic Structures," *Proceeding of the ASCE Journal of the Engineering Mechanics Division*, Paper No. 7496, August 1970.
67. R. N. Iyengar and K. T. S. R. Iyengar, "Probabilistic Response Analyses to Earthquakes,"

Proceeding of the ASCE Journal of the Engineering Mechanics Division, Paper No. 7321, June 1970.

68. I. S. Sokolnikoff and R. M. Redheffer, *Mathematics of Physics and Modern Engineering*, McGraw-Hill Book Company, New York, 1966.
69. L. A. Wainstein and V. D. Zubakov, *Extraction of Signals from Noise*, translated from the Russian by R. A. Silverman, Dover Publication, New York, 1970.
70. Y. K. Lin, *Probabilistic Theory of Structural Dynamics*, McGraw-Hill Book Company, New York 1967.
71. A. Papoulis, *Probability, Random Variables, and Stochastic Processes*, McGraw-Hill Book Company, New York, 1965.
72. A. Papoulis, *The Fourier Integral and its Applications*, McGraw-Hill Book Company, New York, 1962.
73. Fon Mathews and R. L. Walker, *Mathematical Methods of Physics*, W. A. Benjamin, New York, 1964.
74. U.S. Department of Commerce, National Bureau of Standards, *Handbook of Mathematical Functions with Formulas, Graphs and Mathematical Tables*, Applied Mathematics Series 55, June 1964.
75. L. E. Elsgolc, *Calculus of Variations*, Addison-Wesley Publishing Company, Reading, Mass., 1962.
76. R. L. Wiegel, Coordinating Editor, *Earthquake Engineering*, Prentice-Hall, Englewood Cliffs, N. J., 1970.
77. R. Weinstock, *Calculus of Variations*, McGraw-Hill Book Company, New York, 1952.
78. D. G. Fertis, *Dynamics of Structural Systems*, Vol. 1, The University of Akron Press, Akron, Ohio, 1971.
79. D. G. Fertis, *Dynamics of Structural Systems*, Vol. 2, The University of Akron Press, Akron, Ohio, 1972.
80. R. V. Churchill, *Fourier Series and Boundary Value Problems*, McGraw-Hill Book Company, New York, 1941.
81. S. Timoshenko and S. Wainowsky-Kriger, *Theory of Plates and Shells*, McGraw-Hill Book Company, New York, 1959.
82. D. G. Fertis, "Material and Strength Characteristics of Concrete Mixtures by Acoustic Spectra Analysis," Research Report No. APS-201A, Ohio Department of Highways and U.S. Department of Transportation, Federal Highway Administration, February 1971.

INDEX